Lecture Notes in Computer Science 1304

Edited by G. Goos, J. Hartmanis and J. van Leeuwen

Advisory Board: W. Brauer   D. Gries   J. Stoer

T0222932

Springer

Berlin
Heidelberg
New York
Barcelona
Budapest
Hong Kong
London
Milan
Paris
Santa Clara
Singapore
Tokyo

Wayne Luk   Peter Y.K. Cheung
Manfred Glesner (Eds.)

Field-Programmable Logic and Applications

7th International Workshop, FPL '97
London, UK, September 1-3, 1997
Proceedings

Springer

Series Editors

Gerhard Goos, Karlsruhe University, Germany

Juris Hartmanis, Cornell University, NY, USA

Jan van Leeuwen, Utrecht University, The Netherlands

Volume Editors

Wayne Luk
Peter Y.K. Cheung
Imperial College, Department of Computing
180 Queen's Gate, London SW7 2BZ, UK

Manfred Glesner
Darmstadt University of Technology, Institute of Microelectronic Systems
Karlstr. 15, D-64283 Darmstadt, Germany

Cataloging-in-Publication data applied for

Die Deutsche Bibliothek - CIP-Einheitsaufnahme

Field programmable logic and applications : 7th international
workshop ; proceedings / FPL '97, London, UK, September 1 - 3,
1997. Wayne Luk ... (ed.). - Berlin ; Heidelberg ; New York ;
Barcelona ; Budapest ; Hong Kong ; London ; Milan ; Paris ; Santa
Clara ; Singapore ; Tokyo : Springer, 1997
　　(Lecture notes in computer science ; Vol. 1304)
　　ISBN 3-540-63465-7

CR Subject Classification (1991): B.6-7, J.6

ISSN 0302-9743
ISBN 3-540-63465-7 Springer-Verlag Berlin Heidelberg New York

© Springer-Verlag Berlin Heidelberg 1997
Printed in Germany

Typesetting: Camera-ready by author
SPIN 10545913   06/3142 – 5 4 3 2 1 0   Printed on acid-free paper

Preface

This book contains the papers presented at the 7th International Workshop on Field Programmable Logic and Applications (FPL'97), held at Imperial College of Science, Technology and Medicine, London, September 1–3, 1997.

We are delighted to continue the FPL series of workshops. The previous events were held in Oxford in 1991, 1993, and 1995, in Vienna in 1992, in Prague in 1994, and in Darmstadt in 1996. The proceedings of the 1991 and 1993 workshops are available from Wayne Luk (w.luk@doc.ic.ac.uk), and the proceedings of the 1992, 1994, 1995, and 1996 workshops are available as volumes in the Lecture Notes in Computer Science series (LNCS 705, LNCS 849, LNCS 975, and LNCS 1142).

Exciting advances in field programmable logic show no sign of slowing down. New grounds have been broken in architectures, design techniques, and applications for field programmable devices; many of these innovations are reported in this volume. Some readers may also be interested in finding that the topic of run-time reconfigurability of field programmable logic has been receiving growing attention, as reflected by the abundance of contributions on this topic.

We are fortunate in receiving a large number of high-quality papers covering a wide range of topics. From the submitted papers, fifty-two were selected for presentation at the workshop. All selected papers, except one, are included in this book.

We extend our thanks to the authors who submitted their work to this workshop, and to the members of the programme committee for reviewing the submitted papers. We acknowledge the help of the following who contributed to the organisation of paper selection or to the preparation of this volume: D. Choy, S.R. Guo, R.W. Hartenstein, P.I. Mackinlay, S.W. McKeever, U. Nageldinger, M. Rychetsky, R. Sandiford, N. Shirazi, and D. Siganos.

We are grateful to Springer-Verlag, particularly Alfred Hofmann, for their work in publishing this book.

July 1997

Wayne Luk
Peter Y.K. Cheung
Manfred Glesner

General Co-Chairs

Wayne Luk and Peter Y.K. Cheung, Imperial College, UK

Programme Chair

Manfred Glesner, Darmstadt University of Technology, Germany

Programme Committee

Doug Amos, Altera, UK
Jeff Arnold, Independent Consultant, USA
Peter Athanas, Virginia Tech, USA
Stephen Brown, University of Toronto, Canada
Klaus Buchenrieder, Siemens AG, Germany
Bernard Courtois, INPG, Grenoble, France
Keith Dimond, University of Kent, UK
Carl Ebeling, University of Washington, USA
Patrick Foulk, Heriot-Watt University, UK
Norbert Fristacky, Slovak Technical University, Slovakia
John Gray, Xilinx, UK
Herbert Gruenbacher, Vienna University, Austria
Reiner Hartenstein, University of Kaiserslautern, Germany
Brad Hutchings, Brigham Young University, USA
Udo Kebschull, University of Tuebingen, Germany
Andres Keevallik, Tallin Technical University, Estonia
Patrick Lysaght, University of Strathclyde, Scotland
Will Moore, Oxford University, UK
Klaus Mueller-Glaser, University of Karlsruhe, Germany
Wolfgang Nebel, University of Oldenburg, Germany
Peter Noakes, University of Essex, UK
Franco Pirri, University of Firenze, Italy
Jonathan Rose, University of Toronto, Canada
Zoran Salcic, University of Auckland, New Zealand
Mariagiovanna Sami, Politechnico di Milano, Italy
Michal Servit, Czech Technical University, Czech Republic
Mark Shand, Digital Systems Research Center, USA
Stephen Smith, Altera, USA
Steve Trimberger, Xilinx, USA

Table of Contents

Devices and Architectures

Multicontext dynamic reconfiguration and real time
probing on a novel mixed signal programmable device
with on-chip microprocessor 1
*J. Faura, J.M. Moreno, M.A. Aguirre,
P. van Duong and J.M. Insenser*

CAD-oriented FPGA and dedicated CAD system
for telecommunications 11
*T. Miyazaki, A. Takahara, M. Katayama, T. Murooka,
T. Ichimori, K. Fukami, A. Tsutsui and K. Hayashi*

Rothko: A three dimensional FPGA architecture,
its fabrication, and design tools 21
M. Leeser, W.M. Meleis, M.M. Vai and P. Zavracky

Extending dynamic circuit switching to meet the challenges
of new FPGA architectures 31
G. McGregor and P. Lysaght

Performance evaluation of a full speed PCI initiator
and target subsystem using FPGAs 41
D. Robinson, P. Lysaght, G. McGregor and H. Dick

Implementation of pipelined multipliers on Xilinx FPGAs 51
T.-T. Do, H. Kropp, M. Schwiegershausen and P. Pirsch

The XC6200DS development system 61
S. Nisbet and S.A. Guccione

Devices and Systems

Thermal monitoring on FPGAs using ring-oscillators 69
E. Boemo and S. López-Buedo

A reconfigurable approach to low cost media processing 79
I. Kostarnov, S. Morley, J. Osmany and C. Solomon

Riley-2: A flexible platform for codesign and dynamic
reconfigurable computing research 91
P.I. Mackinlay, P.Y.K. Cheung, W. Luk and R. Sandiford

Reconfiguration I

Stream synthesis for a wormhole run-time
reconfigurable platform 101
B. Kahne and P. Athanas

Pipeline morphing and virtual pipelines 111
W. Luk, N. Shirazi, S.R. Guo and P.Y.K. Cheung

Parallel Graph colouring using FPGAs 121
B. Rising, M. van Daalen, P. Burge and J. Shawe-Taylor

Run-time compaction of FPGA designs 131
O. Diessel and H. ElGindy

Partial reconfiguration of FPGA mapped designs with
applications to fault tolerance and yield enhancement 141
J.M. Emmert and D. Bhatia

A case study of partially evaluated hardware circuits:
Key-specific DES 151
J. Leonard and W.H. Mangione-Smith

Run-time parameterised circuits for the Xilinx XC6200
R. Payne 161

Reconfiguration II

Automatic identification of swappable logic units in
XC6200 circuitry 173
G. Brebner

Towards an expert system for a priori estimation of
reconfiguration latency in dynamically
reconfigurable logic 183
P. Lysaght

Exploiting reconfigurability through domain-specific systems 193
B.L. Hutchings

Design Tools

Technology mapping by binate covering 203
M.Z. Servít and K. Yi

VPR: A new packing, placement and routing tool for
FPGA research 213
V. Betz and J. Rose

Technology mapping of heterogeneous LUT-based FPGAs 223
M.K. Inuani and J. Saul

Technology-driven FSM partitioning for synthesis of large
sequential circuits targeting lookup-table based FPGAs 235
K. Feske, S. Mulka, M. Koegst and G. Elst

Technology mapping of LUT based FPGAs for delay optimisation 245
X. Lin, E. Dagless and A. Lu

Automatic mapping of algorithms onto multiple
FPGA-SRAM modules 255
S.J.B. Acock and K.R. Dimond

FPLD HDL synthesis employing high-level evolutionary
algorithm optimisation 265
R.B. Maunder, Z.A. Salcic and G.G. Coghill

A hardware/software partitioning algorithm for custom
computing machines 274
A.V. Chichkov and C.B. Almeida

Custom Computing and Codesign

The Java environment for reconfigurable computing 284
E. Lechner and S.A. Guccione

Data scheduling to increase performance of parallel accelerators 294
R.W. Hartenstein, J. Becker, M. Herz and U. Nageldinger

An operating system for custom computing machines
based on the Xputer paradigm 304
R. Kress, R.W. Hartenstein and U. Nageldinger

Signal Processing

Fast parallel implementation of DFT using configurable devices 314
A. Dandalis and V.K. Prasanna

Enhancing fixed point DSP processor performance
by adding CPLDs as coprocessing elements 324
D. Greenfield, C. Crome, M.S. Won and D. Amos

A case study of algorithm implementation in reconfigurable
hardware and software 333
M. Shand

A reconfigurable data-localised array for morphological algorithms 344
A.S. Chaudhuri, P.Y.K. Cheung and W. Luk

Virtual radix array processors (V-RaAP) 354
B. Bramer, D. Chauham, M.K. Ibrahim and A. Aggoun

An FPGA implementation of a matched filter detector for spread
spectrum communications systems 364
*T. Mathews, S.G. Gibb, L.E. Turner, P.J.W. Graumann
and M. Fattouche*

An NTSC and PAL closed caption processor 374
S. Teerapanyawatt and K. Athikulwongse

Image and Video Processing

A 800 Mpixel/sec reconfigurable image correlator on XC6216 382
T. Kean and A. Duncan

A reconfigurable coprocessor for a PCI-based real time
computer vision system 392
F. Lisa, F. Cuadrado, D. Rexachs and J. Carrabina

Real-time stereopsis using FPGAs 400
P. Dunn and P. Corke

Sensors, Graphics and other Applications

FPGA implementation of a digital IQ demodulator using VHDL 410
C.C. Jong, Y.Y.H. Lam and L.S. Ng

Hardware compilation, configurable platforms and ASICs
for self-validating sensors 418
I. Page

Postscript™ rendering with virtual hardware 428
S. Singh, J. Patterson, J. Burns and M. Dales

P4: A platform for FPGA implementation of protocol boosters 438
I. Hadžić and J.M. Smith

Satisfiability on reconfigurable hardware 448
M. Abramovici and D. Saab

Auto-configurable array for GCD computation 457
T. Jebelean

Structural versus algorithmic approaches for efficient adders on
Xilinx 5200 FPGA 462
B. Laurent, G. Bosco and G. Saucier

Control and Robotics

FPGA implementation of real-time digital controllers using on-line
arithmetic 472
A. Tisserand and M. Dimmler

A prototyping environment for fuzzy controllers 482
T. Hollstein, A. Kirschbaum and M. Glesner

A reconfigurable sensor-data processing system for personal robots 491
K. Nukata, Y. Shibata, H. Amano and Y. Anzai

Author Index 501

Multicontext Dynamic Reconfiguration and Real-Time Probing on a Novel Mixed Signal Programmable Device with On-Chip Microprocessor

Julio Faura[1], Juan Manuel Moreno[2], Miguel Angel Aguirre[3], Phuoc van Duong[4], and Josep Maria Insenser[1]

[1] SIDSA, PTM, C/ Isaac Newton 1, Tres Cantos 28760, Spain (faura@sidsa.es)
[2] Universitat Politècnica de Catalunya, c/ Gran Capità, s/n, edif C4, 08034 Barcelona, Spain
[3] Universidad de Sevilla, GTE, Avda Reina Mercedes s/n, 41012 Sevilla, Spain
[4] MIKRON GmbH, Am Söldermoos 17, 85399 Hallbergmoos, Germany

Abstract. In this paper we present a novel RAM-based field programmable mixed-signal integrated device consisting of a large granularity Field Programmable Gate Array (FPGA), a set of programmable and interconnectable analog cells, and a microprocessor core. Two configuration contexts are available, at least one of them mapped on the microprocessor memory space. The microprocessor can be used for partial (or complete) fast dynamic reconfiguration, to run general purpose user programs, and to probe in real time internal points of the analog and digital programmable hardware. The device can be partially or totally reconfigured *while* it is working by loading the new configuration (including initial states for FFs) in the non-active context without stopping operation, then transferring it to the active one in just one microprocessor write cycle.

1 Introduction

In the past years, the microelectronics industry has shown an increasing interest in flexbility and in-house programmability. FPGAs and field programmable microcontrollers have thrived due to this interest, and recently also analog field programmable arrays are also appearing [1], [2], [3]. Mixed signal field programmable arrays [4] and flexible microcontrollers with A/D, DAC and PWM interfaces try to take a step beyond: That of a field programmable *system*.

Within this framework we introduce the FIPSOC (FIeld Programmable System On Chip) prototyping and integration system. The goal is to bring together the benefits reported by field programmable gate arrays to a system integration level. A novel field programmable integrated circuit has been developed including a FPGA, a set of programmable analog cells, and a standard microcontroller. The power of this approach relies on the optimized communication between the three subsystems: The microprocessor can read the outputs of the programmable blocks of the FPGA as memory locations, the microprocessor bus can be directly interfaced to the FPGA routing channels, the digital side of the analog cells (DAC, ADC, comparators) can be directly connected to the digital blocks and to the microprocessor, and the microprocessor can dynamically change the configuration of the digital and analog programmable hardware in terms of memory accesses.

The ability of the microprocessor to overwrite the configuration of the digital and analog cells makes it specially suitable to be used as a dynamic reconfiguration engine. To further enhance this behavior, the configuration cells are duplicated, and at least one of these cells are mapped on the microprocessor memory space. This way, the microprocessor can interact with the configuration memory while the device is working, without having to stop operation while the new configuration is being loaded.

The other great advantage of having an on-chip microcontroller is the possibility to use it as a development system. For this reason, the (digital) outputs of the programmable blocks have been mapped on the microprocessor memory space, so any output of any programmable block can be read in real time. Furthermore, any point of the analog subsytem can be dynamically routed to the internal ADC block, which is directly interfaced to the microcontroller. Therefore, the microcontroller can be used to probe in real time any point of the analog architecture, thus acting as an emulated digital oscilloscope, and any output of the logic programmable blocks, thus emulating a logic analizer.

This paper is mainly focused on the enhancements that an on-chip microprocessor can bring when included in a field programmable device, especially for the multicontext dynamic reconfiguration mechanism and the real time probing capability. The presented material comes from worst case simulation results of a full custom implementation of the programmable cells. Area results are valid since they come from actual full custom implementations included in the first fabricated chip prototype.

2 System Description

Fig. 1 shows a block diagram of the FIPSOC device.

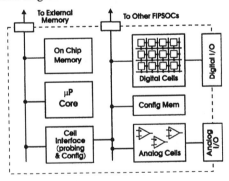

Fig.1. Block diagram of the FIPSOC chip

The chip is a mixed signal field programmable device with an on-chip microcontroller. It includes a Field Programmable Gate Array (FPGA), a set of fixed-functionality yet configurable analog cells, and a microprocessor core with RAM memory and some peripherals. The programmable digital and analog blocks are well defined and are separated from one another due to noise inmunity considerations. Nevertheless, the different interfaces between these blocks themselves and to the microprocessor provide a very powerful interaction between software, digital hardware and analog hardware. A whole family of FIPSOC devices has been planned, with different number of programmable cells in each family member.

The FIPSOC chip includes a two-dimensional array of programmable DMCs (Digital Macro Cell). The DMC is a large granularity, Look Up Table (LUT) based, synthesis targeted 4-bit wide programmable cell, whose structure is depicted in fig. 2. The combinational and sequential outputs of the DMCs are mapped as memory locations in the microprocessor address space. The microprocessor can then *probe* in real time actual digital signals just by issuing memory read commands to these memory locations. Figure 3 outlines the programmable analog tile, a four analog channel section made out of fully balanced differential amplifiers, filters, comparators, and a flexible programmable ADC/DAC structure. More details about the programmable blocks included in the FIPSOC chip can be found in [5].

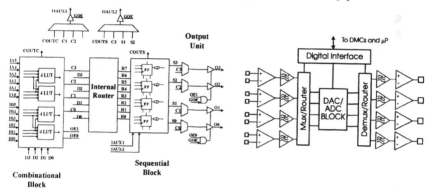

Fig.2. Simplified DMC block diagram   **Fig.3.** Analog block

In particular, it is worth noting that nearly any internal point of the analog block can be routed to the ADC. Then, the microprocessor can use the ADC to probe in real time nearly any internal signal of the analog structure by dynamically reconfiguring these analog routing resources. The ADC/DAC block is especially suitable for reconfigurable applications: it can be configured as one 10-bit DAC or ADC, two 9-bit DAC/ADCs, four 8-bit DAC/ADCs, or even one 9-bit DAC/ADC and two 8-bit DAC/ADCs at the same time. In the latter, two 8-bit DACs can be used to dynamically set the references for the 9-bit DAC/ADC, easily adjusting ranges and offsets on the fly.

As it has been said, the chip contains a standard 8051 microcontroller, which can be used either for general purpose user applications or for configuration management tasks. Apart from FPGA-microprocessor interface issues [6], most of the efforts done for on-chip integration of a microprocessor and programmable hardware have been targeted to enhance the processing power of the microprocessor [7] ,[8], by providing reconfigurable coprocessors or customizable instruction sets, while we have focused a bit more on prototyping needs.

As the commercial device, the microcontroller contains not only a microprocessor core but a serial port (RS232), timers, parallel ports, etc. In particular the serial port is used to allow the chip to communicate to a PC, so the user can download the configuration and debug his applications using a simple monitor program running in the microprocessor and controlling it from the PC. The real time probing of the digital and analog programmable hardware is also done in terms of read data transferred to the PC via the serial port, though it could be also done in a parallel fashion.

3 Configuration Bits and Multicontext Dynamic Reconfiguration

In this section we describe how the configuration memory has been implemented and how the dynamic reconfiguration can be done in the FIPSOC chip. The study that follows is based on a full-custom implementation on real silicon of the. Data about memory stability, noise inmunity, etc. has been extracted from SPICE simulations and validated afterwards in real silicon. Area results are valid since they come from actual full-custom implementations: a small array of DMCs have been included in the fabricated chip.

3.1 Microprocessor Access to the Configuration Memory

As it has been already mentioned, the chip configuration is managed by the internal microprocessor, which can in turn be used as a dynamic reconfiguration engine. To do so, the configuration memory is organized in words which could be mapped onto the microprocessor address space. Such a system may have two important shortcomings: First, a weak memory cell used for hardware configuration would be prone to suffer from noise inmunity problems, so the access and the use of these cells would be tricky; second, a large part of the memory space would be devoted to configuration, which makes inefficient the use of the microprocessor for general purpose user applications: Any user not willing to use the dynamic reconfiguration feature would have to cope with a large and critical part of the memory map devoted to configuration instead of user data or programs.

Mainly due to the first, it was clearly advisable to isolate the *actual* configuration cell from the interfaced memory. Fig. 4.A shows this concept: The actual configuration bit is separated from the mapped memory through a NMOS switch. This switch can be used to load the information coming from the memory bus onto the configuration cell. The microprocessor can only read and write the *mapped* memory, and it could only transfer the information in one way.

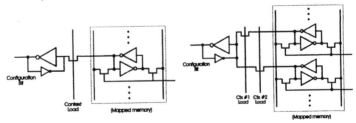

Fig.4.A (left) and 4.B (right). One (left) and two (right) *mapped* context and a *buffered* one.

This technique entails several benefits:

- The mapped memory can be used to run programs and to store general purpose RAM data. After configuration, which would be triggered by a *context load* command after downloading the configuration to the mapped memory, this memory would be free and could be used for something else.

- The configuration bit is isolated from the mapped memory cell. Therefore the memory cell is safer as its output does not have to directly drive the other side of the configuration bit. In our case, the routing architecture, which uses most of the configuration data, is made out of wide multiplexers, so the configuration signals cross with the routing channels and switches and therefore would be prone to be overwritten if they came from relatively weak memory cells.

- As long as these wide multiplexers have been implemented using NMOS switches, it becomes especially interesting to be able to drive them with a voltage level above the power supply. This implementation is especially suitable for this, because the power supply of the *actual* configuration bits can be easily separated from the rest of the circuit.

This implementation is said to have one *mapped* context (the one mapped on the microprocessor memory space) and one *buffered* context (the *actual* configuration memory which directly drives the configuration signals). There was also the possibility of having more than one mapped contexts to be transferred to the buffered context, as depicted in Fig. 4.B. However, as the number of mapped contexts increase, the efficiency of the proposed solution could decrease due to the bigger decoders needed to drive so much memory. The size of the DMC would of course increase, but more user memory would be available to the user.

A final possibility to increase the number of contexts would be to stack them, as depicted in Fig. 5. This technique consists of providing a greater number of contexts, maybe four or eight, but only having access to one of them. Data can be transferred both ways to effectively *push* and *pop* configurations onto the *configuration stack*. This way, a given application would have different smaller subsystems which could be treated as *hardware subroutines* [9]. These subroutines could be hierarchical and call more sub-subroutines, and so on. Everytime a subroutine is released, the actual DMC space upon which it was working would recover the task it was doing at a higher hierarchical level. Of course, the efficiency of this technique would greatly depend on the application and the detailed scheduling of hardware tasks. A further detailed study of this solution would be necessary to assure its feasibility.

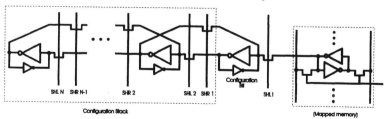

Fig. 5. A configuration stack structure

3.2 What has been implemented

As it has been said, the goal is to provide a familiy of devices. Two full custom implementations have been done, one as depicted in Fig. 4.A and another one as in Fig. 4.B. The DMCs included in the first silicon implementation have two mapped contexts and a buffered one as depicted in Fig. 4.B. Area results given in next section account for this implementation.

3.3 Economics

Table 1 shows an area breakdown of the DMC implemented in silicon:

Table 1. Area breakdown of a multicontext DMC

Function	Area percentage
Routing switches (NMOS multiplexers)	34
Buffered context	9
Mapped context	14
Extra mapped context	9
Four 4-LUTs	20
FFs and their control	4
Output buffers, microprocessor interface logic, etc	3
Internal interconnections and stray area	7

As it can be seen from the table, an extra context is not prohibitive. The area difference between the mapped and the extra mapped context comes from the decoder and its interface to the microprocessor bus - the same decoder is used for both contexts, while it is counted only once in the breakdown. The buffered context is bigger than the mapped one because it uses a larger gate width to be 5V compatible (this is interesting since the routing resources extensively use NMOS multiplexers), and because of noise inmunity considerations. This buffered context should not be completely regarded as an "extra" option to configuration: It provides two polarity signals to the NMOS multiplexers, which would be complicated to route from the mapped memory blocks (which is evenly organized).

It is also important to note that the mapped context, after the configuration load, can be freely used for general purpose use. The memory is distributed along the DMCs, but the microprocessor can use it normally to run programs and to store user data. For a 16x16 DMC array, around 16 Kbytes of RAM memory is provided to the user.

The FFs are also internally duplicated to support the dynamic reconfiguration feature. The area percentage shown in the table accounts for two-context FFs.

An important point is the LUT configuration data. The area required by the LUTs includes the memory cells (which store the configuration), their connections and their interface. As far as this amount of area is somewhat large, the multicontext implementation of the LUTs must be studied separately.

3.4 The LUTs

Four LUTs have been included in the DMC. They have been implemented as 16x1 dual port memory blocks. One of these two ports is connected to the microprocessor interface, while the other one can be explicitly connected to the DMC inputs and outputs - to use the LUT as an actual user memory block with read and write access -, or implicitly connected to perform custom boolean functions - no user write access provided here.

Every LUT has been designed to be used as a single 16-bit LUT or as a two-context 8-bit LUT. A (multicontext) configuration bit decides the LUT mode (static or dynamically reconfigurable) and the context being used in each moment. As far as every two static 4-LUTs share two inputs (that makes a total of six inputs per pair of LUTs), two dynamically reconfigurable 3-LUTs are available in this mode, sharing no inputs. This way, no input pins are wasted, and while dynamic LUTs are smaller, thus supporting smaller functions, they share no inputs, increasing flexibility.

Exactly as the static 4-LUTs in a DMC can be combined to form two 5-input LUTs or one 6-input LUT, in the dynamically reconfigurable mode one can have four 3-LUTs, two 4-LUTs or one 5-input LUT. The rest of the functionality is the same. In particular, the 4-bit adder macro mode fits in the dynamically reconfigured mode, i.e., it can be performed with 3-LUTs.

3.5 Dynamic reconfiguration

An optimized interface between the microprocessor and the configuration memory, suitable for partial dynamic reconfiguration, is provided. Fig. 6 schematically shows how this interface works, not very different from existing ones [6].

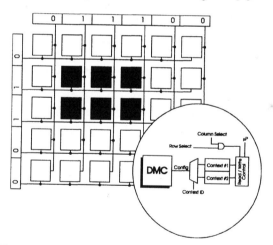

Fig. 6. Microprocessor interface for dynamic reconfiguration

A mask of columns and rows can be selected for any memory write operation, including a context load command. Therefore, any *logical rectangle* of DMCs can be selected for reconfiguration, so any write command from the microprocessor would apply to all members of the selected set at the same time. We define a logical rectangle as a set of DMCs (X,Y) such as if (X1,Y1) and (X2,Y2) are in the set, then (X1,Y2) and (X2,Y1) are also in. This structure is especially efficient for array based applications such as systolic multipliers, array-based FIR filters, etc.

For a DMC structure based on the configuration cell depicted in Fig. 4.A, dynamic reconfiguration would work as follows: Upon reset, the microprocessor would load the initial configuration on the mapped context, and the circuit would start working. Then, while the circuit is working, the microprocessor would fill the mapped context of the set of cells to be reconfigured, which does not interact with the active context (the buffered one) of these or any other cells. The initial states of the FFs in the new context could also be loaded without affecting the actual values stored in the active FFs. When the new configuration is loaded on the mapped context, a mask would be set to design the set of cells going to be reconfigured. Hitherto the circuit is still working with the old configuration. Then, a single write command is performed (the *context load* signal in Fig. 4.A is activated) and the selected part of the circuit starts working with the new configuration, including the desired initial states of the FFs.

Furthermore, the FFs can be used in single context or double context mode. In the single context mode the data is kept upon reconfiguration, just as global variables or pointed parameters in software programs and subroutines. In the double context mode data would be swapped away with the context change, and it could be swapped back upon a new reconfiguration. This second mode is especially useful for storing initial values before the reconfiguration takes place.

A DMC structure with two mapped contexts, as depicted in Fig. 4.B, goes one step forward in reconfiguration: The microprocessor can load two different configuration contexts upon reset, then dynamically change between them on the fly with no extra reconfiguration time (just that of a microprocessor write cycle). If only two contexts are necessary, this approach would be enough for a number of applications.

A related work which should be mentioned here is the DPGA chip [10], a real multi-context device for programmable combinational functions based on dynamic RAM. Finally, it is worth noting how in [11] an interesting study on dynamic reconfiguration and its applications can be found. It can be seen how the improvements suggested there have been already implemented in this new system.

4 The FIPSOC as an Integrated Prototyping Workbench with Internal Probing Capability

Another interesting possibility comes from the optimized interface between the microprocessor and the programmable hardware itself, not its configuration memory as we have already studied. It is the possibility of probing in real time nearly any point of the analog or digital user application mapped onto the programmable hardware. In fact, it is possible to emulate a whole laboratory benchmark using just a FIPSOC Chip and a PC. The communication between them would normally be done through the on-chip RS232 serial port (an external driver is needed for RS232 voltage levels).

Logic data acquisition for this real time probing can be done in terms of memory read operations from the outputs of the DMCs, which are mapped as memory locations on the microprocessor memory space. LUTs configured as memories and dedicated DMCs configured as counters could also be used for fast logic data acquisition like in logic analizers. Note that the microprocessor can look up data from the LUTs while the LUT is in operation.

As it has been said, the internal ADC block can be dynamically rewired to probe nearly any internal point of the analog architecture. An analog data acquisition system, emulating a digital oscilloscope, could then be done using the microcontroller or some dedicated DMCs, as far as the digital side of the ADC can be connected to the digital routing channels and can be directly interfaced to the microprocessor. Even the internal DAC could also be used as a function generator, accepting data from the microprocessor core or from some DMCs configured as counters and memories.

The rest of the laboratory workbench is a matter of software: A digital oscilloscope could be emulated to present the acquired analog data (from the internal ADC) on the PC screen; a logic state analyzer could be provided printing out the digital data obtained from the DMC outputs in real time. The real advantage is the integrated approach that the user can follow to develop his system application: Everything is integrated, and the final hardware and software solution is exactly what the user is measuring when probing in real time. The proposed analog programmable hardware is especially suitable for this operation. In particular, the possibility of severing the ADC/DAC block into a 9-bit ADC and two 8-bit DACs, and then using these two DACs to provide the ADC references, greatly simplifies the emulated digital oscilloscope described: the "offset" and "amplification factor" knobs would directly apply over the reference DACs.

5 CAD Tools for FIPSOC

An integrated set of software CAD tools is being developed for design entry and optimization, technology mapping, placement and routing, device programming, mixed-signal simulation and real-time system probing from a Windows^{TM}-based PC station.

A dynamic reconfiguration management tool, able to handle the multicontext operation of the chip, would constitute a very interesting research area here. Such a tool could analyse HDL code to check the coincidence in time of the processes, their criticallity and their system requirements. Up to now, some experiences on dynamic reconfiguration software have already been reported [12].

The FIPSOC chip is especially suitable for hardware-software co-design techniques due to the flexible interfaces between the programmable hardware areas and the micrprocessor core, which results in a very powerful hardware-software interaction. This interaction is enhanced mainly due to: *a)* Internal signals from the programmable hardware can be probed and read as memory locations from the microprocessor core (analog signals have to be converted with the ADC). *b)* The microprocessor can dynamically reprogram a piece of hardware by overwritting the configuration memory. A co-design CAD tool could then be targeted to this device, putting in hardware those critical processes needing too high a computational speed, and performing with software those tasks which would be prohibitively area-consuming.

We think that FIPSOC would be a suitable becnhmark platform for this kind of tools and methodologies.

6 Conclusions

In this paper we describe some of the benefits of including an on-chip microprocessor within a mixed-signal field programmable device, in terms of enhanced dynamic reconfigurability and integrated system prototyping. A new dynamic reconfiguration technique, based on buffered contexts, is introduced. An actual silicon implementation of the reconfiguration model has already been done and shows the feasibility and effectiveness of this technique: An extra mapped context is not prohibitively large; the reusability of the configuration memory is maximized because the mapped context can be used for general purpose programs and data if no dynamic reconfiguration is desired; the buffered context is isolated from the weak memory cells, prone to be affected by the circuit noise; and the buffered context can be independently powered, which is interesting if voltage boosting is necessary when NMOS routing switches are used.

The optimized interface between the microprocessor and the actual signals mapped on the programmable hardware makes possible to provide a whole development system, including an emulated digital oscilloscope and logic analizer, with just a FIPSOC chip and a host computer, namely a PC (a RS232 driver is also needed to convert the 3V chip signals to RS232 levels). This arises from the fact that the outputs of the programmable logic blocks are mapped on the microprocessor memory space, and the internal points of the analog hardware can also be probed with the internal ADC/DAC block.

7 Acknowledgements

This work is being carried out under the ESPRIT project 21625. The authors would like to thank the European Commission for the financial support.

References

1   D. Anderson, C. Marcjan *et al*, *"A Field Programmable Analog Array and Its Application"*, 1997 IEEE Custom Integrated Circuits Conference, Santa Clara (CA, USA).

2   Hans W. Klein, *"The EPAC architecture: an expert cell approach to field programmable analogue devices"*, FPGA'96, Monterrey CA.

3   *"Analog-Silicon-Breadboard - an analogue FPGA"*, Fraunhofer-Institut für Mikrolektronische Schaltungen un Systeme, IMS2 Dresden (commercial communication).

4   Chengjin Zhang, Adrian Bratt and Ian Macbeth, *"A New Field Programmable Mixed-Signal Array And Its Application"*, The 4th Canadian Workshop on Field-Programmable Devices, 1996 Toronto, Canada.

5   Julio Faura, Chris Horton, Phuoc van Duong, Jordi Madrenas, Miguel A. Aguirre, and Josep M. Insenser *"A Novel Mixed Signal Programmable Device with On-Chip Microprocessor"*, 1997 IEEE Custom Integrated Circuits Conference, Santa Clara (CA, USA).

6   Stephen Churcher, Tom Kean and Bill Wilkie, *"The XC6200 fastMapTM processor interface"*, European FPL'95, Oxford (UK).

7   H.F. Silverman et al, *"Processor Reconfiguration Through Instruction-Set Metamorphosis"*, IEEE Computer, 23 (3) March 1993.

8   Neil Hastie and Richard Cliff, *"The implementation of hardware subroutines on field programmable gate arrays"*, 1990 IEEE Custom Integrated Circuits Conference.

9   Neil Hastie and Richard Cliff, *"The implementation of hardware subroutines on field programmable gate arrays"*, IEEE 1990 CICC.

10   E. Tau, D. Chen, I. Eslick, J. Brown and A. DeHon, *"A First Generation DPGA Implementation"*, FPD'95 -- Third Canadian Workshop of Field-Programmable Devices, 1995 Montreal, Canada.

11   Patrick Lysaght and John Dunlop, *"Dynamic reconfiguration of FPGAs"*, European FPL'93, Oxford (UK), pp 82 to 94.

12   Patrick Lysaght and Jon Stockwood, *"A Simulation Tool for Dynamically Reconfigurable Field Programmable Gate Arrays"*, IEEE transactions on VLSI systems, vol. 4, nr. 3, September 1996.

CAD-oriented FPGA and Dedicated CAD System for Telecommunications

Toshiaki Miyazaki, Atsushi Takahara, Masaru Katayama,
Takahiro Murooka, Takaki Ichimori[†], Kennosuke Fukami[††],
Akihiro Tsutsui[†††], Kazuhiro Hayashi[†††]

NTT System Electoronics Laboratories
A1-329S, 3-1 Morinosato Wakamiya, Atsugi, 243-01 JAPAN
e-mail: miyazaki@aecl.ntt.co.jp

[†] NTT Network Service Systems Laboratories

[††] NTT Science and Core Technology Laboratory Group

[†††] NTT Optical Network Systems Laboratories

Abstract. This paper describes a newly developed FPGA and its dedicated CAD system. The FPGA is an improved version of our previous telecommunication-based FPGA, especially in terms of the routing resource architecture. Thus, in addition to having the good features of our previous FPGA for realizing telecommunications circuits, it enables us to adopt a top-down design methodology for application circuits configured in the FPGA. The architecture is determined based on a quantitative evaluation carried out to balance the FPGA with CAD algorithms.

1 Introduction

Today, more functions are needed in each protocol layer to realize flexible multimedia services in digital telecommunication networks [1]. However, current telecommunication systems are often constructed by dedicated hardware. This is because data transmission requires high throughput and various bit-level manipulations must be performed, which CPUs or DSPs cannot handle well. So, the implementation of telecommunication circuits is far from rich in terms of flexibility. To remedy this situation, we have developed a telecommunication-based FPGA called *PROTEUS* [2] and reconfigurable telecommunication systems utilizing it [3]. *PROTEUS* was developed based on analytical results for telecommunication circuits [4], which have the following characteristics;

- relatively many latches compared to the amount of combinational logic,
- a lot of pattern matching operations, and
- strong directions in the main data streams.

Thus, application circuits related to telecommunications can easily perform real-time operations with the *PROTEUS* FPGA. Unfortunately, the logic vs. routing resource balance in *PROTEUS* was not optimal, and its dedicated CAD system [5], especially the router, suffered from a lack of routing resources. To overcome these problems, we have newly designed a well-balanced FPGA using a FPGA/CAD co-evaluation system called *FACT* [6]. The FPGA architecture has been improved to the point where the CAD system can easily handle it without sacrificing the good features of the original *PROTEUS* when the user designs an application circuit.

In this paper, how we decided upon the new FPGA architecture is described first. Next, we introduce newly developed FPGA, which is called *PROTEUS-Lite*. Finally, its dedicated CAD environment with some experimental results is discussed.

2 Quantitative Analysis

The *PROTEUS-Lite* architecture was determined quantitatively, using *FACT* system [6]. *FACT* is an FPGA/CAD co-evaluation system. It can simulate ordinary circuit-design processes for FPGAs such as technology mapping, placement, and routing. In addition, the user can define a new FPGA architecture using the Architecture Definition Format (ADF) provided in *FACT* system.

A routing bottleneck often occurs at the inputs or outputs of each logic block, which is called *Basic Cell*, or *BC*, in *PROTEUS* and *PROTEUS-Lite*. Thus, we concentrated on improving the routing topology around the BCs.

First, we considered the three architecture types shown in Fig. 1. Each type has a symmetrical array structure. However, the input and output directions to/from each BC differ among the three types. In type *A*, the inputs come from both the left vertical and upper horizontal routing channels, and the outputs go to both the right vertical and upper horizontal channels. In type *B*, the inputs come only from the upper horizontal channel, and the outputs also go to it. In type *C*, both the inputs and outputs are connected to the upper and lower horizontal channels. Six circuits were applied to the above architectures while the numbers of tracks in the vertical and horizontal channels and switch pattern in each switch box were changed.

We evaluated a total of 23 architectures in the three categories. The results are shown in Fig. 1. In the graph, the y-coordinate indicates the number of unrouted nets and the x-coordinate indicates the number of switches in each cyclic pattern in the FPGA, i.e., the switches in several switch boxes. The two numbers in parentheses are the numbers of horizontal and vertical tracks in each routing channel. The target area had less than 450 switches per one cyclic pattern and less than ten unrouted nets. These were not strict constrains; they were decided from the standpoints of hardware cost and CAD aspect.

In one type *A* architecture, all nets are completely routed, i.e., the number of unrouted nets is zero. The architecture has complete connection switches that can connect any two tracks coming into the same switch box. Thus, we concentrate only on the width of each channel, and the routing results indicate that the architecture has enough channel width (in this case, twenty) to route the circuit data. However, this architecture requires too many switches. So, we also

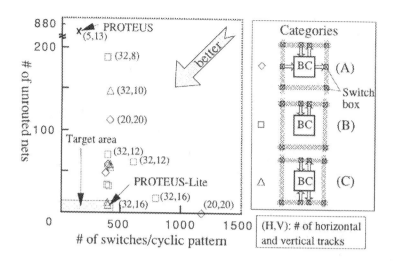

Fig. 1. Architecture Types and Evaluation Results

recognized that it is difficult to realize the desired architecture without reducing the number of switches.

Type B is aimed at row-based routing, because the inputs and outputs of each BC are connected to the upper horizontal channel. Unfortunately, the routing results are not so good. However, there is still room to apply channel routing algorithms developed for ASICs to improve the routing. In fact, the routing tool we used in this evaluation is based on a *Dijkstra*-based algorithm, not a channel routing algorithm.

The type C architecture is a modification of type B. That is, the inputs and outputs of each BC are connected to both the upper and lower horizontal channels. These architectures achieve relatively good routing results compared to types A and B, without increasing the number of switches.

Finally, we chose a type-C architecture with 32 horizontal and 16 vertical tracks. We expected much improved routability compared to *PROTEUS* architecture, which has a type-A architecture minus the upper channel connections.

3 PROTEUS-Lite FPGA

3.1 Chip Overview

PROTEUS-Lite is a Look-Up Table (LUT)-based FPGA [7]. A *PROTEUS-Lite* chip overview is shown in Fig. 2. This chip comprises *Basic Cells* (BCs), *I/O Blocks* (IOBs) and routing resources. The BCs are for circuit logic, and they are placed regularly. As shown in Fig. 2(d), a BC has four 3-input 1-output LUTs (3-LUTs) and one 5-input AND (5-AND) gate. There are 136 IOBs, 28 located on the up side, and 28 on the bottom side, and 40 each on the left and right sides. Here, the 80 IOBs on the sides are grouped by four IOBs, and

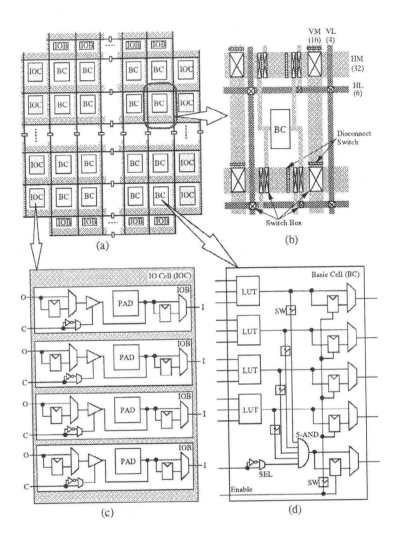

Fig. 2. PROTEUS-Lite Architecture

they structures one I/O Cell (IOC) as shown in Fig. 2(c). The specifications of *PROTEUS-Lite* and the die photograph are shown in Fig. 3. The chip was fabricated using 0.5 μm CMOS technology and packaged in the three types of LSI packages: QFP, CSP, and PGA. There are several configuration modes including "synchronous", "asynchronous", and "memory clear". In addition, the standard JTAG boundary-scan feature is implemented for testing.

In short, the *PROTEUS-Lite* architecture features

- a BC structure that can easily realize basic functions such as pattern matching,
- a lot of regularly placed latches that can perform pipelined data processing for high data throughput,

- local lines for connecting neighboring LUTs, and
- an identical IOC(IOB) and BC connection topology for the surrounding routing resources. This homogeneous wiring topology simplifies the routing algorithm.

All features except the last one are basically inherited from *PROTEUS*, but they are enhanced.

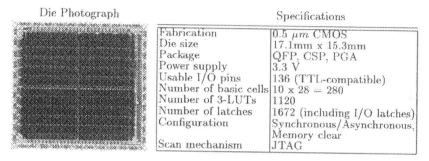

Die Photograph	Specifications	
	Fabrication	0.5 μm CMOS
	Die size	17.1mm x 15.3mm
	Package	QFP, CSP, PGA
	Power supply	3.3 V
	Usable I/O pins	136 (TTL-compatible)
	Number of basic cells	10 x 28 = 280
	Number of 3-LUTs	1120
	Number of latches	1672 (including I/O latches)
	Configuration	Synchronous/Asynchronous, Memory clear
	Scan mechanism	JTAG

Fig. 3. Die Photograph and Specifications of PROTEUS-Lite Chip

3.2 Basic Cell

As shown in Fig. 2(d), there is one 5-AND gate after four 3-LUTs in a BC. This structure makes it easy to implement basic functions for telecommunication circuits such as pattern matching. In addition, a latch is located at each output of the 3-LUTs and the 5-AND gate, and it can be programmed to latch or not latch the output. There is a switch between an LUT output and one input of the 5-AND gate. In the *OFF* state, the input of the 5-AND gate is disconnected from the LUT and is pulled up. So, the 5-AND gate can be used as a one- to five-input AND gate by controlling the switches.

3.3 Data Latch and Clock Signal

As shown in Fig. 2(d), data latches are located at each output of the 3-LUTs and 5-AND gates. Users can select the outputs they want latched when the *PROTEUS-Lite* chip is programmed. Furthermore, the latching of a new signal in the current clock can be controlled by the enable lines. The total number of latches including those in the IOBs is 1672. Compared to commercial FPGAs, the ratio of the number of latches to the logic size realized with the LUTs is about twice, and this is important for realizing practical telecommunication circuits.

A clock signal is delivered to all latches in a *PROTEUS-Lite* chip using dedicated clock lines. The clock lines were implemented in the fabrication process, and, of course, decreasing the clock-skew was considered. It is less than 1 ns. In addition, each latch has a clock enable pin. Thus, we can easily design circuits that can drive a large number of sub-modules using a timing pulse. This kind of circuit is often used in data transmission circuits.

3.4 Routing Resources

Compared to the previous FPGA, the routing resource architecture was drastically changed in order to improve routability. *PROTEUS-Lite* has three kinds of routing resources: *local lines*, *middle lines*, and *long lines*. Each routing resource plays a different role in achieving effective routing.

Local Line As shown in Fig. 4, local lines are used to connect neighboring LUTs directly. A mesh connection of all LUTs in the chip can be realized using the local lines without worrying about the boundary of each BC. The output of an LUT connects five adjacent LUTs; the LUT above, below, to the right, to the upper-right, and to the lower-right. We call this structure *"Sea of LUTs"*. Compared to the other routing resources, the local lines have the smallest propagation delay. Thus, they should be used as much as possible to make high-performance circuits. Our preliminary experimental results showed that 33 percent of all nets in a circuit on average can be routed using only local lines.

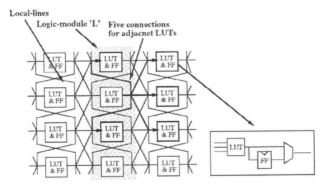

Fig. 4. Local Line Structure: Sea of LUTs

Middle Lines and Long Lines Details of the routing resource architecture are shown in Fig. 2(b). To connect BCs and IOBs, middle lines and long lines are provided. Long lines are used to establish long-distance routes. The length of each middle line is equivalent to the length of one BC, and a long line runs over half of the chip. A long-line segment has several times larger delay than one middle line. Thus, middle lines should be used for rather near connections. However, long lines are better for long-distance routes.

Switch Patterns The switch patterns in each switch box were decided in consideration of the routability determined from the architecture evaluation process described in Section 2. Figure 5 shows simplified but essential switch patterns in *PROTEUS-Lite*. As shown in the figure, each input or output line to/from a BC or an IOC is connected to both upper and lower middle lines through the switch boxes, and the switches in one pair of upper and lower switch boxes are located on different horizontal middle-line tracks. For example, the switches in switch box SB3 are located on odd tracks, and those in SB4 are located on even tracks. With these switch patterns, a signal on the odd track such as "a" in Fig. 5 can be fed to the BC or IOC through the upper switch box, and an even-track signal is fed through the lower box. Furthermore, even-track signals, like "b" in Fig. 5, can be routed to the corresponding track in the lower middle lines via two switch boxes, SB1 and SB2, which connect vertical and horizontal middle lines. This is because the switches are placed diagonally in each switch box. These characteristics of the switch patterns contribute greatly to fast routing because the track candidates can be easily selected by simply checking the switch patterns or looking up some pattern tables. In fact, our router with the *line-search* algorithm reduced routing time by more than 90 percent, compared to the *Dijkstra*-based router in the *FACT* system, and the routing process for a circuit, which occupies more than 50 percent of the FPGA area, often finishes within fifteen minutes.

In addition, wires W1 and W2 in Fig. 5 can be used as *bypass* lines to connect upper and lower middle lines by opening switches SW1 and SW2, if the BC or IOC is unused.

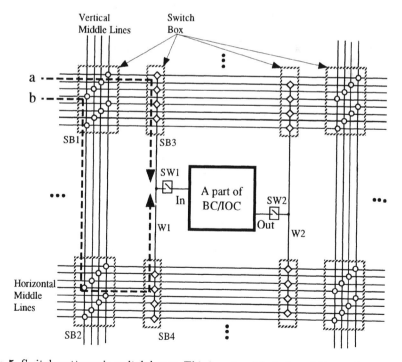

Fig. 5. Switch patterns in switch boxes. This is a simplified view of the routing resource architecture around a BC or an IOC.

4 PLCAD

4.1 System Overview

A dedicated CAD system called *PLCAD* was developed for *PROTEUS-Lite*. An overview of the system is shown in Fig. 6. *PLCAD* supports a top-down design methodology. With the Synopsys' Design Compiler^{TM} or the PARTHENON logic synthesis system [8], the programming data downloaded into *PROTEUS-Lite* can be obtained directly from RTL descriptions.

In addition, a visual design editor was developed to allow the user to interfere freely on any design level. Using the design editor, any part of a designed circuit can be saved at any time, and can be re-used as a *hardmacro* in other design phases. An actual programming data stream is created after design rule checking (DRC). The static delay calculator is based on a modified Elmore model [9] and can estimate critical path delays to an accuracy within 4 % on the average. Furthermore, there is a netlist converter, which produces Verilog-HDL descriptions from our original netlist format. It can also generate delay information in the SDF format if necessary. Thus, we can link to other CAD tools and take the application circuit to an ASIC design process.

5 Experimental Results

Using the *PLCAD* system, some telecommunication circuits were evaluated. The results are shown in Table 1. The results for the same circuits implemented as *PROTEUS* are also shown for comparison. In the table, *#LUTs* and *#Nets* indicate the number of used LUTs and the number of nets that should be routed, respectively. *Time* represents the total execution time for technology mapping, placement, and routing in seconds, and *#UN* is the number of unrouted nets. In *PROTEUS-Lite*, 100 % routing is performed for every circuit, while the execution time is comparable to that of *PROTEUS*. This indicates our architecture is enhanced from points of view of routability or creating a CAD-oriented FPGA.

6 Conclusion

We introduced a telecommunication-based FPGA called *PROTEUS-Lite* and its CAD system, which was developed according to FPGA/CAD co-evaluation results. The FPGA architecture and CAD system are well-balanced. Thus, in spite of the rich routing resources in the FPGA, the routing process finishes rapidly without sacrificing application-circuit performance. This is vital for realizing a top-down design environment.

The FPGA has been applied to a flexible ATM system. In addition, as an application of the *PROTEUS-Lite* FPGA, we are developing a real-time rapid prototyping system, which consists of heterogeneous boards, i.e., FPGA boards, 156Mbps (OC-3) x4 optical I/F boards, 8 x 8 ATM switch boards, and MPU

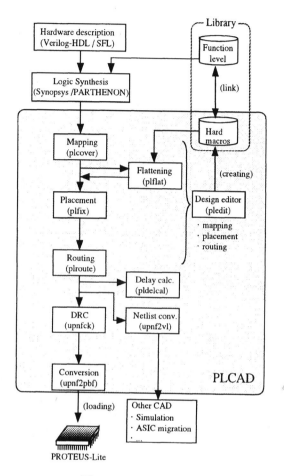

Fig. 6. PLCAD system

Table 1. Experimental results

Circuit	PROTEUS #LUTs	#Nets	#UN	Time	PROTEUS-Lite #LUTs	#Nets	Time
crc2_2p	12	16	0	150.1	10	14	98.5
atm_d_t	47	56	1	131.8	37	46	105.3
add_8	64	81	3	140.9	59	76	108.5
atm_d_g	169	190	7	235.0	154	175	207.7
50msp	185	195	1	198.7	181	191	131.2
n1s	191	193	7	215.4	161	163	149.9
atm_g	254	255	31	346.1	237	237	385.6
aal_1_s	611	639	169	711.1	504	531	659.1

$\#UN$ = the number of unrouted nets. (In *PROTEUS-Lite*, $\#UN = 0$ for every data.)
Time (s) = technology mapping + placement + routing.

boards. It is designed to emulate telecommunication functions such as an IP (Internet Protocol) routing and ATM-cell switching in real time. One of our future tasks is to construct system-design environment for the prototyping system.

Acknowledgment

The authors would like to express their gratitude to Mr. Kenji Ishii and Mr. Hideyuki Tsuboi for providing part of the experimental results. They also wish to thank Mr. Tadanobu Nikaido, and all members of the LSI Architecture Research Group and Programmable Transport Research Group for their helpful suggestions.

References

1. W. Stallings, "Advances in ISDN and Broadband ISDN," IEEE Computer Society Press, 1992.
2. A. Tsutsui, T. Miyazaki, K. Yamada, and N. Ohta, " Special Purpose FPGA for High-speed Digital Telecommunication Systems," Proc. ICCD'95, pp. 486-491, October 1995.
3. K. Hayashi, T. Miyazaki, K. Shirakawa, K. Yamada, and N. Ohta, "Reconfigurable Real-Time Signal Transport System using Custom FPGAs," Proc. FCCM'95, pp. 68-75, April 1995.
4. N. Ohta, H. Nakada, K. Yamada, A. Tsutsui, and T. Miyazaki, "PROTEUS: Programmable Hardware for Telecommunication Systems," IEEE Proc. ICCD'94, pp. 178-183, October 1994.
5. A. Tsutsui, and T. Miyazaki, "An Efficient Design Environment and Algorithms for Transport Processing FPGA," Proc. VLSI'95, pp. 791-798, August 30-September 1 1995.
6. T. Miyazaki, A. Tsutsui, K. Ishii, and N. Ohta, "FACT: Co-evaluation Environment for FPGA Architecture and CAD System," Proc. FPL'96, pp. 34-43, September 1996.
7. S.D. Brown , R.J. Francis, J. Rose, and Z.G. Vranesic, "Field-Programmable Gate Arrays," Kluwer Academic Publishers, 1992.
8. Y. Nakamura, K. Oguri, A. Nagoya, and R. Nomura, "A Hierarchical Behavioral Description Based CAD System," Proc. EURO ASIC, 1990.
9. W.C. Elmore, "The Transient Response of Damped Linear Networks with Particular Regard to Wideband Amplifiers," J. Appl. Phys., Vol. 19, No. 1, pp. 55-63, 1948.

Rothko: A Three Dimensional FPGA Architecture, Its Fabrication, and Design Tools

Miriam Leeser, Waleed M. Meleis, Mankuan M. Vai, and Paul Zavracky

Department of Electrical and Computer Engineering
Northeastern University
Boston, MA 02115, USA

Abstract. We are designing and plan to fabricate a 3-dimensional field programmable gate array. The three dimensional VLSI technology, developed at Northeastern University, is based on transferred circuits with interconnections between layers of active devices. Interconnections are in metal, and can be placed anywhere on the chip. Our FPGA architecture, called Rothko, is based on the sea-of-gates FPGA model first proposed in the Triptych architecture[1] (a 2-D architecture) in which individual cells can be used for routing, logic, or both. We provide 3-D connections to each cell from above and below. This makes our architecture truly 3-D with each cell having connections to cells on other layers. In this paper we present the architecture of Rothko, discuss the 3-D technology we use, and discuss CAD tools for mapping designs onto Rothko.

1. Introduction

To achieve new levels of integration and utilization in field programmable logic requires new FPGA architectures. Problems with existing architectures include low utilization of resources, routing congestion, high delay due to interconnect and insufficient I/O. In this paper we present a novel 3-D FPGA architecture aimed at solving some of these problems.

One of the main obstacles to mapping large designs onto existing FPGA architectures is routing congestion. FPGA routing resources are more expensive than ASICs because programmable interconnections are required. Not only do these programmable interconnections prevent a more efficient routing, they also introduce longer propagation delays. By going to a 3-D design which allows flexible interconnect in every dimension we expect to be able to relieve routing congestion and shorten interconnect lengths dramatically, thus improving speed. The speed of an FPGA is a measure of the delay required to implement a function and to propagate signals to neighboring functions. The FPGA logic speed is often slow due to the interconnect delay. This delay can account for over 70% of the clock cycle period[2].

Another problem with FPGA designs is the number of I/O connections available. According to *Rent's Rule*, the number of I/O pins needed on a given FPGA grows faster than the square root of the number of logic elements. However, the number of perimeter bonding pads that can fit along the die periphery only grows as the square root of the area. This means that for a given pad pitch (about 100 microns) and logic element pitch, there will be a die size beyond which the demand for I/O far exceeds the supply, in which case the device becomes pin limited. Experience with existing FPGAs shows that this results in low logic element utilization.

Others have proposed using multichip module technology (MCMs), area-I/O, and optical interconnections to address some of these issues. These technologies all require that interconnections between chips or layers go through I/O pads and solder bumps. The geometries of solder bumps are on the order of 100 microns, an order of magnitude larger than our interconnections. In addition, I/O pads plus solder bumps are inherently more power hungry and complex than our technology which uses aluminum to interconnect chip layers.

A major advantage of 3-D VLSI technology is shorter interconnections. A 3-D FPGA has the following advantages over an MCM FPGA:

- More logic units are available in the same foot print area.
- The planar long interconnections between FPGAs in an MCM are replaced with significantly shorter vertical interconnections.
- The shorter average interconnect distance in a 3-D FPGA ($O(n^{1/3})$ for an n-block 3-D FPGA vs. $O(n^{1/2})$ in the 2-D case) implies shorter signal propagation delay.
- The increased number of logic block neighbors (e.g., 8 in 3-D vs. 6 in 2-D for our architecture) affords greater versatility and resource utilization.
- The power consumption is significantly lower due to the elimination of I/O pins and planar long interconnections between FPGAs.

Researchers at the University of Virginia have demonstrated the theoretical advantages of 3-D FPGAs[3,4]. Their benchmarks, which assume a standard FPGA architecture and straightforward extension of switch boxes to 3-D technology, demonstrate that a 3-D FPGA can shorten the average net length by 13.8%, reduce the number of switches by 23%, and reduce the radius of logic elements by 26.4% Our architecture, which makes better use of routing resources and 3-D interconnections should improve on these figures.

In the rest of this paper we discuss the Rothko architecture in detail, describe the technology for building 3-D circuits, and present the CAD tools we are developing in order to make use of the Rothko architecture. First we discuss related work.

1.1 Related Work

Our work is closely related to research on multichip modules (MCMs) for FPGAs. There are other groups investigating three dimensional FPGA architectures; most of these architectures depend on multichip module (MCM) packaging techniques. Researchers at the University of California at Santa Cruz have shown the advantages of placing more programmable resources with smaller interconnect, of using area I/O in place of perimeter I/O and of stacking MCMs vertically by making use of solder bump connections and area I/O[5,6]. Even with MCMs stacked vertically, connections must still go through I/O pads and solder bumps. These connections are larger than our vertical vias, and are inherently slower and more power hungry due to the size and the need to drive I/O pads and to use solder for connections. Finally, due to their large size only a small number of such connections can be handled between chips. Our technology will allow more connections between chips due to the smaller size of our interconnections. Others have begun building 3-D FPGAs using MCM-V technology[7]. While this technology does not use solder bumps, it requires that all connections between layers are done at the periphery of a chip. Our technology allows connections between layers to be made anywhere on the chip.

A three dimensional FPGA using optical interconnections has been demonstrated[8]. Their optical interconnections are very complex, requiring drivers, LEDs, detectors and receivers. The technology to integrate optical and CMOS on the same chip is not sufficiently mature, so they mount bare FPGA die on chip carriers with the optical elements. Our interconnections are much simpler to manufacture and are completely integrated with the FPGA circuitry.

Researchers at the University of Virginia have implemented 3-D place and route tools and have shown the improvement by applying their three dimensional tools to a hypothetical 3-D FPGA[4]. Their results show the importance of pursuing an architecture like Rothko.

2. Rothko Architecture

Rothko is based on the sea-of-gates FPGA model first proposed in the 2-D Triptych architecture[1]. The sea-of-gates FPGA architecture integrates logic and routing resources, and has provided a logic density improvement of up to a factor of 3.5 over commercial FPGAs with comparable performance. This improvement is obtained by allowing a per-mapping tradeoff between logic and routing resources, and with a routing scheme designed to match the structure of typical circuits.

2.1 Sea-of-Gates Architecture

FPGAs currently available are based on a strict separation between logic and routing resources. This strict separation is too confining and the resources are often under-utilized. The sea-of-gates approach, first proposed by the Triptych architecture, allows the split between logic and routing area to be made on a per-mapping basis. This scheme addressed two fundamental efficiency problems: increasing routing area and decreasing cell utilization.

In the Triptych architecture, the logic blocks of standard FPGA architectures are replaced with Routing & Logic Blocks (RLBs), which perform both logic and routing tasks. In addition to the capability of routing signals through RLBs, two types of routing resources are provided between RLBs. Short, fast direct connections are provided between neighbor RLBs. Segmented routing channels are provided between columns to connect RLBs in these columns and facilitate larger fanouts.

2.2 Rothko: A 3-D FPGA Architecture

The FPGA layers to be stacked together in the Rothko architecture are based on the sea-of-gates FPGA model. The Triptych RLB is adapted to provide programmable vertical connections between layers. In the process we developed a new routing structure. While the original Triptych routing structure, which has two overlaid arrays of RLBs routed in opposite directions, may work well in a 2-D architecture, we have found that a 3-D technology allows a more flexible routing structure to be used. Fig. 1 shows a layer of this new routing structure in which all the RLBs in the same layer are routed in the same direction. Two adjacent FPGA layers assume opposite routing directions.

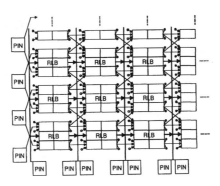

Fig. 1. The routing structure of a Rothko FPGA layer.
(○: Connections to the layer above; ●: Connections to the layer below).

Another important feature of the Rothko architecture is the vertical metal connections between adjacent layers. Simulation results have shown that these vertical metal connections have similar delay properties to the intra-layer direct connections. Fig. 1 shows a logical and cost effective way of providing vertical connections between layers.

In addition, a segmented routing channel is provided between columns to connect RLBs beyond the reach of the direct connections described above. Fig. 2 shows the connectivity of the segmented channel which was adapted from the Triptych architecture. The Rothko segmented routing channel includes 7 tracks, with 5 handling inter-cell RLB routing and 2 carrying pin inputs. There are two tracks provided between 8 RLBs (4 sources and 4 sinks), two tracks provided between 16 RLBs (8 sources and 8 sinks), and one track provided between 32 RLBs (16 sources and 16 sinks). Segmented routing channels have the advantage of not requiring active switching circuits in the routing channel. However, since a small driver is driving the signal a potentially large distance, the delay due to routing in the channels can be significant. The design of a Rothko RLB is shown in Fig. 3.

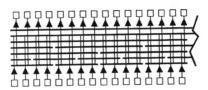

Fig. 2. A portion of the segmented channel.

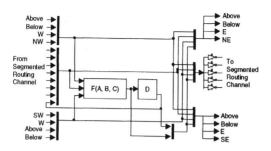

Fig. 3. The Rothko RLB design.

3. Mapping Example

This section describes an example of mapping a 4-bit×4-bit combinational array multiplier, which is an important building block for many signal processing applications. The mapping was performed on both a two layer Rothko architecture (Fig. 4) and the Triptych architecture (not shown) for the purpose of comparison. Table 1 summarizes the comparison results between these two mappings. The criteria we used to judge the quality of a mapping include:

- the footprint;
- the number of unused RLBs, which indicates the effectiveness of resource utilization;
- the number of orphan RLBs, which are unused RLBs located away from the footprint periphery so that they are unlikely to be used for other parts of the circuit; and
- the number of channel tracks used in the critical path, which contribute to delay more heavily than other kinds of routing resources.

Table 1 shows that the Rothko mapping is a very compact design and has advantages over the 2-D architecture in every criterion.

Table 1. A comparison of mapping a multiplier onto a two layer
Rothko architecture and the 2-D Triptych architecture.

	Footprint	Unused RLBs	Orphan RLBs	Channel Tracks
3-D Rothko	56	16	0	6
Triptych	96	28	14	8

4. 3-Dimensional Circuit Fabrication

At Northeastern University, the 3-D Microelectronics group has developed a unique technology which allows us to design individual CMOS circuits and stack them to build 3-D layered FPGAs which can have vertical interconnections placed anywhere on the chip[9]. In the current design rules, the size of a vertical interconnection has a diameter of around 6 microns. Northeastern's 3-D process has several advantages over other 3-D methods:

- The procedure is simple. The process uses conventional VLSI processes.
- Transfer is done at wafer scale, leading to potentially high production rates.
- The capability of fabricating circuits with more than two layers is possible with multiple transfer steps.

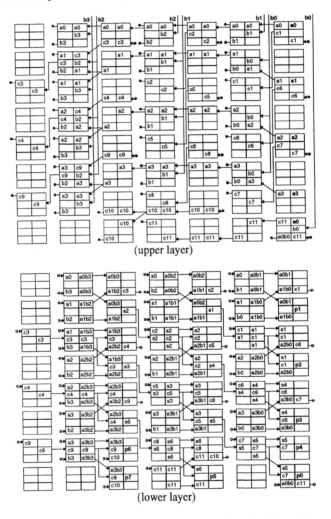

Fig. 4. The mapping result of a multiplier onto a 2 layer Rothko architecture.

The development of a three dimensional microelectronics is based on two fundamentally important processing capabilities: the ability to transfer circuits in thin film form and the availability of an interconnection technology. Kopin

Corporation (695 Miles Standish Blvd, Tanuton, MA. 02780, USA) has been developing their thin film transfer technology for applications in flat panel displays for the past six years. Researchers at Northeastern have teamed with Kopin Corporation on a program to develop an interconnection process. We use the wafer transfer technology developed at Kopin Corporation to transfer fully fabricated silicon circuits from one substrate to another. For a two-level 3-D circuit, the receiving substrate contains a portion of the circuit while a second portion will be transferred and aligned to it. Circuits with more than two layers can be fabricated by repeating the alignment and transfer to an already patterned wafer. By employing transfer techniques, the circuit fabrication can be performed using existing CMOS processing techniques. The second key to the success of this concept is the ability to fabricate electrical interconnections between layers. Researchers at Northeastern have focused on this issue and have successfully fabricated 3-D circuits.

4.1 Transfer Process

For the two level FPGA, a bulk silicon wafer is processed with half of the circuit. A second Silicon-on-Insulator (SOI) wafer is processed using standard CMOS fabrication techniques creating the second half of the circuit. An SOI wafer consists of a bulk silicon substrate with a thin layer of single crystalline silicon on top and separated from the substrate by a silicon dioxide layer (buried oxide). The SOI wafer is used because the buried oxide layer acts as a etch-stop during a subsequent back-etch step. The SOI circuit will be transferred face down onto the top of the bulk wafer as shown in Fig. 5. An adhesive is used to bond the transferred circuit to the bulk silicon wafer. The result is the two layer 3-D circuit shown in Fig. 6. Electrical connections need to be made between the two active device layers after the transfer.

4.2 Interconnection Process

The objective of the 3-D interconnection process is to make electrical connections between bulk devices on the lower layer of the 3-D structure and SOI devices on the upper layer. The interconnection scheme is illustrated in Fig. 7. An extra metal layer (metal 3) positioned on the very top of the 3-D circuit is introduced. Bulk metal 2 (assumed to be the top most metal layer on the bulk CMOS circuit) and SOI metal 2 (upper metal layer on the SOI CMOS circuit) are connected to the metal 3 through separate vias. The via etching is accomplished using an Inductively Coupled Plasma (ICP) to anisotropically etch both oxide and adhesive layers. Via filling is performed using a conventional magnetron sputtering source with a high bias.

5. Design Tools for Rothko

We are developing design tools for mapping digital circuits onto the Rothko architecture. This involves three main steps: mapping (or partitioning) of the input logic onto the FPGA's computational and IO blocks, placement of these blocks within the FPGA, and the selection of a feasible set of routes to connect the blocks.

In the first phase of the configuration process a netlist is subdivided into a set of subcircuits, each of which is mapped onto a specific RLB. We have developed a tool that accepts a specification of such a subcircuit and outputs the necessary configuration bits.

Fig. 5. Transfer process taking the device layer from the SOI wafer and bonding it to the top of a processed bulk silicon wafer.

Fig. 6. A simplified cross-sectional view of a 3-D circuit created using Kopin's circuit transfer technology.

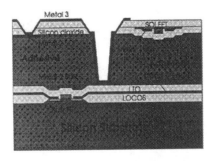

Fig. 7. Schematic cross-section of an ideal deposited metal interconnection.

5.1 Design Tools for Rothko

The placement algorithm for the Rothko architecture uses simulated annealing to place RLBs implementing logic functions in a way to minimize the expected routing cost and maximize the design's routability. As described above, the cost of routing a net is affected by the presence of both direct connections (inter-layer and intra-layer direct connections have similar delay properties) and segmented routing channels in the architecture. An RLB can connect directly to its NE, SW, and W neighbors, as well as the RLBs above and below it on adjacent layers. A lower bound of a routing cost can thus be computed by examining the locations of the source and sink RLBs and assuming that only direct connections are being used. This cost is used as part of the simulated annealing cost function. The actual cost of routing a net may be higher due to the unavailability of routing resources. The presence of segmented routing channels is factored into the expected routing cost by reducing vertical distances by a constant multiplicative factor throughout. This factor is designed to balance the use of direct connections and segmented routing resources.

The placement algorithm improves a design's routability in two ways: by monitoring local routability, and through density smoothing. Local routability refers to the ability to make a small number of required connections near an RLB. During the placement stage, care must be taken not to oversubscribe the routing capabilities of RLBs. The evaluation of local routability can help to detect problems of routing oversubscription and adjust the placement accordingly. Density smoothing is used to discourage placements that completely saturate a group of RLBs with computational logic, making it difficult to route signals in that area. The cost metric encourages placements with enough unused RLB routing capability to allow for later routing. The metric also discourages overly diffuse placements that spread a small design over the entire chip.

The routing algorithm under development for the Rothko architecture consists of a signal router that routes one net at a time, and a global router that makes use of a congestion and delay model to guide the signal router. In the basic algorithm, each net is routed by the signal router using a breadth-first search, under the assumption that all routing resources are equally available. The algorithm iteratively rips up and reroutes individual nets while gradually increasing the cost of using congested routing resources. This increased cost encourages the router to find alternate paths where possible. This process continues until a feasible set of routes has been found.

The signal router also attempts to minimize the delay for each source-sink pair in the circuit by giving higher routing priority to longer connections. For each source to sink connection in the circuit, a slack ratio is computed which is equal to the ratio of the cost of the longest path including the source and sink to the cost of the longest path in the circuit. This quantity gives the relative priority of each connection attempting to use a routing resource. Connections with a large slack ratio contribute more to the overall delay of the circuit and can replace connections whose delay is less critical.

The notion of local routability pays special attention to the connectivity of each RLB with its neighbor RLBs above and below it. These interlayer routing resources are monitored to ensure that they are not oversubscribed. Density smoothing is used to ensure that not all interlayer connections in a region are saturated, making later connections between layers difficult.

6. Summary

We have described an architecture, fabrication process and design tools for Rothko. We plan to fabricate two layer Rothko chips in 1997. We are also building an architectural simulator for Rothko, and will use this to demonstrate results from our design tools on the Rothko architecture. In the future, we plan to fabricate four layer as well as two layer FPGAs. We will use results of our design tool simulations to improve on the proposed architecture.

7. Acknowledgment

We would like to thank Scott Hauck for his help with the Triptych architecture.

References

1. G. Borriello, C. Ebeling, S. Hauck, and S. Burns, "The Triptych FPGA architecture," *IEEE Trans. on VLSI Systems*, vol. 3, no. 4, pp. 491-501, 1995.

2. S. M. Trimberger, *Field-Programmable Gate Array Technology*, S. M. Trimberger, ed., Kluwer Academic Publishers, Boston, MA, 1994..

3. M. Alexander, J. Cohoon, J. Colflesh, J. Karro, and G. Robins, "Three-dimensional field-programmable gate arrays," *Proc. IEEE International ASIC Conf.*, Austin TX, September 1995, pp. 253-256.

4. M. Alexander, J. Cohoon, J. Colflesh, J. Karro, E. Peters, and G. Robins, "Placement and routing for three-dimensional FPGAs," *Fourth Canadian Workshop on Field-Programmable Devices*, Toronto, Canada, May 1996, pp. 11-18.

5. J. Darnauer, P. Garay, T. Isshiki, J. Ramirez, and W. Dai, "A Field Programmable Multi-chip Module (FPMCM)," *in IEEE Workshop on FPGA's for Custom Computing Machines*, pages 1-10, 1994.

6. V. Maheshwari, J. Darnauer, and W. Dai, "Design of FPGA's with Area I/O for Field Programmable MCM," *in Proc. ACM/SIGDA Intl. Symposium on Field-Programmable Gate Arrays*, 1995.

7. The Trimorph project, Electronic Systems Group, University of Sheffield. http://www.shef.ac.uk/uni/academic/D-H/eee/esg/research/mcm.html

8. J. DePreitere, H. Neefs, H. Van Marck, J. Van Campenhout, Baets, Thienpont and Veretennicoff, "An Optoelectronic 3-D Field Programmable Gate Array," in *Proc. 4th Int'l Workshop on Field Programmable Logic and Applications*.

9. P. Zavracky, M. Zavracky, D-P Vu and B. Dingle, *"Three Dimensional Processor using Transferred Thin Film Circuits,"* US Patent Application # 08-531-177 allowed January 8, 1997.

Extending Dynamic Circuit Switching to Meet the Challenges of New FPGA Architectures

Gordon McGregor and Patrick Lysaght

Dept. of Electronic and Electrical Engineering
University of Strathclyde
204 George Street
Glasgow G1 1XW

Abstract

Dynamic Circuit Switching (DCS) was the first CAD tool to allow the specification and simulation of dynamically reconfigurable logic. It was designed to be extensible to cope with new FPGA architectures. The use of DCS to simulate the operation of two new families of dynamically reconfigurable FPGAs is considered in this paper. The top-down strategy of DCS is also compared with an alternative bottom-up technique that has recently been reported.

1. Introduction

Dynamic Circuit Switching (DCS) was the first reported CAD tool to allow the specification and simulation of dynamic reconfigurable logic [1]. Conventional tools allow the simulation of individual FPGA configurations, but were not designed to allow the user to investigate the operation of circuits that are present on a device while it is being reconfigured. This is important, since by definition, parts of a design can continue to operate during the reconfiguration interval. The DCS tool allows the performance of both the static and dynamic sections to be simulated simultaneously.

DCS was designed to integrate with existing CAD tools. It was also designed to be extensible to cope with new, dynamically reconfigurable FPGAs. One of the underlying benefits of the DCS technique is that it is portable to new architectures with modest effort. The recent development of two new families of FPGAs [2, 3] for dynamically reconfigurable applications suggests that it is timely to investigate whether DCS can indeed be adapted easily to support newer device architectures.

This paper discusses the extension of DCS to address the challenges posed by the introduction of these new FPGAs. The work was partly initiated by communication with the creators of the Field Programmable System on a Chip (FIPSOC) [2], a new FPGA architecture that combines analogue and digital arrays with two configuration contexts and a microprocessor. The second new device architecture considered is the Xilinx XC6216 FPGA reported by Churcher [3]. Finally, we compare DCS with the CAD tool developed by Kwiat [4] for simulating dynamically reconfigurable logic.

The paper is organised into six further sections. Section 2 presents a brief review of the DCS tools. Section 3 identifies the new architectural features of FIPSOC and the XC6216 that were not available when DCS was initially developed. The following section investigates how DCS may be used with the new FPGAs and what enhancements, if any, are needed to extend it to cope with new device features. Kwiat's technique for simulating dynamically reconfigurable logic is reviewed in section 5 and compared to DCS. Section 6 offers some conclusions and suggestions for future work.

2. DCS System Overview

If the use of dynamic and partial reconfiguration is to become generally accepted, as its proponents predict [5 - 8], efficient design automation tools are required. Among the areas that need to be considered are design specification, simulation of reconfiguration intervals and design implementation. Most tools support the design of static applications and do not address the additional requirements of *dynamic* reconfiguration. Our approach to this challenge has been to produce a tool that co-exists with current tools and extends them to allow investigation of the various performance trade-offs arising in dynamically reconfigurable designs.

The approach used in DCS is based on an abstract model of the device or devices being simulated. A fundamental premise of our approach is that while a circuit is being reconfigured no assumptions should be made about its behaviour. This results in a three-stage model of the reconfiguration process:

- Circuits to be reconfigured are isolated from other circuits
- The circuits are reconfigured
- The new circuits become operational

In the actual circuit, it is the designer's responsibility to ensure that, before a task is reconfigured, it is isolated from any other tasks with which it may communicate. It is also his responsibility to re-connect newly instantiated tasks to other tasks. We expect that while dynamic reconfiguration is taking place, no assumptions are being made about the behaviour of the tasks that are 'in transit'.

To see how these ideas are used in DCS, consider Fig. 1. The diagram on the left shows a circuit consisting of five blocks. The shaded blocks represent three circuits, labelled TASK_B, TASK_C and TASK_D, that are dynamically reconfigurable. The two other circuits are static. In DCS, a design is entered using conventional schematic capture tools and reconfigurable tasks are identified at this stage. Each reconfigurable task is encapsulated into a hierarchical symbol block. A RECONFIG_ELEMENT attribute is then attached to it, to identify it as a dynamically reconfigurable component of the design. Two further attributes, ACTIVATE and DEACTIVATE, are used to specify the logical conditions under which a section is switched into and out of the active circuit. Finally, the last two attributes, LOADTIME and

REMOVETIME, are used to specify how long it takes to reconfigure circuits. At this stage, the designer has fully specified the logical sequence of the dynamically reconfigurable design. Note, that for clarity, we have not specified the actual values of the attributes, since these are design dependent.

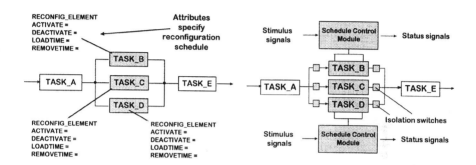

Fig. 1. The Application of DCS to a simple dynamically reconfigurable circuit

The design netlist is then processed by the DCS tools, prior to simulation. The circuit on the left of Fig. 1 is transformed into the circuit on the right of the same diagram. The interpretation of the reconfiguration attributes, attached to the three reconfigurable sections of the design, results in two kinds of virtual components being inserted in the new netlist. These are the schedule control modules (SCMs) and the isolation switches. Both are implemented in VHDL. DCS models the process of dynamic reconfiguration by automatically interpreting the reconfiguration attributes in the netlist and inserting SCMs and isolation switches into the simulation database.

Circuits are switched in and out of the simulation model according to the state of the isolation switches. The state of these switches is controlled by the SCMs. Each SCM monitors the appropriate circuit signals to determine if the conditions for enabling reconfiguration of a circuit block are valid. If they are, it manipulates the state of the isolation switches accordingly. Note that the designer is responsible for specifying the reconfiguration timing to determine how long it takes to remove and load tasks. Status signals are added as outputs of the SCMs to assist the designer in monitoring control flow.

3. New Dynamically Reconfigurable FPGAs

The two FPGA architectures considered in this section are very different. The field programmable system on a chip attempts to integrate all the essential components of a mixed-signal, programmable ASIC: these include a microprocessor, memory, and programmable digital and analogue arrays. In contrast, the Xilinx XC6200 architecture comprises a field programmable digital array with a highly optimised interface for connection to external microprocessors. Despite their obvious dissimilarities, the two architectures share a common design goal in that they both attempt to exploit dynamic reconfigurability more comprehensively than previous FPGAs. They are probably the first FPGAs to have been designed with the explicit intention of exploiting dynamic reconfiguration.

3.1 FIPSOC

A block diagram of FIPSOC is shown in Fig. 2. The system integrates an 8051 microprocessor core, program memory, and field programmable analogue and digital arrays. Control of both of the programmable arrays is via the CPU. A special interface block converts the information on the processor's address, data and control buses to an optimised serial protocol for communication with the two arrays. The serial bus is bi-directional and is a shared resource.

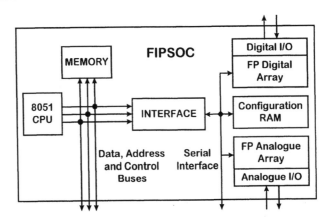

Fig. 2. Block diagram of the Field Programmable System on a Chip

FIPSOC's programmable digital array has the distinction of being the first dynamically reconfigurable, coarse-grained FPGA. Individual logic blocks are called digital macro cells (DMCs). They contain four, four-input look-up tables, that may be configured to provide any combinatorial function of six inputs. Each DMC also contains four flip-flops. A single DMC can implement functions as complex as 4-bit adders and 4-bit counters. The analogue array consists of a set of fixed-function, analogue blocks that can be configured to perform a range of analogue functions whose parameters are controlled programmatically. Configurable analogue-to-digital (ADC) and digital-to-analogue (DAC) converter blocks are also provided. One use of these blocks is to allow the microprocessor access to the analogue array.

FIPSOC is further distinguished by its implementation of context-switched configuration planes, similar to those discussed by DeHon [9]. There are two context planes, either one of which is the active at any time. The CPU can access the inactive context plane while the DMCs carry out their logical functions. The configuration information for individual DMCs is addressable and their flip-flops are duplicated. This allows dynamic reconfiguration, incorporating pre-loading of register state. Digital circuits can be pre-initialised before switching them onto the array so that they become active in one processor clock cycle. A more detailed account of the device architecture can be found in [2].

3.2 Xilinx XC6216

In contrast, the XC6216 is based on a fine-grained, sea-of-gates architecture. It consists of an array of 64 x 64 core cells surrounded by 256 input-output ports. Every logic cell can implement any combinatorial logic function of two inputs. Each cell can also implement a D-type flip-flop which can be used to register the cell's combinatorial function. The device has been optimised for datapath designs, such as bit-level processors, that typically require a large number of registers.

Cells have nearest neighbour connections to their North, South, East and West neighbours. The device has a hierarchical bussing scheme. Cells are organised into blocks of 4 x 4, 16 x 16, and so on, increasing by a factor of four each time. A set of fast busses is associated with each size of block. Access to these fast busses is via routing switches that are adjacent to the cells on the periphery of the respective blocks.

The XC6216 is an FPGA that has been designed to be used in two broad classes of applications. The first class is the conventional role of a general-purpose ASIC device for logic integration. The other role is that of an intelligent peripheral that can operate as a memory-mapped, coprocessor in fast microprocessor systems. It is in the degree of synergy that may be achieved between the operation of the FPGA and its host microprocessor that the XC6216 architecture sets entirely new standards. This is largely due to the introduction of the most advanced FPGA to CPU interface currently available on any FPGA. The design philosophy appears to have been driven by the desire to produce the first FPGA optimised for reconfigurable computing.

The processor interface is a 32-bit wide data bus that may also be configured for 16 or 8 bit operation. The XC6216 has been designed to appear in system as random access memory so a 16-bit address bus is also provided. All data registers on the array are accessible, making it possible to interface with user logic via the processor interface alone. Registers are addressed in columns via a map register. Up to 32 of the 64 registers in a column may be read from or written to by nominating their appropriate row positions in the map register. This is can be extended so that all of the 64 registers in a column may be written using a single 8-bit operation. This is particularly useful in reconfigurable computing applications. Intermediate results may be read from or written to coprocessor registers, before the next stage of computation is configured onto the array.

The FPGA also supports broadcast configuration methods. Techniques such as address 'wildcarding' allow comparatively large amounts of logic to be configured quickly. This is useful for highly structured designs such as bit-slice circuits. Circuit 'slices' can be configured onto the array very efficiently.

Every configuration bit of the XC6216 is randomly accessible and its address location and function are fully documented. The FPGA may be dynamically reconfigured by simply writing data into the configuration data memory space. The speed at which dynamic reconfiguration is possible is a key feature of the architecture. A full, 32-bit write cycle can take place in 40 ns which ensures fast, partial reconfiguration and transfer of data between the processor and FPGA. The reconfiguration control signals of the device have been made accessible to user logic on the array, by routing them through the input-output ports. Thus, for the first time, it is possible for a computation on the array to control directly reconfiguration of the array. This capability will lead to designs that can selectively reconfigure themselves in response to changes in their operating conditions without reliance on external intelligence to provide the necessary control functions.

4. Evaluating DCS support for new FPGA architectures

Several enhancements to DCS are being considered. Extensions to support direct synthesis of reconfiguration controllers are being examined [10]. Additionally, an expert system front-end to aid in the estimation of reconfiguration latency has been reported [11]. The reconfiguration latency is defined as 'the time that elapses between a request for new circuitry to be loaded onto an already active FPGA and the point at which the new circuitry is ready for use'.

Two distinct application areas can be identified for dynamically reconfigurable FPGAs:

- Increasing effective logic density
- Algorithm acceleration

The two application categories place different emphasis on reconfiguration latency. The latency is usually not as critical a parameter for applications of dynamically reconfigurable FPGAs seeking to increase effective logic density. DCS simulation may be used to verify the correct logical sequencing of reconfigurable tasks in the design. Detailed consideration of reconfiguration latency is not as important early in the design cycle. The designer will have access to full timing information after placement and routing.

With continuing increases in array size and faster configuration interfaces, the potential to exploit dynamic reconfiguration to accelerate algorithms is attracting more attention. The accurate estimation of the reconfiguration latency is more important for these types of applications. The aim is to assist the designer in quickly assessing the potential speed-up that can be obtained using dynamic reconfiguration, without having to prototype the final hardware. With aid of the expert system, increasingly accurate estimations of the reconfiguration latency are available as the design progresses. With this iterative refinement, the designer can decide whether or not a particular algorithm will run faster in reconfigurable hardware.

The use of DCS to model the operation of FIPSOC raises several points. Support for context switched, dynamic reconfiguration is already implicit within DCS: as far as DCS is concerned, the only change implied by context switching is faster device reconfiguration. This is reflected in smaller estimates of LOADTIME and REMOVETIME. This observation assumes that the required configuration is in the inactive context, which is not always true. If the task scheduling is sequential, and the interval between task switching is sufficiently large, then there will be time to load the subsequent configuration into the inactive context. However, if more complex dynamic reconfiguration is required, using a non-deterministic scheduling algorithm, the reconfiguration overhead incurred will be significant.

The relatively slow, serial configuration port will reduce the potential speed-up that can be achieved with dynamically reconfigurable applications. The bottleneck appears to be further compounded by the fact that the serial data port that provides access to the digital and analogue arrays is a shared resource. If, after a context switch, data is being transferred to or retrieved from the array, then reconfiguration requests cannot be processed until the transfer is complete. It appears that the microprocessor must poll the digital array to test if a new configuration is required.

The modelling of the analogue circuitry was initially thought to present new challenges. The available literature on FIPSOC, however, does not appear to suggest that the analogue array is dynamically reconfigurable. Nor is there direct connection between the analogue and digital arrays. Instead, communication between the two blocks is via the processor core. If these assumptions are correct, the task of simulating FIPSOC using DCS is greatly simplified.

The use of DCS has not been evaluated in a mixed-mode simulation environment but there is no reason to suspect that this will present problems. The current implementation of DCS is targeted at Viewlogic's Workview CAD suite, which supports mixed-mode simulation. The creators of FIPSOC report the development of their own CAD tools which include mixed-mode schematic capture and simulation. There are two alternatives available for simulating FIPSOC using DCS. The first would require a port of the DCS software to the FIPSOC CAD suite. The second alternative needs the simulation libraries for FIPSOC to be ported to the Workview suite. The latter scheme would require less work.

The serial configuration port and loose coupling of the array and processor indicate that FIPSOC is better suited for increasing logic density in reconfigurable applications and to tailor array logic to different product configurations. The analogue array on the device introduces these capabilities, for the first time, to mixed-signal applications.

In contrast to FIPSOC, the XC6216 is optimised for algorithm acceleration. Support for the XC6216 in DCS requires the ability to predict the use of the broadcast configuration modes. DCS can be used to give an initial estimate for the design performance. However, it is not until very late in the design cycle, when the logic has actually been placed on the device, that the opportunities to use these features can be identified. This new timing information can then be back annotated onto the schematics.

Using the broadcast configuration methods supported by the device, comparatively large amounts of logic can be configured in a very short time period. The speed-up that can be achieved in this manner is very design dependent. It is difficult to obtain an accurate estimate of the reconfiguration time for a task if these features are being exploited. For DCS to cope efficiently with these techniques, early in the design cycle, it will have to rely on direct input from the designer.

5. A Bottom-Up Approach to the Simulation of Dynamic Reconfiguration

Kwiat has reported what is essentially a 'bottom-up' strategy for simulating dynamically reconfigurable logic [4]. His interest in this technology is motivated by research into fault tolerance and testing of digital systems. By using dynamic reconfiguration, he creates a 'virtual' FPGA. He reports achieving the equivalent of a 20,000 gate FPGA by reconfiguring a 5,000 gate device. Simulation is used to verify the operation of the dynamically reconfigurable array.

The bottom-up simulation method is summarised as follows. First, a model of the target FPGA's logic, routing and configuration circuitry is produced. Kwiat uses a proprietary, in-house netlist, called Netlist Interchange Format (NIF), to represent this information. NIF contains structural information only and is supported by a range of netlist converters to other tools and formats (including EDIF and VHDL).

The structural cell models within NIF are replaced by equivalent behavioural models written in VHDL. The authors report 'exhaustive simulation of the behavioural cell' to ensure that it corresponds exactly to the structural model. The function performed by cells can be sequential, combinatorial or constant, or alternatively programmable within the design.

All simulation is performed in VHDL. The structural model in NIF is converted to VHDL for simulation, where both structural and behavioural cell models are used. Automatic tools have been developed to generate the actual design in a format compatible with the NIF structure. The VHDL model is exercised with vectors corresponding to circuit stimuli and device configuration data. To increase the speed of simulation, a number of improvements are made to the design representation. Where appropriate, functions performed by groups of cells are identified and replaced with equivalent behavioural models. Cells that are unused by the design are replaced by 'stubs' which are less complex because their programming logic is removed and their behavioural description is simplified.

We refer to Kwiat's method as a 'bottom-up' technique because the design flow is inverted relative to the conventional flow. Where normally designs are developed from behavioural descriptions and converted into structural representations, his technique begins with structural representations which are then transformed into behavioural descriptions.

Several aspects of the bottom-up method are noteworthy. The technique is very device specific. Indeed, it could be argued that one is required to know *too much* about the internal programming logic of the FPGA. It requires the development and exhaustive simulation of a VHDL model of the entire FPGA to ensure correct operation. The FPGA model that has been developed, is reported as being 'based loosely' on Atmel's AT6000 FPGAs. How this affects the use of design tools such as automatic placement and routing software and bit-stream generators is not discussed. It would be very undesirable if new tools had to be developed. To port the technique to other devices implies that appropriate cell models would have to be constructed and rigorously tested, for each new target architecture. This is clearly a disadvantage. In contrast, DCS uses vendor-supplied primitives and macro cells, allowing new architectures to be supported relatively quickly.

Programming data is required to simulate the device operation when the entire array is being modelled. Kwiat reports that device programming data and the simulation database are generated concurrently. Programming data, however, can only be obtained after a design has been placed and routed, which is one of the last stages in the design cycle.

The bottom-up method simulates both device logic and programming circuitry. Despite the techniques employed to reduce model complexity, simulation run times can be expected to suffer as a result of the size of the final VHDL model. For small arrays of coarse grained cells (e.g. FIPSOC), this approach is certainly feasible. However, when large arrays of fine-grained cells are considered, the technique becomes more impractical as the number of redundant cells increases.

6. Conclusions

DCS represents a significant advance towards automating the development of dynamically reconfigurable systems. The ability to adapt to new dynamically reconfigurable FPGAs, such as the XC6200 devices and FIPSOC, shows the flexibility and generality of the DCS simulation technique. The basic principles of DCS are equally valid in a mixed-mode environment.

With the addition of enhanced reconfiguration timing estimation, the DCS environment will become increasingly useful for investigation of applications targeting algorithm acceleration. The investigation of dynamically reconfigurable analogue circuitry is an area for future work.

7. References

[1]   Lysaght, P. & Stockwood, J. *A Simulation Tool for Dynamically Reconfigurable Field Programmable Gate Arrays,* IEEE Transactions on VLSI Systems, Sept. 1996

[2]   Faura, J., et al *Multicontext dynamic reconfiguration and real time probing on a novel mixed signal programmable device with on-chip microprocessor,* ibid.

[3]   Churcher, S., Kean, T. & Wilkie, B. *The XC6200 FastMap Processor Interface* in Field Programmable Logic and Applications, Proceedings of FPL'95, pp.36-43, Moore, W. & Luk, W., Eds., Springer-Verlag, 1995

[4]   Kwiat, K. & Debany, W. *Reconfigurable Logic Modelling,* Integrated System Design, December 1996 (www.isdmag.com)

[5]   Foulk, P.W. *Data Folding in SRAM Configurable FPGAs,* IEEE Workshop on FPGAs for Custom Computing Machines, pp. 163-171, Napa, CA, Apr. 1993

[6]   Hutchings, B. L. & Wirthlin, M., J. *Implementation Approaches for Reconfigurable Logic Applications* in Field Programmable Logic and Applications, Proceedings of FPL'95, pp.419-428, Moore, W. & Luk, W., Eds., Springer-Verlag, 1995

[7]   Brebner, G *A Virtual Hardware Operating System for the Xilinx XC6200* in Field-Programmable Logic: Smart Applications, New Paradigms and Compilers, Proceedings of FPL'96, pp. 327-336, Springer-Verlag 1996

[8]   Villasenor, J. & Mangione-Smith, W. H. *Configurable Computing,* Scientific American, June 1997

[9]   DeHon, A. *DPGA-Coupled Microprocessors: Commodity ICs for the Early 21st Century* Proceedings of IEEE Workshop on FPGAs for Custom Computing Machines. pp. 31-39, Napa, CA, 1994

[10]   Lysaght, P., McGregor G. & Stockwood, J., *Configuration Controller Synthesis for Dynamically Reconfigurable Systems,* IEE Colloquium, Bristol, February 1996.

[11]   Lysaght, P. *Towards an expert system for a priori estimation of reconfiguration latency in dynamically reconfigurable logic* ibid.

Performance Evaluation of a Full Speed PCI Initiator and Target Subsystem Using FPGAs

David Robinson, Patrick Lysaght, Gordon McGregor and Hugh Dick*

Dept. Electrical and Electronic
Engineering
University of Strathclyde
204 George Street
Glasgow G1 1XW
United Kingdom

* Dynamic Imaging Ltd
9 Cochrane Square
Brucefield Industrial Park
Livingston
Scotland
United Kingdom

Abstract

State-of-the-art FPGAs are just capable of implementing PCI bus initiator and target functions at the original bus speed of 33 MHz. This paper reports on the use of a Xilinx 4000 series FPGA and LogiCore macros to implement a fully compliant PCI card for a specialist data acquisition application. The design required careful performance analysis and manual intervention during the design process to ensure successful operation.

1. Introduction

The Peripheral Component Interconnect (PCI) bus is an important component in high performance, data intensive computer systems [1]. Its maximum bandwidth of 132 Mbytes/s and its support for automatic configuration of peripheral cards have been major forces behind the wide acceptance of PCI in PC and workstation environments.

To achieve 100% PCI compliance, a PCI card must adhere to the very strict electrical, timing and protocol specifications imposed by the bus standard. Although mask programmed ASIC devices can easily meet these specifications, FPGA implementations push current technology to the limit [2]. FPGAs that meet the strict specifications are now available, as are third-party PCI interface macros [3]. However, implementing a PCI interface that achieves optimum performance remains difficult because state-of-the-art FPGAs and careful design are needed to meet the timing requirements of the bus.

This paper describes the performance evaluation of a PCI card for a specialist data acquisition application. The aim of the design was to transfer image data quickly over the PCI bus between a proprietary interface to an ultrasound scanner and a graphics controller card. The PCI interface is implemented in a Xilinx XC4013E FPGA using the LogiCore PCI macros. The design is distinguished from earlier reported cards [2],[4] in that it includes both 32-bit PCI target and initiator capabilities operating at the maximum bus speed of 33 MHz. The card is designed to operate under the

Windows 95 PC operating system via a custom virtual device driver (VxD). The card also uses a Xilinx XC6216 FPGA to emulate the characteristics of the data source prior to its implementation. PCI bus performance analysis was conducted before the design phase to guarantee the data transfer rate under worst case conditions.

2. System Overview

PCI is a high performance, processor independent, local bus standard. Reflected wave signalling is used to allow CMOS ASICs to interface directly to the bus. Each signal wave propagates to the end of the unterminated bus and is reflected back to the point of origin, doubling the voltage on the bus trace. For 33 MHz operation, the round trip delay can last only 10 ns [1]. This places severe restrictions on the length and capacitive loading of the bus and also the electrical characteristics of any device connected to it.

The original PCI bus specification stipulated 32-bit data transfer at 33 MHz, offering a maximum bandwidth of 132 Mbytes/s. (A 64-bit, 66 MHz version of the bus has more recently been defined). Data and address lines are time division multiplexed, reducing the required number of pins to 49 in the 32-bit version. All data is transferred in burst mode, starting with a single address phase and followed by one or more data phases. PCI devices come in two types, Masters and Targets (also referred to as Initiators and Slaves respectively). An Initiator can request access to the bus for data transfer whilst a Target has to wait until it is accessed by an Initiator. A single PCI card can perform both Target and Initiator functions. Access to the PCI bus is request-based rather than time-slot based and is controlled by a PCI arbiter chip. Each initiator device on the bus has its own request and grant lines connected to the arbiter chip. This allows arbitration to occur while the bus is in use by other PCI devices.

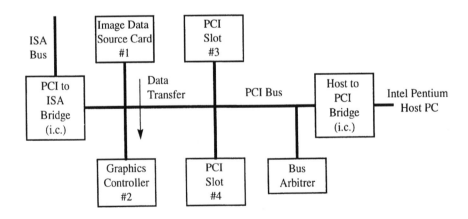

Fig. 1. Image Data Source System Configuration

The PCI specifications are surprisingly vague when describing the algorithm used by the PCI arbiter chip to award bus access:

"An arbiter can implement any scheme as long as it is fair and only a single GNT# is asserted on any rising clock" [1].

GNT# refers to the set of grant lines through which all initiators have a unique connection to the arbiter. The important implication of this statement is that the effective PCI bus bandwidth cannot be calculated directly from the bus specifications. Instead it is dependent on the PCI chipset used in the target system. The following calculations determine the *worst case* per card bandwidth of a PCI system with four initiator cards, based on the 438VX PCI chipset in mode 0 [6]. Fig. 3 shows the bus access priority sequence.

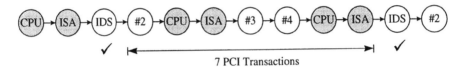

Fig. 3. Bus Access Priority Queue for PCI Arbiter

This scheme grants access priority to one device for every slot in the queue. Each slot lasts the maximum 256 clock cycles defined by the chipset. Under worst case conditions each device will always request the bus and will perform maximum length transfers. Under these conditions bus access effectively becomes time slot based. From Fig. 3 one can calculate that the maximum bandwidth for a single initiator card has dropped from 132 Mbytes/s to 16.5 Mbytes/s. Half the available bandwidth is reserved by the arbiter for use by the CPU and ISA bridges. The remaining bandwidth (66 Mbytes/s) is shared equally between the four PCI masters.

4.   Software Device Drivers

Almost all hardware devices added to the PCI bus depend on software to function correctly. This can result in a substantial decrease in device performance. Software overheads are typically large and non-deterministic. Windows 95 operates a complex virtual machine (VM) environment [7]. Software running in a VM operates as if it has exclusive access to a particular hardware device. Communication between hardware and software is routed through system VxDs to allow resource arbitration between VMs. The non-deterministic delays incurred are one of the disadvantages of this environment.

All PCI device drivers must support interrupt chaining [1]. Multiple VxDs may respond to an interrupt and a subset of these may invoke further interrupt handler routines. A VxD responding to an interrupt may notify several VMs. These in turn may pass an interrupt on to several software applications. This situation is shown in Fig. 4. The delays experienced depend heavily on the system configuration and current activity.

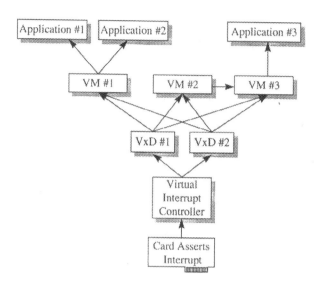

Fig. 4. Possible Interrupt Flow in Windows 95

5. System Design

The image acquisition card is required to transmit a picture of 640 by 440 pixels at 30 frames per second. Each pixel is described by one byte of data. This necessitates a bandwidth of 8.448 Mbytes/s. During burst writes, the LogiCore initiator interface automatically inserts one wait state per data transfer. This doubles the required bandwidth to 16.896 Mbytes/s. Worst case analysis of the PCI environment established that a single card could only achieve an effective bandwidth of 16.5 Mbytes/s, which is less than the required amount.

Under fully loaded conditions, the much quoted maximum PCI bus bandwidth of 132 Mbytes/s and the maximum achievable data transfer rate with a fully compliant card differ by almost an order of magnitude.

Two actions are taken to guarantee the card the required bandwidth.

1. The system is limited to a maximum of three PCI master devices. This allows each card a maximum bandwidth of 22 Mbytes/s.

2. Interaction with software is kept to a minimum to avoid poor performance. The VxD configures the card with the information necessary to start a transfer to the graphics controller. Once a transfer has begun, software intervention is limited to infrequent use of control commands only. The IDS card has no support for interrupts.

6.   FPGA Implementation

Due to increasingly complex designs and faster time to market requirements, individual companies will be unlikely to design all the blocks necessary for their systems in the future. Instead, designers may use system components designed externally [8]. The PCI LogiCore macro from Xilinx is one of the first examples of an intellectual property (IP) building block commercially available for an FPGA to realise the "systems on a chip" concept.

Fig. 5 shows a floorplan of the LogiCore PCI interface after placement and routing.

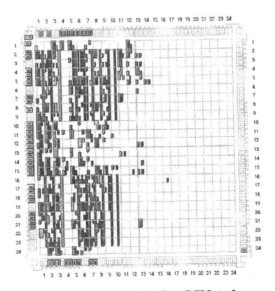

Fig. 5. Floorplan of the LogiCore PCI Interface

The PCI bus is connected to the pins on the left hand side of the chip in Fig. 5. Free space is plentiful on the right of the chip for user logic. Problems arise when connecting user logic on the right to PCI signals on the left. Timing specifications limit the distance that a heavily loaded net can be routed. User logic must be carefully placed next to the PCI logic, as near as possible to the source of the required signals.

Default timing constraints in the LogiCore macro limit the delay between any two flip flops in the user design to a maximum of 30ns. Combinatorial block delays can easily exceed 30ns, causing the design to fail the timing specifications. Routing delays also quickly exceed this specification. Path analysis of user logic can identify components that should be removed from this group. Creating more realistic timings relaxes the overall constraints on the partition, place and route (PPR) software.

In cases where the timing constraints cannot be adjusted, careful floorplanning must be conducted to meet the maximum path delays. Detailed knowledge of the device architecture allows the designer to create a more efficient design. For example, each CLB in the XC4013E contains two flip-flops. Every flip-flop has an associated three-state-buffer (TBUF) which is connected to a horizontal longline. Columns of TBUFs can be driven by a single, vertical longline. These resources can be used to make highly efficient registers that drive a shared bus, Fig. 6.

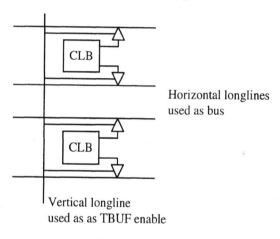

Horizontal longlines
used as bus

Vertical longline
used as as TBUF enable

Fig. 6. Efficient Register in XC4013E

The PPR software does not automatically exploit this feature. Flip-flops are commonly connected to TBUFs in other columns, registers are not aligned with the bus and vertical longlines are not used to control the TBUFs. Floorplanning and using attributes at the schematic level force the PPR software to create a more efficient layout.

Logic duplication was also used in an effort to meet the timing specifications. Duplicating the generation of certain control signals increased the probability of the PPR software satisfying the timing constraints. Register control signals benefited most from this technique. The XC4013E is segmented into quadrants to make efficient use of longline resources. 32-bit registers were split across two quadrants, 16 bits in each quadrant. Providing separate, yet identical, control lines to each half of the register allowed the PPR software to place and route the register within the timing constraints.

A critical point is quickly reached where no additional logic can be added to the circuit without consistently failing the timing specifications. This effect removes one advantage of using FPGAs which is the ability to add and remove extra circuitry for debugging purposes. Simple tasks, such as routing signals to I/O pins for monitoring with an external logic analyser, become impossible.

To achieve a working 33 MHz PCI initiator and target design, the LogiCore macro itself must be modified. The macro generates a data valid control signal, called DATA_VLD. This signal is derived from circuitry in both the Target and Initiator sections of the LogiCore macro. For its operation, the IDS design needs to decompose DATA_VLD back into two separate signals. One of the signals represents DATA_VLD for the Target while the other represents DATA_VLD for the Initiator. The effect of splitting DATA_VLD outside the macro is to heavily load nets that are already critical with respect to the timing specifications. The same goal can be achieved more simply by altering the macro slightly to provide external access to the two signals that are combined inside the macro to create DATA_VLD. The same technique was repeated several times with more complex signals.

7.   Practical Results

The chip resource utilisation of the final design can be seen in Table 1 and the floorplan in Fig. 7.

	Available	Actual Used	% Used
CLBs	576	403	69%
Bonded I/O Pins	160	123	76%
F & G Function Generators	1152	458	39%
H Function Generators	576	118	20%
CLB Flip-flops	1152	395	34%
IOB Input Flip-flops	192	45	23%
IOB Output Flip-flops	192	44	22%
3-State-Buffers	1248	353	28%
3-State Half Longlines	96	32	33%

Table 1 Resource Utilisation

It can be seen from Table 1 that the design uses a large number of three-state-buffers. These are used by both the LogiCore interface and the user application to drive a shared internal bus. To avoid bus contention which could easily destroy the device, extra care must be taken during the design and simulation phases. Treating the LogiCore macro as a complete "black box" design may lead to designs that fail in the long term.

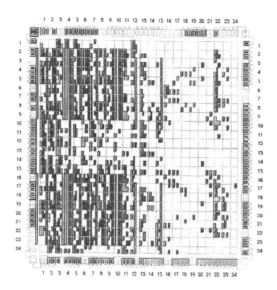

Fig. 7. Floorplan of Complete PCI Design

The frame rate of the final design was measured using a Hewlett Packard 1671D logic analyser and a FuturePlus PCI Preprocessor FS16P64 card. When the IDS was the only initiator device on the PCI bus, and the CPU was performing no background processing, a frame rate of 213 frames per second (fps) was measured. This corresponds to a bandwidth of 119.96 Mbytes/s. This figure is the product of the image width, the image height, the frame rate and the wait states inserted by the LogiCore macro.

8.   Conclusions

Performance evaluation of the PCI bus revealed important issues concerning the bus's bandwidth. The omission of the arbiter algorithm from the PCI specification prevents a complete analysis of the bus's performance. Worst case bandwidth can only be calculated relative to a specific PCI chipset. Bus performance is limited by the quality of an individual vendor's PCI chipset. Analysis of the bus with a specific chipset revealed that the maximum worst case bandwidth for a single card was almost an order of magnitude lower than the headline bus bandwidth.

FPGA implementations of a PCI bus interface remain difficult, even with state-of-the-art devices. As can be seen from Fig. 7, modern, high-capacity FPGAs have ample resources to implement PCI interfaces. The XC4013 is a smaller member of the XC4000 family. PCI design pushes FPGAs very close to the limits of the technology with respect to the operating speed of the logic. Designs must be hand crafted to ensure correct logic operation.

Using the LogiCore interface greatly accelerated the design process. Creating a highly efficient IP building block that is both flexible and easy to use is difficult and trade-offs have to be expected. While not quite a "black box" solution, designing a complete PCI interface would have been more difficult than modifying aspects of the LogiCore interface.

9.   Acknowledgements

The design and development of the PCI system was a joint project between the University of Strathclyde and Dynamic Imaging Ltd. The authors wish to gratefully acknowledge the support of all the staff who helped to make it successful.

10.   References

[1]   PCI Local Bus Specification Revision 2.1, PCI Special Interest Group, USA, 1995

[2]   Fawcett, B K, "Designing PCI Bus Interfaces with Programmable Logic", Eighth Annual IEEE International ASIC Conference, Austin, USA, 1995

[3]   LogiCore PCI Master and Slave Interface User's Guide Version 1.1. Section 8.1, Xilinx, USA, 1996

[4]   Luk, W and Shirazi, N, "Modelling and Optimising Run-Time Reconfigurable Systems", Proc. IEEE Symposium on FPGAs for Custom Computing Machines, USA, 1996

[5]   The Peripheral Component Interconnect Bus X-Note Number 5A, Xilinx, USA, January 1995

[6]   Intel 430VX PCISET 4.6.1, Intel Corporation, USA, 1996

[7]   VtoolsD, Version 2.01.001, Vireo Software, USA, 1995

[8]   Holmberg, Per, CORE-BASED FPGA DESIGNS, Special Report: "Intellectual Property: Reusable Cores & Macros", http://www.pldsite.com/, 1996

Implementation of Pipelined Multipliers on Xilinx FPGAs

T.-T. Do, H. Kropp, M. Schwiegershausen, P. Pirsch

Laboratorium für Informationstechnologie, University of Hannover
Schneiderberg 32, 30167 Hannover, Germany
e-mail: *toan@mst.uni-hannover.de*

Abstract. In this paper we present an approach for handoptimized pipelined FPGA–multipliers, namely carry save array multipliers (CSM). By a detailed adaptation to the underlying architecture of XC4013E–3 FPGAs, we derive high throughput and compact implementation of FPGA–Multipliers. By means of a sophisticated pipelining scheme, clock frequencies of up to 96MHz are achievable for operand's wordthwidth of up to 10 bits.

1 Introduction

Algorithms applied in digital signal processing (DSP), have a steadily increasing computational complexity. In particular, if the processing has to be performed under real time conditions, such algorithms have to deal with high throughput rates. In many cases implementation of DSP algorithms demands application–specific ICs and hardware. This is especially given for image processing applications. Since development costs for such hardware and ASICs are high, algorithms should be verified and optimized before implementation.

Traditionally, software simulation is used for verifying algorithms and circuits. However, due to the enormous throughput rate and huge amount of operations which have to be performed in real time image processing, the software simulation for above purposes takes usually much too long, hours or days for simulating a few seconds of real time processing. Therefore, software simulation is clearly inadequate for verifying image processing algorithms under real time constraints.

Advantages in VLSI technology and, in particular, the increasing complexity and capacity of programmable logic devices (PLDs) have been making hardware emulation possible. The underlying key of the emulation systems is the use of SRAM–based field programmable gate arrays (FPGAs) which are very flexible and dynamically reconfigurable. In recent years, hardware emulation has emerged as a suitable way for verifying, especially digital signal– and image processing systems under real time constraints. Hardware emulation can considerably reduce the time taken in analysis and verification of algorithms and circuits. Furthermore, emulation systems can operate in the target environment with image data and can consequently help to investigate and to ascertain impacts of different parameters of algorithms on the quality of processed images. Thus, algorithms can be optimized before implementation.

Taking a closer look at image processing applications reveals that many iterative algorithms consist of a large number of multiplications and additions. Furthermore, high throughput rates are required, if those applications have to be performed under real time constraints. Hence, fast multiplications are very common in real time DSP– and image processing applications, especially for regular algorithms like filtering or discrete cosine transformation. Unfortunately, multipliers are often very costly parts and, due to the nature of the computation, may be the time–critical path of the system. For an efficient implementation and emulation of DSP–algorithms on FPGA–based emulation systems, an implementation of multipliers on FPGAs with a high throughput rate and low cost, in terms of FPGA resources, is crucial.

Hence, in this paper, we present an approach leading to an efficient realization of different types of FPGA–multipliers on Xilinx XC4013E–3. In sections 2, the general algorithm for multiplication is figured out. Related works for FPGA–implementation of multipliers are discussed in section 3. Based on hardware macros provided by high–level CAD tools, in a first step we synthesized multipliers with automatically inserted pipeline stages in section 4. Analyzing the results, reveals that these modules do not fulfill the requirements of real time image processing. Therefore, in section 5 we present a new approach for handoptimzed pipelined multipliers for Xilinx–FPGAs. The results obtained are compared to the automatically generated multipliers. In Section 6 concluding remarks are provided.

2 Implementation of Multipliers

By the multiplication of two n–digit binary numbers $A = a_{n-1}... a_1 a_0$, called multiplicand, and $B = b_{n-1}... b_1 b_0$, called multiplier, is meant to generate a $2n$–digit product $P = p_{2n-1}... p_1 p_0$, where the relationship between the value of P ($V(P)$) and the values of A ($V(A)$) and B ($V(B)$) is expressed as follows

$$V(P) = V(A) \cdot V(B) = V(A) \cdot \sum_j b_j 2^j = \sum_i \sum_j (a_i b_j) 2^{i+j} \qquad (1)$$

Hence, the multiplication can be most straightforwardly performed by a process of AND operation for generating the partial products $V(A)b_j$ and ADD operations for summing these partial products, as illustrated in figure 1 for $n = 4$. There is a wide range of possible architectures for multipliers which realize (1) in different manners providing different performances at different costs [1]. Generally, regarding the architectures, they may be categorized in array multipliers and tree multipliers. Typical array multipliers are ripple carry array multipliers (RCMs) and carry save multipliers (CSMs). RCMs are regular in structure (figure 1). The excellent repeatability of basic cells in the array and their local interconnections make RCMs attractive and eligible for realizations on Xilinx FPGAs which namely consist of a regular array of function generators. However, due to carry propagation, RCMs have the drawback of a long delay. Partly overcoming that drawback, carry–save–technique can be used to abridge the above delay in array multipliers. In such multipliers, called carry save array multipliers (CSMs), during the multiple additions of the partial products, the carry propagation is saved until all the additions are completed, and a final sum vector and a final carry vector are generated. These carry save additions are performed just using full adders, which are referred to as (3, 2) counters, since each of them reduces three inputs to two outputs, with no interstage carry propagations. Finally, a merging adder is used to merge the sum vector and the carry vector into one final result of the multiplication [1], [2].

In contrast to array multipliers, tree based architectures, e. g., Wallace tree multipliers and Dadda multipliers, utilize the parallelism of computations and can therefore accelerate the multiplication while sacrificing the locality of connections between the computation cells and the regularity. These sacrifices overshadow the speed–up gained by making the computations parallel, especially when the multiplier is implemented on FPGA, where routing delay significantly contributes to the overall delay of the circuit. Because of the sophisticated interconnect structure, tree multipliers are not considered in this paper. In order to derive general multipliers we assume only non constant operands. Therefore, booth recoded multipliers are not considered either.

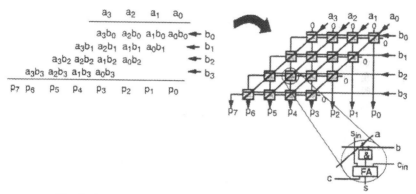

Fig. 1. Partial product matrix and correspondent array architecture

The above nonpipelined multipliers, which are here called combinational multipliers, may be invoked. Throughput of combinational multipliers implemented on FPGAs are apparently far too low to meet the demands of real time processing.

The two well acquainted solutions to the above problem are parallel processing and pipelining. Both of them exploit parallelism in different ways, allow to speed up the processing and thereby increase the throughput. However, parallel processing with merely straight hardware duplication may not be cost effective, especially for implementation on FPGAs, where resources are limited and costly. In contrast to that, by pipelining, data processing is broken into smaller stages, and the processing of successive data is performed in an overlapped manner. Therefore, pipelining can provide significant increase in throughput with only a moderate increase in hardware investment for pipeline registers. Hence, application of pipelining is indispensable for image processing under real time constraints.

Due to the common use of multipliers in DSP–applications, the time critical characteristic of multiplication and the high cost of multipliers, in terms of FPGA resources, compact and pipelined multipliers providing high throughput rates are mandatory.

3   Related Research

There exist several approaches for the implementation of pipelined multipliers on Xilinx FPGAs which utilize a channelled array architecture with so called configurable logic blocks (CLBs) as logic resources [3]. Those approaches have mostly focused on minimizing the number of required CLBs for the multipliers. In [5], multipliers implemented with the number of CLBs proportional to the operand width n have been reported. The multiplication algorithm implemented utilizes a $2n$ linear array to compute, in a bit serial manner, the product of two n–bit numbers in $4n-1$ time steps. The critical path delay was 42ns [5]. Another way is the realization of lookup tables instead of multipliers by distributed arithmetic which also processes operands in a bit serial manner. This approach is reported in [3] and [4], where it is used to implement FIR filters. The achievable data rate is proportional to $1/(n+1)$, e. g., 5.4MHz for $n = 8$. However, this approach is limited to filters with constant coefficients. The above multipliers require only a small number of CLBs by invoking a bit serial approach. However, utilizing such a bit serial approach can only provide a moderate throughput rate. Looking at real time image processing and video applications, higher throughput rates and consequently, faster multipliers are required. In order to derive multipliers providing high throughput rates, our approach is to use bit parallel implementations enhanced by a sophisticated scheme for inserting pipeline registers. The common way developing FPGA design is by using a high–level synthesis tool. Therefore, in a first step we automatically generate carry save array multipliers including pipeline stages as described in the next section.

4   High–Level Synthesis of Pipelined Multipliers

For arithmetic building blocks like adders and multipliers, modern high–level synthesis tools include module generators, with the opportunity to specify, e. g., module type, operand's wordwidth, etc. Furthermore, these tools enable balancing pipeline stages.

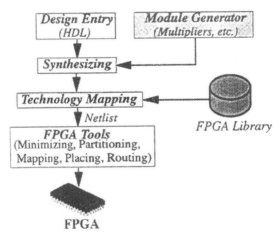

Fig. 2. Typical Design Flow of a High–Level Synthesis Tool for FPGAs

For realization of the circuit on FPGAs, for example Xilinx FPGAs, after synthesizing a technology mapping has to be performed. This means a mapping of logic gates to lookup table based logic blocks. Afterwards, the FPGA netlist can be mapped and routed on the FPGA using partitioning, placing and routing tools. Among the different architectures of multipliers, due to the regularity and the nature of the dataflow in the array, carry save array multipliers can be most conveniently and efficiently pipelined. As an example,

we generated and synthesized 8–by–8 pipelined carry save array multipliers with different number of pipeline stages (# P_stage) by Synopsys tools. Cost, in number of required
CLBs, and performance in term of maximal clock frequency of those multipliers are
given in table 1, assuming Xilinx XC4013E-3 as target FPGA.

From table 1, it is obviously that the multipliers can only provide moderate throughput rates. Table 1 also shows that finer pipelining (i. e., with more pipeline stages) has
not always increased throughput of the multipliers as it is expected. The two potential
causes of it are, on the one hand, the optimization and the inserting of pipelining registers
are made by the high level tool at the gate level and independent of the technology, i. e.,
independent of the structure of the FPGA. So, this insertion of pipelining registers may
be not optimal for the FPGA. On the other hand, the finer the pipelining is, the more
registers and CLBs are required. It becomes harder to place the CLBs, and the multiplier
becomes larger. So, the routing delay increases and degrades the speed–up provided by
pipelining.

Beyond, another important issue, which should be taken into account in implementation of circuits on Xilinx FPGAs, is the heavy dependence of the number of occupied
CLBs and delay of the circuits on placing and routing these CLBs.

Table 1: 8–by–8 Pipelined Carry Save Array Multipliers synthesized by
Synopsys tools and implemented on Xilinx XC4013E-3

# P_stage	2	6	13
# CLB	144	156	240
f_{clk} [MHz]	30	24	54

Beyond, another important issue, which should be taken into account in implementation of circuits on Xilinx FPGAs, is the heavy dependence of the number of occupied
CLBs and delay of the circuits on placing and routing these CLBs.

In order to obtain more compact multipliers with high throughput rate, we have
approached manual optimizations. In the following section, our approach and experimental results of implementation of pipelined carry save multipliers on FPGAs are
presented. A comparison between these implementations and those derived by using the
Synopsys high–level synthesis tools are also included.

5   FPGA–Implementation of Handoptimized Multipliers

For realization of pipelined multipliers providing high throughput rate, we have performed manual optimizations for mapping of these multipliers on FPGAs. This optimization includes the adaptation of the multiplier to the FPGA structure, i. e. combining and
mapping basic cells of the multiplier on CLBs and inserting pipeline registers with
respect to the FPGA structure. Furthermore, positions of occupied CLBs on the FPGA
are prescribed relatively to each other to ensure a minimal routing delay.

a) Fully–Pipelined Carry Save Array Multiplier on XC40xx

The basic structure of a 4–by–4 carry save array multipliers [1] is depicted in figure
3a. In order to minimize the number of pipeline stages and thereby the number of required

flip flops (# *FF*) for pipelining, the computation cells of the multiplier were adapted to the structure of XC4000–function generators and –CLBs before inserting registers. So, since each of such function generators can incorporate up to 4 variables, the cells 2, 3, 4 in the first row can be respectively coupled with the cells 8, 7, 6 in the second row (figure 3a) giving the basic cells drawn in figure 3c, where, for n–by–n multiplication, R_a_i denotes the pipelined bit of a_i. With respect to figure 3b, $c_{i,1}$ and $s_{i,1}$ are the carry–out and sum–out of the basic cells in the first row and $s_{i+1,j-1}$ and $c_{i,j-1}$ are sum–in and carry–in, and $s_{i,j}$ and $c_{i,j}$ denote sum–out and carry–out of the basic cells in the jth row of the CSM.

Therefore, the boolean equations result in:

$$c_{i,1} = a_i b_1 a_{i+1} b_0$$
$$s_{i,1} = a_i b_1 \oplus a_{i+1} b_0 \qquad i = 0, 1, \dots, n-2 \qquad (2)$$

$$c_{i,j} = a_i b_j s_{i+1,j-1} \vee a_i b_j c_{i,j-1} \vee s_{i+1,j-1} c_{i,j-1} \qquad i = 0, 1, \dots, n-2$$
$$s_{i,j} = a_i b_j \oplus c_{i,j-1} \oplus s_{i+1,j-1} \qquad\qquad j = 2, 3, \dots, n-1 \qquad (3)$$

Similarly, the half adder and the full adder of the merging adder in the last row can be coupled together. Afterwards, the registers can be inserted. The most convenient way to insert pipeline stages is by using equitemporal cuts according to figure 3b. In the following, two possible strategies of pipelining will be described.

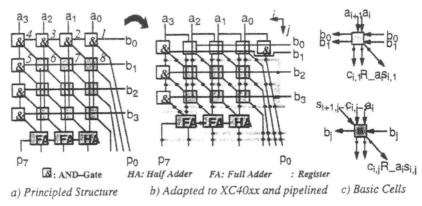

| : AND–Gate | HA: Half Adder | FA: Full Adder | : Register |

a) Principled Structure *b) Adapted to XC40xx and pipelined* *c) Basic Cells*

Fig. 3. 4–by–4 Fully Pipelined Carry Save Array Multiplier

Since the atomic logic delay of XC40xx is the delay of a 4–inputs function generator, the maximal performance for CSMs can be attained with fully pipelined circuit as in figure 3b. Finer pipelining would not improve performance of the circuit. Every two basic cells of figure 3c, which contain six flip flops and four 4–inputs function generators, can be mapped on three CLBs as in figure 4. For a n–by–n CSM of this version, which consists of 2n–3 pipeline stages, its cost can be approximately computed as follows:

The array of basic cells and AND–gates requires $1/2(3n^2 - 2n) + n$ CLBs. Number of flip flops for preskewing inputs of the $(n-1)$ bit merging adder:

$$\#FF_{M-ADD} = 2 + 4 + \dots + 2(n-3) = (n-3+1)(n-3) \qquad (4)$$
$$\#FF_{M-ADD} = (n^2 - 4n + 3)$$

$$\#FF_{M\text{-}ADD} = 2 + 4 + ... + 2(n-3) = (n-1)(n-3) = (n^2 - 4n + 3) \qquad (5)$$

The above FFs can be placed in $1/2\,(n^2 - 4n + 3)$ CLBs. Besides, the logic for the merging adder occupies $n-1$ CLBs. Hence, number of CLBs required for the multiplier, without pre– and deskewing is:

$$\# CLB^* = 2n^2 - 3n \qquad (6)$$

In oder to allow a word parallel multiplication, the operand bits have to be preskewed, whereas the result bits have to be deskewed. Therefore, some more flip flops, called pre– and deskewing flip flops are required. The number of pre– and deskewing flip flops for the above multiplier can be accounted as $1/2\,(5n^2 - 11n)$, which can be placed in $1/4\,(5n^2 - 11n)$ CLBs. Due to that, the minimum number of required CLBs for a n–by–n fully pipelined CSM, including CLB flip flops for pre– and deskewing is:

$$\#CLB = \frac{13}{4}n^2 - \frac{23}{4}n \qquad (7)$$

However, due to limited routing resources additional CLBs for routing signals are required. This leads to a higher amount of total number of CLBs (s. Tab. 2).

Delay from CLB–flip flop to CLB–flip flop (T_{CC}) of CSMs of this version is independent of the word length n and composed of the delay from clock to output of a CLB–flip flop (T_{CKO}), the set–up time for CLB–flip flop from CLB–inputs via function generator (T_{ICK}), see [3], and the routing delay (T_R) between two neighbouring CLBs. Therefore, T_{clk} of the multiplier can be expressed as follows.

$$T_{CC} = T_{CKO} + T_{ICK} + T_R \qquad (8)$$

For an XC4013E–3, $T_R \approx 1.6$ ns (worst case), $T_{CKO} = 2.8$ ns and $T_{ICK} = 3.0$ ns, leading to $T_{CC} \approx 7.4$ ns ($f_{CC} \approx 135$ MHz).

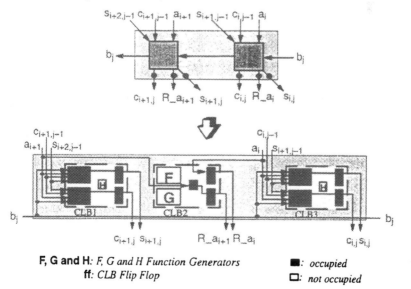

F, G and H: *F, G and H Function Generators*   ■: *occupied*
   ff: *CLB Flip Flop*   □: *not occupied*

Fig. 4. 4–by–4 Fully Pipelined Carry Save Array Multiplier on XC40xx

Our experimental results are summarized in Table 2, where #CLB expresses the cost of fully pipelined CSMs, and f_{CLK} denotes the achievable external clock frequency. As can be seen the achievable external frequency is below f_{CC}. In more detail, it depends on the wordwidth n, due to the pads and the additional routing of the operands bits b_j, since each of them has to be broadcasted to all CLBs of the jth row.

Table 2. Implementation Results of Fully Pipelined CSMs on XC4013E–3

Operand width (bit)	4	5	6	7	8	9	10
#P_stage	7	9	11	13	15	17	19
#CLB	39	67	98	144	184	244	296
f_{CLK} [MHz]	96	96	96	82	75	77	78

As can be seen from Table 2, fully pipelined CSMs are very costly, especially, when flip flops for pre– and deskewing must be included. However, these flip flops can be omitted, if the multiplier is part of a processing chain, where each module processes the bits in an identical skewing scheme. Consequently, only one preskewing circuit at the front, and one deskewing circuit at the end of the processing chain are necessary.

Nevertheless, fully pipelined CSMs require a large amount of CLBs. In order to reduce the hardware expense with respect to flip flops, a coarsely pipelining scheme is more appropriate. Among possible pipeline schemes, we reduce the number of pipeline stages by a factor of two. This leads to so called two–rows–pipelined CSM, four–rows– pipelined CSM, etc.

b) Two–Rows–Pipelined Carry Save Array Multiplier on XC40xx

As an example, an 8–by–8 two–rows–pipelined CSMs is depicted in figure 5, where the basic cells are the same as in figure 3c.

With multipliers of this version, an n–by–n multiplication is performed in $3n/4$ cycles. All the flip flops, exclusive of flip flops for pre– and deskewing, and logic of an n–by–n multiplier of this version can be placed in *(n* columns $\times$ *n* rows + 1) CLBs. So, the cost of the multiplier, without pre– and deskewing flip flops is n^2+1 CLBs. Furthermore, from figure 5, number of all required CLBs, for $n>4$, inclusive of flip flops for pre– and deskewing can be expressed as follows

$$\#CLB = \frac{21}{16}n^2 - \frac{11}{8}n \tag{9}$$

Table 3. Implementation Results of Two–Rows–Pipelined CSMs on XC4013E–3

Operand width in bit	4	5	6	7	8	9	10	11	12
#P_stage	4	6	7	8	9	10	11	12	13
#CLB	26	44	61	87	111	147	175	221	256
f_{CLK} [MHz]	79	70	74	66	66	58	65	56	62

The implementation results are summarized in table 3. Due to the increase of delay between pipeline stages, clock frequency decreases by a factor of 1/4, whereas the number of CLBs is reduced by more than 40% for operand wordwidth $n \leq 10$. It shows, that two–rows–pipelined CSMs represent a good trade–off between cost and performance in terms of achievable frequency.

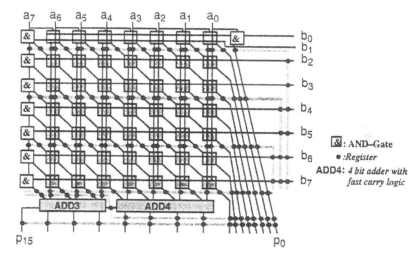

Fig. 5. 8–by–8 Two–Rows–Pipelined Carry Save Array Multiplier

We compare our handoptimized multipliers with automatically generated modules using 8–by–8 pipelined CSMs implemented on XC4013E–3 FPGA. Figure 6 shows the hardware expense in terms of used CLBs and performance in terms of achievable frequencies, depending on the applied pipelining scheme. As can be seen, handoptimized CSMs reduce the amount of required CLBs about 30% – 42% compared to automatically generated CSMs. Furthermore, the fully and two–rows–pipelined 8–by–8 CSMs of our approach are 39% faster than those ones, which are automatically generated. Accordingly, the achievable clock frequency can be increased from 54MHz to 75MHz maximal.

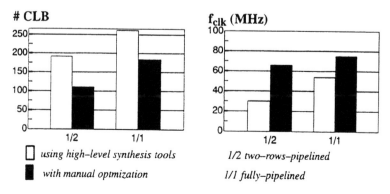

Fig. 6. Comparison between 8–by–8 Pipelined CSMs Implemented on XC4013E–3

6   Conclusion

By a handoptimized mapping of pipelined carry save array multipliers (CSMs) onto Xilinx XC4013E-3 FPGAs, it becomes possible to derive high throughput multiplier modules, working at 75MHz minimum, for operand's wordwidth $n \leq 10$. The number of allocated CLBs for a complete pipeline scheme in case of $n \leq 10$ always remains below 50% of the whole FPGA-area. It is worth to mention that our approach provides one region of adjacent CLBs, thus the remaining area can be efficently used for additional modules. However, if the number of CLBs becomes a limiting factor, e.g., in case of smaller XC4000-FPGAs, it is more appropriate to use a CSM with a coarse pipeline scheme. For example, a two-rows-pipelined CSM working at 56MHz minimum, reduces the number of CLBs by more than 40% for operand's wordwidth $n \leq 10$.

In case of high performance applications like image processing, handoptimized designs are crucial to cope with real time constraints. Hence, handoptimized multipliers, presented in this paper, enable processing of such real time applications on FPGA based hardware platforms.

Acknowledgement

One of the authors, T.-T. Do, receives a scholarship from the German Academic Exchange Service (Deutscher Akademischer Austauschdienst – DAAD). He is grateful to this organization for supporting his research.

References

1. Pirsch, P.: Architectures for Digital Signal Processing, John Wiley & Sons, to be published in (1997)

2. Hwang, K.: Computer Arithmetic, John Wiley & Sons (1979)

3. Xilinx Inc.: The Programmable Logic Data Book (1994)

4. Mintzer, L.: FIR Filters with Field-Programmable Gate Arrays, IEEE Journal of VLSI Signal Processing, vol. 6 (1993) 119 – 127

5. Saha A., Rangasayee E.: Fast and Efficient Implementation of Integer Arithmetic Algorithms Using FPGAs, IEEE Southeast Conference, (1993)

The XC6200DS Development System

Stuart Nisbet[1] and Steven A. Guccione[2]

[1] Xilinx Development Corporation
52 Mortonhall Gate
Edinburgh EH16 6TJ (Scotland)
Stuart.Nisbet@xilinx.com
[2] Xilinx Inc.
2100 Logic Drive
San Jose, CA 95124 (USA)
Steven.Guccione@xilinx.com

Abstract. The *XC6200DS* is a development system for reconfigurable logic design. The hardware features an *XC6216* reconfigurable logic device, a 33 MHz PCI bus interface and up to 2 MB of on-board SRAM. Support software including interface libraries in both *C++* and *Java*, as well as the *WebScope* graphical debug interface and sample applications are also discussed.

1 Introduction

The *XC6200DS* [12] development system is a complete development platform, including hardware and software for producing applications based around the *XC6200* [11] reconfigurable logic device. This system features a PCI bus board containing an *XC6216* reconfigurable logic device and up to 2 MB of SRAM. In addition, software support in the form of the *XACT/6000* placement and routing tool, as well as run-time support libraries in both *C++* and *Java* are available.

2 The System Hardware

Figure 1 shows a diagram of the *XC6200DS* hardware. The board features a 33 MHz *PCI* host interface bus [9] and up to 2 MB of SRAM. In addition, a PCI mezzanine connector is supplied. This permits custom interfaces to the *XC6200DS* hardware to be developed within a high performance, standard framework.

Other support hardware is provided on the board. This includes a programmable clock generator and a current sensing meter. Figure 1 diagrams the *XC6200DS* hardware.

2.1 The XC6200 Reconfigurable Logic Device

The *XC6216* reconfigurable logic device [11] is the first member of Xilinx's XC6200 FPGA family. This device features a microprocessor compatible interface bus for configuration and data access to the device. In addition, all logic,

to support the PCI interface. This leaves the remainder available for other glue logic and interface tasks.

2.3 The Memory Interface

The *XC6200DS* contains 2 banks of RAM, each containing as much as 1 MB. Each bank is organized as 1M 16-bit words of data. Multiplexers and bus switches on the *XC6200DS* permit the data and address paths to the RAM to be dynamically reconfigured. This provides the different configurations suitable for different processing requirements. Figure 2 shows the 4 major modes of memory operation.

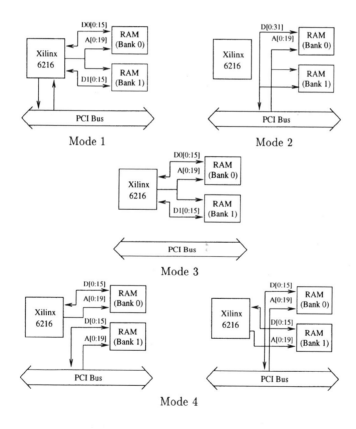

Fig. 2. The memory access paths.

Mode 1 sends the same 20 bit address to both RAM banks, and provides a data path from the RAM to the XC6200. This permits RAM to be accessed as a single large bank. RAM may be accessed as 8-, 16- or 32-bit words by the

XC6200. In addition, the PCI bus is free to communicate in parallel with the *XC6200* while it accesses the RAM. This mode is useful for circuits which wish to process data to/from the PCI bus from/to the RAM, with the *XC6200* in this data path, performing processing.

Mode 2 again groups the 2 banks of 16-bit RAM to appear as a single bank or RAM. Unlike *Mode 1*, the RAM is accessed via the PCI bus. This allows fast loading and unloading of the on board RAM directly from the host processor. In this mode, the RAM may be accessed as 16- or 32-bit words.

Mode 3 is essentially the same as *Mode 1*, but without the PCI bus communicating with the *XC6200*. This permits the *XC6200DS* to run independent of any external bus activity. This mode is useful for using the *XC6200DS* to develop embedded applications which will not use the host processor.

Finally, *Mode 4* permits one bank of RAM to be accessed by the *XC6200* and the other by the PCI bus. The bank attached to the *XC6200* may be switched with the bank attached to the PCI bus. This mode permits the *XC6200* device to perform operations in parallel on one bank of RAM while the PCI bus loads or unloads the other.

2.4   Other Features

In addition to the PCI interface, the RAM and the *XC6200* device, the *XC6200DS* board has additional support hardware. An analogue to digital converter is connected to a current sensing resistor on the *XC6200* power supply. This permits real-time monitoring of device current. Because of the unidirectional routing on the *XC6200* device, excessive current due to internal bus conflicts is not a problem. The current meter is provided simply for those concerned with power consumption, not as a safeguard feature.

A programmable clock generator is also supplied. The clock generator permits circuits in the *XC6200* to be clocked over a wide range of frequencies. This permits circuits in the *XC6200* to be run at their maximum speed without requiring the user to produce clock divider circuits. The *XC6200* may also be clocked in single or multiple clock mode. This mode is useful when debugging circuit designs.

A standard PCI mezzanine interface permits the user to attach suplementary designs to the *XC6200DS*. All *XC6200* I/O pins are attached to this mezzanine, giving full flexible access to the reconfigurable hardware. This mezzanine connector permits daughtercards to perform functions such as video or real-time control interfacing.

Finally, the *XC6200DS* provides support for serial downloading of data. Support for serial PROMs for both the XC6200 and XC4013 are on the *XC6200DS* board. In addition, the *XC6200DS* contains a serial *Xchecker* cable connection. This supports the development of embedded designs intended to run independent of the host processor.

3 The Support Software

3.1 The XACTstep Series 6000 Software

XACTstep Series 6000 is a graphical tool for 6200 Family designs. This system is a back-end tool with *EDIF* as its primary input. Because of the use of standard *EDIF* for design input, design data can be generated with any existing CAD tool which produces EDIF. Existing schematic capture and hardware design language tools, when combined with libraries for the *XC6200* can be used with the *The XACTstep Series 6000 software*.

The *XACTstep Series 6000* editor preserves the hierarchy of the input design. This hierarchy information is used to support both top down design through floorplanning and bottom up design through either manual or automatic techniques. In addition, fully automatic place and route is supported. The graphical editor gives full access to all resources on the 6200 Family architecture.

While the The XACTstep Series 6000 software is the primary means of producing design data for the *XC6200DS*, detailed description of the software is beyond the scope of this paper.

3.2 Run-time Support

The *XC6200DS* system comes with run-time support software for both *C/C++* and *Java* programming languages. The basic support is in the form of a class name *Pci6200*. This class contains the interface to the *XC6200DS* hardware. Figure 3 shows a diagram of the *C/C++* interface library.

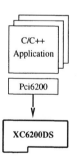

Fig. 3. The C/C++ support software.

The *Pci6200* class contains all functions necessary to read and write the *XC6200* device, the on-board RAM and all support devices, such as the clock generator and current meter.

The *Java* interface class contains identical functionality to the *C/C++* class. In fact, the *java* class was derived directly from the *C/C++* version of *Pci6200*. Figure 4 shows the *Java* interface.

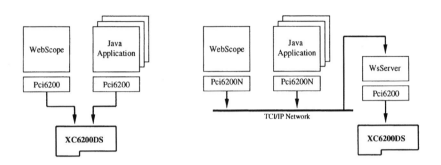

Fig. 4. The Java support software.

While the *Java* support software is based on the original *C/C++* code, there are some differences. First, the *Java* code defines two implementations of the interface, one for direct connection to the hardware, the other for remote access.

The remote access class takes advantage of the sophisticated network support in *Java* and allows remote execution of any function in the *Pci6200* class. While this is usually substantially slower than direct access over the PCI bus, it permits hardware to be easily shared in a networked environment.

Finally, the *Java* support software includes *WebScope* a *Java*-based debug tool for the *XC6200*.

3.3 The WebScope Debug Tool

WebScope is a portable graphical debug interface for the *XC6200* device implemented in *Java*. This tool interacts with the *XC6200* device to aid in the design and debug of circuits.

WebScope provides a point and click interface which permits the logic value of each cell in the *XC6200* to be probed. In addition, all control registers in the *XC6200* are graphically displayed.

In addition to displaying the state of the *XC6200* device, *WebScope* also provides symbolic access to data in the *XC6200*. Selected cells in columns of the *XC6200* may be read back from the microprocessor bus. These groups of cells may function as "variables" such as those in a software program. *WebScope* permits such variables to be defined for a given design. These variables are then displayed either textually or graphically by *WebScope*.

Finally, *WebScope* provide all basic control functions for the *XC6200* device, such as *reset* and *clock*ing. As with the *XACTstep Series 6000* software, details of *WebScope* are beyond the scope of this paper.

4 Conclusions

The *XC6200DS* provides a complete development platform for reconfigurable logic development. Support for activities such as coprocessing applications, embedded applications and research into reconfigurable computing are all possible uses for this system. Application development both inside and outside of Xilinx is continuing. Applications developed to date include; a 32x16 reconfigurable correlator, a CIC Filter, and a fax decoder.

In addition, Xilinx has made the design information for the *XC6200DS* an open standard. Several third party vendors are already producing hardware which is compatible with the *XC6200DS* specification.

Acknowledgements

Thanks to Tom Kean and Nabeel Shirazi for early work on the *XC6200DS* system. Thanks also to the engineering staff of the Xilinx Development Corporation for their efforts in tool and application development for the XC6200DS.

References

1. Gordon Brebner. A virtual hardware operating system for the Xilinx XC6200. In Reiner W. Hartenstein and Manfred Glesner, editors, *Field-Programmable Logic: Smart Applications, New Paradigms and Compilers*, pages 327–336, 1996. Proceedings of the 6th International Workshop on Field-Programmable Logic and Applications, FPL 96. Lecture Notes in Computer Science 1142.
2. Gordon Brebner and John Gray. Use of reconfigurability in variable-length code detection at video rates. In Will Moore and Wayne Luk, editors, *Field-Programmable Logic and Applications*, pages 429–438, 1996. Proceedings of the 5th International Workshop on Field-Programmable Logic and Applications, FPL 95. Lecture Notes in Computer Science 972.
3. Stephen Churcher, Tom Kean, and Bill Wilkie. The XC6200 FastMap TM processor interface. In Will Moore and Wayne Luk, editors, *Field-Programmable Logic and Applications*, pages 36–43, 1996. Proceedings of the 5th International Workshop on Field-Programmable Logic and Applications, FPL 95. Lecture Notes in Computer Science 972.
4. J. P. Heron and R. F. Woods. Architectural strategies for implementing an image processing algorithm on XC6000 FPGA. In Reiner W. Hartenstein and Manfred Glesner, editors, *Field-Programmable Logic: Smart Applications, New Paradigms and Compilers*, pages 317–326, 1996. Proceedings of the 6th International Workshop on Field-Programmable Logic and Applications, FPL 96. Lecture Notes in Computer Science 1142.
5. Tom Kean, Bernie New, and Bob Slous. A constant coefficient multiplier for the XC6200. In Reiner W. Hartenstein and Manfred Glesner, editors, *Field-Programmable Logic: Smart Applications, New Paradigms and Compilers*, pages 230–236, 1996. Proceedings of the 6th International Workshop on Field-Programmable Logic and Applications, FPL 96. Lecture Notes in Computer Science 1142.

68

6. Stefan H.-M. Ludwig. Design of a coprocessor board using Xilinx's XC6200 FPGA – an experience report. In Reiner W. Hartenstein and Manfred Glesner, editors, *Field-Programmable Logic: Smart Applications, New Paradigms and Compilers*, pages 77–86, 1996. Proceedings of the 6th International Workshop on Field-Programmable Logic and Applications, FPL 96. Lecture Notes in Computer Science 1142.
7. W. Luk, S. Guo, N. Shirazi, and N. Zhuang. A framework for developing parameterized FPGA libraries. In Reiner W. Hartenstein and Manfred Glesner, editors, *Field-Programmable Logic: Smart Applications, New Paradigms and Compilers*, pages 24–33, 1996. Proceedings of the 6th International Workshop on Field-Programmable Logic and Applications, FPL 96. Lecture Notes in Computer Science 1142.
8. Wayne Luk, Nabeel Shirazi, and Peter Y. K. Cheung. Modelling and optimising run-time reconfigurable systems. In Kenneth L. Pocek and Jeffrey Arnold, editors, *IEEE Symposium on FPGAs for Custom Computing Machines*, pages 167–176, Los Alamitos, CA, April 1996. IEEE Computer Society Press.
9. Edward Solari and George Willse. *PCI Hardware and Software*. Annabooks, 11848 Bernardo Plaza Court, Suite 110, San Diego, CA 92128 (USA), 1994.
10. R. Woods, A. Cassidy, and J. Gray. Architectures for field programmable gate arrays: A case study. In Kenneth L. Pocek and Jeffrey Arnold, editors, *IEEE Symposium on FPGAs for Custom Computing Machines*, pages 2–9, Los Alamitos, CA, April 1996. IEEE Computer Society Press.
11. Xilinx, Inc. *The Programmable Logic Data Book*, 1996.
12. Xilinx, Inc. *XC6200 Development System*, 1997.

Thermal Monitoring on FPGAs Using Ring-Oscillators

Eduardo Boemo and Sergio López-Buedo

Lab. de Microelectrónica, E.T.S. Informática, U. Autónoma de Madrid,
Ctra. Colmenar Km.15, 28049, Madrid - España.
e-mail: eduardo.boemo@ii.uam.es

Abstract. In this paper, a temperature-to-frequency transducer suitable for thermal monitoring on FPGAs is presented. The dependence between delay and temperature is used to produce a frequency drift on a ring-oscillator. Different sensors have been constructed and characterized using XC4000 and XC3000 chips, obtaining typical sensibilities of 50 kHz per °C. In addition, the utility of the Xilinx OSC4 cell as thermal transducer has been demonstrated. Although a complete temperature verification system requires a control unit with a frequency counter, the use of ring-oscillators presents several advantages: minimum FPGA elements are required; no analog parts exists; the additional hardware needed (multiplexers, prescaler, etc.) can be constructed using the resources of an FPGA, the thermal-related signals can be routed employing the standard interconnection network of the board, and finally, the sensors can be dynamically inserted or eliminated.

1 Introduction

Lower operating temperature on CMOS devices reduces the intrinsic delay and interconnection resistance. It also produces important reliability improvements, considering that electromigration and other failure effects rise exponentially with the temperature [1]. In the area of FPGAs, the gate density and speed of recent devices have appended thermal considerations to the traditional design trade-offs. Applications that make intensive use of chip resources at high speed can dissipate beyond current packaging limits. Miniature heat sinks and fans originally developed for the high-end microprocessor market are becoming familiar in the area of fast-prototyping.

The thermal considerations presented above results enlarged on FPGA-based systems like custom computers (FCCMs) and logic emulators. Their exhaustive utilization of dynamic reconfiguration increases the risk of configuration errors and signal contention. These situations may cause a significant increment of temperature and can produce a permanent chip damage. Moreover, like occurs on a single FPGA, the consumption associated to a given machine configuration is a priori unknown; thus, the particular features of an implementation (fine-grain pipelined datapaths, heavily loaded buses, etc.) can produce an unforeseen power overhead. Consequently, the implementation of a thermal monitoring unit allows several failures in FCCMs to be detected. For example, this strategy has been adopted in the XMOD board [2]: an 8-bit CPU examine the both the temperature and current at each FPGA.

Considering that the processing tasks on a multiple-FPGA board are performed in several chips, the detection of hot-spots requires to sense the temperature in each FPGA that composes the system. However, if the number of chips is relatively high, it is difficult to use discrete thermal transducers, as is common on current PC boards. Thermocouples or integrated sensors require both extra wiring and hardware that must be immune to the influence of the high-frequency signals usually present on the board. Moreover, the designer must also pay attention to topics beyond the scope of the fast-prototyping area, like sensor positioning, thermal coupling, or analog instrumentation.

The implementation of on-chip thermal transducers allows the designer to avoid the inconveniences described above. Main techniques to construct temperature sensors on CMOS technology make use of analog effects like the temperature dependence of the junction forward voltage, or the Seebeck effect [3]. Although these ideas can be useful to FPGA architects, they appear inadequate to the end-users of commercial chips. In this paper, this limitation is overcome by using ring oscillators as temperature transducers. This type of circuits can be easily implemented using few FPGA elements. The advantages of this approach are multiple:

a. Like other on-chip sensors, the junction temperature instead of the package one is measured.
b. All signals are digital; thus, they can be routed using the general interconnection network of the board.
c. The sensor itself is small: practical circuits make use of one or two logic blocks, and a minimum-size sensor can be fitted in just an I/O block.
d. The hardware needed to centralize the thermal status of the machine (basically a multiplexer and a prescaler counter to reduce the frequency) can be mapped in the FPGA, meanwhile the remaining low-speed tasks can be performed by the host or using a low-cost microcontroller.
e. A sensor or even an array of them can be placed in virtually in any position of the chip, making possible to construct a thermal map of the die.
f. The sensor can be dynamically inserted or eliminated.

Several researchers have proposed the use of on-chip thermal transducers. In [4], ring oscillators are used to measure both the temperature and power supply fluctuations. The oscillator is activated during a fixed period, and a counter with an scan path is used to read back the resulting frequency. In [5], an approach based on a "thermal-feedback oscillator" have been developed, whose main advantage is the small dependence between frequency and power supply fluctuations. At PCB level, a thermal monitoring method based on the measurement of a copper trace resistance has been proposed in [6]. In a different context, the use of thermal testing to detect gate oxide short failures have been proposed in [7].

Table 1. Ring-oscillators constructive characteristics

Test Circuit	Chain of inverters	Wiring
s1 and s5	Three inverters. Mapped in two CLBs.	General interconnection. Long delays.
s2 and s6	Three inverters. Mapped in two CLBs.	General interconnection. Short delays.
s3	Three inverters. Mapped in two CLBs.	Three long-lines plus one direct-line.
s4 and s7	One inverter. Mapped in a IOB output buffer.	General interconnection. Short delays.
OSC4	Internal cell.	General interconnection.

Table 2. Ring-oscillators features

Test Circuit	Experiment goals
s1 and s5	Medium-size, low-frequency sensor.
s2 and s6	Compact-size, medium-frequency sensor.
s3	Long-line based sensor. Suitable for clocking a synchronous counters without using the dedicated clock lines.
s4 and s7	Minimum-size. Worst-case sensor (maximum allowable frequency).
OSC4	XC4000 internal 5-frequency clock-signal generator cell

Table 3. Ring-oscillators timing characteristics

Test Circuit	Net delays (*Xdelay* tool)	Combinatorial delays (*Xdelay* tool)	Chip sample
s1	50.7 ns	22 ns (four LUTs)	XC3030PC84-125
s2	14.4 ns	22 ns (four LUTs)	XC3030PC84-125
s3	17.9 ns	22 ns (four LUTs)	XC3030PC84-125
s4	12 ns	8 ns (one obuf + one ibuf)	XC3030PC84-125
s5	47.8 ns	24 ns (four LUTs)	XC4005PC84-6
s6	20.1 ns	24 ns (four LUTs)	XC4005PC84-6
s7	10.8 ns	9 ns (one obuf + one ibuf)	XC4005PC84-6

2 Ring-oscillators on FPGAs

A ring-oscillator basically consists on a feedback loop that includes an odd number of inverters (Fig.1). Thus, the necessary phase shifting to start the oscillation is produced. The oscillation period is twice the sum of the delays of all elements that compose the loop.

Fig. 1. A ring-oscillator scheme

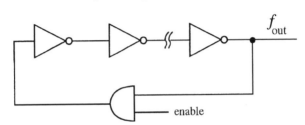

Ring-oscillators can be mapped on FPGAs using the look-up tables or the programmable inverters included on the I/O blocks. Considering that different interconnection elements can be inserted in the loop, the number of possible implementations are extremely large. In order to restrict the experiments, in this work just four different circuits, called s1, s2, s3 and s4, were characterized in the XC3000 family, and three circuits versions (s5, s6 and s7) where selected for the XC4000 family. In addition, the thermal response of the 8-MHz output of the built-in clock signal generator OSC4 [8] was measured. Main circuit features are summarized in Tables 1, 2 and 3.

An external control signal was *ANDed* with the loop in all CLB-based test circuits s1, s2, s3, s5 and s6, in order to allow the oscillators to be stopped. As a consequence, although these circuits have three inverters, their loops include four LUT delays. In the IOB-based oscillators s4 and s7, the loop was opened by using the 3-state control of the output buffer. These IOB versions were constructed to analyze the performance of the minimum allowable sensor size. All circuits were placed in the chip border in order to minimize the wiring capacitance between the oscillators and the corresponding output pads. As example, the layouts of the XC4000 oscillators s5, s6, s7 and osc4 are depicted in Fig.7

3 Experimental Results

The frequency-to-temperature response of each sensor was obtained by introducing each FPGA in a temperature-controlled oven. An Iron-Copper/Nickel (*Iron-Constantan*) thermocouple probe was placed in the center of the package, and was fixed to it with a heat conductive silver epoxy. An study about mechanical details of thermal sensors can be found in [9].

Fig. 2. Output frecuency vs. Temperature.
XC3090-125 CLB-based oscillators s1, s2 and s3

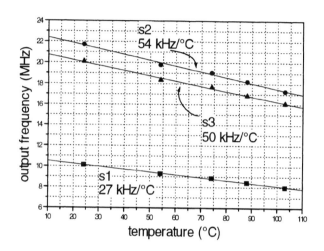

Fig. 3. Output frecuency vs. Temperature. XC4005-6
CLB-based oscillators s5, s6 and OSC4 cell

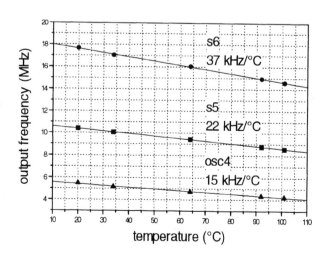

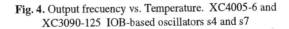

Fig. 4. Output frecuency vs. Temperature. XC4005-6 and XC3090-125 IOB-based oscillators s4 and s7

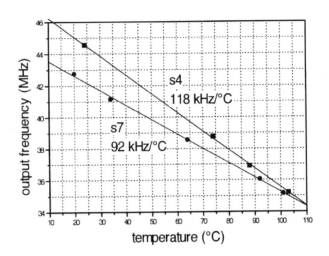

A long ribbon cables (near 0.8 meters) were utilized to carry both output and control signals outside the oven. In order to prevent an excessive sensor power consumption due to these high off-chip loads, a driver 74HC125 was inserted to isolate the FPGA from the cables.

Each FPGA was configured with all oscillators versions, but just one was enabled during the short period of time necessary to accomplish the frequency measurement. After that, the corresponding circuit was stopped again in order to maintain uniform the chip temperature. It allowed the error produced by self-heating to be minimized. An x-t curve tracer was utilized to verify the thermal equilibrium in the system after each temperature step. The error in the temperature measurement was maintained near 1 °C.

In Figs. 2, 3 and 4, the main experimental results are shown. All sensors exhibit a quite linear dependence with the temperature in the normal range of operation. The temperature sensitivity (in percentage per °C) also is very similar for all circuits. However, the IOB-based circuits, s4 and s7, present a high frequency oscillation, over 40 MHz, and should be discarded for practical applications.

The CLB-based sensors have relatively high speed (between 10 and 20 MHz), although their frequencies can be easily managed by a low-cost microcontroller if a prescaler is used. For example, a popular 68HC11 can be employed for counting if all these frequencies are previously divided by ten.

The best results corresponded to the built-in OSC4 cell. This oscillator, not only runs at lower speed and do not make use of extra FPGA resources, but also exhibits a small sensitivity to power supply fluctuations. The use of this cell as thermal transducer have not be reported in the manufacturer data books.

The power supply dependence of all sensors resulted linear in the operation range. This is depicted in Fig.5 for the XC 4005-6 oscillators. Thus, errors caused by power supply fluctuations can be corrected if the voltage of the board also is monitored. However, the sensibility was smaller for the OSC4 cell, as is depicted in Table 4. In addition, was observed that sensors whose loop delay is mainly caused by wiring are slightly less susceptible to power supply fluctuations.

Fig. 5. Output frecuency vs. power supply voltage.
XC4005-6 oscillators s5, s6 and OSC4 cell.

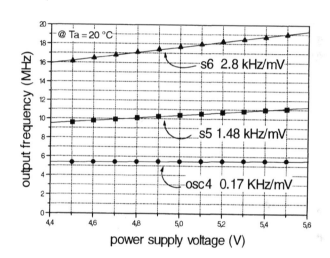

An alternative method for the temperature calibration of a given sensor can be carried-out if the approximate CMOS delay coefficient given by the manufacturer is utilized. This value is situated between 0.3 % per °C [10], and 0.35 % per °C [11]. In this way, the designer must first to construct a particular oscillator, and then to measure its output frequency at a known room temperature. After that, the remaining pairs (T,f) can be calculated by applying the delay coefficient to the measured point. Two examples of this method are shown in Fig.6.

Table 4: Output frequency reduction at Vcc=4.5 V

Test circuit	Frequency reduction at Vcc=4.5V in relation to normal operation
osc4	-1.8 %
s1	-7.3 %
s5	-7.8 %
s2	-7.9 %
s3	-8.1 %
s6	-8.7 %

Fig. 6. Measured (square points) and predicted oscillation frequencies (lines) vs. temperature using the CMOS delay coefficient. Circuits s1 and OSC4. Output frequency at room temperature as reference point.

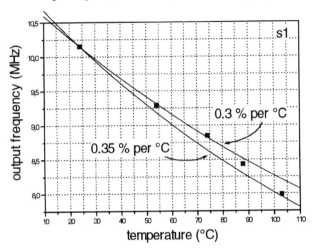

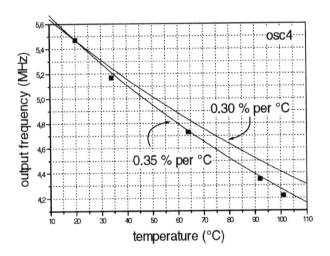

4 Conclusions

A group of experiments to demonstrate the feasibility of on-chip temperature transducers based on ring-oscillators have been presented. The proposed circuits allow the junction temperature of an FPGA to be easily measured. All prototypes analyzed showed a linear response with the temperature.

Although two methods for sensor calibration have been described, they can be simplified if the goal is just to detect a peak power value. In that case, the adjustment can be done in terms of power consumption, by measuring both chip input current and sensor output frequency during the normal operation of a given application. Thus, the correct thermal status of the machine can be described by a range of expected frequency values in each FPGA.

Best results in frequency range, resource occupation, and power supply sensitivity corresponded to the built-in XC4000 oscillator. However, the main disadvantage of this circuit is their fixed position in a corner of the chip. On the contrary, CLB-based ring oscillators can be situated in virtually any position.

The combination of temperature transducers and FPGAs could be also a powerful tool for researchers interested in thermal aspects of integrated circuits and packaging. Just the possibility of "moving" a sensor (or an array of them) from one point of the die to other, in a simple, fast and inexpensive way, is almost unthinkable in any other VLSI technology.

Future work will include a comparative study of die thermal maps using ring-oscillator sensors and an IR microscope.

Fig. 7. Layout of the s5 (top. left), s6 (bottom, left), s7 (bottom, right) and osc4 (top, right) oscillators (shaded areas represent used resources of the FPGA)

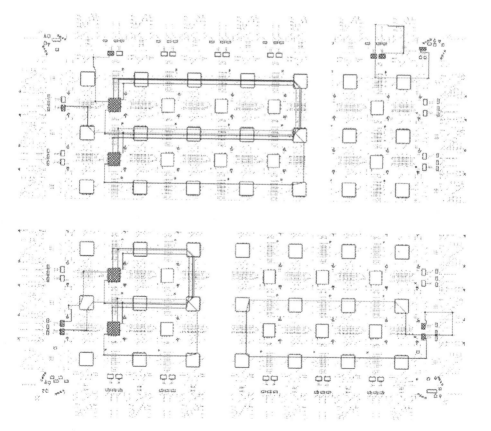

Acknowledgements

This work has been supported by the CICYT of Spain under contract TIC95-0159. The authors wish to thank Javier Garrido for his valuable contribution during the setup of the experiments.

References

1. Bakoglu, H., *"Circuits, Interconnections, and Packing for VLSI"*, Reading, Massachussets: Addison-Wesley Publishing Co. 1992
2. Giga Operations Corp., "XMOD Features", 1997.
3 Wolffenbuttel, R., (ed.), *"Silicon Sensors and Circuits. On-chip compatibility"*, London: Chapman & Hall, 1996.
4. G. Quenot, N. Paris and B. Zavidovique, "A temperature and voltage measurement cell for VLSI circuits", *Proc. 1991 EURO ASIC Conf.*, pp-334-338, IEEE Press, 1991.
5. V. Székely, Cs. Márta, M. Rencz, Z. Benedek and B. Courtois, "Design for thermal testability (DfTT) and CMOS realization", submitted to *Sensors and Actuators*, Special Issue on Therminic Workshop.
6. W. Sjursen, "Sense PCB Thermal Equilibrium", *Electronic Design,* pp.115-116, March 18, 1996.
7. Altet, J. and Rubio, A., "Dynamic Thermal Testing of Integrated Circuits", *Proc. DCIS'96*, pp.537-541, Barcelona: Universitat Politècnica de Catalunya, 1996.
8. Xilinx Inc.,"Xact Libraries Guide", pp. 3-388, April 1994.
9. Steele, J., "Get Maximum Accuracy from Temperature Sensors", *Electronic Design,* pp. 99-110, August, 1996.
10. Xilinx Inc., *"Programmable Logic Breakthrough '95. Technical Conference and Seminar Series"*, pp.6-11, 1995.
11. Xilinx Inc., "Trading Off Among the Three Ps: Power, Package & Performance", *XCELL*, N°22, 1996.

A Reconfigurable Approach to Low Cost Media Processing

Igor Kostarnov, Steve Morley, Javed Osmany and Charlie Solomon
[iak|stm|jo|chas]@hplb.hpl.hp.com

Appliance Computing Department, Hewlett-Packard Laboratories,
Filton Road, Stoke Gifford, Bristol BS12 6QZ, UK

1. Introduction

Future products retailing for $300 or less will need to support high performance implementations of media algorithms such as image and video decompression. The design of such "media appliances" imposes many constraints, not only on cost and performance, but also on development time and hence time-to-market.

This paper proposes a design approach using an "off the shelf" CPU supplemented by application-specific acceleration implemented in a single Field Programmable Gate Array (FPGA). We describe the operation of the experimental board used investigate strategies for mapping algorithms onto such systems and, for two representative algorithms, determine the performance that can be achieved with this modest amount of FPGA.

This paper first contrasts existing design approaches with the features of the CPU/FPGA approach. The following section describes the evaluation board, called Riley, which was developed to investigate a few system alternatives and their utility in accelerating media algorithms. The two specific algorithms are then presented in detail and finally some conclusions are drawn.

2. Background

2.1 Existing Design Approaches

The various approaches to low-cost media processing include general purpose micro-processors, augmentation with packed arithmetic instructions, digital signal processors (DSPs), application-specific instruction processors (ASIPs), and application-specific integrated circuits (ASICs). These make significant tradeoffs between processing speed, time-to-market, non-recurring engineering costs (NREs), and flexibility.

2.2 CPU/FPGA Approach

Our approach supplements an "off the shelf" CPU with a modest amount of FPGA for hardware acceleration. This combines the performance benefits of an application-specific hardware solution with the flexibility of a general programmable solution [1].

Consistent with the low-cost criteria only a single FPGA is considered (i.e. a Xilinx XC4013). The advantages offered are:

1. No NREs (with respect to those of an ASIC). The hardware accelerator is a programmable device and development time is significantly reduced.

2. Flexibility. Algorithm changes (such as occur through changes in standards), the can be reprogrammed without having to do expensive hardware modifications.

3. Dynamic reuse. The same silicon (FPGA) can be reconfigured for different modes of use during the normal life cycle of the appliance.

However a number of new design challenges for CPU/FPGA systems remain:

• How much FPGA is required to achieve significant performance improvement? More specifically can we meet our performance targets given the limited FPGA budget? (Note that this contrasts with much existing work that uses arrays of FPGAs for custom computing e.g. [2]).

• How is the scarce reconfigurable resource used effectively? The partitioning of algorithms into hardware and software is obviously crucial to achieving good performance.

Crucial to the effective utilisation of the FPGA is the consideration given to:

1. The relative latencies for implementing the same operation sequences on the CPU versus the FPGA.

2. Exploitation of parallelism both within the FPGA as well as between CPU and FPGA. Both the algorithm and the available FPGA resources will constrain this.

3. Match the data feed and store rates to the particular FPGA configuration's requirements.

4. Reduce the synchronisation overhead between the CPU, FPGA and other system components.

5. Use internal data stores to hold intermediate results. This reduces the requirement for off chip I/O.

We have applied these criteria in the investigation of a few implementations of two representative algorithms an evaluation platform called *Riley*, which we describe in Section 3. The results from these implementations are presented in Section 4.

3. Riley Board Description

3.1 Design Philosophy

Riley is a board designed for the investigation of system architectures comprising a RISC CPU and FPGA. Much of the board space is given over to debug and measurement ports to enable detailed analysis of results.

The Riley hardware permits the major system components to be interconnected in a variety of ways. This also allows experimentation with the way different algorithms are mapped to the board.

3.2 Architecture

Riley comprises an i960J CPU [4], a Xilinx 4013 FPGA [5] and shared system memory. There is also a boot EPROM containing MON960 and an I/O block containing both serial and parallel interfaces. The connectivity is shown in Figure 1. The FPGA is memory-mapped into the i960 address space.

The i960J has a multiplexed address and data bus that is de-multiplexed in a bus control unit. The FPGA has direct access to both the multiplexed and de-multiplexed busses. It can arbitrate for bus mastership using the i960 bus hold/grant protocol or can grab the de-multiplexed busses directly using the "Abus_get" and "Dbus_get" signals. FPGA configurations are stored in the configuration RAM. The FPGA can disconnect this memory from the main system bus and use it as a local working memory completely independent from the main system memory.

Zero wait state SRAM has been used for both the system and configuration memories. This permits the simulation of on-chip storage. The i960 has on-chip I and D caches which can be enabled and disabled as required by an application

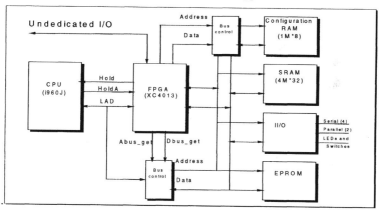

Fig. 1: Riley architecture

3.3 Modes of Operation

We chose to investigate 3 modes of interaction between the CPU, FPGA and memory.

In *functional unit mode* (Figure 2), the CPU is always bus master. The CPU issues instructions to the FPGA, writes operands, and reads results. All data therefore passes to and from the FPGA via the CPU register file. There are two types of instruction issued. 1) An *asynchronous instruction* allows the CPU to continue execution after the instruction has been issued. A signal back from the FPGA (either an interrupt or a polled memory-mapped register) indicates that the instruction has completed. 2) A *synchronous* instruction tells the FPGA to suspend the CPU bus interface unit by use of a "READY" signal. The FPGA asserts this signal when the instruction has completed.

In *lockstep mode*, the FPGA can read and write directly to the de-multiplexed bus using addresses generated by the CPU. On reads, the FPGA traps the address generated by the CPU and reads data from the de-multiplexed data bus. On writes, the FPGA uses the "Dbus_get" signal to isolate this bus from the CPU. The FPGA writes data directly to the bus during the CPU write cycle.

In *datapath mode* the FPGA can arbitrate to become bus master. The CPU and FPGA execute as two independent execution units and most of the information that passes between them is control and synchronisation.

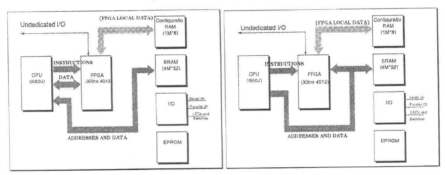

Fig. 2: Functional unit mode   Fig. 3: Lockstep mode

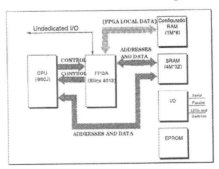

Fig. 4: Datapath mode

The mode of operation strongly influences the partitioning options for a particular algorithm and therefore the performance. An important part of the Riley investigation was to evaluate these modes across a range of algorithms.

3.4 Tools

All applications were implemented using C via the gnu960 compiler and VHDL for the FPGA. Though optimisation switches were used to generate efficient object code, occasionally, inline assembler was used. The VHDL source was synthesised to a Xilinx netlist using Synopsys synthesis tools and Xilinx libraries. Some structural coding was required to instantiate certain components, e.g. on chip RAM.

4. Results

This section describes the findings of two experiments that were conducted using the Riley board. Both performance and partitioning issues are discussed for each experiment.

4.1 Multiply/Accumulate Operations

It is generally possible to build more than one multiplier or accumulator in the FPGA and execute some of these operations in parallel. The capacity of the FPGA dictates how much parallelism can be exploited. In order to make partitioning decisions for the architectures being considered it is necessary to know the cost and performance of MAC operations in both the CPU and the FPGA.

The i960 used on Riley has an on chip multiply divide unit which generates a 32 bit result in 2 to 4 cycles. Addition takes one cycle.

The Synopsys tools synthesise a VHDL "*" operator to a combinatorial multiplier. The delay of a 16*16=32 bit multiplier in the FPGA is 5 cycles worse case at the 20Mhz clock frequency used on Riley. However, this one component uses around half of the available gates. An accumulator in the FPGA will execute in a single cycle. The gate count is insignificant in comparison to the multiplier.

Savings can be made in the FPGA by being efficient on bit-width i.e. only using the precision required. For example an 8*8=16 bit multiplier uses only 1/4 of the gates and executes in 3 cycles worse case. Further savings can also be made if one of the operands is fixed and known at compile time, for example the filter coefficients. Typically, a multiplier with a fixed coefficient uses half as many gates as an arbitrary multiplier.

Table 1 shows the parallel multiply/accumulate operations that fit into the Riley FPGA for various operand precisions. Both fixed and arbitrary multipliers are shown.

No. bits	8	12	16
Arbitrary	5	2	1
Fixed	12	5	2

Table 1: No. of taps in FPGA for filters

To realise the extra performance obtainable by parallelism in the FPGA, the system architecture must support dataflows that ensure the multipliers/accumulators are kept supplied with data. The location of data in the system is key to partitioning.

4.1.1 Partitioning and Results

All the issues just described affect the partitioning process for a DSP algorithm. This is now illustrated using the example of a Finite Impulse Response (FIR) filter.

The absolute performance of a single multiply/add is between 3 and 5 cycles in the CPU and is 6 cycles in the FPGA. In an N-tap filter however, the N multiplies can execute in parallel in the FPGA, so only one extra cycle for the accumulate is added for each tap. Assuming an average of 4 cycles per tap for the CPU, speedups can be achieved if two or more taps are executed in parallel in the FPGA. The speedup increases as taps are added, subject to the resource limits of the FPGA.

The performance of this partition is offset by the interface overhead in reading the new data sample and storing the result. This overhead depends upon the mode of operation. In *functional unit mode*, a load and a store are required for getting the sample from the memory to the FPGA, and a further load and store are required for returning the result. Each CPU load and store takes 3 cycles. In *lockstep mode*, only one load and one store are required as the data goes direct from memory to the FPGA. *Datapath mode* also only requires one load and one store but this time the FPGA can access memory on each cycle. Datapath mode offers the best performance then but requires more gates to implement the address generation.

Figure 6 shows the performance of the three operating modes and of the optimised CPU-only implementation for a 5-tap filter, as measured on Riley. A clock for clock comparison with a DSP, the Motorola DSP56000, is also shown. In all cases, code on the CPU runs entirely from I-cache, the FIR filter code requires only 20 instructions or so. All data is held in D-cache.

The performance in this example is dictated by the CPU/FPGA interface overhead. In functional unit mode, the FPGA is only in use 20% of the time with the CPU working flat out supplying data. In lockstep mode, the FPGA is in use 50% of the time.

When the number of taps in the filter exceeds the number that can be fitted into the FPGA, a different partition must be used. The calculation is broken down into stages so an N-tap filter is calculated in N/M stages (rounded up) of M taps each. An intermediate result must be accumulated between each stage. Fixed multipliers cannot be used in this partition as the coefficients change between sections. This will limit the value of M as shown in Table 1.

The coefficients and filter memory must now be stored either in a memory local to the FPGA or in the main system memory. M-way parallel access to this data is required to exploit the parallelism in the FPGA. This can be achieved for the coefficients by having M memories, one for each multiplier. Each memory has N/M elements. This is more problematic for the filter memory however as this data must be shifted by one place in memory at each sample interval.

In this case we choose to store the filter memory in main memory. To calculate one filter output therefore, N-1 samples are read from filter memory in addition to the new data sample. Figure 7 shows the results obtained for a 100-tap filter running on Riley.

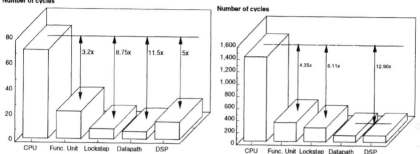

Fig. 6: Performance of a 5-tap filter   Fig. 7: Performance of a 100-tap filter

Comparing these results, the impact of the extra data marshaling can be seen when considering the DSP result (whose performance increases linearly as a function of the number of filter taps). Datapath mode now only just achieves parity with the DSP. Again, the CPU/FPGA interface limits the performance achieved in each mode.

4.2 The DPMatch Algorithm

Dynamic programming [7] is used in many "pattern matching" media algorithms. The algorithm has many different attributes to the digital filtering previously described.

A Dynamic Programming Match (DPMatch) engine (below) evaluates the "edit distance" between two code strings.

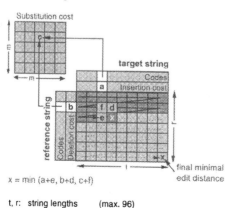

$x = \min(a+e, b+d, c+f)$

t, r: string lengths   (max. 96)
m : no. possible codes (max. 20)

Fig. 8: DPMatch performance

Each cell in the "t" x "r" match-matrix represents the "edit distance" between the strings at that point of the comparison. The cell value is computed by first calculating three "distances" that result from:

1. **inserting** a particular target string code. This is calculated by adding the insertion cost ("a") of the target string to the combined "distance" thus far ("e").

2. **deleting** a particular reference string code. This distance is calculated by adding the deletion cost ("b") to the combined distance ("d").

3. **substituting** the particular target string code for the relevant reference string code. The cost is defined by the pair-wise mapping of the relevant target and reference codes. The distance value is calculated by adding the substitution cost to the combined string distance *before* the last insertion and deletion operations ("f").

The resulting distance for the current cell ("x") is determined by **assigning the lesser** ("min.") of these three values. The final edit distance is the value that results in the last (lower right-hand) cell. More discussion of this algorithm can be found [8].

4.2.1 The DPMatch Partitions

The DPMatch algorithm was investigated in three different partitions that were compared with the CPU-only version running on Riley. The definition of these partitions was derived from a visual inspection of the "C" code. Starting from the smallest partition and working outward we selected:
1. A cell-by-cell computation. This is a minimal partition in the sense that the computation of one cell in the "t" x "r" matrix in Figure 8 is performed in the FPGA. The CPU performs all the data marshalling.
2. A row-by-row computation. In this partition the FPGA computes one row of the matrix in a single instruction.
3. The full match. Here the FPGA performs a complete match between the target and reference strings.

4.2.2 Cell-by-Cell

In this partition the FPGA hardware consists only of the DPMatch's cell computation. This is the addition of the three costs and the selection of the minimum. This partition uses *functional unit mode*. The insertion, deletion and substitution costs are all calculated by the CPU and passed as operands to the FPGA.

This minimal partition didn't quite reach parity with the CPU-only version of the DPMatch. It used 12% of the available FPGA gates. The obvious observation is that the CPU/FPGA interaction overhead is slightly larger than the savings had from the faster computation on the FPGA. Nonetheless, there are a few important observations to be made:
1. Because this partition is close to the computational core, the marshalling is complex.

2. It is difficult to derive much concurrency between the CPU and FPGA owing to the current cell's dependency on the results from the previous one.
3. On Riley, as the bus overhead was significant, operands (i.e. costs) were packed into a single bus write to the FPGA. The extra instructions used for packing in the CPU proved to be more efficient than using separate bus cycles.

4.2.3 Row-by-Row

This partition allows the CPU to load blocks of "pre-costed" target and reference strings to the FPGA at the beginning of the match. "Pre-costing" is the translation of the target and reference strings into their respective insertion, deletion and substitution costs.

As with the cell-by-cell partition, this operates solely in *functional unit mode*. Because on-chip storage is limited in the FPGA, the large matrix of "pre-costed" substitution costs required for a match could not all be stored simultaneously. Therefore the CPU performs an initial load of the strings' insertion and deletion costs, and after each successive computation of a row, the CPU loads the next row of substitution costs. The final result is read back after the loading of the final row. The row-by-row implementation performed 1.6 times faster than the CPU-only implementation and uses 42% of the available FPGA gates. The performance is limited by the overhead of the CPU writing the substitution costs for each row.

The row-by-row partition provides some interesting features over the cell-by-cell:
1. The provision for some internal intermediate state obviated the interactions required to retrieve and re-feed these values.
2. Data marshalling is much simplified because word-aligned blocks of operands were passed hence the interaction overhead per computation was reduced.
3. Some concurrency was obtained while the CPU marshalled for the next row of substitution costs the FPGA calculated the current row.
4. Limited capacity in the FPGA meant that the full block of substitution costs for a match could not be allocated. To simplify the management of this only the capacity for one row of substitution cost was provided.
5. The bottleneck in achieving better performance is the marshalling and transfer of the substitution costs from the CPU to the FPGA.

4.2.4 Full Match

In this partition, as much of the algorithm as possible has been placed in the FPGA This includes the code-to-cost mapping (*"costing"*), obviating the row-by-row's requirement to down-load the large cost matrix for each match.

The interface between the CPU and the FPGA is divided into two phases:
1. The initial set-up phase, when the substitution cost look-up tables (LUTs) are downloaded into one the FPGA configuration memory pages. This uses datapath mode. After this phase, all costing is done in the FPGA.
2. The run-time passing of the strings. This uses functional unit mode.

The CPU has two ways of reading the result. It can poll a ready register and read only after the result is ready, or it can read directly via a *synchronous read* instruction (see Section 3.3). In this case the read cycle will be suspended until the DPMatch matrix computation is complete.

This partition achieves a 21x performance increase over the CPU-only implementation. It uses 50% of the available gates in the FPGA. Note that the substitution costs are now held in the configuration memory not in the FPGA; which is why this partition does not use significantly more CLBs than the row-by-row.

The following factors contribute to this result:
1. The costing of the coded strings (i.e. converting the code strings into their representative insertion/deletion cost strings) is being done much more effectively by the FPGA. In the case of the insertion and deletion costs, combinatorial logic is used to generate the cost from the string codes instead of a table lookup.
2. The substitution costs are being accessed via the FPGA's own configuration memory. These accesses are faster than CPU accesses (2 vs. 3 cycles) and the FPGA does not have to arbitrate with the CPU for memory access.
3. Functional unit mode is only used for passing the target and reference strings.
4. All registers and data paths are carefully tailored to the specific minimal requirements of the algorithm.

4.2.5 DPMatch Conclusions

Figure 9 summarises the performance of the CPU-only version of the DPMatch compared with that of the three partitions.

The CPU/FPGA interface overhead limits the performance of the DPMatch algorithm in the cell-by-cell and row-by-row partitions. The important lesson learned from the full match partition is that the main performance gain was achieved *not* because of parallel computation (in fact, there is only one non-pipelined computing unit) but because of the concurrent data access and careful design of the control state machine.

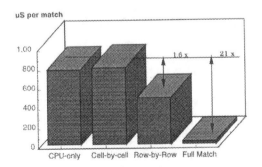

Fig. 9: DPMatch performance

5. Conclusion

In this paper we have proposed a new approach to the design of media intensive appliances using a CPU and a modest amount of FPGA for hardware acceleration. From implementations of representative algorithms (a digital filter and a dynamic programming match), we have demonstrated that a small amount of reconfigurable logic can be used to achieve high performance. For these algorithms the equivalent of around 500 Xilinx CLBs, plus a small local memory, can increase performance 12 to 21 times when compared to an optimised CPU-only implementation.

The FPGA is able to exploit parallelism available in an algorithm. The FPGA can also be "bit-width efficient", using no more bits of precision than are necessary. Internal store can be used for local data, state and control. The system architecture must however be able to support this increased computation rate. We have explored three architectures that enable this.

In functional unit mode the FPGA is a slave coprocessor to the CPU. This is the simplest mode for partitioning but performance can be limited by marshalling overheads on the CPU. It can be used where FPGA resources do not enable direct memory access or where only portions of an algorithm computation can be accommodated.

Lockstep mode is very similar to functional unit mode but increases performance by having a direct data path between the FPGA and memory.

Datapath mode offers the best potential performance but, as the CPU and FPGA execute as two independent units, there are many more issues that must be considered when partitioning. Datapath mode is particularly effective when the FPGA can process a large independent part of the algorithm without need for complex control and synchronisation with the CPU.

These architectures provide templates for the synthesis of CPU/FPGA systems. They provide contexts within which partitions can be evaluated. Although the process is manual at the moment, we expect the architectures and techniques discussed in this paper to form the basis for future tools which will automate this process. Research at HPLabs Bristol is targeted at providing these design tools.

References

[1] W.Sharpe, D.M.McCarthy, "Why can't hardware be more like software?", IEE Colloquium on Partitioning in Hardware-Software Codesigns, Feb 1995.

[2] D.A.Buell, J.M.Arnold, and W.J.Kleinfelder, "Splash 2 : FPGAs for Custom Computing Machines", FCCM 95, Los Alamitos, IEEE Computer Society Press, 1995.

[3] Rahul Razdan, "PRISC: Programmable Reduced Instruction Set Computers", PhD dissertation at the Dept. Of Computer Science, Harvard University, May 1994.

[4] *"1960® Jx Microprocessor User's Manual", Sept.1994, Intel Corp.; ISBN 1-55512-228-0.*

[5] *"The Programmable Logic Data Book", Xilinx Inc., 1994. San Jose, CA*

[6] *Steve Morley, "Multiply-Accumulate Intensive Algorithms on the Riley Experimental Board", Hewlett Packard Laboratories, Bristol, November, 1995; HPL-95-127.*

[7] *Igor Kostarnov, Javed Osmany, Charles Solomon, "Riley DPMatch - An exercise in algorithm mapping to hardware", Hewlett Packard Laboratories, Bristol, November, 1995; HPL-95-128.*

[8] *Robert Sedgewick, "Algorithms", Addison-Wesley Publishing Company, Reading, Massachusetts, 1983: ISBN 0-201-06672-6; "Chapter 37: Dynamic Programming", pp 483ff.*

Riley-2: A Flexible Platform for Codesign and Dynamic Reconfigurable Computing Research

Patrick I. Mackinlay[1], Peter Y.K.Cheung[1], Wayne Luk[2], Richard Sandiford[2]

Imperial College of Science, Technology and Medicine,
Exhibition Road, London SW7 2BT.

Abstract. The paper proposes a number of requirements for an ideal platform for codesign research. Riley-2, a new platform developed at Imperial College, is shown to meet these requirements. It is a PCI based board consisting mainly of four dynamically reconfigurable FPGAs and an embedded processor. A VHDL model of the Riley-2 system, including all major components except the PCI interface, has been produced. Two design routes, one based on VHDL with parametrised hardware libraries and the other based on a codesign language called Cedar, have been developed for Riley-2. Finally, an image processing application running on a PC with the Riley-2 and a Quickcam camera, is described.

1 Introduction

Electronic systems containing application specific hardware working alongside an embedded microprocessor are now common place. However, to design such systems quickly and reliably, while exploring the design space sufficiently to ensure near optimal solutions presents many new challenges. The problem is further complicated by reconfigurable hardware such as FPGAs, some of which can be dynamically reconfigured. This paper describes the design of an experimental platform, known as Riley-2, that supports our research in codesign and reconfigurable computing at Imperial College. It will also describe the tools that we have been developing to assist the design and analysis of such systems. We hope that the collection of tools would form a framework that facilitates rapid design exploration, evaluation and validation.

---

[1] Department of Electrical and Electronic Engineering,. (p.cheung@ic.ac.uk, p.i.mackinlay@ic.ac.uk)
[2] Department of Computing. (w.luk@ic.ac.uk, r.sandiford@ic.ac.uk)

2   Platform requirements

An ideal platform for codesign and reconfigurable computing would allow us to investigate both embedded and PC acceleration applications. Such a platform would comprise the following:

1. A general purpose processor (possibly used in embedded systems), with a modest amount of memory.
2. An adequate amount of reconfigurable resources, comprising a number of FPGAs with local memory. This would allow investigation of partitioning across FPGAs.
3. A flexible interface to the reconfigurable resources, allowing fast, partial and runtime reconfiguration of the reconfigurable resources.
4. A flexible and extensible external IO interface.
5. A fast host interface.
6. A complete model of the system, which allows detailed examination of the interaction between the software and the dynamically changing hardware.

There are many FPGA-based computing machines [1] which could be used in codesign research, however most fall short of the above requirements. For example the Riley board (Riley-2's predecessor) [2] only has one FPGA, has a slow interface to its host and does not support partial configuration. A more recent PCI-based XC6216 board, whose architecture is outlined in a previous paper [3], only has a single FPGA and no processor.

3   Riley-2 overview

Riley-2 is a successor to an earlier system, known as Riley [2], designed by Hewlett Packard laboratories, Bristol. It is designed to be used in a PC with a PCI interface. A photograph of Riley-2 can be seen in Figure 1.

Fig. 1. Riley-2 board

Figure 2 shows a block diagram of the Riley-2. The design has two main buses, the microprocessor's local bus and the reconfigurable resource (RR) bus.

The local bus connects the components meeting requirement 1 of the above list. The i960JF core is a 33 MHz integer RISC core used in many embedded system designs. The shared memory is implemented with a single 72 pin SIMM, currently seating a 16 Mbyte 50 ns BEDO DRAM, allowing a burst throughput of 88 Mbytes/sec. The boot ROM, not shown in the diagram, is a 256 Kbyte FLASH ROM, which contains the boot-up sequence and initialises the PCI interface.

In accordance with requirement 2, the RR bus connects four reconfigurable resource units, each unit containing a Xilinx 6216 FPGA and a 512 Kbyte fast local memory with a 32 bit data bus.

The RR interface along with the 6216s [4] provide the i960JF core with a flexible interface to the reconfigurable resources (requirement 3). The 6216 has the following features:

- Advanced microprocessor interface with direct read/write access to the reconfigurable resources. All registers (6216 internal and user defined), SRAM control store memory and local SRAM are mapped onto the microprocessor address space.

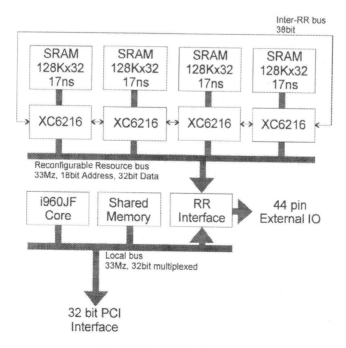

Fig. 2. Riley-2 block diagram

- Advanced dynamic reconfiguration capability with a high speed CPU interface, unlimited and partial reprogrammability.

The implementation of the PCI interface gives the host programmer flexibility by mapping the i960JF's shared memory and the reconfigurable resources into the host processor's address space. The processor, Intel's i960RP [5], contains a 32 bit PCI interface in the same chip as the i960JF core. It has inbound and outbound transfer queues which can support sustained burst transfers of up to 132 MBytes/sec (requirement 5). Actual speeds will depend on the host's PCI clock and on its chipset.

The RR interface also has a 44 pin external IO connector, which provides access to any external device, such as a video source. Since the RR interface is implemented with an FPGA, the IO interface can be changed and is very flexible (requirement 4).

4 Cosimulation

One important aspect of a codesign environment is the possibility of cosimulation, as outlined in requirement 6. Both hardware and software can be simulated together with a proper interface between them. To facilitate cosimulation of the Riley-2 System, we have developed a functional simulation model in VHDL for the i960 microprocessor. The model handles i960 machine code level instructions produced by the gnu *gcc960* compiler. In addition, we have also developed a VHDL model for the rest of the Riley-2 system (except for the PCI interface), including models for all the memory modules and the 6216s. A commercially available VHDL simulator is used for cosimulation.

The i960JF core model from the Riley system [2] was modified for the Riley-2. This was a relatively simple task since the processor core is the same. Due to the complexity of the microprocessor core, the following compromises were made. Only the functional behaviour of the instruction is modelled. Not all instructions and processor modes are implemented, however these can easily be added if needed. Pipeline timing and architecture is only approximate, since Intel does not publish the detailed internal architecture of its chips. The local bus model does not take account of setup and hold times. However, the following functionality was modelled correctly: the local bus protocol, the internal data and instruction caches, the register file and scoreboarding hardware, and the on-chip memory. Hence, the model only gives an approximate cycle-by-cycle model, but the bus traffic is modelled correctly.

The RR interface is actually an XC4013E whose functionality is synthesised from VHDL. The model is the actual code used to create the 4013E configuration.

The following 6216 functionality is modelled: Microprocessor interface; State accesses; Map register; Cell hierarchy; Cell functionality. The 6216 model can be run on its own with a given 6216 configuration (CAL file), or the entire Riley-2 system can be simulated, with i960 code parsing a CAL file. Wildcarding has not yet been fully tested, since the current tools do not create any CAL files that use this feature. However, the VHDL code has all the necessary features to support wildcarding. The model allows us to look in detail at the bus traffic during execution of a program. The exact details of configuring and executing FPGA designs can also be easily extracted. The detail in the 6216 model allows us to simulate FPGA designs that change dynamically. However, because of the complexity of the model, simulation is very slow and memory hungry. In most cases it is only practical to simulate part of the 6216.

5 Design tools

In developing design tools for Riley-2, we aim to meet several challenges, including:

1. Efficient utilisation of FPGA resources;
2. Systematic description and compilation of hardware, software and host programs;
3. Versatile facilities for simulating system operation at different stages of the development process;
4. Appropriate strategies for partitioning data for local and global memories, and for partitioning hardware and software;
5. Performance enhancement by reconfiguring FPGAs at run time.

The following provides an overview of how the first three challenges can be met using parametrised libraries and the Cedar system. Some work on meeting the remaining challenges, such as design tools for exploiting run-time reconfigurability, can be found elsewhere [6].

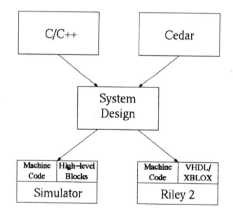

Fig. 3. High level compilation route

5.1 Overview

The first challenge is to enable efficient utilisation of FPGA resources. Our approach allows the user to optionally guide placement in a high-level language; for instance we have employed user-defined attributes in VHDL for this purpose. This development route is supported by a package of commonly-used library components that have been optimised for the 6200 series [7]. They can be customised by parameters such as the size of input and output operands, the separation between the bits of an operand and the number of pipeline stages. In particular, the ability to specify the separation between bits allows the interfaces of connected components to be aligned. This alignment leads to a more predictable placement and reduces the amount of routing resources required.

The second challenge is to systematise description and compilation of hardware, software and host programs. Our approach supports a style of hardware/software codesign, in which both the hardware and software parts of an application are defined in a single source file. This infrastructure is built on top of the VHDL route explained above, and there is a simple but effective mapping from Cedar to the VHDL libraries or to Xilinx XBLOX components (Figure 3).

The software components running on the I960 and the PC host are usually written in C or C++, and the hardware components are described in a language called Cedar. Cedar is loosely based on C, but has extensions to deal with parallelism and communication; its semantics clearly defines the cycle-by-cycle behaviour of a design and is similar to that of Handel [8]. Cedar achieves portability by using a target-specific interface, itself written in Cedar, to provide a standardised set of channels with which to communicate with the processor and other components of the board. At the most abstract level, therefore, a design can be specified in a completely imperative manner using C and Cedar. This method treats the software and hardware as parallel processes of equal status that communicate using a message passing scheme.

Our third challenge is to provide facilities for simulating system operation at different stages of the development process. At the early stages, a design for Riley-2 involving C and Cedar can be tested using a high-level simulator that we have developed. If necessary, sections of a design may be optimised by redefining them using register transfer level Cedar or by the direct use of Riley-specific features such as Well Known Address accesses. Further refinement is possible via the creation of custom VHDL libraries, which can then be included into the main Cedar design. When all hardware descriptions have been compiled into VHDL, their behaviour can be obtained using commercial VHDL simulators. After FPGA programming files have been generated, detailed interactions of the Riley-2 system can be explored using the VHDL model of Riley-2 outlined in the previous section.

Further details of our tools will be given in the next section, when we consider a simple example.

5.2 Example

We illustrate the design route with an example which performs a threshold transformation on an image. In this example, the threshold value is specified at compile time and the hardware process stores just one pixel. Two fragments of the codesign file are shown in Figure 4 and Figure 5, which contain Cedar and C code respectively. The Cedar part enters an infinite loop in which a pixel is read from the 'original_channel' channel and the processed value is returned via the 'processed_channel'. The code is encased in a 'using' directive which tells the compiler to use the interface called 'riley6216channels' to provide these channels. The software in Figure 5 uses the matching communication primitives to send each pixel of an image to the hardware and reads back the processed result.

```
typedef unsigned int pixel : 8;
using riley6216channels
{
  pixel original_pixel;
  while (true)
  {
    input  (processor.out [original_channel],
            original_pixel);
    output (processor.in [processed_channel],
            original_pixel<threshold?
            black_pixel:white_pixel);
  }
}
```

Fig. 4. Simple Cedar threshold program

```
for (y=0; y<height; y+=1)
{
  for (x=0; x<width; x+=1)
  {
    output (threshold_fpga.in  [ original_channel],
            picture[y][x]);
    input  (threshold_fpga.out [processed_channel],
            newpicture[y][x]);
  }
}
```

Fig. 5. Inner loop of software program

The implementation of Cedar is based on the one-hot encoding method, which relies on a small set of token-passing control components [8]: the presence of a token in a circuit indicates that the corresponding statement is being executed. A library-based strategy is adopted in the design of the Cedar compiler. From the source code, an abstract syntax tree is generated which is then transformed into modules independent of the chosen implementation technology [9]; these modules provide the interface to the back-ends which target technology-specific libraries.

The Cedar compiler currently possesses two back-ends: one converts the high-level modules into Xilinx XBLOX format, while the other translates it into VHDL that uses the 6200 libraries. The language provides a mechanism to include designs written in other languages into the output of these back-ends. Such designs are treated by Cedar as external functions that can be called in the same way as true Cedar functions. These functions can be bound, at the user's request, to built-in operators such as addition and multiplication. This external function facility is especially flexible, since it supports polymorphic arguments and return values, multi-cycle functions and, if necessary, allows the function to be managed by a dedicated Cedar process. For example, the language can describe interfaces to multiplier libraries that are pipelined or combinational, that operate on signed or unsigned numbers, that are parametrised or of fixed size, and that require the two inputs to be the same size or allow different bit widths.

The other components of the high-level modules are mapped onto standard 6200 libraries when applicable, and onto Cedar-specific VHDL entities otherwise. In particular, control logic is mapped using a combination of gates and custom libraries such as the IOPRIM block, which implements the (symmetrical) input and output message passing primitives. Similarly, parallel statements are implemented by the PAR block, which collects tokens from parallel processes and releases a token when they have all finished.

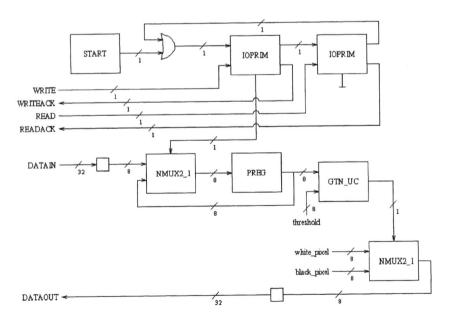

Fig. 6. Pictorial representation of VHDL circuit description produced from Figure 4

The VHDL code for the threshold example is illustrated in schematic form in Figure 6. The riley6216 channel interface uses a wrapper that consists mainly of buffers and a small amount of bus control logic, and has the input and output ports shown on the left of the diagram. The control logic that was generated for the design is shown in the top half. The START block is used to generate a token after reset, while the two instances of IOPRIM are responsible for controlling the transfer of data, and the OR gate implements the loop. The data path is shown below the control logic; the PREG and leftmost NMUX2_1 together form the 'original_pixel' variable, while the other NMUX2_1 implements the condition 'original_pixel<threshold ? black_pixel : white_pixel'. GTN_UC (Greater ThaN Unsigned Combinational) is the library that was chosen to implement the comparison operator.

Other applications, mainly for image and video processing, are currently being developed using the design tools described above. The Cedar compiler has also been used in producing designs for other FPGA systems, such as the EVC-1 board from Virtual Computer Corporation.

6 Case study

Support for Windows 95 has been developed in the form of a plug-and-play device driver and a simple monitor program. Using these, a demo program was written which performs some image processing form a video source. The application setup is shown in Figure 7. The design continuously grabs frames from the QuickCam, processes them on the Riley-2 and displays the result in a window.

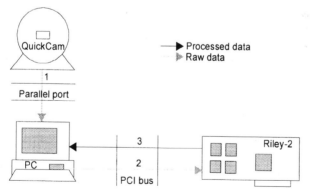

Fig. 7. Image processing application setup

The application can display either a normal picture (Steps 2 and 3 in Figure 7 are not performed), an inverted picture, a Gaussian filter or various edge detection pictures. This is achieved by configuring three of the 6216s with the following designs: a simple inverter, a filter/edge detector, a filter/edge detector followed by a thresholding function. The latter two are based on a design outlined in a previous paper [3]. All four designs could have been fitted in a single 6216, but for test purposes three chips were used. Currently the frame size is only 320x240, 64 gray levels and the application achieves a relatively low frame rate. This is mostly due to the limitations of the QuickCam, since there is no visible difference in frame rate when the application is displaying a normal or a processed picture.

7 Conclusions

Many experimental platforms have been designed to support research in reconfigurable computing in the past. Riley-2 is perhaps unique in providing the necessary architecture to experiment with practical issues concerning dynamic partial reconfiguration and hardware/software codesign. For example, accessing the static memory on each 6216 from the PCI or local bus provides a number of interesting alternatives. One could configure each 6216 such that the SRAM device appears permanently on the memory map of the local bus. Alternatively, one could use the demand-driven model where the 6216 are dynamically configured to provide access to the SRAM when necessary. These two alteratives represent a possible trade-off between hardware resource and reconfiguration time. A number of demanding real-time image processing applications are currently being implemented on the Riley-2 in order to evaluate the strength and weakness of such a computing model, the efficiency of the cosimulation software, and the effectiveness of our design tools.

Acknowledgements. The support of Hewlett Packard Laboratories Bristol, the UK Engineering and Physical Sciences Research Council (research studentship and contracts GR/L24366 and GR/L54356) and the UK Overseas Research Student Award Scheme is gratefully acknowledged.

References

1. Steve Guccione, "List of FPGA-based Computing Machines", on the web, URL: http://www.io.com/~guccione/HW_list.html
2. P.Y.K. Cheung, W. Luk, Patrick Mackinlay, "Hardware Cosynthesis for the Riley System", IEE Colloquium Disgest on Hardware-software cosynthesis for reconfigurable systems, Febuary 22, 1996.
3. W. Luk, N. Shirazi and P.Y.K. Cheung, "Modelling and optimising run-time reconfigurable systems", in Proc. IEEE Symposium on FPGAs for Custom Computing Machines, 1996.
4. The programmable logic data book published by XILINX.
5. i960RP Users Manual.
6. W. Luk, N. Shirazi and P.Y.K. Cheung, "Compilation tools for run-time reconfigurable designs", in Proc. IEEE Symposium on Field-Programmable Custom Computing Machines, K.L. Pocek and J. Arnold (editors), IEEE Computer Society Press, 1997.
7. W. Luk, S. Guo, N. Shirazi and N. Zhuang, "A framework for developing parametrised FPGA libraries", in Field-Programmable Logic, Smart Applications, New Paradigms and Compilers, R.W. Hartenstein and M. Glesner (editors), LNCS 1142, Springer, 1996, pp. 24-33.
8. I. Page and W. Luk, "Compiling occam into FPGAs, in FPGAs", W. Moore and W. Luk (editors), Abingdon EE&CS Books, 1991, pp. 271-283.
9. G. Brown, W. Luk and J.W. O'Leary, "Retargeting a hardware compiler using protocol converters", Formal Aspects of Computing, Vol. 8, 1996, pp. 209-237

Stream Synthesis for a
Wormhole Run-Time Reconfigurable Platform

Brian Kahne and Peter Athanas
Virginia Tech
Department of Electrical and Computer Engineering
Blacksburg, Virginia (USA), 24061-0111

Abstract. Configurable Computing is a technology which attempts to increase computational power by customizing the computational platform to the specific problem at hand. An experimental computing model known as *wormhole run-time reconfiguration* allows for partial reconfiguration and is highly scalable. The *Colt/Stallion* project at Virginia Tech implements this computing model into integrated circuits. In order to create applications for this platform, a compiler has been implemented which utilizes a genetic algorithm in order to map dataflow graphs to the physical hardware of the *Colt/Stallion* chip. A description of the language is presented, followed by experimental results.

1. Introduction

Rapid partial run-time reconfiguration (RTR) extends the available physical computing resources through resource swapping and paging [Luk96,Sin96]. Applications designed with run-time reconfiguration in mind make use of a corollary to the Principle of Locality for the computing structures within a program – recently or frequently used structures may reside on the computing platform, while unmapped structures may be brought into the platform for subsequent execution. The performance of these applications depends upon many factors, including the size of the swapped overlay structure, the reconfiguration bandwidth (the number of configuration bits per second that can cross the chip boundary), and the controllability of the configurable resources (how easy it is to change an internal configuration bit).

Platforms that accommodate rapid partial run-time reconfiguration typically use FPGAs featuring random access support to the configuration resources (XC6200, CLAy, and ATMEL 6000, for example). Offering random access to the configurable resources is *the best* one can do in terms of configuration controllability. However, random access to the configurable resources does not come without a price. Physically, random access techniques are expensive in terms of global row/column signals needed in the physical design (analogous to the BIT and WORD lines associated with RAM designs). In addition, the global row/column signals are heavy capacitive loads and consume power when switched. Furthermore, it is typically assumed that a microprocessor of some sort is used to manage the context switching/swapping of the configurable media. If the platform consists of many FPGAs, the controlling microprocessor can prove to be the bottleneck in reconfiguration throughput [Bit97].

An experimental alternative called *wormhole run-time reconfiguration* attempts to address some of the shortcomings of the random access methods. Wormhole RTR is area and power efficient (in terms of VLSI), relies on only one shared global signal among the configurable computing resources $(CLOCK)^1$, and lends itself to distributed control. With wormhole RTR, the unit of computing is a *stream* -- a long ordered sequence of data. Streams possess the ability to tunnel through the reconfigurable media, creating custom operators, pathways, and pipelines for a given application. Once a pathway is created by a (set of) stream(s), the computation proceeds with no further overhead.

The purpose of this paper is to present a programming approach for wormhole RTR platforms and contrast it with conventional design flows for FPGA design. The *Tier2* compiler presented here transforms an input structural specification of an application into one or more streams that can be used to (partially) configure the distributed resources in a wormhole RTR platform. The streams synthesized by this compiler can be used for computing on a platform consisting of *Colt/Stallion* configurable computing integrated circuits [Bit97].

This paper begins with a brief introduction to stream-based configurable computing (Section 2). The structural synthesis "compiler" called Tier2 forms the stream headers. Section 3 covers the compiler process and contrasts this process with traditional FPGA design flows. Section 4 provides results obtained from using this compiler.

2. Wormhole RTR and Stream-based Computing

In order to place the theme of this paper into proper context, this section provides a brief overview of the stream processing paradigm and how wormhole RTR is used. A more detailed account can be found in [Bit97].

Wormhole[2] run-time reconfiguration provides a framework for implementing large-scale rapid run-time reconfigurable CCM platforms. It is intended as a method for rapidly creating and modifying custom computational pathways using a distributed control scheme (*data-driven* partial run-time reconfiguration). Similar to the data movement process in data-flow machines, the onset or completion of computational streams is used to instigate the process of reconfiguration. The essence of the

---

[1] *Reset* also is obviously a shared global signal, but does not participate in computation.
[2] The term *wormhole* is used here, as it is used in the computer networks community, to describe the method in which data are passed from one computational node to another (wormhole routing).

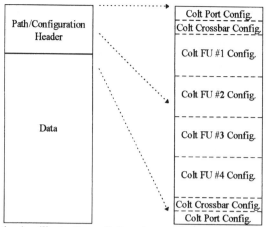

Fig. 1: An illustration of the stream format. A stream is composed of a header segment (expanded on the right) and a data segment.

wormhole RTR concept is formed from independent self-steering streams of programming information and operand data that interact within the architecture to perform the computational problem at hand. The method of computation itself can be compared to that of Pipenets, as discussed by Hwang [Hwa93], in which multiple intersecting pipelines (streams) are formed within the processor to perform a task. One application of wormhole RTR would be as a high-speed configuration methodology for such a system.

For wormhole RTR, a stream is a concatenation of a programming header and operand data (refer to Fig. 1). The programming header is used to configure a computational pathway through the system as well as to configure the operations to be performed by the various computational elements along the path. The stream is self-steering and, as it propagates through the system, configuration information is stripped from the front of the header and is used to program the unit at the head of the stream; thus, the size of the header diminishes as the stream propagates through the system. The stream header is composed of an arbitrary number of packets of programming information. Each packet contains all the information needed to configure a designated unit in the system. The composition and length of the packets are variable so that different packet types may coexist within the same stream header, and hence a given stream may traverse heterogeneous unit types. The stream data section can contain zero to an infinite number of data words for subsequent processing. Example stream sources are video cameras, A/D converters, and antennas. Intermediate streams may contain, for example, filter weight updates, partial computational results, or new computing contexts. Details on how wormhole RTR and streams are supported at the physical level can be found in [Bit96].

3. A Tiered Approach towards Stream Synthesis

The stream concept can be extended to work in a heterogeneous computing environment [Aco95]; however, additional hardware is needed to transform unconventional stream components (such as microprocessors) to properly process streams. This stream synthesis discussion focuses on the *Colt* prototype chip, which contains native support for wormhole RTR. The architecture of the *Colt* chip is similar in concept to conventional FPGA organizations in that *Colt* has a mesh of computational elements that are interconnected by fast local programmable switches and supports longer chip-wide communication through a segmented skip bus. Similar to other FPGA-like devices geared towards computation, *Colt* utilizes a coarse-grain architecture and favors processing of both 1- and 16-bit pathways. In addition, Colt has an on-chip "smart" crossbar for signal distribution, an integer multiplier, and six configurable 16-bit I/O ports. Streams independently entering through any of the six I/O ports can dynamically allocate all resources in the Colt.

In order to synthesize streams, two compilers have been developed and a third one is planned. Each level of compiler development, or *tier*, adds a level of abstraction to the process of mapping an algorithm to the final platform. In the first case, the *Tier1* "compiler" acts as an assembler, allowing the user to have explicit control over all aspects of *Colt*. The *Tier2* compiler only requires the user to construct a data-flow graph of the algorithm, where each vertex represents a functional unit in the *Colt* chip. It then takes care of finding a valid placement for the algorithm and routes each of the data flow edges. When using the final compiler, *Tier3*, the user will write in a more mathematical language, which is not constrained to dealing with individual functional units. The *Tier3* compiler will partition applications into one or more sub-tasks (in the form of a hierarchy of graphs). From there, each sub-task will be processed by the lower tiers and eventually placed and routed onto individual chips.

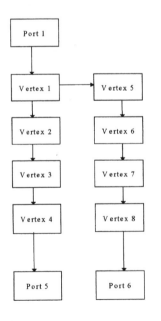

The *Tier1* compiler uses a custom syntax for describing the functionality of each of the functional units (FUs) within the *Colt* chip. Each program consists of a series of streams, port definitions, and macros. The port definition maps data port numbers (currently one through six) to symbolic names, which are then used throughout the stream definitions. For example, the

Fig. 2: A sample of a data flow graph for a computation. For simplicity, no loops or conditional execution paths are shown.

following port definition statement specifies that physical data port one will be mapped to the name of "indata" and that physical data port two will be mapped to "outdata."

```
ports
        indata = 1;
        outdata = 2;
end ports;
```

Macros allow the user to directly manipulate configuration fields of functional units. A macro is evaluated at compile time and allows for basic parameter passing and limited flow control. This example macro, called "rightop," takes one parameter, named "bit," and checks to make sure that its value lies between 0 and 4 inclusive. If not, then the error message is displayed and execution of the macro stops. If it is a valid value, then the configuration field "FNOutSel" receives the value of "bit."

```
macro rightop (bit)
        assert (bit = [0,4]) "Invalid right operand bit specified!";
        FNOutSel = bit;
end macro;
```

Finally, stream definitions describe the flow of information through the computing platform. The outer definition defines what data port is to be used for input and what data ports are for output. Within the body of the statement lie crossbar routing instructions and declarations for functional-unit behavior. For example, this simple stream takes a value from an input data port, adds a constant to it, then places it back onto the crossbar and then to an output data port.

```
stream COLUMN (in indata, out outdata)

        port indata => ifu ifu1 at COL;

        block ifu1
        rightreg = from local north;  // Right operand takes from local north
        out = ADDVAL, add;        // left operand set to ADDVAL, ALU set to
    add operator
        end, go south to ifu2;
        //
        // Definitions for ifu2, ifu3, and ifu4
        //
        ifu ifu4 => port outdata;

end stream;
```

The *Tier2* compiler removes the constraint of having to explicitly describe the path that programming information and data must take through the chip. The user works at a more abstract level -- one that simply describes the flow of data through a graph, where each vertex represents a functional unit (FU) within the *Colt* chip.

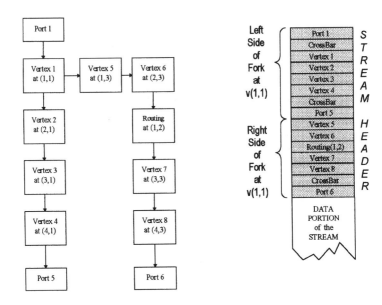

Fig. 3: Depiction of the synthesized programming stream for the sample data flow graph in Figure 2. The left side depicts the stream pathway through the configurable resources, while the right side shows the actual stream structure for this example. Note that in this example, the stream must split at Vertex (1,1).

The main task of the *Tier2* compiler is to fit this graph into the configurable resources on the wormhole RTR platform. The top of the *Colt's* computational mesh can receive data from the smart crossbar network, which also connects to the chip's data ports. The left and right sides of the mesh are connected together to form a cylindrical structure. Data can pass from an FU to an immediate neighbor or can pass over an FU without disturbing its operation.

3.1. Genetics for Placement and Routing

Because a fairly large number of routing resources exist, a brute force technique is not feasible. For instance, if a graph contains 16 vertices, then the first vertex can be placed in one of 16 physical locations, the next vertex can be placed in any one of 15 locations, and so forth. The result is 16!, or 2.092×10^{13}, a rather large number to have to deal with.

In order to solve this problem in a reasonable amount of time, a genetic algorithm was used to search through the solution space. The allele set consists of each of the vertices of the data flow graph, and a particular placement for an algorithm forms a gene. The encoding is simply done by listing out, row by row, the data flow vertices contained within the four-by-four mesh of functional units. For instance, the gene

1	5	9	10	2	6	11	12	3	7	13	14	4	8	15	16

describes the data flow graph in Figure 2 as having Vertex 1 placed in the Colt FU at row 1, column 1. Vertex 5 is located in row 1, column 3, and Vertex 3 is located at row 3, column 1.

The genetic algorithm starts with an initial random population of placements. Each successive generation is created by either mutation, by swapping elements in the gene, or by creating children, in which two parent genes are combined to form a child. In order to judge the fitness of a particular placement, the compiler attempts to route all of the data flow edges contained within the graph. The number of edges that could not be routed serves as a measure of fitness. Because a solution that uses fewer routing resources is considered a better solution, than one that uses more, the fitness function returns either the count of routing resources used or a count of unrouted edges multiplied by a bias value.

Once valid solutions have been found, and the values returned by the fitness functions have converged to a relatively stable value, the genetic algorithm exits. The *Tier2* compiler then generates programming streams describing how the configuration and data information is to flow through the *Colt* chip. For instance, using the placement gene

19	5	10	2	11	6	12	3	13	7	14	4	15	8	16

the compiler might place Vertex 1 through Vertex 4 into the first column on the chip, then place Vertex 5 through Vertex 8 into the third column. The remaining functional units are unused, but must be specified by the gene anyway.

The resulting programming streams, graphically depicted, would look like Figure 3. In Figure 3, an extra routing resource is required to connect Vertex 6 to Vertex 7. The compiler could have placed Vertex 5 through Vertex 8 in the second column to produce a better solution.

The results of the *Tier2* compiler are generally pretty good compared to the alternative of manual placement and routing. For instance, a floating-point multiplier [Bit97] that was laid out by hand ended up requiring 20 routing resources and approximately two weeks worth of time. The *Tier2* compiler was able to place and route the same design in approximately 30 seconds on a Pentium Pro computer. The longest signal paths for both the manual and automatic place-and-route are comparable. In its current form, the *Tier2* compiler tends to use about 10% more routing resources than manual routing.

4. Experimental Results

In order to judge the efficacy of the compiler and its genetic algorithm, several programs were written in the *Tier2* language and then compiled. By far the most complex of these was a floating-point multiplier [Bit97]. This particular algorithm accepts two 32-bit floating point numbers, in a custom format, multiplies them together, and produces a 32-bit output. In addition to the multiplication, various normalization and overflow checking operations are performed. The program itself

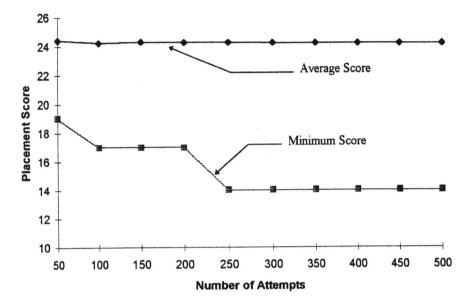

Fig. 4: Average and minimum score (cumulative) for 500 placement attempts.

requires all 16 FUs of the *Colt* chip, the hardware multiplier, and all six data ports. Manual placement took approximately two weeks and resulted in a score of 20.

An experiment was performed in order to attempt to judge the efficacy of the *Tier2* compiler. A total of 500 independent routing attempts were performed using the floating-point multiplier as input. The resulting cumulative average and minimum scores are presented in Fig. 4.

As can be seen, the average performance of the compiler, for 500 independent trials, is worse than manual placement. However, the minimum score was far better. Because the minimum score is what is kept by the compiler as the final placement to be used, the end result is that the system outperforms what a human can easily do and takes far less time doing it. The graph also demonstrates a feature of this compiler: The quality of the solution can be traded for execution speed. In other words, fewer trials can be done in order to obtain a less impressive placement result, versus longer run times to produce an extremely high quality score.

Although it would be optimal if the genetic algorithm could obtain the best solution for every run, this is unlikely to occur. The problem of premature convergence, before the optimal solution has been found, is frequently observed in genetic algorithms, due to the exponential reproduction of the best chromosomes combined with the functioning of the crossover operator. Once a population has converged to the point that crossover operations simply cycle through a reoccurring set of genes and a large mutation would be required to break out of this predicament, some literature recommends that the run be stopped [Fog94]. This is the technique used with *Tier2*: Convergence is detected and the algorithm stops. Another run is begun in the expectation that it will converge in an even better solution. Another method for handling this is to use a hill-climbing method, or greedy method, to search for a minimum, which lies in close proximity [Fog94]. This might be a promising area of research for future compiler versions.

5. Conclusions and Future Research Directions

The *Tier2* compiler effort was successful in automating placement and routing for *wormhole run-time reconfiguration* on the *Colt* architecture. The result is that it is now far easier for an applications designer to map an algorithm to this chip than it was using the original, entirely manually operated tools [Bit97]. Not only has the development time been reduced, but the quality of the placements, as judged by the number of routing resources used, has also been increased.

Many features must still be added to the compiler before it could be considered to be complete. One of the primary issues to be dealt with is the ability to map arbitrarily large data flow graphs to a group of *Colt* chips. This first-generation chip is very limited in the number of resources it contains and will soon be replaced by a far more powerful version called the *Stallion*. However, it will still be necessary to deal with large graphs that cannot be contained within a single device. Two approaches might be taken towards implementing this feature: (1) use a genetic algorithm to map all of the nodes to an aggregate of chips or (2) use an initial step that partitions the graph into subgraphs, then use the existing genetic algorithm to route each of the subgraphs.

As the *Colt/Stallion* architecture matures, the complexity of these chips will increase dramatically, as will the algorithms intended for use. The *Tier2* approach, that of using a custom language combined with a genetic algorithm for placement purposes, represents a scalable solution for the programming of run-time reconfigurable software.

6. References

[Aco95] E. Acosta, V. Bove, Jr., J. Watlington and R. Yu, "Reconfigurable Processor for a Data-Flow Video Processing System," *Field Programmable Gate Arrays (FPGAs) for Fast Board Development and Reconfigurable Computing*, J. Schewel, ed., SPIE – The International Society for Optical Engineering, Bellingham, Wash., 1995, pp. 83-91.

[Bit96] R. Bittner, M. Musgrove, and P. Athanas, "Colt: An Experiment in Wormhole Run-Time Reconfiguration," *High-Speed Computing, Digital Signal Processing, and Filtering Using Reconfigurable Logic*, (SPIE), Nov. 1996, pp. 187-194.

[Bit97]   R. Bittner, "Wormhole Run-time Reconfiguration: Conceptualization and VLSI Design of a High Performance Computing System," doctoral dissertation, Virginia Tech, Bradley Department of Electrical and Computer Engineering, Jan. 1997.

[Eld94]   J. Eldredge and B. Hutchings, "Density Enhancement of a Neural Network Using FPGAs and Run-Time Reconfiguration," *Proceedings of IEEE Workshop on FPGAs for Custom Computing Machines*, D. Buell and K. Pocek, eds., Napa, Calif., April 1994, pp. 180-188.

[Fog94]   D. Fogel, "An Introduction to Simulated Evolutionary Optimization", *IEEE Transactions on Neural Networks*, Vol. 5, No. 1, Jan. 1994, pp. 3-8.

[Har97]   R. Hartenstein, J. Becker, and R. Kress, "Custom Computing Machines vs. Hardware/Software Co-Design: from a Globalized Point of View," *Sixth International Workshop of Field Programmable Logic and Applications*, Lecture Notes in Computer Science, no. 1142, Springer-Verlag, Oxford, England, Sept. 1996, pp. 65-76.

[Hwa93]   K. Hwang, *Advanced Computer Architecture*, McGraw-Hill, New York, N.Y., 1993, pp. 442-446.

[Luk96]   W. Luk, S. Guo, N. Shirazi, and N. Zhuang, "An Integrated Framework for Developing Parametrised Macros," *Sixth International Workshop of Field Programmable Logic and Applications*, Lecture Notes in Computer Science, no. 1142, Springer-Verlag, Oxford, England, Sept. 1996, pp. 24-33.

[Sin96]   S. Singh, J. Hogg, and D. McAuley, "Expressing Dynamic Reconfiguration by Partial Evaluation," *IEEE Symposium on FPGAs for Custom Computing Machines*, D. Buell and K. Pocek, eds., Napa, Calif., April 1996, pp. 188-194.

Pipeline Morphing and Virtual Pipelines

W. Luk, N. Shirazi, S.R. Guo and P.Y.K. Cheung

Department of Computing, Imperial College, 180 Queen's Gate,
London SW7 2BZ, UK

Abstract. Pipeline morphing is a simple but effective technique for reconfiguring pipelined FPGA designs at run time. By overlapping computation and reconfiguration, the latency associated with emptying and refilling a pipeline can be avoided. We show how morphing can be applied to linear and mesh pipelines at both word-level and bit-level, and explain how this method can be implemented using Xilinx 6200 FPGAs. We also present an approach using morphing to map a large virtual pipeline onto a small physical pipeline, and the trade-offs involved are discussed.

Introduction

Pipeline architectures are commonly used in high-performance designs. This paper introduces morphing, a technique for enhancing the efficiency of reconfigurable pipelines at run time. We shall also describe the use of morphing in the emulation of large virtual pipelines by small physical pipelines, and explain how temporary storage can be used to improve performance.

Implementing pipeline architectures using reconfigurable devices is attractive for several reasons. Many FPGAs facilitate the realisation of pipelines, since they have a regular structure and an abundance of registers. Moreover, leading-edge FPGAs often provide built-in support for fast reconfiguration. An example is the wildcarding facility for Xilinx 6200 devices [1], which allows concurrent reconfiguration of a block of FPGA cells. With such facilities, it is possible to reconfigure each pipeline stage rapidly at run time to implement multiple functions. For a system operating in an unpredictable environment, this possibility enables the selection of functions adaptively.

Partial reconfiguration [4] is a powerful method of exploiting the flexibility of FPGAs such as the Xilinx 6200: one part of the FPGA can be reconfigured while other parts are continuing to function. Pipelines provide a simple but effective scheme for partial reconfiguration, since pipeline registers isolate one pipeline stage from another so that computation and reconfiguration can take place at the same time without interference. The regular structure of pipelines also simplifies the development of hardware operators which can be relocated to different regions of a pipeline, maximising the re-use of design effort.

An obvious method for reconfiguring an n-stage pipeline involves three steps. First, one needs to complete the current computation and clear the pipeline, which takes n cycles. Then reconfiguration can take place. Finally one has to wait for n cycles for the result to flow through the newly-configured pipeline. This method of reconfiguring a pipeline leads to a latency of $2n$ cycles, in addition to the time for reconfiguring all the pipeline stages. In highly-pipelined systems when n is large, the pipeline latency cycles and reconfiguration time will have a significant impact on performance.

Pipeline Morphing

We present a method, called pipeline morphing, for reducing the latency involved in reconfiguring from one pipeline to another. The basic idea is to overlap computation and reconfiguration: the first few pipeline stages are being reconfigured to implement new functions so that data can start flowing into the newly-configured stages of the pipeline, while the rest of the pipeline stages are completing the current computation. Instead of changing the entire pipeline at once, our method involves morphing one pipeline to another, just as one can morph two images by interpolating them. In our case, the pipeline registers isolate one pipeline stage from its neighbours, enabling computation and reconfiguration to take place concurrently in different stages.

Figure 1 shows how a five-stage pipeline F can be morphed into a pipeline G in five steps. It should be clear from this example that during morphing, the flow of reconfiguration is synchronous with the flow of data, and hence the pipeline latency cycles are eliminated. If the time for reconfiguration is longer than the pipeline processing time, the pipeline will need to include flow control mechanisms to slow down the rate of data flow while morphing is taking place. We shall explain later how this can be achieved in Xilinx 6200 FPGAs.

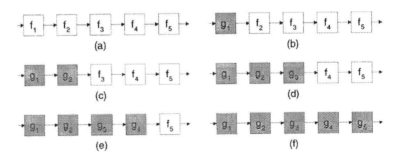

Fig. 1. Given the pipeline F with stages $f_1, \ldots, f_5$ and the pipeline G with stages $g_1, \ldots, g_5$, the diagram shows how F can be reconfigured into G in five steps using the morphing technique. Only one stage of the pipeline is being reconfigured in each step.

Whether morphing is used or not, a designer's task is to ensure that the slowing down due to run-time reconfiguration will not affect system performance. For instance in video processing, there may be sufficient time for reconfiguration between the end of one image frame and the beginning of the next frame.

Because of the elimination of latency cycles, pipeline morphing will improve the performance of systems that reconfigure at run time. It is particularly suitable for devices supporting rapid reconfiguration, and it works best when reconfiguration time is comparable to the pipeline computation time. To meet this condition, the user can build single-cycle reconfigurable structures in an FPGA [7]. FPGAs specially-designed for supporting rapid reconfiguration [10], [11] will particularly benefit from pipeline morphing. Morphing can also be applied to systems with multiple FPGAs arranged as a pipeline.

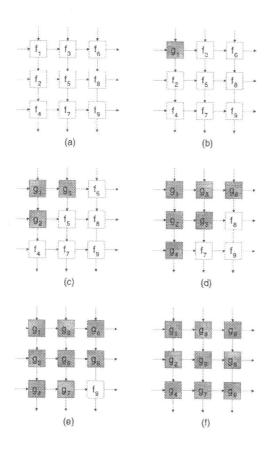

Fig. 2. Given the square mesh F with components $f_1, \ldots, f_9$ and the mesh G with stages $g_1, \ldots, g_9$, the diagram shows how F can be reconfigured into G in five steps using the morphing technique.

Our method is not confined to linear pipelines. It can be applied to pipelines of other shapes, such as two-dimensional meshes or tree-shaped designs. Figure 2 shows the steps of morphing from one mesh to another, given that every component in the mesh has a pipeline register at each of its two outputs. This approach can also be applied to bit-level operators. In the next section, we shall explain how a pipelined adder can be morphed into a pipelined subtractor.

Morphing Pipelines on Xilinx 6200 Devices

We illustrate the morphing reconfiguration technique by showing how it can be applied to reconfigure a six-bit, three-stage pipelined adder to a pipelined subtractor of the same size on a Xilinx 6200 FPGA. The pipelined adder is shown in Figure 3a and the resulting pipelined subtractor is shown in Figure 3d.

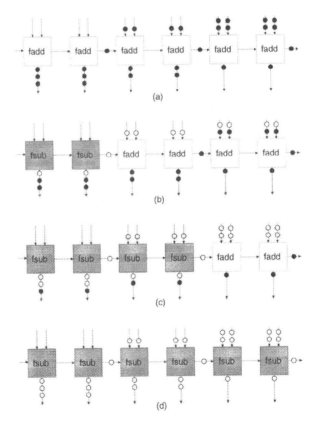

Key: ● Input/Output Register for Addition
     ○ Input/Output Register for Subtraction

Fig. 3. Morphing a pipelined adder to become a subtractor.

As explained above, if all n stages of a pipelined adder ($n=3$ in the above example) are reconfigured into a pipelined subtractor at once, an additional $2n$ clock cycles would be needed in order to flush the pipeline and to refill it. In order to avoid this delay, the reconfiguration is performed in three steps where each stage of the pipelined adder is reconfigured followed by one clock cycle of computation. These three reconfiguration steps are shown in Figures 3b, 3c and 3d. The partial configuration information involved in these steps was obtained using tools described in [8].

A dual clocking scheme is used in order not to clock invalid data values into the pipeline registers during reconfiguration. The two operand values for the adder are stored in two six-bit registers. When using the processor interface [1] on the Xilinx 6200 FPGA, a pulse is produced whenever a register is written with a value, and this pulse can be used as a clock for the registers within the design. The input registers are set up so that a clock pulse is generated when the second operand is written into the register. An additional configuration clock is used to control the reconfiguration of the logic in each pipeline stage. The reconfiguration sequence is therefore broken down into three steps. First, a stage of the pipeline is reconfigured; on completion the operands are written into the input registers; the write action then triggers the clock for the pipeline registers. This sequence is continued until all the stages are reconfigured.

In the above example, it takes four cycles to reconfigure the pipelined adder to a pipelined subtractor. Without morphing it takes an additional three cycles to flush the pipeline and three cycles to refill it; hence a total of ten cycles are needed to perform the reconfiguration and to begin producing correct results. The morphing technique therefore improves the reconfiguration time by 2.5 times; Table 1 summarises these results. Clearly the higher the degree of pipelining, the larger the improvement that can be obtained using the morphing technique, since it takes more cycles to evacuate and to refill the pipeline.

Table 1. Comparing morphing and non-morphing reconfiguration of a pipelined adder into a pipelined subtractor. N/A is short for "not applicable".

	With morphing (number of cycles)	Without morphing (number of cycles)
Reconfigure Figure 3a to Figure 3b	2	N/A
Reconfigure Figure 3b to Figure 3c	1	N/A
Reconfigure Figure 3c to Figure 3d	1	N/A
Reconfigure Figure 3a to Figure 3d	N/A	4
Time to flush and refill the pipeline	N/A	6
Total reconfiguration time	4	10
Speedup factor	2.5	1

Virtual Pipelines

An advantage of adopting a pipeline structure is the ease of mapping a large virtual pipeline onto a small physical pipeline. Our approach involves feeding back partial results to the physical pipeline which morphs between different sections of the virtual pipeline. The performance of such a system can often be enhanced by a temporary storage (Figure 4), as we shall discuss later.

Let us begin with an example: the mapping of a six-stage virtual pipeline onto a three-stage physical pipeline (Figure 4). The first three stages of the virtual pipeline are time-multiplexed with the last three stages. Morphing is used to replace one of the two pipeline configurations by the other.

This design operates as shown in Figure 4. Note that the physical pipeline operates in two modes: the "fill" mode and the "feedback" mode. In the fill mode, the first pipeline stage is connected to the external input and data start filling up the pipeline. Once the pipeline is filled up, partial results will emerge and will be stored in the temporary storage. When all input data have been processed by the first three stages of the virtual pipeline or when the temporary storage is full, the pipeline will operate in the feedback mode.

When an n-stage physical pipeline first starts in the feedback mode, its first stage will be reconfigured to become the $(n + 1)$-th stage of the virtual pipeline (Figure 4b). After reconfiguration is completed in the first pipeline stage, it is provided with the partial results from the temporary storage. When the computation is completed, the second stage of the physical pipeline will be reconfigured to become the $(n + 2)$-th stage of the virtual pipeline (Figure 4c), and so on. For the example in Figure 4, eventually the partial results will flow through the physical pipeline configured to be the last three virtual pipeline stages (Figure 4d).

When the virtual pipeline has been emulated once, the result will start to emerge at the external output. When new data can be accepted, the physical pipeline will operate in fill mode again and will morph back to the first three virtual pipeline stages. Note that adequate flow control is necessary to stop the external input while the pipeline is operating in feedback mode. The next section will describe the use of the temporary storage to optimise pipeline performance.

Temporary Storage

First, note that if the physical pipeline only supports global reconfiguration, the temporary storage shown in Figure 4 is needed to hold partial results while the entire pipeline is being reconfigured. If the pipeline can be partially reconfigured at run time, then the temporary storage is not necessary as the pipeline itself can provide storage of partial results.

However, a small temporary storage will result in frequent reconfiguration, since the physical pipeline has to operate in "feedback" mode (see previous section) once the temporary storage is full. It will remain in the "feedback" mode until outputs are ready which will then free up space for further inputs. If the combined storage in the pipeline and the temporary storage is large enough

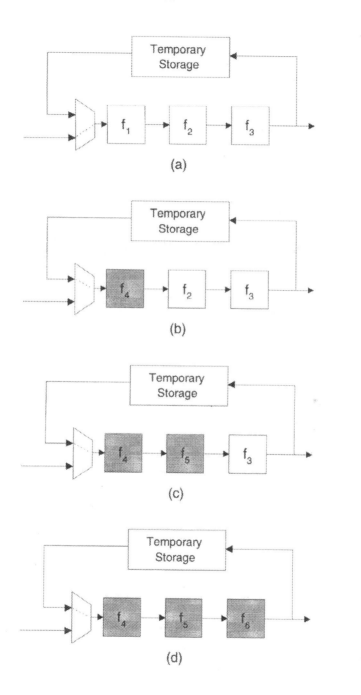

Fig. 4. Emulating a six-stage virtual pipeline $f_1, \ldots, f_6$ using a three-stage physical pipeline. The pipeline is in the fill mode for Step (a), and it is in feedback mode for Steps (b) to (d). The control to the switch that selects the external or the feedback data is not shown.

to contain all input data, then each virtual pipeline stage will only need to be emulated once. Having sufficient temporary storage is particularly important for pipelines which require a long reconfiguration time, since it would be desirable to minimise the frequency of reconfiguration for these pipelines.

Let us now consider different ways of implementing the temporary storage. If a large amount of temporary storage is required, then external memory can be used; otherwise on-chip registers or embedded memories within the FPGA may be sufficient. As explained above, pipelines supporting rapid reconfiguration can afford a small temporary storage. When this happens, the feedback connections can be made entirely on-chip, possibly using global connections in the FPGA so that output data from the last stage can be fed back to the first stage in the feedback mode. Global connections are provided in most FPGAs; such connections can themselves be pipelined to ensure high performance.

Another way of implementing a physical pipeline with little or no temporary storage is to partition the physical pipeline into half, and "fold" one half of the pipeline onto the other half by interleaving the components (Figure 5). This method avoids global connections at the expense of requiring an efficient integration of non-neighbouring elements in a pipeline structure.

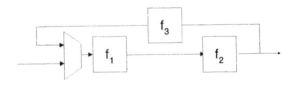

Fig. 5. A folded version of the physical pipeline in Figure 4a, which does not have temporary storage and avoids using global connections.

Virtual Pipelines on Xilinx 6200 PCI System

The viability of virtual pipelines has been demonstrated by a PCI board supplied by Xilinx Development Corporation, which contains a Xilinx 6216 or a Xilinx 6264 device and four 8-bit wide memories organised into two banks [7]. Each bank of memory can be accessed from either of the two separate address busses (Figure 6), and each of the four memories can be controlled individually. This memory architecture allows multiple modes of operation to be set-up by selecting multiplexers and bus switches for flow control in the desired manner.

This system provides a flexible platform for implementing virtual pipelines. One possibility is to use the two memory banks to provide the temporary storage (Figure 4) for a virtual pipeline. Partial results can be stored into one memory bank, and they can be used later when the FPGA has been reconfigured to

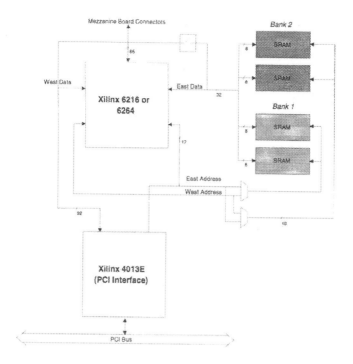

Fig. 6. Xilinx 6200 PCI system.

implement a different region of the virtual pipeline. Another possibility is to use the on-chip registers of the Xilinx 6200 FPGA to implement the temporary storage. If registers or global connections are at a premium, the folded structure (Figure 5) may prove to be an appealing alterative in implementing a physical pipeline.

Summary

This paper introduces morphing, an effective technique for reconfiguring pipelines. We explain how morphing can be applied to linear and mesh pipelines at word-level and bit-level, and how it can be implemented in Xilinx 6200 FPGA technology. We also describe the mapping of large virtual pipelines onto small physical pipelines, and how the resulting implementations can benefit from morphing.

Current and future work includes extending the scope of morphing to cover various architectural templates; this extension will enable us to morph between pipelines with different number of pipeline stages, or to morph a linear pipeline into a tree-shaped architecture. Frequently there are multiple ways of morphing between designs, and it will be important to evaluate the trade-offs involved.

The use of languages such as Ruby [2] and VHDL [9] for modelling hardware morphing is also being explored; we expect such work to result in techniques and tools for automating the implementation of morphing and virtual pipelines.

Acknowledgements

The authors are indebted to John Gray, Douglas Grant, Hamish Fallside, Tom Kean, Steve McKeever, Stuart Nisbet and Bill Wilkie for their constructive comments. The support of Xilinx Development Corporation, the UK Engineering and Physical Sciences Research Council (Grants GR/L24366 and GR/L54356) and the UK Overseas Research Student Award Scheme is gratefully acknowledged.

References

1. S. Churcher, T. Kean and B. Wilkie, "The XC6200 FastMap Processor Interface", in *Field Programmable Logic and Applications*, W. Moore and W. Luk (eds.), LNCS 975, Springer, 1995, pp. 36–43.
2. S. Guo and W. Luk, "Compiling Ruby into FPGAs", in *Field Programmable Logic and Applications*, W. Moore and W. Luk (eds.), LNCS 975, Springer, 1995, pp. 188–197.
3. J. Hadley and B. Hutchings, "Design Methodologies for Partially Reconfigured Systems", in *Proc. FCCM95*, P. Athanas and K.L. Pocek (eds.), IEEE Computer Society Press, 1995, pp. 78–84.
4. B. Hutchings and M.J. Wirthlin, "Implementation Approaches for Reconfigurable Logic Applications", in *Field Programmable Logic and Applications*, W. Moore and W. Luk (eds.), LNCS 975, Springer, 1995, pp. 419–428.
5. W. Luk, "A Declarative Approach to Incremental Custom Computing", *Proc. FCCM95*, P. Athanas and K.L. Pocek (eds.), IEEE Computer Society Press, 1995, pp. 164–172.
6. W. Luk, S. Guo, N. Shirazi and N. Zhuang, "A Framework for Developing Parametrised FPGA Libraries", in *Field-Programmable Logic, Smart Applications, New Paradigms and Compilers*, LNCS 1142, Springer, 1996, pages 24–33.
7. W. Luk, N. Shirazi and P. Y. K. Cheung, "Modelling and Optimising Run-Time Reconfigurable Systems", in *Proc. FCCM96*, K.L. Pocek and J. Arnold (eds.), IEEE Computer Society Press, 1996, pp. 167–176.
8. W. Luk, N. Shirazi and P. Y. K. Cheung, "Compilation Tools for Run-Time Reconfigurable Designs", in *Proc. FCCM97*, K.L. Pocek and J. Arnold (eds.), IEEE Computer Society Press, 1997.
9. P. Lysaght and J. Stockwood, "A Simulation Tool for Dynamically Reconfigurable Field Programmable Gate Arrays", *IEEE Transactions on VLSI*, Vol. 4, No. 3, September 1996.
10. H. Schmit, "Incremental Reconfiguration for Pipelined Applications", in *Proc. FCCM97*, K.L. Pocek and J. Arnold (eds.), IEEE Computer Society Press, 1997.
11. S. Trimberger, D. Carberry, A. Johnson and J. Wong, "A Time-Multiplexed FPGA", in *Proc. FCCM97*, K.L. Pocek and J. Arnold (eds.), IEEE Computer Society Press, 1997.

Parallel Graph Colouring Using FPGAs

Barry Rising, Max van Daalen, Peter Burge and John Shawe–Taylor

Department of Computer Science, Royal Holloway, University of London
Egham, Surrey TW20 0EX, UK

Abstract. In [13] a freely expandable digital architecture for feedforward neural networks was described. This architecture exploits the use of stochastic bitstreams representing real valued signals.

This paper proposes a hardware implementation of a recurrent neural network to form part of a novel system for solving the graph colouring problem [5], particulary relevant to frequency assignment in mobile telecommunications systems [7].

The core elements of the design are multi-state bitstream neurons, arranged in a pipelined architecture that allow for probabilistic connections between nodes, a fixed temperature cooling schedule, and control over the degree of parallelism used in the network update rule.

We also introduce the concept of hardware paging permitting the size of graph to be unconstrained and reducing the hardware overhead.

The highly regular and compact nature of the proposed circuitry, makes it an ideal candidate for utilizing the flexibility provided by FPGA's. In such a reconfigurable system, efficient problem specific hardware is easily generated.

1 Introduction

This paper describes a technique designed to solve the graph colouring problem, using a stochastic recurrent network. Initially the project will investigate the colouring of a set of Leighton graphs, as these are known to be difficult to colour. These graphs are often used for benchmarking purposes, and are freely available in the public domain [8].

The proposed colouring system consists of a novel algorithm [6] governing a custom designed Multi–State BitStream recurrent Neural Network (MSBSN), based on the Generalised Boltzmann machine, generating candidate solutions to the problem.

In this paper, we present the hardware architecture for an appropriate MS-BSN network designed to take advantage of the flexibility provided by current FPGAs [2] [3].

We also present a set of simulation results, obtained from the proposed network architecture, which show the efficiency gains of operating the MSBSN network at differing levels of parallelism. These results illustrate that the optimum degree of parallelism is directly related to the edge density of the graph to be coloured.

1.1 The graph colouring problem

A graph consists of a set of nodes and a set of edges between the nodes. The graph colouring problem is as follows; Given an undirected graph $G = (N, E)$, a *colouring* of an undirected graph is an assignment of a label (from a small finite set) to each node. The labels on the pair of nodes incident to any edge must be different. Graph colouring can be applied to many real–world applications [9]:

- *Frequency Assignment for mobile radios:* Two people who are close enough together must be assigned different frequencies, while those who are distant can share frequencies. The problem of minimizing the number of frequencies is then a graph colouring problem [7].
- *Register Allocation in compilers:* This problem has to assign variables to a limited number of hardware registers during program execution. The goal is to assign variables that do not conflict so as to minimize the use of non–register memory.
- *Printed Circuit Board Testing:* For unintended short circuits, where the nodes in the graph correspond to the nets on the board and an edge corresponds to a potential short circuit between the nets.

1.2 The Boltzmann machine

The Boltzmann machine is a recurrent neural network whose neurons have binary states and update stochastically using the simulated annealing algorithm. The connections between neurons are bi–directional and symmetrically weighted.

Definition 1. Let $G = (N, E)$ be an undirected graph with vertex set N and edge set E with no self connections. (ie. a set of N neurons and a non–empty set $E \subset N \times N$ of connections). A Boltzmann machine $\beta(G)$ on G has a symmetric weight associated with each edge, $w : E \to \mathbb{R}, (i, j) \in E \Rightarrow w_{ij} = w_{ji}$, a threshold θ_i with each vertex i, and a stochastic binary update function which sets the activation of X_i of a randomly chosen vertex as follows: $X(i) = 1$ with probability $p_i = 1/(1 + exp(-\Delta H_i/T))$ where T is the temperature parameter for simulated annealing and ΔH_i is the reduction in energy (or error) $= \sum w_{ij}x_j - \theta_i$. The energy function of $\beta(G)$ is given by $H = -\frac{1}{2} \sum_{(i,j) \in E} w_{ij}x_i x_j + \sum_{i \in E} \theta_i x_i$.

1.3 The generalised Boltzmann machine

Each neuron in a Generalised Boltzmann Machine (GBM) can be in one of a finite number of possible states (i.e. more than two states). The GBM was first introduced in [12] and is formally defined as follows:

Definition 2. Let A be a finite alphabet of r symbols and $G = (N, E)$ be any graph with vertex set N and edge set E. A GBM $\beta(G, A)$ on G over A is specified by a mapping ω from the set E into the set of matrices with real entries indexed by $A \times A$ and a mapping z from the set N to vectors indexed by A. The matrices will be referred to as weights and the vectors as thresholds. A state of the machine is an assignment σ which specifies for each node a symbol from A. The state of a node v is its value under the assignment σ. The update rule generalises the standard Boltzmann machine by choosing state i for selected neuron v with probability proportional to $exp(-H_{[v \rightarrow i]}/T)$, where $H_{[v \rightarrow i]}$ is the energy when neuron v is moved to state i.

1.4 The multi–state bit stream neuron

A network of Multi-state bit stream neurons (MSBSNs) is essentially an extension of the GBM idea. A network of MSBSNs can be used to find good solutions for the graph colouring problem [5]. In such a system, each node in the graph is represented by a single MSBSN and the number of allowed states correspond to the maximum number of possible colours. The edge connections within the graph are represented by the network weight values (these are implemented as stochastic bit streams).

An individual MSBSN contains a number of counters, one to represent each of the available colours. Whilst updating, a MSBSN looks at all of its adjacent nodes and increments the relevant counter according to; the colour of the adjacent nodes and a stochastic weight value (a stochastic bit stream encoded with a bit probability providing a fixed temperature. [1] showed that annealing at a fixed temperature is as good as applying a cooling schedule. An initial value of 0.92 has been determined by experiment, however this is likely to be problem dependent). The MSBSN's new colour is then chosen as the state represented by the lowest internal MSBSN counter value. In the event of a tie, the new colour is chosen randomly from the lowest values.

```
initialise bitstream temperature
WHILE (graph not coloured correctly) DO
  Randomly pick a node to update
  FOR (all adjacent nodes) DO
    IF (bitstream = 1)
      Increment counter that has same state as the adjacent node
    ELSE
      Ignore this connection
    END IF
  END FOR
  Choose colour whose counter is the lowest
END WHILE
```

Fig. 1. Algorithm for graph colouring with MSBSN's

Figure 1 gives the outline of the low-level MSBSN algorithm described in this paper for finding candidate solutions to the graph colouring problem [5]. This algorithm is based on earlier work by [10].

2 Hardware design of the network

Each MSBSN has a counter for every possible colour or state. This colour information is passed to each MSBSN in the form of an n-bit code bussed around the system (where n is the number of possible colours). Within a given MSBSN, each of the n bit signals directly controls its respective colour counter enable line.

Codes passed into the MSBSN only ever contain a single '1'. The position of this '1' defines a particular colour. For this reason, on a given iteration only one bit in the n–bit code is ever set to '1'.

To represent and store the structure of the network, an m by m *connection matrix* is created (where m is the number of nodes in the graph). A '1' at position (i, j) in this matrix, represents a connection between node i and node j. As all connections are bidirectional, the matrix must be symmetrical.

When a given MSBSN updates, the colour information of all other MSBSNs in the system, in the form of n-bit codes, are sequentially transmitted to it. If the colour of the updating MSBSN is the same as one of its adjacent neighbours (i.e. this is indicated by the presense of a '1' in the appropriate place of the weight connection matrix) and the stochastic weight bit stream representing the temperature is also set to '1', then the internal counter whose enable line has been selected by the n-bit colour code is incremented. i.e. The updating MSBSN node totals up the colours of its connected neighbours.

Once the colour states of all MSBSNs have been transmitted to the updating MSBSN, the counter with the lowest value is selected, and loaded into the parallel output shift register. This is then shifted out on the next update cycle thus generating the next n-bit colour code (or state).

2.1 Choosing the new colour

Once a given MSBSN has aggregated the colours of all its adjacent MSBSNs, the lowest valued internal counter must be selected. This determines the new colour of the given MSBSN. It is possible for several of the internal counters to have the same minimum value, in which case a random selection between these values must be made. The above process can be split into two disjoint parts: Selecting the minimum valued colours and then randomly choosing between them.

Selecting the candidates for the new colour Selecting the lowest count values within a given MSBSN can be trivialised if up/down counters are used. At the start of a given MSBSN update cycle, all of its internal counter bits are set to '1' except for the top most bits which are set to '0' (i.e. 8 bit counters would be initialised to the value $7F_{16}$).

During the operational cycle of a given MSBSN, colour codes from adjacent MSBSNs and the stochastic temperature bits will cause the appropriate internal counters to increment. If at some stage during this update cycle all of these internal counters get 1's in their top most bit positions (i.e. every counter has been incremented at least once), then the counters are all decremented by 1.

This is achieved by holding the counter that is about to be incremented while decrementing all the other counters. This allows the counters to increment in just one clock pulse (instead of two) but means there must be an extra clock pulse at the end of the iteration cycle.

The result of this procedure will be that the lowest valued counters always have a '0' in their top bit positions (in fact these counters will contain their initialisation values of $0111 \cdots 111_2$); with the remaining counters containing 1's in their top most bit positions.

Once the update cycle is complete, the top bits of the counters are loaded into the output shift register of the MSBSN. These codes are now passed to the minimum value selection circuitry (remember, it is possible to have multiple minimum values at this stage).

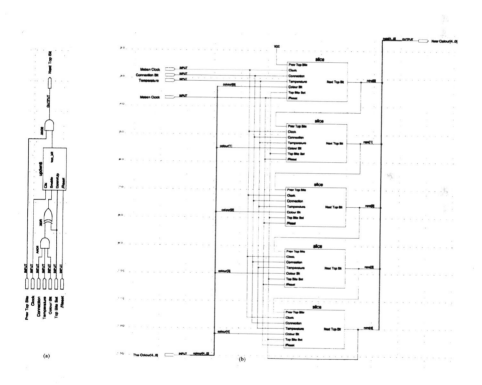

Fig. 2. (a) Multi State Bit Stream Neuron Slice (b) Five State MSBSN

Figure 2a shows a MSBSN slice which can be cascaded to form an arbitary number of states. Figure 2b shows five of these slices cascaded together, forming a five state MSBSN.

Using this "bit slice" design graphs with any number of colours may be processed. Here, the flexibility and reconfigurability of FPGAs are a significant advantage as the hardware can be configured by the host computer to be problem specific (i.e. by downloading the appropriate MSBSN design).

Randomly selecting multiple minimum values Once the lowest valued counters have been selected, one of them (if there are ties) must be chosen to become the new colour for the particular node being considered. In order for the network to function correctly this choice has to be completely random, such that no one colour has any bias. Picking a value randomly with simple logic is a difficult task.

The best way to implement this function for a small number of colours is to use a large appropriately initialised RAM/ROM lookup table. This simplifies the design of the individual MSBSNs, but also introduces a potential bottleneck as all raw colour codes must pass through this external lookup table. If each node in the graph has its own MSBSN, then the look–up table can be accommodated into a single pipeline stage thus removing the potential "bottleneck" [11].

A five colour system, can produce 31 different possible colour codes, (11111 should never occur as there always has to be at least one lowest valued counter). The incoming colour code, potentially containing multiple minimum values, is used as an address input for the lookup table. After a fixed lookup delay, it will output an appropriately modified colour code. A five bit input address containing one '0' is inverted, an address containing multiple '0's is randomly translated into an output colour code containing only one '1'. However, this '1' will correspond to one of the original '0's in the input address. The randomness of this translation is governed by the values loaded into the lookup table. This function would be performed by the host system.

To increase the randomness of this procedure, multiple randomly generated lookup tables would be loaded into a larger RAM or ROM. The additional random address bits required to select a particular lookup table can be obtained by appropriately tapping a feedback shift register wired to generate a maximal length m-sequence [13]. In fact these extra bits may be obtained from the feedback shift register used in the generation of the stochastic bit streams used throughout the design.

If a large colour set is required, the incoming colour can be binary encoded to reduce off–chip wiring (ie. 32 colours can be encoded onto 5 wires).

2.2 Overall architecture of the network

The regular, repetitve nature and modular design of the MSBSN (a simple processing element) makes FPGA's an excellent implementation technology. The full hardware design has been simulated in software (see section 3) sucessfully, and currently a prototype system is being built using Altera FLEX8000 chips on a RIPP10 development board [14] donated by Altera as part of their University research programme.

If the entire network can fit into the processing elements available then an elegant pipelined design can be implemented [11]. However, to solve graphs with different numbers of nodes a more general approach is required.

The number of FPGA's required to implement large graphs completely becomes prohibitive. Eight 5 colourable MSBSN's can fit onto a single FLEX8000. Therefore a fully populated RIPP10 board can implement 64 processing nodes. This does not include controlling state machines for the connection matrix and look–up table.

Hence a 450 node Leighton graph would require 56 FLEX8000 chips. At any given clock cycle most of this hardware is actually redundant as only a small proportion of the nodes are required to update in parallel (see section 3).

Hardware paging The approach taken is to use a hardware paging technique and limit the number of nodes that can update in parallel. With this technique a fixed number of processing elements are implemented and extra circuitary is used to randomly choose which nodes will update on a given cycle

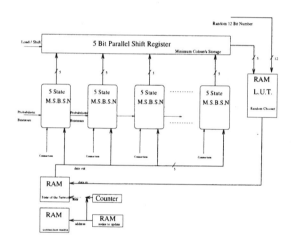

Fig. 3. Overall Design Enabling Hardware Paging

Each processing element has an 8 bit shift register to store a byte of the connection matrix relevant to the current node being updated. This is loaded with continuous bytes from the relevant row of the connection matrix until each processing element has seen every node in the graph and hence can choose a candidate solution.

This strategy makes the system much more flexible allowing any number of "virtual nodes" to be coloured with minimal modification to the hardware. These modifications could be made "on the fly" by reconfiguring the FPGA's appropriately by the host computer.

Figure 4 shows all the data paths to each unit in the system. The design is modular and so each processing unit (containing say, four or eight MSBSN's) could be cascaded to increase the parallelism of the network updates. This can dramatically increase the efficiency of the algorithm (see figure 5).

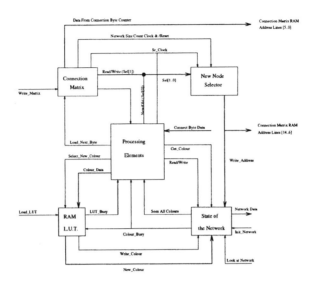

Fig. 4. Communication Paths of Entire Hardware System

3 Simulation results

The simulation written in C++ allows variable numbers of processing nodes in the network, thus varying the amount of parallel processing. This *fast* simulation took several days to solve each graph many times (The final hardware takes a matter of seconds to produce the same results).

The results are based on 450 node graphs that are known to be 5–colourable. Initially, the simulator processed two 5–colourable graphs with edge densities (the number of connections between the nodes, expressed as a fraction of the total number of possible connections) of 0.5 and 0.1 respectively using a sequential update. This is based on the pseudo–code given in figure 1.

The results in this section show the average number of iterations taken to solve the graph in 100 attempts. This is sufficient to smooth out any random pertubations introduced by the stochastic elements of the network, i.e. the probabilistic temperature of 0.92 and the random colour choice made by the lookup table when more than one counter achieves the minimum value within a MSBSN.

The mean iteration figures obtained for the two graphs are show below:

Edge Density	Average Iterations
0.5	3120.00
0.1	7018.77

The efficiency of the colouring algorithm, implemented by this architecture, can be improved by allowing nodes to update with a certain degree of parallelism.

The total number of nodes updating at any one time is an important parameter, this is illustrated by the results shown below. The number of iterations taken to colour a particular graph is dependant on not only the percentage number of nodes updating in parallel, but also on the edge density within the graph. It is this parallel update that allows the possibility of hardware paging.

For the purpose of these simulations, three graphs were generated, each with edge densities of 0.5, 0.1 and 0.05 respectively.

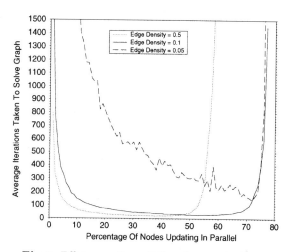

Fig. 5. Efficiency of parallelism for graph colouring

A high level of parallelism is not desirable as the phenomenon of *Parallel Interference* comes into effect [11]. This occurs when two adjacent nodes update in parallel and so increases the probability that they will update to the same colour.

From the graphs it appears that a good fixed parallel update percentage for the network is somewhere in the region of 30–40%, as this yields good results for all levels of edge density. Therefore the paging hardware needs approximately 30–40% of the total nodes. ie. for a 450 node graph 135 – 180 nodes would be required.

An improved alternative, suggested by [4] is to use a more adaptive update schedule. Here the network updates with a high level of parallelism and then as time progresses reduces the amount.

3.1 Hardware results

We have implemented a protoype system in hardware to confirm the above test results. This system contains a four node (four MSBSNs) processing unit and currently can solve upto 5 colourable 512 node graphs. The results from the hardware are encouraging and give similar results to the above graph. The system runs at 10MHz and performs approximately 500 iterations per second.

4 Conclusion

This paper has introduced a hardware accelerator system for processing graph colouring problems efficiently. Currently we are using the prototype system to solve Leighton graphs with the assistance of a novel high level algorithm [6]. Thus providing an effective solution to a difficult problem. This research is on–going, and further results will be produced.

Altera have produced a successor to the FLEX 8000 family, namely the FLEX 10000 series [3]. This family of FPGA's have a higher logic capacity and small amounts of on–chip RAM. Utilising this RAM does not reduce the amount of logic available within the device. This new architecture would be advantageous to use for a MSBSN network, as the localised on–chip RAM can be utilized as an independent weight connection matrix for each neuron.

References

1. E. Aarts and J. Korst. *Simulated Annealing and Boltzmann Machines*. John Wiley & Sons, 1989.
2. Altera Corp. FLEX8000 programmable logic device family, 1993.
3. Altera Corp. FLEX10000 embedded programmable logic device family, 1995.
4. P. Burge. Parallel pruning. *in private conversation*, 1996.
5. P. Burge and J Shawe-Talyor. Bitstream neurons for graph colouring. *Journal of Artificial Neural Networks*, pages 443 – 448, 1995.
6. P. Burge, J. Shawe-Talyor, and J. Zerovnik. Graph colouring by maximal evidence edge adding. presented at the Udine Workshop on Approximate Solution Of Hard Combinatorial Problems.
7. B. Chamaret, S. Ubeda, and J. Zerovnik. A randomized algorithm for graph colouring applied to channel allocation in mobile telphone networks. In *Operational Research Proceedings KOI'96*, pages 25 – 30, 1996.
8. Dimacs colouring benchmarks. available via ftp at `rutgers.dimacs.edu`.
9. DIMACS Research Group. *Clique and Colouring Problems, A Brief Introduction*. Rutgers University, December 1992.
10. A. Petford and D. Welsh. A randomised 3–colouring algorithm. In *Discrete Mathematics*, volume 74, pages 253 – 261, 1989.
11. Barry Rising. Hardware design of multi–state bit stream neurons for graph colouring. Technical Report CSD–TR–96–12, Royal Holloway, June 1996.
12. J. Shawe–Taylor and J. Zerovnik. Generalised boltzmann machines. Technical Report CSD-TR-92-29, Royal Holloway, University of London, 1992.
13. M. van Daalen, P. Jeavons, and J. Shawe-Taylor. A stochastic neural architecture that exploits dynamically reconfigurable FPGAs. *IEEE Workshop on FPGAs for custom computing machines*, pages 202 – 211, 1993.
14. D. van den Bout. RIPP10 user manual, 1993.

Run–Time Compaction of FPGA Designs

Oliver Diessel[1] and Hossam ElGindy[2]

[1] Department of Computer Science and Software Engineering
[2] Department of Electrical and Computer Engineering
The University of Newcastle, Callaghan NSW 2308, AUSTRALIA

Abstract. Controllers for dynamically reconfigurable FPGAs that are capable of supporting multiple independent tasks simultaneously need to be able to place designs at run–time when the sequence or geometry of designs is not known in advance. As tasks arrive and depart the available cells become fragmented, thereby reducing the controller's ability to place new tasks. The response times of tasks and the utilization of the FPGA consequently suffer. In this paper, we describe and assess a task compaction heuristic that alleviates the problems of external fragmentation by exploiting partial reconfiguration. We identify a region of the chip that can satisfy the next request after the designs occupying the region have been moved. The approach is simple and platform independent. We show by simulation that for a wide range of task sizes and configuration delays, the response of overloaded systems can be improved significantly.

1 Introduction

Recently, FPGA architectures have become partitionable and dynamically reconfigurable — chips have become capable of supporting several independent or interdependent tasks/designs at a time, and parts of the chip can be reconfigured relatively quickly while the remaining tasks continue to execute [6, 1]. These new capabilities promise exciting new application areas and pose several challenging engineering problems, including the design of suitable controllers [4]. When the sequence of tasks to be performed by the chip is known in advance the designer can optimize the use of resources off–line and design an appropriate static controller. However, when the sequence is not predictable, or the task designs are not fixed, the controller needs to make allocation decisions on–line. Unfortunately, on–line allocation schemes that allocate contiguous resources suffer from external fragmentation as variously sized tasks are allocated and deallocated. Tasks end up waiting in a queue despite there being sufficient, albeit non–contiguous resources available to service them. The time to complete tasks is consequently longer, and the utilization of resources is lower than it could be, thereby contributing to the selection of larger, less utilized chips.

We are interested in alleviating this problem. Our aim is to satisfy the next allocation request to a dynamically reconfigurable FPGA that is executing a set of tasks when it is not possible to satisfy the request without compaction. Two subproblems naturally arise:

1. how to identify a good *allocation site*, a sub–array of the requested size efficiently, and
2. how to schedule the compaction so as:
   (a) to free the allocation site of other executing tasks as quickly as possible,
   (b) to delay the tasks that are to be moved as little as possible, and
   (c) to complete compacting the tasks as quickly as possible.

On a grid, it is NP–complete to decide whether or not a set of *rectangular* tasks can be placed orthogonally without overlap [3]. Efficient heuristics are therefore sought. Partial task compaction to reduce fragmentation on meshes has been investigated by Youn et al [7]. However, effective heuristics needed to carry out the arbitrary rearrangements generated by Youn's approach with minimum delay to FPGA tasks are still being sought. In this paper we present a more structured heuristic that is simple, effective, and platform independent. We describe a method that has the effect of sliding the set of tasks to be compacted in a single direction along a single dimension while preserving their relative order. Without loss of generality, we describe compacting the tasks to the right along the rows of the FPGA cells. Fig. 1 contains an example of such a one–way one–dimensional order–preserving compaction [2].

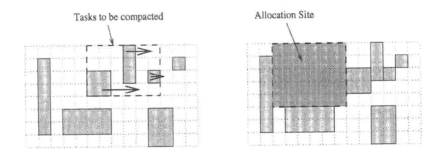

Fig. 1. An example of a one–way one–dimensional order–preserving compaction. The initial arrangement on the left shows the tasks to be compacted so as to allocate a task of size 5 × 6. The final arrangement on the right indicates the allocation site.

In the following section we describe the assumptions leading to the FPGA task allocation model. In Section 3 we describe and analyze the algorithms used by the controller to compact FPGA tasks at run–time. We show that it is possible to reduce the number of potential allocation sites from $O(H \times W)$ to $O(n^2)$, which is a considerable saving when the number of tasks is relatively small. Thereafter, we describe the construction of a visibility graph over the executing tasks that allows us to determine the feasibility and cost of freeing the executing tasks from each candidate site in $O(n)$ time. In the worst case, we therefore spend $O(n^3)$ time determining whether the incoming task can be allocated with compaction. The site that can be freed of executing tasks in minimum time, and thus satisfies the request quickest of all, can be identified at the same time. In Section 4 we demonstrate by simulation the improvement in performance achievable by use of the compaction heuristic. Our conclusions appear in Section 5, which also mentions areas for future research.

2 Model

For the remainder of this paper we use the terms "task" and "design" synonymously. At the cost of possibly introducing internal fragmentation, we assume that a task and the used routing resources surrounding its perimeter, which may or may not be associated with the task, can be modeled as a rectangular sub-array of arbitrary yet specified dimensions. We assume that the time needed to configure a sub-array (place a design) is proportional to the size (area) of the sub-array since at worst, cells are configured sequentially. Since the delay properties of commercially available chips are isotropic and homogeneous, we assume that the time needed to configure a task and route I/O to it is independent of the task's location.

Moving a task involves: suspending input to the task and waiting for the results of the last input to appear, or waiting for the task to reach a checkpoint; storing register states if necessary; reconfiguring the portion of the FPGA at the task's destination; loading stored register states if necessary; and resuming the supply of input to the task for execution. We do not consider tasks with deadlines, and therefore assume that any task may be suspended, with its inputs being buffered and necessary internal states being latched until the task is resumed. The time needed to wait for the results of an input to appear, or for the task to reach a checkpoint, is considered to be proportional to the size of the task, which, in the absence of feedback circuits, is the worst case. However, since the time to configure a cell and associated routing resources is typically at least one order of magnitude greater than the signal delay of a cell or the latency of a wire, the latency of the design is considered negligible compared with the time needed to configure the task. We investigate the effectiveness of reconfiguring the destination region of a task by reloading the configuration stream with a new offset. This approach naturally re–incurs the cost of placing the task, but is applicable to any device. In this paper we do not address the problem of rerouting I/O to a task that is moved. If I/O to tasks is performed using direct addressing, then tasks not being moved may be delayed by reloading the configuration stream of tasks being moved. We ignore this phenomenon here.

Overall management of tasks is accomplished in the following way. Tasks are queued by a sequential controller as they arrive. A task allocator, executing on the controller, attempts to find a location for the next pending task. If some executing tasks need to be compacted to accommodate the task, then a schedule for suspending and moving them is computed by the allocator. The allocator coordinates the partial reconfiguration of the FPGA according to the compaction schedule, and associates a control process with the new task and its placement. If a location for the next pending task cannot be found, the task waits until one becomes available following one or more deallocations as tasks complete processing.

We use the following notation in this paper: The FPGA of size $F(H, W)$ consists of H rows and W columns of configurable cells arranged in a grid. A task $T_i(s_i, b_i)$ of size $s_i = (r_i, c_i)$ and base $b_i = A_{y_i, x_i}$ is allocated to a submesh of r_i rows and c_i columns of cells with bottom–leftmost cell A_{y_i, x_i}, $1 \leq y_i, 1 \leq x_i$ and top–rightmost cell $A_{y_i + r_i - 1, x_i + c_i - 1}$ with $y_i + r_i - 1 \leq H$ and $x_i + c_i - 1 \leq W$. The task T_i is said to be based at b_i. We denote the ith row of cells R_i and the jth column of cells C_j. The intersection of R_i and C_j is the cell $A_{i,j}$. The interval of cells $A_{i,k}, A_{i,k+1}, \ldots, A_{i,k+m}$ in the ith row is denoted $R_i[k, k+m]$. A similar definition applies to $C_j[l, l+n]$. The intervals $R_i[k, k+m]$ and $C_j[l, l+n]$ intersect at cell $A_{i,j}$ iff $l \leq i \leq l + n$ and $k \leq j \leq k + m$.

3 Algorithms

3.1 Identifying Potential Allocation Sites

Definition 1. For the request of size $s_{n+1} = (r_{n+1}, c_{n+1})$ we define a *top cell interval* for each executing task $T_i((r_i, c_i), A_{y_i, x_i})$, $1 \leq i \leq n$, that consists of the set of possible base locations for T_{n+1} were the bottom edge of T_{n+1} to abut the top edge of T_i. The existence and extent of the top cell interval for T_i is constrained by the boundaries of the chip but disregards the intersection of T_{n+1} with other executing tasks. A top cell interval is also defined with respect to the bottom edge of the chip. We thus define the set of *top cell intervals for s_{n+1}* to be the set $T = \{R_{r_i + y_i}[\max(1, c_i - c_{n+1} + 1), \min(c_i + x_i - 1, W - c_{n+1} + 1)] : 1 \leq i \leq n, r_i + y_i \leq H - r_{n+1} + 1\} \cup R_1[1, W - c_{n+1} + 1]$.

We similarly define for each executing task T_i a *right cell interval*, consisting of the the set of possible base locations for T_{n+1} were the left edge of T_{n+1} to abut the right edge of T_i. A right cell interval is also defined with respect to the left edge of the chip. The set of *right cell intervals for s_{n+1}* is thus defined to be the set $R = \{C_{c_i + x_i}[\max(1, r_i - r_{n+1} + 1), \min(r_i + y_i - 1, H - r_{n+1} + 1)] : 1 \leq i \leq n, c_i + x_i \leq W - c_{n+1} + 1\} \cup C_1[1, H - r_{n+1} + 1]$.

The set of cells at the intersection of the set of top and right cell intervals is denoted $B = T \cap R$.

These intervals, which define the minimum cost locations for placing the base of the incoming task T_{n+1} if it is to be allocated in the neighbourhood of T_i, are illustrated in Fig. 2.

Theorem 2. *If T_{n+1} can be allocated by means of compaction, then the cost of freeing the executing tasks is minimized for an allocation site based at some cell in B.*

Proof. The proof considers the time needed to free the space for the incoming task for all of its possible base positions as it is shifted along a row from the left edge of the FPGA to the right. Our assumption is that the cost to free an

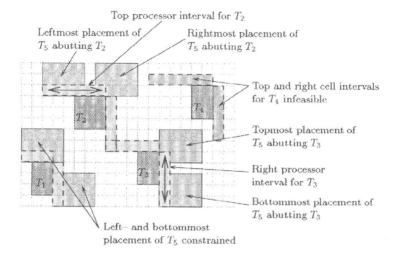

Top processor interval for T_2

Leftmost placement of T_5 abutting T_2

Rightmost placement of T_5 abutting T_2

Top and right cell intervals for T_4 infeasible

Topmost placement of T_5 abutting T_3

Right processor interval for T_3

Bottommost placement of T_5 abutting T_3

Left- and bottommost placement of T_5 constrained

Fig. 2. Definition of top and right cell intervals for four executing tasks and an incoming task T_5 of size 3×4.

allocation site is at least proportional to the area of tasks that need to be moved out of the allocation site.

Let us consider the allocation site based at $A_{r,1}, 1 \le r \le H - r_{n+1} + 1$. Executing tasks lying within the allocation site are to be compacted to the right. Assume the leftmost allocated cell(s) within the allocation site are in column C_c. As the base of the allocation site is shifted to the right from $A_{r,1}$ to $A_{r,c}$, additional allocated tasks potentially become covered by the right edge of the allocation site, thereby increasing the time to free the site of occupying tasks. However, it is not until the first task occupying the allocation site, T_l say, is completely uncovered by the left edge of the allocation site that the time needed to free the site of occupying tasks potentially decreases, since while any cells of T_l remain within the allocation site T_l must be moved to the right of it. Thus it is only necessary to check allocation sites based in the columns of cells $C_{x_i+c_i}$ immediately to the right of executing tasks T_i. The base is constrained from moving to either side with the potential of reducing the cost to free the allocation site by the presence of T_i to the left, and the possibility of covering additional tasks to the right. However, the columns $C_{x_i+c_i}$ need only be checked over the interval in which the task T_i potentially intersects the allocation site, namely $C_{x_i+c_i}[\max(1, r_i - r_{n+1} + 1), \min(r_i + y_i - 1, H - r_{n+1} + 1)]$.

By a similar argument it follows that it is only necessary to check sites in the rows of cells $R_{y_i+r_i}$ immediately above executing tasks T_i. These rows $R_{y_i+r_i}$ need only be checked in the interval in which the task T_i intersects the allocation site, $R_{y_i+r_i}[\max(1, c_i - c_{n+1} + 1), \min(c_i + x_i - 1, W - c_{n+1} + 1)]$.

Consider the top cell interval associated with a task T_i. The allocation site is constrained from moving below or above it without potentially increasing the cost to free the site. Allocation sites based to the right or left of the interval are potentially more costly than sites based within a column that intersects other top cell intervals. A similar argument applies to a right cell interval. Therefore if we consider potential bases within the top interval, the cost to free the allocation site is least where it intersects right intervals. These intersections are guaranteed to exist due to the fact that T_{n+1} cannot be allocated without compaction. □

Constructing the set $\mathcal{B}$ of potential bases for the incoming task requires $O(n^2)$ time if each member of the set of right cell intervals is used to check for intersections against each member of the set of top cell intervals. Since $O(n^2)$ potential base locations have to be identified, this is optimal in the worst case.

3.2 Assessing Allocation Site Feasibility

Allocation sites based at cells in $\mathcal{B}$ are not guaranteed to be feasible since it may not be possible to compact the executing tasks within the allocation site to the right due to a lack of free cells. An efficient way of answering this question for the $O(n^2)$ possible sites is to build a visibility graph of the executing tasks thereby allowing the feasibility of a site to be determined in $O(n)$ time.

Definition 3. (After [5]) Task V *dominates* a task T if, for some cell A_{r_V, c_V} of V and some cell A_{r_T, c_T} of T, $r_V = r_T$ and $c_V > c_T$. Where V dominates T, we say that V *directly dominates* T if there is no task U such that V dominates U and U dominates T. A *visibility graph* is the directed graph having the collection of executing tasks as vertex set; for each pair of tasks T and V it contains an edge from T to V iff V directly dominates T.

We build the visibility graph in $O(n^2)$ time from its roots to its leaves in the following way. The list of executing tasks is sorted into increasing base column order, where if two or more tasks share a column, they are sorted into increasing row order. For each task we create a graph vertex and insert it in sorted order. A vertex already in the graph has associated with it the bottom- and topmost rows covered by tasks in its subgraph. Vertex insertion can therefore be done in linear time by a depth first search of vertices not visited before to determine whether the task is to the right of the subgraph or not. For each edge inserted, we associate the distance from the parent to the newly added child. After the graph has been built, we compute and store at each vertex the maximum distance the task can be moved to the right by summing the edge distances in a bottom–up fashion. Note that the distance the terminal nodes can be moved is given by their base columns and their widths. This final step, which takes $O(n)$ time, saves time during searching by eliminating the need to determine the cost of compaction for allocation sites that cannot be freed of executing tasks. Fig 3 depicts the visibility graph for our example.

For each potential base $b \in \mathcal{B}$, those subgraphs whose covered rows intersect the allocation site based at b are searched depth first down to the leftmost task(s) that intersect the allocation site. The possibility of moving each of these out of the way of the incoming task is then checked in $O(n)$ time per base b.

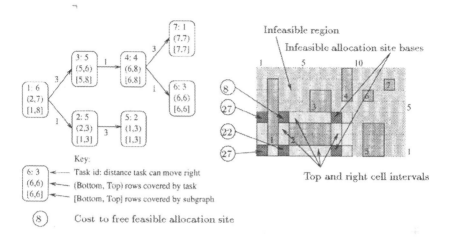

Fig. 3. Visibility graph for arrangement of tasks on the right, with cost to free feasible allocation sites for an incoming task of size 5×6.

3.3 Selecting the Best Allocation Site

We can identify the set of tasks to be compacted and the distance each is to be moved by traversing each subgraph rooted in the potential allocation site. This traversal is needed to accumulate the cost of freeing the site as well since it is assumed to be proportional to the sum of the areas of the tasks that have to be moved. These traversals, which take $O(n)$ time, are required for each feasible allocation site. The cost of freeing each of the potential allocation sites for our example is illustrated in Fig. 3.

3.4 Scheduling the Compaction

Given a set of tasks that are to be compacted, the scheduling policy we investigate moves tasks according to the visibility graph: a task is not moved until all tasks in its subgraph that must move have moved. The policy attempts to minimize delays to executing tasks by suspending each task that is to be moved for the period needed to reload the configuration, and by moving a task onto a region of the FPGA that does not overlap any other executing tasks. Note that the scheduling policy moves the tasks occupying the allocation site last of all, and therefore does not minimize the time needed to free the allocation site. Scheduling the compaction is straightforward and requires time linear in the number of tasks to be compacted.

4 Experimental Evaluation

We evaluated the performance of a simulated FPGA chip operating with and without compaction. The simulator queued and then allocated a set of random requests for service using two allocation methods:

1. With Compaction — attempted to satisfy the next pending request by compaction whenever the task could not be allocated by the Bottom–left method, and

2. Bottom–left Allocation — allocated the next pending task to the bottom-leftmost free subarray whenever possible.

Experiments were conducted to measure the performance under varying load, varying configuration delays, and varying task sizes. All experiments involved generating 10,000 requests for service to an $F(64, 64)$ chip. The service period generated for each task ranged uniformly between 1 and 1,000 time units. The task size per side and the intertask arrival period were independently generated random variables. Each experiment involved fixing two of the maximum task size, maximum intertask arrival period, and configuration delay per cell, and varying the third. The amount of time a task spent waiting at the head of the queue to begin entering the chip, the elapsed time between a task arriving and completing processing, and the chip utilization were measured.

Effect of System Load on Allocation Performance Task sizes were allowed to range up to 32 cells per side and the configuration delay per cell was set to 1/1,000 of a time unit, giving a mean configuration delay per task of about 0.3 time units. See Fig. 4(a). At maximum inter–task arrival periods below 50 time units the FPGA was saturated with work as tasks arrived more frequently than they could be allocated. Compaction resulted in a reduction in mean allocation delay of approximately 19% at saturation, which caused a reduction in mean response time. The reduction in mean response time due to compaction increased from approximately 26% in the saturated region to over 75% at loads where the chip was coming out of saturation before falling to zero in the unsaturated region. Fig. 4(b) illustrates that the utilization was roughly 25% higher in the saturated region due to compaction. This benefit rapidly decreased to zero as the chip came out of saturation because compaction cannot influence the inter–task arrival period. Significantly, when compaction is used, the system has a higher load–bearing capacity, as evidenced by decreases in response times and utilization at greater task arrival frequencies.

Effect of Configuration Delay on Allocation Performance The performance benefits due to compaction, though significant in saturated systems, are greatest when the system is coming out of saturation. We investigated the response times in this operating range as the configuration delay per cell was increased from 1/1,000 time unit to 1,000 time units. Task sizes were again allowed to range up to 32 cells per side and the maximum intertask arrival period was set to 120 time units. Fig 4(c) indicates that at configuration delays of less than 50% of the mean service period, the performance benefit due to compaction was significant.

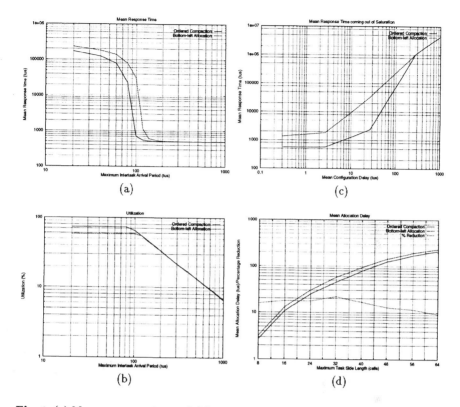

Fig. 4. (a) Mean response time and (b) chip utilization for 10,000 tasks of size $U(1,32)$ $\times$ $U(1,32)$, service period $U(1,1000)$ time units (tus), and uniform intertask arrival periods on $F(64,64)$. (c) Dependency of response time on configuration delay for above tasks with intertask arrival period of $U(1,120)$ tus. (d) Dependency of mean allocation delay on task size for above tasks arriving with intertask arrival period of 1 tu.

Effect of Task Size on Allocation Performance To give some indication of the relative performance benefits as the maximum task size changes, Fig. 4(d) plots the mean allocation delay at saturation as the maximum task size was increased from 8 to 64 cells per side. The configuration delay per cell was set to 1/1,000 time unit, and tasks arrived at intervals of 1 time unit. The figure indicates that the performance benefit due to compaction increased as the maximum task size was increased from 8 since small tasks are more easily accommodated without compaction. The benefit decreased again as the maximum task size approached the chip size, suggesting that less opportunities for free fragment combination by one–dimensional ordered compaction presented themselves.

5 Concluding Remarks

In this paper we described an effective heuristic for alleviating fragmentation by task migration in partially reconfigurable FPGAs. The one–dimensional ordered partial task compaction method is generally applicable because it operates independently of the task scheduling and allocation method. It is platform independent since task are moved by reloading their configurations. Simulations indicate that significant performance benefits can be gained as the FPGA becomes saturated with work even when configuration delays are large. Further evaluation under real run–time conditions is desired.

Many research problems remain to be solved. These include: examining the possibility of improving the time complexity of compaction algorithms; developing methods to relocate the tasks occupying an allocation site to arbitrary locations on the FPGA, which has the potential of improving the performance of compaction; taking into account tasks with deadlines; and, determining practical means of rerouting I/O to migrated tasks.

Acknowledgments

This research was partially supported by grants from the Australian Research Council, and the RMC at The University of Newcastle.

References

1. Atmel: AT6000 FPGA configuration guide
2. Diessel, O., ElGindy, H.: Run–time compaction of FPGA designs. Technical report 97–02, Department of Computer Science and Software Engineering, The University of Newcastle (1997). Available by anonymous ftp:
   `ftp.cs.newcastle.edu.au/pub/techreports/tr97-02.ps.Z`
3. Li, K., Cheng, K. H.: Complexity of resource allocation and job scheduling problems on partitionable mesh connected systems. Technical report UH-CS-88-11, Department of Computer Science, University of Houston, Houston, TX (1988). Proceedings 1st IEEE Symposium on Parallel and Distributed Processing (1989) 358–365
4. Lysaght, P., McGregor, G., Stockwood, J.: Configuration controller synthesis for dynamically reconfigurable systems. IEE Colloquium on Hardware–Software Cosynthesis for reconfigurable systems, London, UK, (1996)
5. Sprague, A. P.: A parallel algorithm to construct a dominance graph on nonoverlapping rectangles. International Journal of Parallel Programming **21** (1992) 303–312
6. Xilinx: XC6200 field programmable gate arrays. Technical report, Xilinx, Inc., (1996)
7. Youn, H.-y., Yoo, S.-M., Shirazi, B.: Task relocation for two–dimensional meshes. Proceedings of the Seventh International Conference on Parallel and Distributed Computing Systems (1994) 230–235

Partial Reconfiguration of FPGA Mapped Designs with Applications to Fault Tolerance and Yield Enhancement *

John M. Emmert and Dinesh Bhatia

Design Automation Laboratory
Department of Electrical and Computer Engineering and Computer Science
P.O. Box 210030
University of Cincinnati
Cincinnati OH 45221-0030, USA

Abstract. Field-programmable gate arrays have the potential to provide reconfigurability in the presence of faults. In this paper, we have investigated the problem of partially reconfiguring FPGA mapped designs. We present a maximum matching based algorithm to reconfigure the placement on an FPGA with little or no impact on circuit performance. Experimental results indicate the algorithm works well for both fault tolerance and reconfigurable computing applications. We also present the motivation and feasibility of using a similar approach for dynamic circuit reconfigurability.

1 Introduction

Field Programmable Gate Arrays (FPGAs) are becoming extremely popular for implementing ASICs and other applications. Recent advances in fabrication technology and device architecture have resulted in tremendous growth in FPGAs, both in terms of density and performance. Commercially announced FPGA parts can easily implement up to 60K gate equivalent designs, and it is predicted that very soon we will see devices that can possibly implement more than one million gate equivalent designs [14]. On the application side, the potential use of FPGAs has grown from applications like "glue logic" to reconfigurable (hardware) computing. While growth in both arenas will continue, significant advancement in CAE tools technology is needed to keep up with the growth [10]. The mapping of a digital design to an FPGA takes place in three steps: *Technology mapping, Placement,* and *Routing.* Normally, if a new circuit configuration is to be generated, the entire task of placement and routing has to be repeated; however, there is a limited capability for incremental change that is available in commercial CAD tool packages[8].

At times, it is desired to partially partially reconfigure the FPGA mapping. If the change from one configuration to another is small, and only involves remapping of certain logic functions, it may not be proper to re-accomplish the entire

---

* partially supported by contract number F33615-96-C-1912 from Wright Laboratories of the US Air Force

physical design cycle. In this paper, we have investigated the problem of rapidly reconfiguring FPGA mapped designs with the following potential applications:

- *Fault Tolerance:* FPGAs are ideal for addressing fault tolerance in electronic circuits. In the case of logic or interconnection fault(s), it is possible to isolate the fault(s) by not using the locality of the fault(s). Also, during the runtime life of an application, this will require the assignment of functionality of a faulty location to a non-faulty location.
- *Yield Enhancement:* Partial reconfiguration can increase the process yield for FPGA circuits [7]. When FPGAs are produced at the foundry, some of the dies contain random defects rendering the chip unusable. As the density of FPGAs continues to grow, the process yield is expected to fall very rapidly due to the increase in the area of silicon. In the case where these defects only occur on a limited number of FPGA resources, the chip may still be rendered usable. By partially reconfiguring the successfully mapped circuit, defects on the silicon die can be avoided. This methodology makes use of imperfect silicon die, thereby artificially enhancing the process yield.
- *Reconfigurable Computing:* Dynamic reconfigurable computing implies the hardware is reprogrammed to process data at the time of execution. In this way, one multi-purpose processor may accomplish the work of several specific processors. Partial reconfiguration can provide the space necessary to place highly optimized applications. For example, it may become necessary to add a new application to the already executing FPGA. A new process can be mapped to the hardware resources if it can fit in the available space. In some cases, this new application may come in the form of a highly optimized macro requiring a specific configuration of LBs. It may be necessary to reconfigure the existing applications on the FPGA to make room for the new application. However in order to efficiently use the FPGA in such applications, an effective means to rapidly update the mapping of the circuit on the FPGA is required.

In all of the cases above, partial reconfiguration of FPGA mapped circuits is necessary, and the problem becomes one of moving the functionality of programmed logic resources from one location to other locations, leaving unwanted (due to fault or otherwise) locations empty or unused.

Hanchek and Dutt developed a method to address fault tolerance and increase FPGA yield [3, 5]. Their method uses node covering and reserved routing resources to replace the functionality of faulty cells. Mathur, Chen, and Liu developed a method to re-engineer placements for regular architectures [12]. Their method uses an affinity matching algorithm to minimize the timing degradation during placement of new modules. Kelly and Ivey developed a redundancy based method to improve the yield for commercial FPGAs [9]. They add a spare element at the end of each row and an on-chip dedicated reconfiguration processor. Durand and Piguet presented an FPGA architecture with self-repair capabilities [2]. They present a new cell architecture that uses a special multiplexer with self diagnosis and self repair capabilities. Chapman and Dufort use techniques developed for wafer scale integration to correct wiring defects on FPGAs. They use a laser to form connections and bypass fabrication time defects found on FPGA die.

2 FPGA Architecture

We use a simplified technology independent model to represent the architecture of the FPGA. Figure 1 depicts the architecture model with a 4×4 array of *logic blocks*(LBs). We assume that the LBs are arranged in a regular $\sqrt{N} \times \sqrt{N}$ array. Two adjacent rows of LBs are separated by a horizontal wiring channel. Similarly, two adjacent columns of LBs are separated by a vertical wiring channel. The LBs are connected to the wiring channels through *connection boxes* (CBs), and the horizontal and vertical wiring channels are connected together using *switch boxes* (SBs). More details of such a generic model can be found in [1]. We use the Manhattan distance metric to measure the distance between between any two LBs.

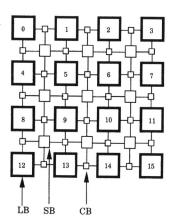

Fig. 1. Example FPGA architecture consisting of a 4×4 array of LBs.

3 Problem Formulation

The overall problem is *Partial Reconfiguration of FPGA Mapped Circuits*. This entails moving or reallocating resources on the FPGA such that impact on circuit performance is minimized and circuit functionality is maintained. In order to carry out reconfiguration, we must identify the resources (LBs) that need to be moved. In order to reconfigure, we must determine new locations for the functions programmed in these LBs, and we must move these functions to unused locations. Third, we must update the circuit's *routing* to account for displaced circuit resources. The reconfiguration problem can be restated as,

Let S be the set of LBs in a mapped circuit C and, let S_m be the set of LBs in a mapped circuit that are to be relocated. Let S_a be the set of unused LBs in

FPGA. Determine a one to one mapping of elements (LBs) of $\mathcal{C}$ to the elements in $S - S_m + S_a$.

$\square$

Clearly, the above mentioned problem can easily be solved by generating a new placement for the design where locations associated with elements of S_a are declared as unusable. However, the overhead of new placement and routing makes this solution unattractive. We would like to generate a new mapping of $\mathcal{C}$ from the original mapping of $\mathcal{C}$. In doing so, we are particularly interested in the preservation of performance while generating the new mapping very quickly.

4 Solution

In this paper we address the problem of partial reconfiguration of FPGA LB mapping. We use a matching technique to determine empty, non-faulty LB locations suitable for reconfiguring the mapped design, and then we use a *shift* move strategy to shift the functions of the LBs between matched locations, toward the matched locations.

4.1 Assumptions and Related Terminology

While reconfiguring a mapped design on an FPGA, we make the following assumptions:

- Only the LBs within the FPGAs can become faulty and routing architecture is always available. In our model, we do not represent interconnect faults. However, interconnect faults that can be emulated using LB faults.
- The location of cells to be reconfigured is known.
- The number of unused, non-faulty LBs must be greater than the number of LBs to be reconfigured.

An *optimally mapped* FPGA circuit is a digital circuit whose logic functions are mapped to an FPGA in such a way that the propagation delays are satisfied as per user's specifications. The mapped circuit consists of a set of LBs and interconnections that have been programmed to perform the required circuit function.

Depending on the LB *utilization*[2], a set S_a of free LBs is available at any time[3]. Similarly, we define a set of LBs that are either known to be faulty (must be re-mapped to non-faulty LB locations) or *tagged* for re-mapping as the set S_m. Clearly, $|S_m| \leq |S_a|$ for re-mapping/reconfiguration to be successful. Here, for simplicity, we have assumed the logic mapped or programmed to each LB must remain grouped. We construct a bipartite graph with two sets of nodes, S_a and S_m.

---

[2] utilization means the percent of LBs on the FPGA used to map the circuit

[3] the utilization must be $< 100\%$ for S_a to be non-empty

4.2 Minimax Grid Matching with Applications to FPGA Reconfiguration

The problem of determining new locations to map the members of S_m becomes one of matching the members of the set S_m to members of the set S_a in such a way that circuit functionality is maintained and circuit performance is minimally impacted. We use general minimax grid matching to accomplish this match. The problem of grid matching is stated below:

Instance[11]: A square with area N in the plane that contains N *grid points* arranged in a regularly spaced $\sqrt{N} \times \sqrt{N}$ array and N *random points* located independently and randomly according to a uniform distribution on the square as shown in figure 2.

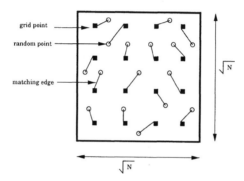

Fig. 2. An Instance of Grid Matching.

Problem: Find the minimum length L such that there exists a perfect matching between the N grid points and N random points where the distance between matched points is at most L. L is also called the *minimax* matching length.

The problem of finding L for a given distribution of random points is solvable in polynomial time since the length L has an upper bound of $O(\sqrt{N})$. An algorithm that solves this problem constructs a bipartite graph between the grid points and random points. Let S_m be the set of grid points and S_a be the set of random points. The edges of the bipartite graph will be $< i, j >$ such that $i \in S_m$ and $j \in S_a$, and i, j are at most distance L apart. We construct a bipartite graph $G = (S_m, S_a, E)$. The members of E are derived as follows,

$\exists$ *an edge* $e = < i, j >$, $i \in S_m$, $j \in S_a$, *iff* $d(i, j) \leq L$, $L = constant$.

where $d(i, j)$ is the Manhattan distance between the LBs i and j on the FPGA.

The algorithm starts with the construction of a bipartite graph for some initial L. It repetitively updates L, adding more edges, until a perfect matching is found in the bipartite graph. The L found by such an algorithm is also the *minimax* length. Leighton and Shor have proved a bound on the expected length of L for a random distribution of points which, with very high probability [4] is shown

---

[4] very high probability means probability exceeding $1 - 1/N^\alpha$ where $\alpha = \Omega(\sqrt{logN})$.

to be $\Theta(log^{3/4}N)$[11]. We use this tight and small bound to attain a reconfiguration that results in minimal impact on circuit performance. In practice, L is set equal to 1 so only adjacent LBs can be connected. If a completely matched graph is not found, L is incremented by 1 until a completely matched graph is found.

Minimax grid matching is performed on the sets S_m and S_a which returns a matched list, or a list of LBs from the set S_a that match the LBs in the set S_m. We use the matched list to update the LB mapping. The functionality of the LBs from the set S_m are moved to other LB locations related to the set S_a.

It should be noted that the bipartite graph construction need not be carried out completely, every time L is updated. The graph can be stored and only new edges may be added. Similarly, the bipartite matching can be found using augmenting paths[13] and every time new edges are thrown in, only the augmenting paths may be updated. In the worst case we expect a runtime complexity of $O(N^3)$ for an FPGA with $O(N)$ cells. This is because matching can take at most $O(N^{2.5})$ time [4] and we will never execute more than $O(\sqrt{N})$ iterations of matching. If however, a binary search on the matching distance L is performed, then the overall time complexity is at most $O(N^{2.5} \cdot logN)$. We also expect the procedure to be faster than $O(N^{2.5} \cdot logN)$ because in practice we are constructing random sparse graphs with very few edges per node. In fact, experimental data indicates that the actual running time on random examples is nearly linear in N.

4.3 Move Strategy

Initially to update the circuit map, we directly moved the functionality of the elements in S_m to their matched locations from S_a. This *direct* swap methodology led to poor circuit performance which is explained as follows. Assuming that highly connected LBs are placed in close proximity to one another (to enhance performance), implies that directly moving the functionality of a LB in the set S_m to its matched LB from the set S_a can be detrimental to circuit performance (see figure 3). In order to minimize the impact of updating the FPGA circuit, we use a *shift methodology* (see figure 4). As can be seen in figure 3. the direct swap methodology can cause unnecessary circuit delays by increasing the length of nets connected to reconfigured LBs. As shown in figure 4 the shift methodology limits the amount of delay introduced during partial reconfiguration by spreading the net length increase across several nets. This methodology shifts not only the functionality of the LB in the set S_m, but also the functionality of the LBs between the LB in the set S_m and the corresponding matching LB in the set S_a. To limit complexity, one of two paths is chosen for the shifts: *horizontal then vertical* or *vertical then horizontal*. (The degenerate case when S_m LB and S_a LB are in the same row/col, only horizontal/vertical shifts take place.) The method for choosing between the two paths is limited to summing the number of nets attached to the LBs in each of the two paths and choosing the path with the fewest nets. Once a path is chosen, the functionality of the LBs in the path are shifted starting at the LB from the set S_m and ending on the matching LB from the set S_a.

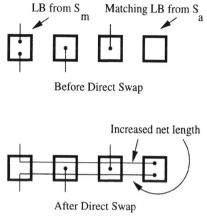

Fig. 3. Net length increase from swap.

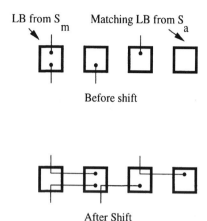

Fig. 4. Move by shifting.

5 Results and Analysis

We used empirical data to test the reconfiguration methodology presented above. The benchmarks were based on circuits that were mapped to the Xilinx XC4000 family, in particular the XC4013 device, of FPGA circuits. We derived the data by reconfiguring several circuits that have been optimally partitioned, placed, and routed using the Xilinx **XACT 5.2 PPR** utility, and we used the Xilinx **XDELAY** static timing analyzer to find the delay of the circuits before and after reconfiguration.

We used several mapped circuits in 130 different configurations to test the methodology described above. The mapped circuits used from 16% to 92% of the available LB resources on a Xilinx XC4013 device; however, we focused our effort on circuits with ≈ 90% LB utilization. In all cases, circuits were 100% routable both before and after reconfiguration. This is expected as our methodology does not account for routing architecture related faults. Also, since the routing is performed using the Xilinx router, it is not possible to block out routing resources very easily.

We applied the reconfiguration algorithm to the test circuits using two methods of selecting LB locations for reconfiguration: *random* and *user* input. The random method used the unix **rand** function to randomly pick LB locations for reconfiguration. The second method gives the user the opportunity to supply the LB locations for reconfiguration. In our experimentation we declared a contiguous block of LBs to be faulty. This scenario resembles that of an area fault and the ability to swap a large function that may be mapped on a collection of contiguous LBs.

The reconfiguration program read in the **ckt.lca** file which was the output of the Xilinx XACT PPR utility. After circuit reconfiguration, the program wrote out a new LCA file, **ckt_2.lca**, which included placement and programming information about the new circuit placement. The new LCA file included only information on routing that was not affected by the movement of LBs. This allowed the new LCA file to be incrementally routed to account for routing of resources that were affected by reconfiguration.

5.1 Results

The left part of figure 5 shows the results of the XDELAY program for randomly supplied faulty LB locations. The x-axis gives the number of faulty LBs inserted into the FPGA circuit. The y-axis indicates the maximum circuit path delay (in n sec) after reconfiguration. This graph indicates the performance of the circuit degrades gradually as the number of faulty LBs on the FPGA increases with little or no degradation for low numbers of faulty LBs. The right part of figure 5 shows the execution time of reconfiguration (minus time for incremental reroute) for randomly supplied faulty LB locations. The left part of figure 6 shows the maximum circuit path delay determined by XDELAY for clustered LB reconfiguration locations. The x-axis gives the number of LBs mapped to the FPGA that were reconfigured. The y-axis gives the normalized XDELAY of the reconfigured circuit. The normalized XDELAY is determined by dividing the XDELAY value of the reconfigured circuit by that of the optimally mapped FPGA circuit (original FPGA circuit). This graph shows the performance of the circuit degrades gradually as cluster size increases with little or no degradation for small clusters. The right part of figure 6 shows the execution time of the reconfiguration (minus time for incremental reroute) for user supplied, clustered LB locations.

Additionally two methods were implemented to update the placement of LBs that were to be re-mapped: *direct* swap and *shift* swap. Both methods may be considered simple, but the second method displayed a quantifiable improvement over the first method. In order to analyze the performance of the *shift* methodology over that of the *direct* swap methodology, we analyzed six test cases. The *shift* methodology showed more than a 6 times improvement in worst case timing as reported by XDELAY over the *direct* swap methodology with a standard deviation of only 2.8.

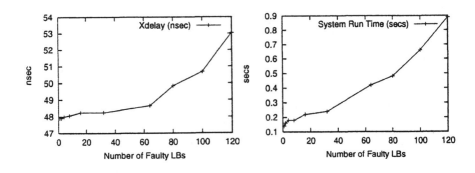

Fig. 5. Reconfiguration of random faulty LBs.

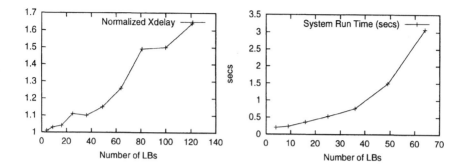

Fig. 6. Reconfiguration of specific LB blocks.

The original placement time for the optimized circuits took between 53 and 55 minutes on a SPARC 10. Reconfiguration time for the same circuits was on the order of seconds with little or no impact on circuit performance.

Typically, redundancy has been adopted as the primary means of fault tolerance for FPGA circuits [6]. However, this methodology adds the additional overhead of extra rows/columns and selector circuits. The additional overhead leads to speed or performance degradation, and the area overhead due to redundancy results in small net yield gain. Our reconfiguration methodology requires no changes to the FPGA hardware resources, and when an FPGA circuit is used in a remote location (e.g. space applications) reconfiguration can be used to keep the circuit functional if certain areas of the die become faulty.

6 Concluding Remarks

We have shown that minimax grid matching can (with good results) be used to determine locations for moving the functionality programmed in LBs to other LB locations. And, we provide a basis for minimizing the affect of moving the functionality of LBs to other LB locations through the use of a *shift* rather than a *directswap* methodology.

The next step is to incorporate the incremental routing step into the methodology. In order to improve the methodology presented above, routing should be included with the movement of LBs. By combining the idea of the "neighborhood" with the idea of local routing channels, routing can be localized, and therefore, the impact to circuit performance minimized.

The method presented above was accomplished off-line for test purposes. The intention is to eliminate the down time required to reload the bitstream by embedding the methodology into the system, either through on-chip processing or through an off-chip co-processor (or other FPGA). Then, the cache style architectures will allow rapid reconfiguration of part of the FPGA mapped circuit without requiring the entire circuit to be disabled.

This still leaves identifying which LBs are to be reconfigured. For fault tolerance, this entails fault detection/location. For yield enhancement, a method for identifying faulty LBs that is transparent to the user is required.

References

1. S. D. Brown, R. J. Fancis, J. Rose, and Z. G. Vransic. *Field-Programmable Gate Arrays*. Kluwer Academic Publishers, 1992.
2. S. Durand and C. Piguet. FPGA with Selfrepair Capabilities. In *ACM Second International Workshop on Field-Programmable Gate Arrays*, pages 1–6, Feburary 1994.
3. S. Dutt and F. Hanchek. REMOD: A New Methodology for Designing Fault-Tolerant Arithmetic Circuits. *IEEE Transactions on Very Large Scale Integration (VLSI) Systems*, 5:34–56, March 1997.
4. S. Even. *Graph Algorithms*. Computer Science Press, 1979.
5. F. Hanchek and S. Dutt. Node-Covering Based Defect and Fault Tolerance Methods for Increased Yield in FPGAs. In *Proceedings of the IEEE International Conference on VLSI Design*, pages 225–229, January 1996.
6. F. Hatori, T.Sakurai, K. Sawada, M. Takahashi, M. Ichida, M. Uchida, I. Yoshii, Y. Kawahara, T. Hibi, Y. Sacki, H. Muraga, and K. Kanzaki. Introducing Redundancy in Field Programmable Gate Arrays. In *Proceedings of the IEEE International Conference on Custom Integrated Circuits*, pages 7.1.1–7.1.4, 1993.
7. N. J. Howard, A. M. Tyrrell, and N. M. Allinson. The Yield Enhancement of Field-Programmable Gate Arrays. *IEEE Transactions on VLSI Systems*, 2:115–123, March 1994.
8. Xilinx Inc. *The Programmable Logic Data Book*. Xilinx, 2100 Logic Drive, San Jose, CA 95124-3400, 1995.
9. J. L. Kelly and P. A. Ivey. A Novel Approach to Defect Tolerant Design for SRAM Based FPGAs. In *ACM Second International Workshop on Field-Programmable Gate Arrays*, pages 1–11, Feburary 1994.
10. K. Keutzer. Challenges in CAD for the One Million Gate FPGA. In *ACM Fifth International Symposium on Field-Programmable Gate Arrays*, pages 133–134, Feburary 1997.
11. F. T. Leighton and P. W. Shor. Tight Bounds for Minimax Grid Matching with Applications to Average Case Analysis of Algorithms. In *Proceedings of the Symposium on Theory of Computing*, pages 91–103, May 1986.
12. A. Mathur, K. C. Chen, and C. L. Liu. Re-engineering of Timing Constrained Placements for Regular Architectures. In *IEEE/ACM International Conference on Computer Aided Design*, pages 485–490, November 1995.
13. C. Papadimitriou and K. Steiglitz. *Combinatorial Optimization*. Prentice Hall Pubblishers, 1982.
14. J. Rose and D. Hill. Architectural and Physical Design Challenges for One-Million Gate FPGAs and Beyond. In *ACM Fifth International Symposium on Field-Programmable Gate Arrays*, pages 129–132, Feburary 1997.

A Case Study of Partially Evaluated Hardware Circuits: Key-Specific DES

Jason Leonard and William H. Mangione-Smith

Electrical Engineering Department, University of California
Los Angeles, CA 90095-1594
http://www.icsl.ucla.edu/~billms

Abstract. FPGA based data encryption provides greater flexibility than ASICs and higher performance than software. Because FPGAs can be reprogrammed, they allow a single integrated circuit to efficiently implement multiple encryption algorithms. Furthermore, the ability to program FPGAs at runtime can be used to improve the performance through dynamic optimization. This paper describes the application of partial evaluation to an implementation of the Data Encryption Standard (DES). Each end user of a DES session shares a secret key, and this knowledge can be used to improve circuit performance. Key-specific encryption circuits require fewer resources and have shorter critical paths than the completely general design. By applying partial evaluation to DES on a Xilinx XC4000 series device we have reduced the CLB usage by 45% and improved the encryption bandwidth by 35%.

1. Introduction

As the volume of digital communications increases, reliable encryption becomes more important in order to ensure secure communications. In 1977 the U. S. National Bureau of Standards certified the Data Encryption Standard (DES) for use within the United States [1]. We will present a configurable computing implementation of DES.

Configurable computing offers a great degree of flexibility in system design due to the programmability of FPGAs. Circuit resources can be used for multiple purposes and refined to perform highly specific functions, thus combining generality with high performance. Secure communications systems often require the capacity to encrypt messages with several different algorithms in addition to the need to change keys regularly. ASICs lack the flexibility to implement multiple encryption algorithms efficiently. On the other hand, software implementations suffer from low speed. Fortunately, certain properties shared by many encryption algorithms make them well suited to FPGA implementations [2]. In particular, the reliance on many parallel bit-level operations generally matches the strengths of many FPGA architectures. However, the implementation of some other common functions, in particular wide permutations, is often problematic on FPGA architectures. This paper will present one technique for improving the performance of an encryption engine.

Encryption algorithms require a significant amount of hardware resources due to their inherent complexity. Our approach alleviates part of this burden by generating key-specific circuitry. This technique has the benefit of improving the speed of the circuit, as FPGA performance is strongly effected by routing complexity. The approach is termed partial evaluation; the effects of the encryption key are evaluated once rather than continuously during execution. FPGA configuration time is often not a problem as encryption keys change at a lower rate than encryption operations.

Session-based configuration has some additional advantages as well. In particular, it is more difficult for a third party to compromise the system integrity. With an ASIC implementation, an attack could involve writing a new value into the key register. For the FPGA implementation developed here such an attack would involve replacing an entire FPGA configuration, which would be difficult to do unobtrusively. There has also been much discussion in the cryptography community about the security of DES because of the 56-bit key length. A commonly presented solution to present or future attacks on DES is to increase the key length. To go to an expanded DES key in an ASIC requires a new design. In a configurable design, however, it is possible to reuse the same physical devices.

The remainder of this paper is organized as follows. Section 2 discusses the relevant background work. Section 3 describes the DES algorithm in detail. Section 4 presents the approach used for specializing DES to a particular encryption key. Section 5 compares the resource requirements and performance of the full DES algorithm against the specialized implementation using partial evaluation. Finally, section 6 discusses some technology issues related to implementing this approach.

2. Related Work

The use of FPGAs for DES has attracted attention primarily in code-breaking machines, through the use of a "known plaintext" attack. These attacks require the possession of ciphertext and a part of the corresponding plaintext. Known plaintext attacks search the entire set of valid keys and hunt for a key which maps some know plaintext to some known ciphertext. FPGAs have received attention for this application because, while being slower than custom hardware, their cost is generally much lower and they outperform software approaches [3-5]. A large number of FPGAs can therefore form a system that searches the key-space in parallel. ASIC implementations suffer in comparison due to high non-recurring engineering cost.

In 1996 an ad hoc group of cryptographers and computer scientists claimed that a small business could produce an FPGA board that would break a key in 556 days, at a cost of $10,000. As a consequence, they suggested that the DES key be expanded to 90 bits. Goldberg and Wagner disagreed with the ad hoc group and implemented DES in an Altera FLEX8000 part [4]. They extrapolated their results for the algorithm to project a cost of $45,000 for a full cryptanalytic machine that would search through the entire key-space in one year. There is much published research on custom hardware cryptanalytic machines for DES, dating as far back as 1977 [6]. Wiener is currently considered the best estimate of custom hardware performance [7].

Gray and Kean implemented DES on the fine-grained Configurable Array Logic device. This design faithfully implemented the standard DES architecture, thus exploiting the bit-level parallelism but not partial evaluation.

Villasenor et al. used partial evaluation to accelerate an automatic target recognition system [8]. Singh, Hogg, and McAuley discuss the use of partial evaluation for the dynamic reconfiguration of FPGAs [9]. They made several observations about what would constitute a good application for partial evaluation of hardware in FPGAs. The first was that the relevant inputs should be pseudo-static, i.e., remain constant for a long period of time. Partially evaluated hardware will obviously bring about an increase in the number of configurations and, because configuration times can be long, designers should be sure that the circuit improvements justify incurring the time overhead. Additionally, the circuit changes resulting from the partial evaluation should be small, so as to minimize the configuration time. This fact calls for a partially reconfigurable FPGA device.

Finally, in the context of CAD support for dynamic reconfiguration, they made the point that an efficient method of creating/modifying the file that programs the FPGA is an important design challenge for partially evaluated hardware. Perhaps the most systematic use of partial evaluation and hardware circuits involves the PECompiler developed by Wang and Lewis, which attempts to automatically apply these techniques without relying on domain- or application-specific information [10]. Their work is in a very early stage, however.

This work makes two specific contributions. While a number of researchers have used partial-evaluation techniques, none have go as far as the work here in removing subsequent control hardware. Thus, we believe that the DES design used here is the most complicated and advanced application of partial-evaluation to date. Secondly, this is the first known design of a cryptography system that capitalizes on key-specific hardware to achieve a cost or performance advantage.

3. DES Algorithm

DES is a 56-bit private key encryption algorithm based on a block cipher with a 64-bit block. Block ciphers deal with blocks of data as separate units, rather than operating on a continuous stream. The input is known as the plaintext while the encrypted output is called the ciphertext. The algorithm contains 16 rounds (iterations) of identical operations performed with a set of 48-bit subkeys. In typical ASIC implementations, a single circuit with registers and a feedback loop implements the iteration control and a second circuit generates each 48-bit subkey from the 56-bit key. The same datapath is used to perform both encryption and decryption. A single key is used for both encryption and decryption, but the subkeys are generated in the reverse order. Because of these properties, DES is called a symmetric private key cryptosystem. Encryption begins with an initial permutation (IP), which scrambles the 64-bit plaintext in a fixed pattern. The result of the initial permutation is sent to two 32-bit registers, called the right-half register and left-half register. These registers hold the two halves of the intermediate result through the succeeding 16 iterations.

The contents of the right-half register are permuted (Permutation E) and sent to an exclusive-OR unit along with the subkey for each iteration. Note that some bits are selected twice, allowing the 32-bit register to expand to 48 bits. The product of the exclusive-OR block is used to address a set of eight substitution memories (S-boxes). The inputs to S-box S1 are generated through the exclusive-OR of: bit 32 from the right-half register with bit 1 from the subkey, bit 1 of the right-half register and bit 2 of the subkey, etc. The subkey is directly applied to the exclusive-OR block of the S-box inputs, while some bits from the 32-bit right-half register must be selected twice to obtain the 48 bits of input for the eight S-boxes. The result of the exclusive-OR forms the inputs to eight 6-bit input to 4-bit output S-boxes. The S-box outputs are permuted (Permutation P) and fed into an exclusive-OR block along with the contents of the left-half register. The output of this block is written into a temporary register, concluding the first iteration. At the next clock cycle, the contents of the temporary register are written into the right-half register and the previous contents of the right-half register are written into the left-half register. This process repeats through 16 iterations. After the sixteenth iteration, the right-half and left-half register contents are subjected to a final permutation (IP^{-1}), which is the inverse of the initial permutation. The output of IP^{-1} is the 64-bit ciphertext.

The 56-bit key is subjected to a permutation (PC-1) and the result of the permutation is stored in two 28-bit registers, C and D. The 56-bit key is often transmitted as a 64-bit block with every eighth bit acting as a parity bit. At the beginning of each iteration the contents of these two registers are rotated by either one or two bit positions, based on the iteration count. Finally, 48-bits from the two registers undergo a permutation (PC-2) to generate the subkey for each iteration. The first 24 bits of the subkey come from the C register, while the last 24 are from the D register. Four bits from each of the registers are unused during each iteration.

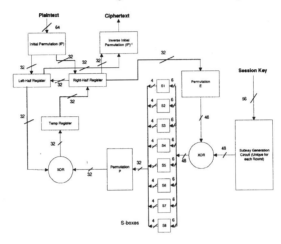

Figure 1. DES Algorithm

4. Partial Evaluation and DES

The key technique presented here for improving DES is to apply partial evaluation to the circuit design. Partial evaluation is a general-purpose optimization technique which involves transforming a function $F(A,B)$ into a new function $F_A(B)$. This new function is optimized under the assumption that some of the inputs to the function, namely the set A, are known. Such a transformation is useful if the designer can make some valuable optimizations as a result of knowing the value of A. DES has three inputs: the plaintext, the session key, and the encrypt/decrypt select line.

In many secure communications systems, the encryption key will remain constant during the complete communication session. The encryption key is sometimes referred to as a session key for this reason. In an ASIC or non-adaptive FPGA implementation, the key will be stored in a 56-bit register and subjected to the shifting and permutations to generate the 48-bit subkey for each of the 16 iterations of the algorithm. The subkey bits then become the inputs to the exclusive-OR block along with bits from the intermediate message halves.

However, because for most usage scenarios the session key is pseudo-static for a relatively long period of time, there exists an opportunity to specialize the DES circuitry. The possibility exists to generate the sixteen subkeys once and then select between them with a multiplexer. This basic approach is particularly well suited to

FPGAs because it removes several of the permutations and much of the logic and routing. These functions are trivial with ASIC technology, where routing is very flexible, but they are expensive for FPGAs.

To understand the nature of this specialization, note that an exclusive-OR gate with inputs A and B can be thought of as a 2:1 multiplexer controlled by B: it selects either input A or its inverse. In the case of DES, the inputs to the multiplexers are inverted and non-inverted versions of the intermediate message half bits. The control lines to the multiplexers are functions that reflect the subkey bits at specific iterations, i.e. hard-wired functions of the iteration count. By optimizing the circuit for a specific session key, we were able to remove the following resources:

- the 56-bit session key register
- the 48-bit subkey register
- all of the shifting circuitry
- permutation routing
- the 48 exclusive-OR gates

This hardware is replaced by 48 2:1 multiplexers that are controlled by the iteration counter. This optimization assumes that the encrypt/decrypt line is kept constant, i.e. the subkeys are for either encryption or decryption but not both. The design can encrypt and decrypt if the control lines to the multiplexers become functions of five bits. We will present results for both approaches.

To see the effects of the partial evaluation more clearly, consider the VHDL code for the two designs (Figure 2). We will refer to the full DES design, with all of the original circuitry for generating the sixteen subkeys, as the static design. The dynamic design has been specialized for a known session key. Recall that the output of the subkey generation circuitry leads to 48 exclusive-OR gates which are used to address the S-boxes. The static design has four processes for addressing the S-boxes. Two of the processes are used to implement permutations. The hardware in this portion of the static design is represented by the "ks" process (key scheduling), which performs the shifting function and the "lut_in" process (lookup table input), which implements the exclusive-OR block. By contrast, the dynamic design contains only the "lut_in" process for addressing the S-boxes. This process consists of a "case" statement, with the address bits conditionally inverted as a result of the intermediate message half bits. The selection of "Sb_inp(*index*) <= rh(*bit*)" or "Sb_inp(*index*) <= not rh(*bit*)" reflects the value of the subkey *index*-th bit. The subkey bits in turn reflect select session key bits, with selection dependent upon the iteration count.

The code shown is for a design that assumes either encrypt or decrypt but not both. For a design that would retain the option to encrypt and decrypt, an "if" statement would proceed the "case" statement, followed by an "else" "case" statement, making the S-box inputs functions of five variables instead of four. Due to the width of the buses, the processes are all abbreviated in the figure.

5. Results

A static DES design and two dynamic designs with 32-bit interfaces were placed and routed into Xilinx XC4000 series parts. Synthesis was initially done with Synopsys and the XACT tools were used for place-and-route. The first dynamic design assumed that the session key and value of the encrypt/decrypt signal are

```
PERM1 : PROCESS (KEY)              LUT_IN : PROCESS (COUNT)
BEGIN                              BEGIN
  SK_CPERM(1)<=KEY(57);             CASE COUNT IS
  SK_CPERM(2)<=KEY(49);              WHEN "0001" =>
        ...                           SB1_IN(1)<= RH(32);
END PROCESS;                          SB1_IN(2)<= RH(1);
--                                    ...
KS : PROCESS (C, ENCRYPT)           WHEN "0010" =>
-- SUBC & SUBD ARE 28-BIT VARS       SB1_IN(1)<= NOT RH(32);
BEGIN                                 SB1_IN(2)<= RH(1);
 SUBC := SK_CPERM;                    ...
 SUBD := SK_DPERM;                    SB8_IN(5)<= RH(32);
 IF (C'EVENT AND C='1') THEN          SB8_IN(6)<= NOT RH(1);
  IF ENCRYPT='1' THEN                 ...
   IF COUNT = "0000" THEN           WHEN OTHERS =>
    SKC<=SC(2 TO 28)&SC(1);           NULL;
    SKD<=SD(2 TO 28)&SD(1);         END CASE;
   ELSIF COUNT = "0001" THEN       END PROCESS;
    SKC<=SC(3 TO 28)&SC(1 TO 2);
    SKD<=SD(3 TO 28)&SD(1 TO 2);
             ...
   END IF;
   ELSE
    IF COUNT="0000" THEN
     SKC<=SUBC;
     SKD<=SUBD;
   ELSIF COUNT="0001" THEN
    SKC<=SC(28)&SC(1 TO 27);
    SKD<=SD(28)&SD(1 TO 27);
             ...
   END IF;
  END IF;
 END IF;
END PROCESS;
--
PERM2 : PROCESS (SKC, SKD)
BEGIN
 SK(1)<=SKC(14);
 SK(2)<=SKC(17);
 ...
END PROCESS;
--
LUT_IN : PROCESS (RH, SK)
BEGIN
 SB1_IN(1)<=RH(32) XOR SK(1);
 SB1_IN(2)<=RH(1) XOR SK(2);
             ...
 SB8_IN(5)<=RH(32) XOR SK(47);
 SB8_IN(6)<=RH(1) XOR SK(48);
END PROCESS;
        Static Design                      Dynamic Design
```

Figure 2. Code Comparison for Static and Dynamic Design

constant, while the second merely assumed the session key was constant. Figure 3 shows the floorplan of the static design implemented in the XC4025 part. Subkey generation hardware, both the function generators and the registers for the 56-bit session key and the 48-bit subkey, occupy much of the right side of the device. By contrast, Figure 4 shows the floorplan for the first dynamic design implemented in the same XC4025 part. There are far fewer occupied configurable logic blocks (CLBs) than for the static design, thus decreasing the burden on routing and logic resources.

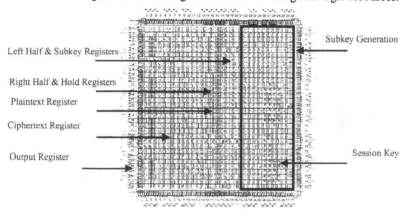

Figure 3: Floor plan of static design, with subkey scheduling circuits.

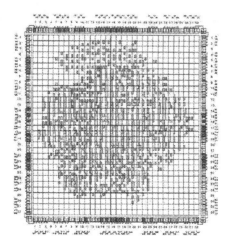

Figure 4. Floorplan of Dynamic (i. e. Partially Evaluated) Design

The resource and performance results are shown in Table 1. The static design required 938 CLBs, while the first dynamic design required 520 and the second 640. The static design was composed of 1103 FG function generators and 197 H function generators. The numbers were 702 FG function generators and 196 H function generators for the key and encrypt/decrypt specific design. The total number of function generators for the first dynamic design was 69% of that for the static design, while the number of occupied CLBs was 55% of the static design. The total number of flip-flops was 365 for the static design and 261 for the first dynamic design.

By inspecting the static design, it was found that 480 FG function generators and 48 H function generators, plus 104 flip-flops for the key and subkeys, went to the subkey generation hardware alone. By removing this circuitry the critical path was shortened and routing constraints loosened. This factor allowed the placement tools greater flexibility, thus producing a more efficient route for the remaining circuits. By reducing the number of occupied CLBs, both of the dynamic designs could be placed in the 4013 while the static design required a 4025. Results from the second dynamic design are reported for both 4013-4 and 4025-4 parts (-4 indicates a speed grade for the device). Interestingly, the performance is significantly better in the larger 4025 part. This indicates that the routing requirements are an important factor for the smaller part. This phenomenon did not occur with the first dynamic design, which uses even fewer resources: performance was nearly identical for 4013-4 and 4025-4. The clock speed for the static design was 4.9 MHz, or 2.45 MB/sec in the 4025-4 (16 clocks per result, 8 bytes per result). The clock speed for the first dynamic design was 6.6 MHz in the 4013-4, which corresponds to a rate of 3.3MB/sec. This rate is a 45% speed improvement over the static design. The clock speed for the second dynamic design was 6 MHz, corresponding to a data rate of 3MB/sec. The speed numbers are based on the Xdelay program, which is conservative: implemented speeds are possibly higher. Also note that these numbers are based on a specific part and speed grade. To compare these numbers to ASIC and software implementations see the *Applied Cryptography* text by Bruce Schneier [1]. He lists VLSI Technology's 6868 chip as the highest performance ASIC at 64 MB/sec. The fastest listed software implementation is on an HP 9000/887 with a clock speed of 125 MHz, and a DES bandwidth of approximately 1.6 MB/sec.

	Full Static Design Without Decrypt	Dynamic With Decrypt	Dynamic With Decrypt	Dynamic Design With Decrypt
Part	4025-4	4013-4	4025-4	4013-4
Occupied CLBs	938	520	640	527
FG Funct Gen	1103	702	711	711
H Funct Gen	197	196	181	181
Flip-Flops	365	261	261	261
I/O Pins	75	73	74	74
Clock Speed	4.9 MHz	6.6 MHz	6MHz	5.2MHz
Data Rate	2.5 MB/sec	3.3 MB/sec	3 MB/sec	2.6 MB/sec

Table 1. Static vs. Dynamic Designs on Xilinx XC4000 Parts

To obtain a better idea of the difference in resource usage for the three designs, they were each synthesized with Synplify from Synplicity and targeted to the ORCA 2C family from AT&T. To understand how the results should be compared considering the change in synthesizers, the previous three designs were also synthesized with Synplify and targeted to the Xilinx 4000. The smaller dynamic design has about the same percentage of resources relative to the static design at 62%, but the absolute number is much greater for all three designs. Synplify has fewer options than Synopsys and produces lower quality circuits (for these cases) in less time. The difference for the ORCA part are less than on the Xilinx part, but still significant. For instance, in the dynamic design with encrypt mode, the number of 4-input look-up tables is 80% of the number for the complete static design. This result indicates that the resource savings can be extended across FPGA architectures.

6. FPGA Technology Considerations

Key-specific hardware requires an efficient way of generating the FPGA configuration bitstreams for each of the 2^{56} possible session keys. In order for partial evaluation to be used in a deployed system, these bitstreams must reflect the effects of the changes in the VHDL code discussed in the last section without actually generating new code and running the backend tools. Obviously, the full set of bitstreams cannot be stored in memory; they must be generated in response to activating a specific session key. Fortunately, a large portion of the circuit remains the same for all keys. Keeping the same placement for those parts of the circuit reduces the changes in the bitstreams from one circuit to another. We are in the process of measuring the number of bits that must be modified as the session key changes. The next step in our research will be to generate a software tool that reprograms the affected bits directly as a function of a specific key.

While reducing the amount of circuitry needed to generate bitstreams for the desired keys is obviously important, there is another component of circuit creation that contributes to delay. In the Xilinx XC4000 parts and many other FPGAs, the entire configuration must be loaded into the device in order to cause even the smallest modification. This application, and in fact most configurable computing applications, benefits from low configuration time. In fact, configuration time will almost certainly determine whether or not the partial evaluation approach of DES is right for a given operating scenario. For instance, if this approach were used in a code-breaking machine the configuration time would have to be added to the processing time, as each key is cycled through once. In more typical communications applications, the frequency of key changes would determine the applicability of partial evaluation. Recalling that blocks are only 64 bits and considering the size of transmissions under a single key, the configuration time can actually be seen to be negligible in many cases. In those cases, the speed advantage of the partial evaluation design over the static FPGA design can be an important consideration.

Partial reconfiguration would be extremely useful for avoiding large configuration times when large portions of a circuit are truly static. In these cases, the whole configuration need not be loaded. Instead, just those portions that change are loaded. As stated earlier, the changes to the partially evaluated DES circuit are small. Section 4 showed the changes in the VHDL code necessary to go from key to key. A "case" statement invoked a function of four bits. A signal, the S-box input, was assigned one of two values depending on the key. In the circuit, such changes result in different configurations for function generators. Thus, partial evaluation of DES would greatly benefit from partial reconfiguration of the FPGA. The changes to the circuit are small enough to make it a good option, and reducing the configuration time makes the proposed way of implementing DES in an FPGA appropriate for a greater number of applications.

7. Conclusion

The advantages of partial evaluation for DES have been described. In the Xilinx XC4000 family FPGAs, one example of a dynamically generated design was shown to consume 69% of the function generators of a fully static DES design. Additionally, the dynamic design improved the speed of encryption from the 2.45 MB/sec of the static design to 3.3 MB/sec. To extend the results beyond the Xilinx part, resource utilization in the AT&T ORCA 2C part was also compared. The dynamic design consumed 80% of the 4-input look up tables that the static design consumed. The necessity for dynamic configuration generation was also discussed for this design.

Acknowledgements: This work was supported by the Defense Advanced Research Projects Agency of the United States of America, under contract DAB763-95-C-0102 and subcontract QS5200 from Sanders, a Lockheed Martin company.

References

[1] B. Schneier, *Applied Cryptography: Protocols, Algorithms, and Source Code in C*. New York: John Wiley & Sons, 1996.

[2] J. P. Gray and T. A. Kean, "Configurable Hardware: A New Paradigm for Computation," presented at Decennial Cal Tech Conference on VLSI, Pasadena, CA, 1989.

[3] M. Blaze, W. Diffie, R. L. Rivest, B. Schneier, T. Shimomura, E. Thompson, and M. Wiener, "Minimal Key Lengths for Symmetric Ciphers to Provide Adequate Commercial Security: A Report by an ad hoc Group of Cryptographers and Computer Scientists," http://www.bsa.org/policy/encryption/cryptographers, 1996.

[4] I. Goldberg and D. Wagner, "Architectural Considerations for Cryptoanalytic Hardware," http://www.cs.berkeley.edu/ ~iang/isaac/hardware, 1996.

[5] S. Son, "Feasibility Study: DES Cryptanalysis Using Reconfigurable Hardware," ftp://ftp.et.byu.edu/papers/ desfpga.ps, 1996.

[6] W. Diffie and M. E. Hellman, "Exhaustive Cryptanalysis of the NBS Data Encryption Standard," in *IEEE Computer*, vol. 10, 1977, pp. 74-84.

[7] M. J. Wiener, "Efficient DES Key Search," presented at CRYPTO '93, Santa Barbara, CA, 1993.

[8] J. Villasenor, B. Schoner, K. Chia, C. Zapata, H. J. Kim, C. Jones, S. Lansing, and W. H. Mangione-Smith "Configurable Computing Solutions for Automatic Target Recognition," in *Proceedings of IEEE Workshop on FPGAs for Custom Computing Machines*, J. Arnold and K. L. Pocek, Eds. Napa, CA, 1996, pp. 70-79.

[9] S. Singh, J. Hogg, and D. McAuley, "Expressing Dynamic Reconfiguration by Partial Evaluation," presented at Proceedings of IEEE Workshop on FPGAs for Custom Computing Machines, Napa, CA, 1996.

[10] Q. Wang and D. M. Lewis, "Automated Field-Programmable Compute Accelerator Design Using Partial Evaluation," presented at Proceedings of IEEE Workshop on FPGAs for Custom Computing Machines, Napa, CA, 1997.

Run-Time Parameterised Circuits for the Xilinx XC6200

Rob Payne

rep@dcs.ed.ac.uk

Department of Computer Science, the University of Edinburgh

Abstract. Current design tools support parameterisation of circuits, but the parameters are fixed at compile-time. In contrast, the circuits discussed in this paper fix their parameters at run-time. Run-time parameterised circuits can potentially out-perform custom VLSI hardware by optimising the FPGA circuit for a specific instance of a problem rather than for a general class of problem. This paper discusses the design of run-time parameterised circuits, and presents a study of run-time parameterised circuits for finite field operations on the Xilinx XC6200. The paper includes a comparison with implementation on a self-timed version of the XC6200 architecture, which illustrates the potential benefits of self-timing for dynamically reconfigurable systems.

1 Introduction

Parameterisation is a common part of many FPGA design tools. For instance, many graphical design tools allow the definition of a bit-sliced component of arbitrary width, such as a N-bit wide adder. More comprehensive parameterisation can often be achieved through use of a Hardware Design Language, such as VHDL.

However, the parameterisation of these designs is fixed at compile time. Often it would be useful for an application to specify the parameterisation of a hardware accelerator at run-time rather than compile-time. For example, a constant multiplier is quicker and more compact than a general multiplier circuit. If a large number of data values are to be multiplied by a constant value, then it is beneficial to configure an instance of a parameterised circuit at run-time, rather than using a general-purpose multiplier circuit. Indeed, run-time parameterised circuits could potentially out-perform a dedicated VLSI implementation, since a dedicated VLSI solution is only optimised for solving a class of problems, whilst a run-time parameterised circuit is optimised to solve a particular instance of a problem. Similar ideas are being explored by other researchers in the context of partial evaluation in functional HDLs [1, 2]

The generation of run-time parameterised circuits poses several new challenges for designers. The key issue is that the generation of the configuration must be done quickly, otherwise the speed-up of using a run-time parameterised circuit is lost in the time taken to generate the FPGA configuration. As a result, neither timing-consuming place and route algorithms, or standard delay analysis algorithms can be used in generating the run-time parameterised circuits.

As standard place and route algorithms are too complex for run-time parameterisation, the configuration has to be constructed using a regular pattern of placement and routing. However, fixed (non-parameterised) components of a run-time parameterised circuit can be placed and routed using standard design tools at compile-time. The second problem of insufficient time for delay analysis at run-time requires that the delays in a class of parameterised circuit have to be analysed at design-time. Given the parameters of a circuit it should be possible to produce a relationship to the delay of the final circuit. Again, a regular structure to the circuit configuration assists in deriving this relationship.

An alternative solution to the problem of delay analysis would be to move away from using synchronous circuits, where the global clock period has to be determined. Circuits using self-timed (asynchronous) blocks to build the run-time parameterised circuit would not require delay analysis, as there is no global clock constraint to be met.

The rest of the paper illustrates techniques for run-time parameterisation using the example of circuits for finite field operations on the Xilinx XC6200 FPGA [3]. Such circuits have application in a number of areas such as Reed-Solomon error correcting. The circuits presented can be used to implement the dynamically reconfigurable communications system proposed by Klindworth [4], where the power of the error-correcting code is varied according to the current noise conditions.

The paper is structured as follows. Section 2 discusses techniques for regular placement and routing for the Xilinx XC6200, by identifying regular subsets of the XC6200 architecture. These techniques are illustrated in the next two sections; Section 3 discusses the design of a constant multiplier over $GF(2^k)$, which illustrates a simple level of parameterisation, and Section 4 describes a circuit for division by a fixed polynomial over $GF(2^k)$, which illustrates a circuit with more complex parameterisation. Section 5 discusses how the circuits can overcome the problems of delay analysis, and describes the benefits of using a self-timed FPGA architecture for the construction of run-time parameterised circuits. Finally, Section 6 summarises the paper and looks at possible future work.

2 Run-time Mapping Architecture Subset

As discussed in the introduction, a key issue in creating run-time parameterised circuits is that the circuits must have a regular structure to allow rapid parameterisation. Irregular routing resources, i.e. those only useable by certain cells cannot be used. As a consequence, before designing the run-time parameterised circuit, a regular subset of the logic and routing resources of the FPGA being mapped to must be identified. For many FPGA architectures, several such regular subsets can be defined depending on the basic building block of the architecture used for the construction of the run-time parameterised circuit. This section identifies regular subsets of the XC6200 for run-time routing and placement.

The Xilinx XC6200 is constructed of a hierarchy of cells that are grouped into 4 × 4, 16 × 16 and 64 × 64 blocks of cells that each have their own routing resources called *flyovers* (*fast lanes* in later versions

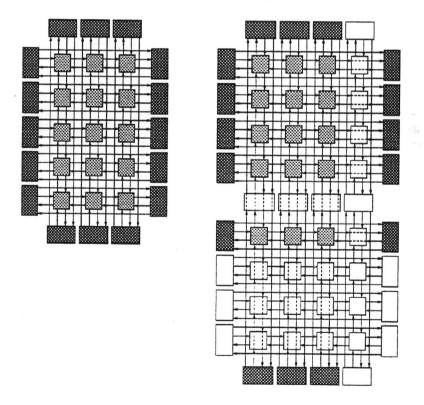

(a) Abstract 3 × 5 Block   (b) Actual 4 × 4 Blocks

Fig. 1. Converting from an Abstract $N \times M$ Block to 4 × 4 Blocks

of the datasheets [3]). Run-time parameterised circuits can be defined using any of these block sizes as the basic building block of the circuit. A problem with using any of these units as the basic building blocks, is that it does not allow the use of the higher level routing resources. For example, a run-time parameterised circuit defined in terms of the basic cells of the architecture, cannot use flyover or Magic routing resources, as these are not a regular routing resources with respect to single cells.

To allow the limited use of higher level routing resources, the concept of an abstract $N \times M$ block of cells is introduced. The abstract block just fits the run-time parameterised circuit that is to be instantiated, and except for its dimensions, it is similar to a 4×4 block of cells, consisting of a nearest neighbour array of cells with boundary multiplexors driving flyover signals that cross the block. Fig.1(a) illustrates an abstract 3×5 block. Such blocks can quickly be mapped to the 4×4 blocks in the architecture, by directly mapping cells to those in an array of 4×4 blocks large enough to contain the circuit, as illustrated in Fig.1(b). Unused cells and boundary multiplexors (unshaded in the figure) are configured so that signals pass through them unaffected and continue in the same direction. Unused cells can be rapidly configured using the XC6200 wildcard configuration interface.

The technique of defining a circuit using an abstract block can be generalised to any of the basic building blocks. For example, a circuit defined in terms of 4×4 cell blocks can utilise level 16 flyovers by mapping an abstract $N \times M$ array of 4×4 blocks to the 16×16 blocks of the architecture.

3 $GF(2^k)$ Constant Multiplier

The section describes the design of a run-time parameterised circuit for multiplying by a constant in $GF(2^k)$. The circuit illustrates the construction of a simple run-time parameterised circuit using the technique of mapping from an abstract $N \times M$ block of cells. The circuit has two basic parameters, the size of the base field, k, and the constant in $GF(2^k)$ to be multiplied by. Operations in $GF(2^k)$ can be define as operations over polynomials in $GF(2)$ modulo an irreducible polynomial. Thus, multiplication of A by a constant B in $GF(2^k)$ can be represented by the multiplication of an input polynomial $A(x)$ by a constant polynomial $B(x)$ over $GF(2^k)$, as shown in Equation 1.

$$A(x).B(x) = \sum_{i=0}^{k-1} a_i x^i.B(x) \mod H(x) \qquad (1)$$

In Equation 1, let $C_i(x)$ equal the term $x^i.B(x) \bmod H(x)$. The terms $C_i(x)$ are constant polynomials over $GF(2)$, since $B(x)$, x^i and $H(x)$ are constant. Thus, the multiplication can be expressed as a sum of the products $a_i.C_i(x)$. Creating products of a polynomial with a term in $GF(2)$ is trivial, since only the multiples one and zero are possible.

Fig.2(a) illustrates the pattern of data flow in the circuit. The input polynomial is split into its component terms a_i, which are multiplied by $C_i(x)$ to give $a_i.C_i(x)$. These terms are then summed to give the output. An example of the circuit layout as generated by the Xilinx tools is shown in Fig.2(b), for $GF(2^5)$. In the figure, the input terms are distributed to the cells using the level-4 flyovers. Additional information to assist in interpretation of the figure has been superimposed on the output of the Xilinx tools. The breakdown into 4×4 blocks is shown by the thick dotted lines. The direction of signal flow is indicated by small arrows next to the signals. Arrows with filled heads are used for local routing, whilst arrows with unfilled heads are used for level-4 flyovers.

The example configuration for $GF(2^5)$ was chosen to illustrate how circuits defined in terms of cells may not map well to 4×4 blocks. In Fig.2(b), the 4×4 block boundaries are marked by the thick dashed lines. Only the bottom left 4×4 block is completely used, the others are mostly full of unused cells. Little can be done to solve this problem without losing the regular structure of the circuit. Other $GF(2^k)$ fields give better mappings; if k is a multiple of 4, then the circuit maps exactly to 4×4 blocks.

The example also illustrates the potential savings from using run-time parameterised circuits. The constant $GF(2^k)$ multiplier is much smaller than a general $GF(2^k)$ multiplier, approximately by a factor k, and is also approximately twice as fast (See [5] for details).

4 Fixed Polynomial Division

This section discusses the construction of run-time parameterised circuit for division of a polynomial over $GF(2^k)$ by a constant polynomial over $GF(2^k)$. The circuit can be used for the generation of Reed-Solomon error correcting codes. By varying the length of the generator polynomial and the choice of field $GF(2^k)$ representation, the error correcting power of the code can be varied.

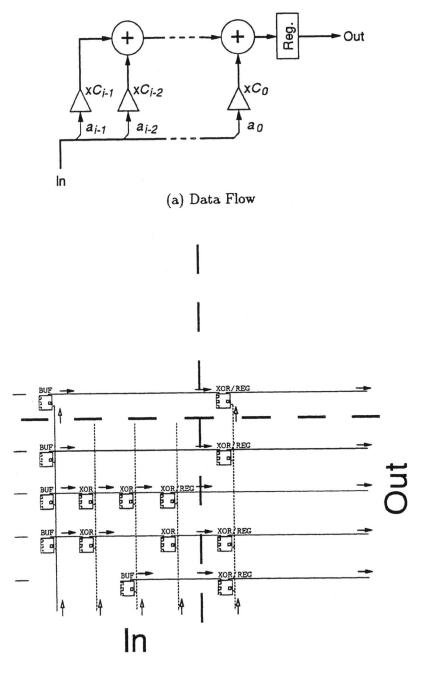

(a) Data Flow

(b) Example Circuit for $GF(2^5)$

Fig. 2. Constant Multiplier in $GF(2^k)$

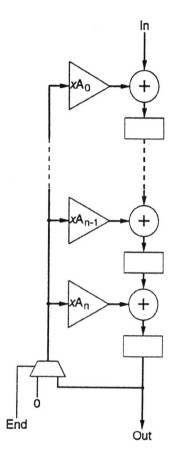

Fig. 3. Data Flow for Fixed Polynomial Division

The circuit illustrates the construction of a complex run-time parameterised circuit with a large number of parameters. The circuit's parameters are the size of the field $GF(2^k)$, the length of the input and fixed divisor polynomials, and the coefficients of the fixed divisor polynomial. In contrast to the last example, the circuit is built using 4×4 blocks of cells as the basic building blocks for the circuit, rather than individual cells.

Fig.3 illustrates the data flow for the fixed polynomial division circuit. It is similar to the linear feedback shift registers used for division by a fixed polynomial over $GF(2)$ in the generation of CRCs (Cyclic Redundancy Codes). The terms of the dividend are shifted down the shift register on the right. On each shift, a multiple of the divisor is subtracted from the value in the shift register. Adders are shown in Fig.3, since addition and subtraction are identical over $GF(2^k)$. In CRC circuits, generating the multiples is simple, since the only multiples possible in $GF(2)$ are zero and one, but in $GF(2^k)$, constant multiplier circuits must be used for each term in the divisor polynomial. When the whole dividend has been shifted in, the End flag is set, which zeros the feedback path and allows the remainder to be shifted out.

Fig.4 illustrates the configuration for a small divisor polynomial of length two with terms in $GF(2^4)$ as generated by the Xilinx Tools. The layout is similar to the layout of the data flow diagram in Fig.3. The 4×4 blocks which are used to construct the circuit are marked with the function that they perform. The Adder blocks integrate the addition and shifting stages in the data flow. The Scale blocks generate the multiples of the constant divisor using the constant $GF(2^k)$ multiplier circuits discussed in Section 3. The Tjunc block is used to route signals to the output and the feedback path. The Fifo block routes the output value to the edge of the circuit. The Corner block zeros the feedback when the remainder is shifted out (i.e. it generates the End signal in Fig.3), and is parameterised according to the length of the divisor polynomial. The Counter block generates the count used by the Corner block, to determine when to zero the feedback and is parameterised according to the length of the input polynomial.

5 Timing Analysis and Self-Timing

So far, the paper has described how the configuration of a run-time parameterised circuit can be generated rapidly using regular routing and placement. However, for synchronous circuits, once the configuration has been generated, some method is required to derive the delay of the circuit, to ensure that the circuit will meet the global clock constraint, or that the clock speed can be changed to match the delay of the circuit.

Since there is insufficient time to perform a delay analysis at run-time, a worst case delay bound for the circuit must be derived from the value of the parameters for the circuit. Thus during design, the class of run-time parameterised circuits, must be analysed to derive an equation for its worst case delay given its parameters. Circuits with a simple regular construction simplify the process of deriving this relationship. For example, it is clear from the structure of the $GF(2^k)$ constant multiplier of Fig.2(b) that the worst case delay of the class of circuit is the path from the bottom left corner to the top-right output via up to k cells configured as XOR and BUF gates.

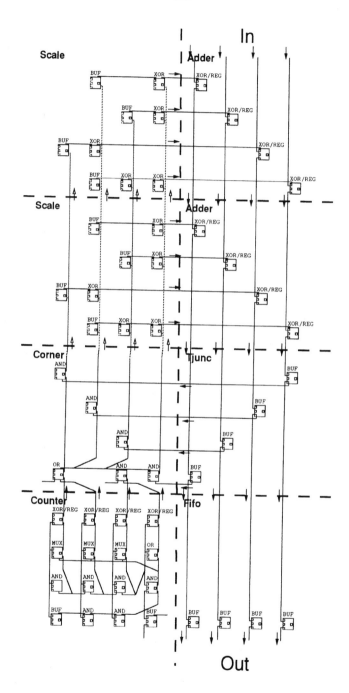

Fig. 4. Fixed Division by a Polynomial of Length Two over $GF(2^4)$

Run-time parameterised circuits constructed from a hierarchy of parameterised parts, such as the fixed divisor circuit, could have their worst case delay calculated knowing the worst case delays of the components. If each parameterised component of the circuit is a pipeline stage, then the worst case delay is simply the largest delay of each stage. If each component is not a pipeline stage, then the components have to be grouped into pipeline stages first.

An alternative approach is to avoid delay analysis by using self-timed circuits. In this case, the circuit only has to meet the local timing constraints of the self-timed protocol used rather the global clocking constraints of the synchronous protocol. This means that circuits can be rapidly constructed from self-timed parts with no need for delay analysis of the whole circuit. Other benefits of the self-timed approach, is that their is that parts of the circuit may be separated from other parts, potentially on separate chips, and the self-timed protocols will accommodate the longer routing delays. Also, self-timed circuits can be designed to exploit data-dependent delays.

The circuits described here have been simulated on a self-timed version of the Xilinx XC6200 architecture [5] based on the STACC model[6] for self-timed FPGA architectures. Self-timed FPGA architectures in general are discussed in [7]. The STACC model involves replacing the global clock of an FPGA with an array of *timing cells* that generate local timing control. The model is general enough to apply to a wide range of FPGA architectures. An important feature of the STACC model is that it preserves the structure of an FPGA's logic array, so that designs can be readily transferred between synchronous and self-timed architectures.

The run-time parameterised circuits were simulated on two versions of the self-timed XC6200, which used different delay techniques. The first model used a reconfigurable fixed delay in each timing cell to provide the local delay. This delay technique, still required the local delay of each block to be derived from its parameters to set the reconfigurable delay element in each timing cell, but did not need any further analysis, for circuits composed from these parameterised circuits.

The second delay methodology used was Current Sensing Completion Detection (CSCD) [8]. CSCD generates a completion signal for each block by monitoring the current drawn by a CMOS circuit. The circuit has completed evaluating when this current becomes negligible. CSCD can exploit data-dependent delays for each block. For example, if the same data is presented in succession then the block will complete immediately. The CSCD run-time parameterised circuits needed no delay analysis, since the delay is generated by the architecture.

The performance of run-time parameterised circuits on different architecture are presented in Table 1. The figures quoted are the delay between results being produced by the circuits. For the synchronous versions this could involve multiple clock cycles. The self-timed figures are averages as the self-timed circuits' delays are data dependent.

The self-timed fixed delay architecture is slower on smaller examples due to additional pipelining overheads (see [5]) but gains on larger examples (lower in the table) due to being able to exploit some data-dependent delays between self-timed regions. The self-timed CSCD architecture consistently out-performs the synchronous architecture by exploiting data-dependent delays within the local self-timed regions. The figures quoted for the CSCD architecture are conservative as they are based on worst case delay figure for the cells from Xilinx. Actual CSCD implementations, can exploit the actual delays of individual cells.

Table 1. Performance Figures

Circuit	Sync.	Self-Timed Fixed Delay		Self-Timed CSCD	
	/ns	/ns	diff%	/ns	diff%
General $GF(2^k)$ Multiplier	50.5	60.2	+19.2%	45.9	-9.1%
Fixed Polynomial Division	57.5	73.3	+27.5%	54.4	-5.5%
Fixed Polynomial Evaluation	162.0	213.6	+31.9%	137.0	-15.4%
General Poly. Evaluation	314.0	387.6	+23.4%	267.8	-14.5%
$GF(2^k)$ Division	330.0	323.7	-1.9%	272.1	-18.5%
Polynomial Division	98.5	94.7	-3.9%	85.1	-13.6%

6 Conclusions

This paper has described the generation of run-time parameterised circuits for the Xilinx XC6200. Such circuits can potentially out-perform a custom VLSI solution as each circuit is optimised to solve a particular problem rather than a general class of problem. Run-time parameterisation requires techniques that allow the rapid generation of configurations at run-time. The examples illustrated that for rapid placement and routing, circuits must be constructed with a regular structure. Also, for synchronous circuits, a bound on the delays in the circuit must be determined, without sufficient time for a full delay analysis. Self-timing simplifies the problem, as only local timing constraints have to be met. Results also highlighted how the self-timed architectures, especially the CSCD version, can exploit data-dependent delays.

The future development of run-time parameterised circuits will require the development of specialised design tools. Such tools will have to analyse a class of circuits rather than a specific instance of a circuit. Unlike current design tools, the output of the tool will not be an FPGA configuration, instead, the design tools will produce a software program that can generate an instance of the parameterised circuit quickly at run-time.

Acknowledgements: Thanks to Gordon Brebner and Iain Lindsay for their advice and guidance.

References

1. S. Singh, J. Hogg, and D. McAuley. Expressing Dynamic Reconfiguration by Partial Evaluation. In *FCCM96: Proceedings of IEEE Workshop on FPGAs for Custom Computing Machines*, 1996.
2. W. Luk, N. Shirazi, and P. Y. K. Cheung. Compilation Tools for Run-Time Reconfigurable Designs. In *FCCM97: Proceedings of IEEE Workshop on FPGAs for Custom Computing Machines*, 1997.
3. Xilinx. *XC6200 Datasheet*, 1996.
4. A. Klindworth. FPLD-Implementation of Computations over Finite Fields $GF(2^m)$ with Applications to Error Control Coding. In W. Moore and W. Luk, editors, *Field-Programmable Logic and Applications*, volume 975 of *Lecture Notes in Computer Science*, pages 261–271. Springer-Verlag, 1995.
5. R. E. Payne. *Self-Timed Field Programmable Gate Array Architectures*. PhD thesis, University of Edinburgh, 1997.
6. R. E. Payne. Self-Timed FPGA Systems. In W. Moore and W. Luk, editors, *Field-Programmable Logic and Applications*, volume 975 of *Lecture Notes in Computer Science*, pages 21–35. Springer-Verlag, 1995.
7. R. E. Payne. Asynchronous FPGA Architectures. *IEE Proceedings on Computers and Digital Techniques - Special Issue on Asynchronous Processors*, 143(5):282–286, September 1996.
8. M. E. Dean, D. L. Dill, and M. Horowitz. Self-Timed Logic Using Current-Sensing Completion Detection (CSCD). In *Proc. International Conf. Computer Design (ICCD)*, pages 187–191. IEEE Computer Society Press, October 1991.

Automatic Identification of Swappable Logic Units in XC6200 Circuitry

Gordon Brebner

Department of Computer Science
University of Edinburgh
Edinburgh EH9 3JZ, Scotland

Abstract. A Swappable Logic Unit (SLU) is an FPGA-based logic circuit that can be managed by a virtual hardware operating system. Since the SLU concept is still in its infancy, it is unreasonable to expect circuit design systems to be enhanced to include SLU awareness at this stage. Therefore, this paper describes a tool that can be used as a back end to any XC6200 circuit design system. It analyses XC6200 circuitry, to detect SLUs and then collect information about their external interfaces. This information is needed by an operating system so that it can manage the SLUs when they are in use.

1 Introduction

In [1], the author described the components of a simple operating system able to support two virtual hardware models using physical FPGA resources, specifically the Xilinx XC6200 family [5,7]. The central feature of the virtual hardware models was the **Swappable Logic Unit** (SLU) — an FPGA-based logic circuit capable of performing a function in terms of specified input(s), and then supplying results on specified output(s). An SLU is the virtual hardware analogue of a page or segment in a virtual memory system. In [3], the nature of SLUs was explored in more detail: their hardware environment; the software environment from which they are harnessed; their basic operational properties; and their input/output interfaces. The central theme was the need to make virtual hardware available efficiently in a normal general-purpose system environment with standard compilers, run time systems, operating systems and hardware architectures.

In this paper, the production of SLU circuitry itself is considered. The central theme is akin to that of [3] but, in this case, to investigate how SLU circuitry can be produced using normal general-purpose hardware design tools. The aim in both cases is to avoid the need to construct new, and different, tool sets, and so make it easier to entice designers from either hardware or software backgrounds into a world of hardware/software co-design. In this case, a further consideration is that it is not possible or desirable to import the notions associated with SLUs into conventional tools until the concept is stable and accepted. Therefore, in this paper, one new tool is introduced — a technology-dependent back-end useable after any particular circuit design system.

The tool analyses a piece of circuitry, and identifies one or more SLUs within it. Information about each SLU found is then stored for use by a virtual hardware operating system. The information required is explained in detail in [1] but, in summary, must include:

- a rectangular bounding box; and
- details of input/output interfaces

and, preferably, to allow efficient swapping:

- replicated areas within the SLU (to allow parallel configuration); and
- similarities between the SLU and others (to allow partial reconfiguration).

A further piece of information required by the operating system is an indication of the position dependence of the SLU — ranging from placeable anywhere on the configurable array through to requiring a fixed position on the array.

The detailed operation of such a tool, for the specific case of the XC6200 FPGA family, is discussed in this paper. A main issue thoughout is: how feasible is implementing a fully-automatic back-end approach? Unfortunately, for the XC6200 at least, full automation is not possible in some cases, and so a little human assistance is required. The essential difficulty is one of *reverse engineering*: trying to find structure that existed at an earlier stage of design but has been flattened out in the circuitry being analysed. Given that the need for human guidance is not particularly desirable, a conclusion is that the incorporation of SLU awareness at higher (technology-independent) levels of hardware design will be very useful, and indeed necessary, in the future.

Section 2 of the paper discusses the general problems associated with finding Swappable Logic Units in XC6200 circuitry described in bit stream form. The two sections following are concerned with the detail of finding input and output interfaces for SLUs. Section 3 deals with the more straightforward case of finding perimeter input/output wiring; Section 4 deals with the trickier case of finding mapped input/output registers within the circuitry. Finally, Section 5 contains some conclusions drawn from the work.

2 Finding SLUs in XC6200 circuitry

The tool makes use of the internal graph-based representation of XC6200 circuitry described in [2]. Circuitry consists of **ports**, represented by graph vertices, and wires between ports, represented by graph edges. There are four types of port: compute, input/output, multiplexor and source. The overall representation style is suitable for technologies other than the XC6200 as well, and so the overall algorithms described here are applicable in more generality. However, some of the detail parts of algorithms are technology-specific, and so would require reworking for other technologies.

Given a circuit description, the first step for the SLU finding tool is to identify unconnected sub-graphs, since each sub-graph corresponds to a potential SLU. From this stage on, each SLU candidate is treated separately so, without loss of generality, only a single SLU is considered in the description that follows. In practice, the expectation would be that a single SLU candidate is presented to the tool; however, finding multiple SLUs is straightforward using a standard algorithm for finding unconnected components of a graph.

Obtaining a minimum bounding box for the SLU is not too hard, given the coordinate information stored in the internal circuit representation. A little care is necessary however, since the exact perimeter is not necessarily marked by the coordinates of the most extreme configured cells. For example, flyover wires, driven by configured cells, may protrude beyond the apparent perimeter. Although contributing nothing to function, inclusion of such wires may be unavoidable because of the nature of the XC6200 technology, and so must be considered as being within the bounding box.

Finding input/output interface information is less straightforward. The work required depends on SLU type. In [3], three different types of SLUs are identified:

- accelerator SLUs, which have register-style interfaces;
- PE (processing element) SLUs, which have perimeter wiring interfaces; and
- system SLUs, which have interfaces that can use any available mechanisms on the host FPGA.

A fully-automatic tool must decide what type of SLU has been found in a piece of circuitry, in order to then obtain appropriate interface information.

In fact, the nature of a putative SLU might be ambiguous. A simple heuristic to help make the right decision is:

1. if there are inputs from (non-clock) global signals, or if there are fixed-position inputs or outputs from or to FPGA pins, it is a system SLU
2. if there is wiring to the perimeter of the circuit, it is a PE SLU
3. if there is at least one input and one output register in the circuit, it is an accelerator SLU
4. otherwise, it is not an SLU candidate

Of course, there is scope for making a wrong decision, and this makes a case for a small amount of checking interaction with the user. However, this can be very high level, for example, just asking the question: 'is this an accelerator, part of a parallel harness or something else?' Alternatively, the tool user can supply a hint initially, in which case the heuristic above is used to check that the circuitry is consistent with the hint.

Once the type of SLU is ascertained, appropriate interface information can be gathered. In the case of system SLUs, no further work is required, since the operating system does not need to be aware of the SLU's specific input and output arrangements — software directly accesses the SLU through the external interface of the FPGA. For PE SLUs, only perimeter interface information is sought; the existence of registers within the SLU is of no relevance. For accelerator SLUs, register interface information is sought; however, perimeter interface information is also sought. In principle, the existence of perimeter interfaces on a putative accelerator SLU is an error. However, depending on location, the tool may be able to convert such interfaces into register interfaces by extension of the circuitry under user control. Sections 3 and 4 discuss the search process for perimeter interfaces, and for register interfaces, in detail.

In general, care is necessary in order to avoid finding spurious interfaces. For example, there may be redundant perimeter wiring, perhaps as a harmless side-effect of an automatic design system. Alternatively, there may be registers present for purely internal use in sequential logic, and these might be wrongly identified as interface components. One solution to this is further interaction with the user, to check whether potential interfaces are intended to be externally visible. A different solution, and one which is to be investigated in future work, is to use typing information from the software side of the system being designed. For each SLU, there will be a description of a corresponding function, giving its parameter and result types. This information can be used to guide the search for SLU interfaces. When external guidance is used, either through interaction with the user, or from software environment information, there is scope for the tool to be proactive in extending the circuitry it is analysing.

Apart from capturing the basic SLU information, analysis of each potential SLU can find replicated patterns within, in order to assist parallel reconfiguration. The search must be done with respect to any actual parallel mechanisms available for the FPGA technology being used. In the case of the XC6200, the mechanism is 'wildcarding', which allows FPGA cell addresses to be wild-carded so that multiple cells can be written with the same configuration data simultaneously. In [6], Luk et al describe an algorithm for making optimal use of the wildcarding mechanism, which has been incorporated into their ConfigDiff tool. This involves enumerating all possible wildcarded addresses, to find which matches a required addressing pattern most closely.

Here, a slightly more abstract approach, consistent with keeping SLU information technology-independent, is adopted. The search is guided by the organisation of the XC6200 address space, but seeks patterns in the cell structure rather than patterns in the required FPGA addressing structure. It looks for increasingly large patterns replicated decreasingly often. That is, the ideal is initially to find that one cell is replicated everywhere; if unlucky, the search finds that one group of cells occurs once, i.e., there is no replication. Note, however, that the methods of Luk et al are a useful contribution to this work, but at the stage when the operating system must instantiate SLUs on the XC6200 array.

Potentially, some assistance in finding replicated patterns has been ignored. In a circuit that has significant structural replication, this fact is likely to be reflected in the original technology-independent design, for example, in a VHDL description. This is a drawback of the back-end approach, in contrast to infiltrating the SLU concept into the higher-level tools. It may also be the case that placement and routing manages to disguise replication so that the back-end tool fails to spot it in the circuit configuration data it is analysing. Note, however, that spotting replication is not *essential* to make virtual hardware work, so here there is an example of a trade-off: improved performance versus generality of design tools; such a trade-off is not uncommon, of course.

A further way of improving performance is by finding similarities between different SLUs. The purpose of this is to assist the use of partial reconfiguration. Clearly, the analysis of a single SLU, as described so far, is not capable of doing this, because it has no family context. However, the tool is able to find one class of information that can lead to partial reconfiguration: information about circuit state, when sequential logic is included. This can be exploited if the same circuitry is being used, but with different internal states.

The simplest way for identifing family similarities, in general, is for the tool to be supplied with at least two examples of family members; on the basis of these examples, it can identify internal bounding boxes of rectangular blocks that differ. This requires some connivance with the designer. A more general alternative is, for each SLU under investigation, to compare it automatically with an existing collection of SLUs, in order to place it in a family context. This seems more promising, being an automatic approach, as long as the collection is not too large. In [6], Luk et al describe their algorithm for calculating incremental configurations, which is at the heart of the ConfigDiff tool mentioned previously. This compares two configurations, and finds the incremental reconfiguration needed to move from one to another. The approach used is directly applicable here for finding SLU family resemblances.

As for finding replicated patterns, so finding similarities from the low-level circuit descriptions is likely to be losing potential benefits from higher level descriptions. These may make clear structural commonality between SLUs. However, as for replication, spotting similarities is not essential to making virtual hardware work, but involves the trade-off between performance and generality.

Finally, the tool can be used to find one further piece of information that is necessary for the operating system. This is to determine the degree of position dependence of the SLU. Essentially, this involves analysing the circuitry to find use of asymmetric features of the XC6200 technology. For example, use of length-4 flyovers or magic wires means that the SLU can only be placed at positions on a four-cell pitch, or use of input/output blocks means that the SLU can only be placed at the FPGA perimeter. Techniques for increasing freedom of placement have been suggested by Burns et al [4]. These include the use of standard two-dimensional geometrical transformations, where irregularities and asymmetries in the XC6200 architecture do not invalidate the circuitry. This offers the possibility of extra proactive action by the tool described here.

3 Finding perimeter input/output interfaces

Finding perimeter input/output interfaces is more straightforward, and has distinctly less ambiguous results, than finding register input/output interfaces. A main aspect of the work is already done when finding the minimal bounding box for the SLU. The search for interfaces then involves an exploration of the perimeter of the box. A complication arises when the circuitry is not a pure rectangle in terms of its external interfaces — this may arise from the arrangement of the configured cells or, as mentioned in Section 2, from wiring that overhangs the extreme configured cells. Since SLUs are inherently rectangular (to aid run time placement by the operating system), an optional proactive facility for the tool is in extending perimeter wires so that all interfaces are on the perimeter of the bounding box.

Given a perimeter that is uniformly rectangular, searching for interfaces involves examining the ports located on the perimeter, to check for wires entering from, or departing to, the area beyond the perimeter. By definition, these are the potential perimeter input/output interfaces. As discussed in Section 2, there is the possibility of the tool finding 'bogus' interfaces. However, the fact is that the tool fails safe: all possible interfaces are identified and, if nothing else, finding unexpected potential interfaces gives the user a warning that the circuitry is not completely self-contained.

For SLU purposes, the presence of input/output pads on the perimeter should not be an issue here, since pads cannot present on non-system SLUs. However, there are two related issues worth mentioning. First, some design tools might only produce bit stream data for circuitry that is complete with input/output pads. Thus, an optional role for this tool would be in stripping away perimeter pads, to give circuitry suitable for inclusion in a parallel harness. Second, a possible option, not particularly relevant in this context but perhaps of use elsewhere, would be automatically to add perimeter input/output pads to a circuit once its perimeter input/output interfaces had been identified.

A more interesting proactive role for the tool is to add additional perimeter registers, so that circuitry with perimeter input/output interfaces can be used as an accelerator SLU. Figure 1 shows an example of how this can be done. In Figure 1(a), a circuit with three input and five output wires at its right-

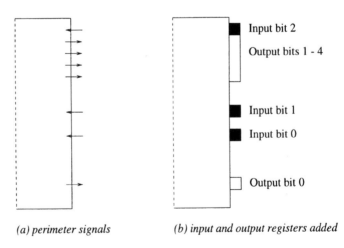

(a) perimeter signals   (b) input and output registers added

Figure 1. Adding input/output registers at right-hand side

hand edge is shown. Figure 1(b) shows extra cells added in an extra right-hand column, to give a three-bit input register and a five-bit output register. The mechanism for implementing the extra registers with interleaved bit positions hinges on the XC6200 mapped register interface [5].

As the example indicates, adding registers at the right-hand edge, or the left-hand edge, is relatively easy, since bit-mapped registers may be placed in single columns. The only issue is to ensure that the total number of input, or of output, bits is less than or equal to the bus width used for register access. Given that this can be no more than 32 bits wide, it would be necessary to use more than one register if there were 56 input bits, for example. Also, of course, an extra column has to be appended to the original circuit. In some cases, it would be feasible to avoid this, if unconfigured registers were available in all the correct positions in the rightmost column of the original circuit.

In contrast, adding registers at the top or bottom edges is not so feasible, because bit-mapped registers cannot be placed in single rows. An additional one or two bit register would be needed for each column that contains an input/output perimeter wire — this is not particularly desirable. In fact, responsibility for avoiding this issue lies with the original circuit designer. It is necessary for XC6200 SLU designers to ensure that registers (or perimeter wires that might be converted) are placed vertically, to use the memory-mapped addressing efficiently.

4 Finding mapped register input/output interfaces

Finding register input/output interfaces is distinctly harder than finding perimeter interfaces, largely because there are no geometrical hints. The situtation is complicated by the fact that registers may be present for internal reasons, unrelated to input and output. In fact, this rules out a totally automatic approach in general, because *any* register can be accessed using the bus-mapped interface, and so there is a need for extra information to identify those registers that are explicitly intended for external access.

As mentioned in the last paragraph of the previous section, the nature of registers relates to the address mapping mechanism of the XC6200. Registers must have a vertical orientation to allow reading and writing with single operations. A subsidiary heuristic, for possible use in finding registers, is that reads and writes can operate in 8, 16 and 32 bit modes, and so these are a natural size to use for registers. Unfortunately, the address mapping mechanism of the XC6200 makes it very straightforward to use register sizes of any number of bits up to 32, so this heuristic is rather weak. The more serious complication introduced by the sophistication of the mechanism is that register bits may be placed in arbitrary column positions, i.e., they need not be contiguous. Not only does this hamper searching for a single register, but it opens up the possibility of two or more registers being interleaved in the same column.

Of course, before a complete register can be identified, it is necessary to locate individual flip-flops that can form part of registers. The XC6200 architecture allows a D-type flip-flop to be configured in any cell of the array. In order to discuss ways of finding register bits, it is necessary to understand exactly how the function unit of a cell is configured. A (slightly simplified) schematic is shown in Figure 2. The two configurable multiplexors labelled 'Y2 mux' and

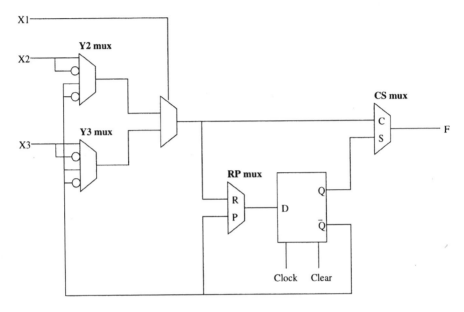

Figure 2. Schematic of XC6200 function unit

'Y3 mux' are concerned with the function computed in the cell, in terms of the input values X1, X2 and X3. The multiplexor labelled 'CS mux' determines whether the logic is combinational or sequential, as can be seen from the figure. Finally, the multiplexor labelled 'RP mux' determines whether the stored bit value is protected or not, that is, whether it can be over-written by the function result.

One preliminary check to determine whether a flip-flop is configured is to confirm that its clock input is connected up. If not, it can be assumed that the flip-flop is not in use. A further check, to rule out flip-flops that contain constant values, is to confirm that the circuit configuration data does not include writing an initial value in the flip-flop. Apart from these two checks, the search heuristics depend on whether an input flip-flop or an output flip-flop is being sought; it is assumed that no flip-flop is designed for both input and output use.

For an input flip-flop, the following are all possible features that are plausible aids to detection, and can be ascertained from the stored circuit representation:

- X1, X2 and X3 are not connected;
- Y1 and Y2 have default settings;
- CS mux is set to S;
- RP mux is set to P.

The first two make sense in that computation of a function seems pointless if the cell is being used to accommodate an input flip-flop. However, they are not totally reliable, in that default or harmless wiring might lead to X1, X2 or X3 being connected, and that default Y2 and Y3 settings might in fact be intentional settings. Moreover, the setting of the CS mux to S seems an essential at first sight. However, looking at Figure 2 more closely reveals that the flip-flop output may act as an input to combinational use of the cell. Therefore, a more apt heuristic is to check that the CS mux is set to S *or* the Y2 or Y3 mux selects the flip-flop output as an input. Note that the second condition cannot arise from the default settings for the Y2 or Y3 mux, so must result from a conscious design decision.

If this new CS/Y2/Y3 condition holds, in addition to the RP mux being set to P, it is reasonable to conclude that the flip-flop is either used for input, or holds a constant value. The latter case arises when the value is zero, and there is no explicit initialisation in the configuration data; it is the only possible cause of erroneous input flip-flop detection. A final point to note is that the RP mux being set to P is not the default setting, and so this fact on its own is usually enough to identify an input flip-flop.

For an output flip-flop, the situation is distinctly less clear cut. The following are possible features that might be of some assistance:

- X1, X2 and X3 are not not connected;
- F output is not connected;
- CS mux is set to C;
- RP mux is set to R.

Unfortunately, none of them yield definitive heuristics. The first one is weak anyway, and may be true just through default settings. The second is the most tempting, but suffers on two grounds. First, an output flip-flop might legitimately be located within a function unit that generates an output signal, sequentially or not; this also makes checking the setting of the CS mux not much help. Second, the default setting for a cell's nearest-neighbour connections selects the function unit output, so it appears that the F output is connected. The RP mux being set to R is a necessary feature, but does not help in differentiating output flip-flops from internal register flip-flops.

As this discussion may suggest, identification of output flip-flops is the weakest feature of the automatic tool. It is possible to find all of the candidates, but the set also includes all internal register flip-flops, and so user interaction is needed. In essence, this problem is inevitable, because the XC6200 bus-mapped interface makes it possible to read all configured registers. Unfortunately, there is no direct way for a circuit designer to flag registers in a way that can be back-deduced from the configuration data. One possible convention might be to mark output flip-flops specially by including initial values for non-protected flip-flops.

5 Conclusions

The tool described in this paper demonstrates that it is feasible for hardware designers to produce SLU circuitry, without the need for awareness of the existence of the SLU concept, or of the virtual hardware model within which the SLU fits. The only constraint on the design process is that the interface to the circuitry must obey certain rules, of types described in [3]. However, circuit designers are well used to interface constraints, so nothing new and unnatural is being imposed. Thus, hardware components of a hardware/software system can be produced in a familiar manner, just as software will be able to use them via function calls (see [3]), also in a familiar manner. Overall, this indicates that hardware/software co-design need not involve a special-case approach. Designers can focus on the interesting questions, like how best to partition systems between hardware and software, rather than the mechanisms for doing so.

As indicated in previous sections, the process of SLU extraction for the XC6200 cannot be entirely automated, and human assistance is needed to help with ambiguous cases. However, this allows the introduction of specialist SLU knowledge at a stage after the hardware design is complete. To illustrate the concept, the tool was tried on a large, complex circuit designed by Tom Kean, which made full use of all available XC6200 features; the test was 'blind', i.e., with no prior knowledge of the nature of the circuit. The tool correctly found the eight input registers in this circuit, and identified a collection of candidate registers that included the eight output registers.

6 References

1. Brebner, "A Virtual Hardware Operating System for the Xilinx XC6200", Proc. 6th International Workshop on Field-Programmable Logic and Applications, Springer LNCS 1142, 1996, pp.327–336.
2. Brebner, "CHASTE: a Hardware-Software Co-design Testbed for the Xilinx XC6200", Proc. 4th Reconfigurable Architecture Workshop, ITpress Verlag, 1997, pp.16–23.
3. Brebner, "The Swappable Logic Unit: a Paradigm for Virtual Hardware", Proc. 5th Annual IEEE Symposium on Custom Computing Machines, IEEE Computer Society Press 1997.
4. Burns, Donlin, Hogg, Singh and Watt, "A Dynamic Reconfiguration Run Time System", Proc. 5th Annual IEEE Symposium on Custom Computing Machines, IEEE Computer Society Press 1997.
5. Kean, Churcher and Wilkie: "XC6200 Fastmap Processor Interface," Proc. 5th International Workshop on Field Programmable Logic and Applications, Springer LNCS 975, 1995, pp.36–43.
6. Luk, Shirazi and Cheung, "Compilation Tools for Run-Time Reconfigurable Designs", Proc. 5th Annual IEEE Symposium on Custom Computing Machines, IEEE Computer Society Press 1997.
7. Xilinx Inc., XC6200 advance product specification, Xilinx 1996.

Towards an Expert System for a *priori* Estimation of Reconfiguration Latency in Dynamically Reconfigurable Logic

Patrick Lysaght

Department of Electronic and Electrical Engineering
University of Strathclyde, 204 George Street,
Glasgow G1 1XW

Abstract

Dynamically reconfigurable FPGAs are increasingly being used to speed up algorithms that would previously have been executed on computers. Reconfiguration latency is the time that elapses between a request for new circuitry to be loaded onto an FPGA and the point at which the circuitry is ready for use. It is a critical parameter in the design of dynamically reconfigurable systems used for algorithm acceleration and needs to be evaluated early in the design cycle. This paper reports on the development of expert system techniques for the *a priori* estimation of reconfiguration latency.

1. Introduction

One of the motivations for using dynamically reconfigurable logic is to reduce the execution time of algorithms that would otherwise be executed on computers. Other applications include using it to increase the effective logic density of FPGAs and its use in fault tolerant systems. A common feature of systems based on dynamically reconfigurable logic is that they are more complex because of the need to control the sequencing of different tasks at different times on to a shared resource.

The reconfiguration latency of dynamically reconfigurable logic is defined as the time that elapses between a request for new circuitry to be loaded onto an already active FPGA and the point at which the new circuitry is ready for use. (The definition extends easily to arrays of reconfigurable FPGAs: in such cases, an array as a whole is dynamically reconfigurable though its component FPGAs need not be).

Consider the class of applications in which FPGAs are used to accelerate algorithms that would otherwise be implemented in software. It is possible that the dynamic swapping of circuits on and off an FPGA will consume significant time relative to the execution time of the algorithm that is being accelerated. The technique, if used inappropriately, could potentially offset any speed-up gained in using FPGAs in the first place.

This paper proposes a new approach to estimating reconfiguration latency. For convenience, reconfiguration latency will be denoted by $t_{RECONFIG}$. There are no reported tools to assist the designer in assessing $t_{RECONFIG}$ when working with high-level design representations, such as VHDL. Instead, he must complete the automatic placement and routing of his design on the target architecture. Only after using the bit-stream generation tools provided by the vendor, can he estimate the size of files that must be loaded to effect a dynamic reconfiguration. As will be shown later, this only provides him with an initial estimate of $t_{RECONFIG}$. It should be pointed out that it is possible to work at much lower levels of design abstraction where the latency can be more accurately predicted. This advantage is sustained, however, at the cost of reduced productivity associated with the use of low-level design techniques.

The approach introduced here is based on first decomposing $t_{RECONFIG}$ into its constituent latencies and analysing each component individually. The results of the analysis are used to develop expert system techniques for the *a priori* estimation of reconfiguration latency. The expert system is intended to be used in conjunction with Dynamic Circuit Switching (DCS), a new technique for specifying and simulating dynamically reconfigurable logic [1]. The goal is to provide the designer with the means to evaluate, as early as possible in the design cycle, the potential for the deployment of dynamically reconfigurable logic. It is expected that a tool of this kind would be particularly valuable in applications where dynamically reconfigurable logic is being used for algorithm acceleration.

Two assumptions are made to contain the scope of the analysis and to simplify its presentation. It is assumed that the device controlling reconfiguration of the FPGA is a microprocessor. The cases of FPGAs that control their own reconfiguration, or FPGAs whose reconfiguration is controlled by dedicated hardware, are not addressed. The second assumption is that the system uses a single dynamically reconfigurable FPGA. The assumption implies no loss of generality since the observations made extend naturally to systems consisting of multiple reconfigurable FPGAs.

The rest of the paper is organised as follows. In section two, the reconfiguration latency is decomposed into five constituent latencies. These latencies are considered in turn in sections 3 to 7. The structure of an expert system to predict $t_{RECONFIG}$ is then described in section 8. Conclusions and suggestions for future work are presented in section 9.

2. Constituent Latencies

On first inspection, it is common to associate $t_{RECONFIG}$ with the time taken to load the new configuration bit-streams onto the array. However, after more consideration, $t_{RECONFIG}$ can be decomposed into a number of smaller, constituent delays, as shown in Fig. 1. They are shown in the order in which they will be encountered as the reconfiguration interval progresses. The relative lengths of the components, as shown, are not intended to be representative of the ratios of the actual latencies.

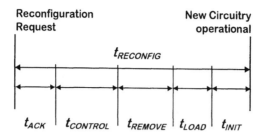

Fig. 1. Constituent Latencies

Each of the latencies is defined in turn:

- The time that elapses before the reconfiguration control subsystem recognises the reconfiguration request, (t_{ACK})

- The time it takes for the reconfiguration controller to execute the control algorithms. These include tasks such as interrupt service routines and procedures for determining what circuits to remove and what circuits to load in response to the reconfiguration request, ($t_{CONTROL}$)

- The delay associated with clearing an area on the array for the new task, i.e. removing a previous configuration by overwriting it with the default cell configuration, (t_{REMOVE})

- The time taken to load the bit-streams for the new circuitry, (t_{LOAD})

- The delay while the newly loaded circuitry is initialised, (t_{INIT})

The following sections consider each of these latencies in more detail.

3. Request Acknowledgement Latency

A request for reconfiguration can occur in a number of ways. It may be the result of a change in status of a signal external to the system, corresponding to a change in state of some "real-world" parameter. The signal may be monitored by circuits resident on the FPGA or directly by the microprocessor. In either case, it will have to be propagated to the microprocessor to trigger execution of the control algorithms. A reconfiguration request may also result from the execution of an algorithm on the FPGA or the microprocessor. For example, the microprocessor may complete one phase of an algorithm and reconfigure the FPGA to execute a part of the next phase.

In all of these cases, t_{ACK} will be measured in units of the microprocessor clock. The exact number is system dependent. At best, t_{ACK} it is unlikely to exceed a couple of clock periods. For example, if the reconfiguration interrupt is presented directly to the microprocessor, and provided it is of sufficiently high priority, it will be acknowledged after the current instruction has executed.

An interrupt or polling scheme may be used to monitor external signals or the results of computations emanating from the FPGA. Interrupt schemes are usually faster and have the advantage of allowing more efficient use of the processor. They have the disadvantage of being more complex and can have a wider range of response times according to the complexity of the priority scheme adopted. In a complex system, with many interrupts and differing levels of priority, t_{ACK} can become quite long, especially if other interrupts of higher priority are serviced before it. The possibility of pre-emption also occurs but this does not change t_{ACK} since it is measured to the point of initial acknowledgement of the reconfiguration request.

4. Control Algorithm Latency

The control algorithm latency is measured from the clock period in which the reconfiguration request is acknowledged by the microprocessor. One implication of this decision is that any interrupt service routines that may execute in response to a reconfiguration request contribute to $t_{CONTROL}$ rather than to t_{ACK}. Delays due to interrupt pre-emption also contribute to $t_{CONTROL}$. The control algorithm latency depends on the details of both the microprocessor hardware and software. There are so many possible permutations that it would be impractical to try to enumerate them all and analyse them individually. For all cases, however, it is clear that once a reconfiguration has been requested, any delay in effecting it represents wasted time. This is likely to reduce the advantage of using the FPGA for algorithm acceleration.

Once the microprocessor starts executing the control algorithm to decide what to do in response to the reconfiguration request, more permutations emerge. In simpler systems, the processor would know exactly what to do in response to a reconfiguration request and would start altering the state of the FPGA. However, several authors have already proposed systems that are more complex. Burns [2], for example, discusses the possibility of generating the configuration files for circuits at run-time. He also suggests that floorplans of circuits may be transformed by operations such as rotation before reconfiguration is performed. An algorithm for run-time compaction of FPGA tasks, to make space available for tasks being reconfigured dynamically, is presented by Diessel [3]. Brebner's "swappable logic unit" [4] will also incur run-time overheads when the FPGA is being dynamically reconfigured. These systems and algorithms are at present the subject of ongoing research and are still at the formative stage. They are indicative, nonetheless, of how systems of the future are developing.

It is possible to identify some important generalities at this early stage in the development of dynamically reconfigurable systems. The use of a microprocessor to control the dynamic reconfiguration of the FPGA gives rise to a classic example of a hardware-software, co-design problem. If the system is being designed to accelerate the execution of an algorithm, the problem is clearly a real-time one. As $t_{RECONFIG}$ increases, the performance of the system as a whole degrades. Consequently, except for the most trivial systems, the use of a real-time operating system (RTOS) would appear to be vital. The RTOS would contribute to reducing run-time overheads that would otherwise compromise overall system operating speeds. The new hardware-software co-design problem is unique in that includes the added complexity of managing dynamically reconfigurable hardware tasks. It is not clear whether conventional RTOS products, in their present form, are adequate for these new applications. They may need further specialisation or new products variant may emerge.

The combined delays associated with interrupt handling, context switching, and executing task relocation and transformation algorithms are likely to be measured in hundreds of microseconds at best, even on fast microprocessors. It is possible that these latencies will be significantly longer than those associated with task removal, loading and initialisation. Thus, optimisation of the control algorithms is likely to be as least as important as the process of task reconfiguration and initialisation.

5. Task Removal Latency

Task removal corresponds to the requirement to clear an area of the device by loading it with the default block configuration. The default logic block configuration is the state of an unconfigured block after power has been applied to an FPGA. Configuring a new task onto an active FPGA may involve:

1. The removal of no other logic

2. The removal or replacement of a single task

3. The removal or replacement of multiple tasks

For the first case, t_{REMOVE} is equal to zero. It implies that there is redundant space on the FPGA that was not in use immediately prior to the current reconfiguration.

It is likely that at least one task would need to be removed from the array to make space available for the new task. The most obvious reason to remove a task is that its area is needed in order to run a new task. When one phase of an algorithm ends, a number of related tasks may become redundant simultaneously. It may be necessary to remove all of these circuits to free sufficient area for a new task. Alternatively, the new task may not require as much area as is released by the removal of all of the tasks, so some will be removed only because they have become redundant.

One might assume that t_{REMOVE} would be equal to the sum of the values of t_{LOAD} for each of the tasks that are being removed. However, it can be substantially smaller, especially with newer FPGAs that incorporate multiple configuration contexts [5] or wildcarding mechanisms for configuration memory loading such as the Xilinx XC6216 FPGA. Since removing a task corresponds to returning its logic blocks to their default configuration state, only a single configuration vector needs to be loaded into the blocks concerned. This provides an excellent opportunity to employ broadcast configuration mechanisms to deliver the same vector to multiple cells simultaneously. Thus, the potential exists to significantly reduce t_{REMOVE}.

6. Task Load Latency

The time taken to load new circuitry, t_{LOAD}, is in turn dependent on a number of factors. It is clearly a function of the new circuitry being loaded onto the FPGA. In particular, it is a function of the complexity of that circuitry. Larger and less structured circuits typically require more time to configure onto an FPGA.

1. It is also dependent on the target FPGA architecture. The technology mapping stage will determine the representation of the circuitry in terms of the logic primitives and macros of the target FPGA. The device architecture and the mapping tools will determine the efficiency of the design representation.

2. t_{LOAD} is also a function of the reconfiguration protocol of the target FPGA. The reconfiguration protocol encompasses the structure of the reconfiguration bit-streams, including preamble, postamble and error correction data. It also includes the programming interface with the device configuration memory. The latter includes consideration of special capabilities such as the use of wildcard configuration mechanisms.

For FPGAs with random access to configuration memory, the worst-case value of t_{LOAD} cannot exceed the time taken to configure the whole FPGA. In the case of FPGAs, such as the Atmel AT6000 series, it is possible that t_{LOAD} can exceed the time to reconfigure an entire array.

7. Initialisation Latency

The last operation before the newly loaded circuitry can be operated is circuit initialisation. It is possible that no initialisation at all is required. This occurs when the circuits that are being loaded have no state information, i.e. they are purely combinatorial. Examples of this kind of circuit include the reconfigurable crossbar switch reported by Eggers [6]. Other authors have recorded the opportunity to preserve state between successive task reconfigurations, e.g. [7]. This is one example of a technique that can be used to minimise t_{INIT}.

Typically, in sequential circuits, one subset of the registers in a circuit will need to be set and another subset will need to be reset. It is also possible that certain registers will not need to be initialised at all and can safely be ignored. The problem is simpler if the state of the registers in a dynamically reconfigurable design can be assumed to automatically default to a known state, such as state zero. Then only the subset of registers to be set needs to be initialised explicitly.

With entirely static designs, the use of one or more global resets is a common method of initialising circuitry. This facility is unlikely to be available after dynamic reconfiguration. A newly loaded task could only be initialised with a global reset line if no other active task on the array was connected to the reset net. This is unlikely to be the case unless the designer consciously arranges for it to be possible.

The initial state of a circuit after dynamic reconfiguration need not be its initial state before it first executes on the FPGA. If a task is being re-started after having been interrupted, then the task of initialisation is to restore it to its state immediately before interruption. Some FPGAs allow the state of device flip-flops to be configured via the device's configuration plane. The microprocessor algorithm would be required to know what flip-flops to access and the level to which they should be set. If no use is made of broadcast addressing techniques to exploit any regularity that may exist in a design, the processor will have to perform as many write accesses to the FPGA as there are registers to be initialised explicitly. The worst-case value of t_{INIT} is clearly a linear function of the number of registers in the particular task. It is guaranteed to be less than or equal to the time needed to initialise all circuit registers independently.

8. Expert System

The original DCS system allows the designer to specify and simulate dynamically reconfigurable logic. It relies on him to enter values for task reconfiguration latencies in the form of two attributes (REMOVETIME and LOADTIME) that are associated with each dynamically reconfigurable task. At present, DCS relies on the designer to calculate these delays and provides no support for their estimation. The decomposition of $t_{RECONFIG}$ into five constituent latencies is the first step towards extending DCS by providing new computer aided design tools to help the designer in estimating system timing. The latencies, and the knowledge of what influences them, comprise the basis of a set of rules that will be used to formulate an expert system for *a priori* estimation of reconfiguration latency.

The focus on *a priori* estimation reflects the critical importance of the timing information. It is entirely possible that simple calculations performed with the assistance of the expert system will be sufficient to establish whether a particular system is practical. The designer needs to know as early as possible in the design cycle if the speed-up that he is hoping to achieve is realistic given his proposed system configuration. This will allow him the earliest opportunity to modify his approach and avoid wasted effort.

The first contribution that the expert system will make is to provide a structured framework within which the estimation of reconfiguration latencies may be undertaken. Given that the best source of design information is often the design engineer, the expert system will capitalise on this in a number of ways. It will present a structured tool for estimating $t_{RECONFIG}$ via a common interface based on the use of a set of consistent *templates*. For each dynamically reconfigurable task, the five constituent latencies have to be considered.

The proposed system will use a separate template to manage the interaction with the designer when specifying the conditions pertaining to each of the individual latencies. The template is the graphical user interface to the expert system's database of knowledge about that latency. The user will be guided through a series of choices that help him to identify the characteristics of his design that influence a particular latency. It is anticipated that the design process will be an iterative, interactive one that successively refines the precision with which the system latencies are being approximated. For example, worst-case values may be accepted as default characterisations of those latencies that are initially unknown. These may be modified directly by the user or in response to calculations based on post-processing of the circuit representations.

Designers will typically be familiar with a few microprocessors and FPGAs, rather than the whole spectrum of products. Similarly, the expert system cannot be expected to have internal knowledge of all possible FPGAs and microprocessors. Instead, it will offer an extensible template that the designer can interact with in order to specify his particular system. For example the latencies, t_{ACK} and $t_{CONTROL}$, are determined in the main by the controlling microprocessor and are thus very system specific. The designer will enter details, such as his estimates of the worst-case response time for t_{ACK} and $t_{CONTROL}$, based on his knowledge of the microprocessor type and its configuration.

The template will provide the interface to codify this information. The dialogue between the designer and the templates will be also be structured to assist different levels of user. Templates will be stored so that they can be re-used or modified as a design progresses or a new one starts. The system will thus promote re-use of valuable expertise. As chipsets with integrated microprocessors and FPGAs [5] become more commonplace, templates can be pre-supplied by the vendors and customised by the designers. For an individual chipset, the range of options diminishes and more pre-compiled information is available.

The expert system will estimate system latencies by manipulating circuit descriptions directly. This approach relies on access to structural design information. Designs consist of instances of FPGA logic primitives (cell configurations) and macros from the vendor libraries. The loading times, t_{LOAD}, for logic primitives is simply the time taken to configure a single device cell. The corresponding time for macros can be calculated easily from the design library files that specify the layout of the macros.

Ultimately all designs will be converted into primitives and macros by the technology mapping process.

This approach can be generalised by encompassing the use of macros from the Library of Parameterised Macros (LPM). The use of LPM is promoted to achieve technology independent design while exploiting optimal realisation of common circuit instances. It is proposed that the existing standard format would be extended so that each LPM macro would contain an indication of its own logic block usage and dimensions. The primitives and macros within each dynamic circuit could be enumerated along with their loading times. From this information, an estimate of t_{LOAD} would be derived for each reconfigurable task.

The estimate will take into account the circuit structure and the reconfiguration protocol. The reconfiguration protocol is sufficiently important to merit its own template since t_{REMOVE}, t_{LOAD} and t_{INIT} are so dependent on it. Further correction factors account for routing and layout efficiency. For example, the target figure for logic utilisation used by the APR tools for the Motorola MPA1000 FPGAs defaults to 40%. A simple rule-of-thumb derived from this fact is that circuits will typically require 2.5 times as many cells after APR as there are logic cell primitives in the design.

The removal of tasks is quite different from task loading. Usually, all logic blocks involved will be reconfigured to the default configuration. The software's initial estimate of t_{REMOVE} will be the worst case possibility; the product of the number of blocks involved (or all blocks on the FPGA, if no other number is available) and the time to reconfigure a single block. This will be refined if more details on the area of the circuit become available. For example, t_{REMOVE} can either be the time needed to remove one task, or the sum of the times needed to remove multiple circuits. If the expert system has established estimates of t_{LOAD} for each of the tasks already, these may be used in the calculation of t_{REMOVE}. The use of broadcast reconfiguration mechanisms will reduce t_{REMOVE}. This would require some quantification of the circuit shape such as an estimate of its bounding block area from the macros used in the circuit.

The expert system can also contribute to estimating t_{INIT}. The best source of design information is again the circuit designer. As part of the structured dialogue with the system, he may be able to classify a task as purely combinatorial or not. In the event that it is not, he may be able to estimate the number of registers that must be initialised. Failing these two options, the system can initially default to the worst case assumption, which is that every flip-flop on the array needs to be initialised.

The user may have previously estimated the area of the circuit as a percentage of the device when interacting with the t_{LOAD} template. The expert system will use this information to automatically refine its estimate of t_{INIT}. The software will also be able to prompt the user for information on the use of special device features such as additional configuration contexts or wildcarding. The user can be prompted to consider the use of special design techniques such as preserving state information between configurations if he requests advice on minimising t_{INIT}. A similar facility would be available on all templates.

Improved estimates can also be derived from automatic parsing of structural representations of the circuit. For example, after technology mapping but before APR, the number of flip-flops in a task may be extracted automatically from the netlist and used to refine t_{INIT}.

9. Conclusions

The design of dynamically reconfigurable systems constitutes an emerging class of real-time, hardware-software, co-design problems. Reconfiguration latency is an important parameter that can be decomposed into several smaller latencies associated with different phases of the reconfiguration interval. Analysis of the five constituent latencies shows that expert system techniques can be formulated to assist designers in managing the complexity of dynamically reconfigurable systems. They can be used to provide initial estimates of $t_{RECONFIG}$ that can be successfully refined by interaction between the designer and the CAD tool. Future work will involve the further refinement of the analysis and expert system techniques by their application to actual systems and real applications and the development of exemplars.

10. References

[1] Lysaght, P. and Stockwood, J. *A Simulation Tool for Dynamically Reconfigurable Field Programmable Gate Arrays*, IEEE Transactions on VLSI Systems, Sept. 1996

[2] Burns et al *A Dynamic Reconfiguration Run-time System* D. Buell and K. Pocek (eds.), Proceedings of IEEE Workshop on FCCMs, USA, 1997.

[3] Diessel, O. *Run-time compaction of FPGA designs*, ibid.

[4] Brebner, G. *A Virtual Hardware Operating System for the XC6200* In, R. W. Hartenstein, M. Glesner (eds.) Field-Programmable Logic, pp. 327-336, Springer Verlag, Germany, 1996.

[5] Faura, J., et al *Multicontext dynamic reconfiguration and real time probing on a novel mixed signal programmable device with on-chip microprocessor,* ibid.

[6] Eggers, H., Lysaght, P., Dick, H., and McGregor, G., *Fast, Reconfigurable, Crosspoint Switching in FPGAs*, In, R. W. Hartenstein, M. Glesner (eds.) Field-Programmable Logic, Springer Verlag, Germany, 1996, pp. 297-306

[7] Hutchings, B.L. & Wirthlin, M., J *Implementation Approaches for Reconfigurable Logic Applications* in Field Programmable Logic and Applications, pp.419-428, Moore, W. & Luk, W., Eds., Springer-Verlag, 1995.

Exploiting Reconfigurability Through Domain-Specific Systems

Brad L. Hutchings
hutch@ee.byu.edu

Department of Electrical and Computer Engineering
Brigham Young University
Provo, UT 84602

1 Introduction

The main thesis of this paper is that reconfigurability is best exploited through the reuse of previously-designed circuit elements. To demonstrate this point, this paper will first compare FPGAs with other computing devices especially with regard to the atomic computing primitives they implement and their usage in practical applications to show how FPGAs might fit into a general computing spectrum. Domain-specific systems are then presented as the part of the computing spectrum best situated to exploit the reconfigurability of FPGAs. Finally, the paper will discuss what kinds of CAD tools are necessary for designing domain-specific systems.

2 Atomic Computing Primitives

The term, atomic computing primitive, refers to the smallest user-controllable programming primitive of a computing device, where a computing device is broadly defined to be a programming device such as a microprocessor, Digital Signal Processor (DSP), or FPGA. Atomic computing primitives are important because they are what ultimately defines the functionality of a device; the device just being a physical realization that implements these primitives. More importantly, these primitives are the target for any compilation tools.

The vendors of programmable devices, be they vendors of processors or FPGAs, have carefully selected their respective set of computing primitives with two basic goals in mind. First and most important, primitives are selected such that they are generally applicable over the broadest possible range of desired computations. This is extremely important because general applicability is one of the prime factors that determines whether a device will ultimately command a market large enough to justify the cost of its design and fabrication. Second, where possible, vendors try to select primitives that closely parallel or are specialized in some way to support the native operations typically used to implement the computations of interest. This is primarily done to improve performance. Of course, these two goals almost always collide to create a fundamental engineering paradox. Generality provides reusability and flexibility at the cost of performance. Conversely, specialization achieves higher performance at the cost of reusability and flexibility. Thus the devices that are currently available are a result of a tradeoff between these two extremes.

For microprocessors and DSPs, the atomic computing primitives are the instructions that comprise the instruction-set. When programming a microprocessor, specific computations are implemented by selecting specific instructions from the instruction set and ordering them sequentially such that they implement the desired computation. As microprocessors are applied in all sorts of general computing problems, instruction sets directly support simple arithmetic operations such as addition, subtraction, multiplication and division and in some cases floating point operations. In addition, instruction sets also include basic control primitives that make it possible to implement any desired computation as a linear stream of instructions. Note that DSPs are similar to processors with further specializations for multiply-accumulate operations, highly iterative loops, and fast address generation. Compilers targeting microprocessors take as input a high level description and automatically generate an ordered set of these primitive instructions that implement the desired algorithm.

On the other hand, FPGA vendors are targeting the implementation of random logic: logic that is typically used to "glue" other components such as microprocessors, memory, I/O, etc., together to form a system. Random logic consists primarily of wires and gates; FPGAs provide a very flexible set of corresponding computing primitives: wires segments with programmable switches for implementing wires and function blocks that implement gates. Function blocks are essentially primitive ALUs that can be configured by the user to implement some arbitrary boolean function. These function blocks typically accept 2-5 single-bit inputs and generate a single-bit output. Wire segments are available in various fixed lengths; programmable switches are used to connect these segments to the inputs or outputs of function blocks, and to other segments when it is necessary to route to function blocks or I/O locations that are inaccessible via a single wire segment. In addition, nearly all modern FPGAs provide some dedicated logic and routing resources for accelerating basic addition, subtraction and some wide fan-in logic functions. Compiling the description of an algorithm into an FPGA implementation is an extremely complex and time-consuming process due primarily to the primitive nature of the function blocks and wire segments. Unlike conventional compilation where the only real task is to schedule instructions in time, FPGA compilers must also spatially place and route the final implementation, an extremely complex and time-consuming task that in some cases actually has no solution because of resource constraints on the FPGA.

Because each device has been carefully optimized for a specific domain, each device serves its own set of applications quite well. However, because of these optimizations, each device is also a poor fit outside of its respective domain. For example, microprocessors do an excellent job of executing operating-system software, applications such as spreadsheets and word-processing, and other general computing applications. However, microprocessors cannot be used to implement random logic because the instruction-level organization does not provide the necessary functionality, I/O and performance. FPGAs are excellent in areas where random, programmable logic is indicated. They are regularly used to implement glue logic and in some cases, complete ASICs. Indeed, because FPGAs essentially consist of uncommitted gates and wires, they can be used to implement *any* digital system subject to their size and speed –even a simple microprocessor. However, one would be ill-advised to implement a microprocessor on an FPGA unless cost and performance are irrelevant: microprocessors implemented on FPGAs will typically operate at around 1/100 the speed and cost at least 10 times that of their fixed-function counterpart.

3 The Computing Spectrum

To further demonstrate both the strengths and weaknesses of various computing devices, consider the computing spectrum depicted in Figure 1. In some sense, this spectrum ac-

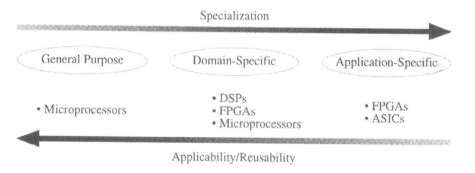

Fig. 1. Computing Spectrum for Programmable Devices

tually has less to do with actual devices and more to do with how these devices are generally used and programmed and thus depicts the range of programming methods for implementing computing applications.

General-purpose computing is defined here as *maximal reuse* computing; this is the realm of the microprocessor. When using general-purpose programming tools and devices, nearly everything is completely reusable across the widest possible range of computations. Reusable items include, of course, the device itself, but more importantly a vast library of previously developed software, including: run-time libraries, operating systems, *high-level* compilers, assemblers, linkers, etc. Moreover, much of this software is relatively portable and can be used with many different processors. All of these tools and software represent a vast amount of leverage that can be used to rapidly and cost-effectively implement complex computations. The ability of general-purpose approaches to reuse nearly everything is very important and is one of the exponential factors feeding modern computing. The primary reason for this level of reusability stems from the general applicability of microprocessor instructions across a wide range of computing tasks and the relative high performance achieved by their highly optimized implementation.

In contrast, application-specific computing is *minimal reuse* computing. This is the sphere of the fixed-function silicon device and the FPGA which replaces it in some situations. In this case, nearly every part of the application is a custom circuit element that has been carefully tailored to apply only to a single, specific computational task. In addition, the tools used in these situations (schematic capture, VHDL synthesis) are *low-level* design tools where the bulk of the design is described in low-level terms (relative to the microprocessor) with little reuse of library elements, at least relative to the microprocessor. In addition, specific architectural differences of modern devices tend to result in VHDL or other descriptions that are typically non-portable across devices from different vendors or even different families of devices from the same vendor. Recently some reuse of core circuit elements has been demonstrated, to implement PCI, etc., however, such reuse has been limited primarily to I/O elements of a design. Cores rarely find their way into the computational kernel of an algorithm and when this occurs it typically requires major modification at either the structural or physical level, or both. The key idea is that each new design starts nearly from scratch and there is little existing "software" that can be exploited across all computations. The lack of reuse occurs because of the need to completely specialize each operation to secure the highest possible performance at the lowest possible cost.

Domain-specific systems lie between these two extremes and are based on a library of highly-optimized routines that are reused across a range of applications within a given domain. These primitives are typically developed by highly skilled experts who have a thorough understanding of the device architecture and all of its otherwise arcane features. These experts thoroughly optimize these primitives such that the primitives achieve the highest performance possible on a given architecture. These primitives are then used by application developers who combine elements from this library with other general-purpose code to create comprehensive applications. Because of their highly specialized design, library elements are not generally applicable outside their area of focus –hence the name, domain-specific systems. However, the approach is still very effective as it allows programmers to rapidly develop extremely high-performance implementations of algorithms with only a limited knowledge of the device architecture. A key idea that we will return to later is that this library of highly optimized domain-specific primitives essentially serves as another set of atomic computing primitives at a higher level of abstraction that can be used by application programmers and potentially targeted by CAD tools.

Modern domain-specific programming approaches are largely a result of the existence of the DSP but these approaches have been used in the past with microprocessors as well. As its name implies, a DSP is a processor that has been customized for signal-processing applications especially with regard to cost and power. It contains several architectural features (multiply-accumulate units, zero-overhead branches, address generators, etc.) that cost effectively achieve high performance in signal-processing applications. However, these arcane features make DSPs notoriously difficult to program –as anyone who has programmed them extensively can attest– especially when you are trying to achieve high performance at low cost[1]. In order to access many of the performance-enhancing features of a DSP, programmers must resort to assembly language and manually manage arcane architectural features such as pipeline interlocks, data dependencies, etc. C compilers cannot typically be used because they deliver lower performance, don't access the special architectural features effectively and create executables that require

---

[1] which is just about all the time as DSPs are only indicated when performance and cost are important.

too many run-time resources. This has lead to a development approach where programmers create large domain-specific libraries of carefully optimized, reusable assembly code. Because these routines are reusable, substantial programming effort is spent thoroughly optimizing their performance. Complete applications are implemented by selecting routines from these libraries and "gluing" these routines together, either with more assembly code, or in some cases, C code. Much of the actual software is actually implemented by experts employed by the DSP vendors as most commercially-available DSP vendors provide extensive libraries of optimized assembly code that implement FFTs, convolutions, etc. The richness of the vendor-provided libraries is a prime consideration of any designer selecting a DSP for a new project.

4   Where FPGAs Fit in the Computing Spectrum

Perhaps the real question to ask is: which one of these three design approaches, application-specific, domain-specific, or general-purpose, can best exploit the reusability and reconfigurability of SRAM FPGAs? This section will consider this question in turn for each of the approaches already presented.

4.1   FPGAS and Application-Specific Approaches

Generally speaking, FPGAs are well-suited to application-specific approaches because their basic computing primitives were designed to be a reasonably good match for the kinds of CAD tools used in ASIC design. They provide the necessary low-level flexibility that allows the implementation of nearly any digital function. And, while FPGAs cannot achieve the same performance, cost or power levels of a custom fixed-function device, these disadvantages are often ameliorated by quicker time to market, no NRE costs, and the benefits of prototyping. However, because of the high level of customization that occurs at all levels of application-specific implementations and because of the low-level nature of the CAD tools used to develop these applications, the potential for reuse is low making it inherently difficult to exploit the reconfigurability of SRAM-based FPGAs.

4.2   FPGAs and General-Purpose Approaches

The primary problem facing FPGAs in relation to general-purpose systems is the existence of the microprocessor. Microprocessors are the best known devices for implementing the operations used by general-purpose programming tools because that is exactly what they are designed to do. The low-level flexibility of FPGAs just becomes an impediment in these kinds of systems, complicating and dramatically lengthening the compilation process and only providing marginal (if any) speedup in most situations. Although there have been some cases where FPGAs have been applied in general-purpose systems, speedup has generally been marginal and it is not at all clear whether the final system will be cost effective when all factors are taken in to account.

4.3 FPGAs and Domain-Specific Approaches

Domain-specific approaches occupy a middle-ground between general purpose and application-specific approaches and within this middle-ground FPGAs can simultaneously achieve, to some extent, both the reusability of general-purpose systems and the high performance of application-specific systems. This is done by using libraries of reusable, domain-specific elements that have been carefully optimized by *FPGA experts* to achieve high levels of application-specific performance and that have also been designed to be reusable across a selected set of applications. For example, in the domain of image processing, a domain-specific library developed for an FPGA would consist of various circuit modules that implement commonly used image-processing functions such as convolution and basic filtering operations. Applications are developed by selecting operations from this domain-specific library and combining these with other user-supplied circuitry, or in some cases, code that executes on a nearby general-purpose processor.

To achieve maximum performance, each circuit in the library is manually optimized down to the physical level (manual place and route, etc.) by FPGA-design experts and is also designed to mesh efficiently with other circuits in the library. Physical optimizations are extremely important with FPGAs; manually-placed and-routed circuits are smaller and often operate several times faster than their automatically placed-and-routed counterparts. In addition, it is also important that any transformations applied by any CAD tools used in the design process preserve the physical organization of circuits in the library. If this is not done, most if not all of the performance advantages of this approach will be lost.

Given these limitations, domain-specific approaches are an excellent way to exploit the low-level flexibility and reconfigurability of FPGAs. The low-level flexibility of an FPGA can be exploited to create a carefully optimized and specialized circuit module that implements the desired function. These circuit modules will typically be developed by FPGA-design experts and because they will be reused in many different applications, substantial design effort can be employed to achieve highly optimized circuits. Once a large library of domain-specific circuits becomes available, it forms a new level of computing primitives that hides the underlying complexity of optimized FPGA implementations. This is extremely important for FPGA-based systems because FPGA devices themselves are even more arcane than DSPs; developing high performance applications (or even low-performance applications for that matter) on FPGAs requires low-level circuit-design skills not usually associated with application programmers. By hiding the FPGA details, where possible, behind higher-level domain-specific primitives, FPGA-based systems will be come more accessible to a wider group of programmers. Finally, this approach directly exploits the reconfigurability of FPGAs through reuse of these domain-specific library elements.

Domain-specific design approaches also bring the following general advantages for FPGA-based systems.

– *Ease of use.* Because the application is described in terms of higher-level operations that are selected from a library, the application can be described at a higher level. For example, instead of describing a convolution in terms of nested for-loops, the programmer instantiates a convolution circuit similar to calling a subroutine.

– *Rapid compilation times.* Because much of the application consists of calls to pre-compiled library elements, the compilation process just involves the assembly of precompiled elements, a much simpler process than full-blown synthesis.

– *Deterministic performance.* In general FPGA synthesis systems the ultimate performance of the system is determined by the final physical layout which is unknown until the final place and route is performed by the FPGA-specific tools. Domain-specific systems have the potential to provide more deterministic performance because the library that comprises much of the application is composed of precompiled circuits that can be fully characterized prior to their final use in an application. The use of precompiled elements makes it possible to more accurately predict the ultimate performance of the application prior to compilation.

– *Processor-FPGA integration.* Domain-specific systems are also amenable to merged FPGA-processor configurations. General-purpose operations, where necessary, can be implemented on the processor while application-specific operations are implemented on the FPGAs.

5 Tools for Domain-Specific Approaches

In order to illustrate many of the problems involved in designing domain-specific systems, this section will refer to the Dynamic Instruction Set Computer (DISC) developed at BYU with which the author has direct experience [4]. DISC was implemented as a domain-specific system for image processing and was based on a library of reusable, domain-specific image-processing circuit modules, referred to as custom instructions, that achieved high-performance when compared to a single microprocessor. Each custom instruction was manually designed and laid out to achieve application-specific performance and to be reusable for a broad range of image-processing applications. Examples of some custom instructions include: mean-filter, skeletonize, histogram, threshold, etc. Although this discussion will focus exclusively on the DISC architecture, most of the concepts discussed here are likely to be encountered when developing libraries of circuit modules for domain-specific systems.

The general DISC architecture as shown in part (a) of Figure 2 consists of static and dynamic elements. The static part of the circuit is downloaded onto the FPGA just prior to execution and remains unchanged throughout operation. The dynamic elements are the custom instructions that are loaded into DISC and that implement the various image-processing functions. The internal static architecture of DISC was organized to support the notion of *relocatable hardware* and is referred to as the context, referring to the "context of the custom instructions." This context provides the necessary static circuitry that supports the execution of any custom instruction at any vertical physical location within the instruction space (horizontal positioning is fixed). The context is shown in the figure as three bundles of wires: address, control and data, wires that span the entire FPGA so that circuit modules can communicate with system resources independent of where they may be located. In DISC all library modules are carefully designed so that the physical location of all ports meshed with this context; each circuit module is designed to span the full width of the FPGA and the full vertical height of the context ensures that application-specific circuit modules will be able to access all control lines independent

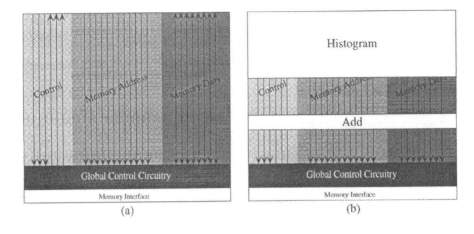

Fig. 2. Internal Architecture of DISC

of their vertical location. This is demonstrated in part (b) of the figure where the FPGA is shown with two circuit modules (histogram and add) loaded. Note that with current implementations of DISC, only one custom instruction can be active at any time.

5.1 High-Level CAD Tool Issues

The compilation process for DISC (and domain-specific systems in general) is dramatically simplified, relative to general FPGA synthesis, for two basic reasons. First, the primitives (elements from the domain-specific library) are defined at a much higher level than the simple function blocks and wire segments that comprise the actual atomic computing primitives of FPGAs. This reduces both development and compilation time. Second, these primitives have all been pre-compiled and have been fully placed, routed, debugged and characterized. In cases where only one library element is active at a time[2], the compiler can be a simple C compiler that simply schedules elements from the library. For example, in DISC all application development was performed using using a modified version of the lcc compiler [2, 1] with special code annotations for custom instructions. This simple compilation process was fast and effective. In more complex situations where more than one library element may be active at a time, for example when library elements are organized as a pipeline, something different than a 'C' compiler may be required to create the final organization. However, even in this case, the compilation phase remains relatively simple because the compiler need only organize fixed physical circuit modules into a structure that was largely predetermined when the library was designed.

---

[2] DISC was implemented in this manner.

5.2 Interfacing to the Context

Many of the problems encountered in the design and implementation of domain-specific systems are related to how circuit elements from the library interface with the context. It should be noted that all domain-specific systems are likely to contain a context –even completely static systems. The context exists in systems of this type because whenever you are dealing with a large set of reusable circuit modules, it is fundamentally important that they all communicate in a standard way. If not, then the design of the library rapidly degenerates into an exercise in frustration: each change to some circuit element may impact another by changing the communication and physical interface. At BYU, we experienced this problem directly in an earlier project, RRANN-2 [3] where we tried to carefully optimize the way each circuit module communicated within the system. RRANN-2 quickly became unmanageable and it only consisted of three configurations! Making sure that all modules communicate in a standard manner is the only way to ensure that the library will be reusable and extensible.

None of the CAD tools known to the author supported the idea of a static context that was used to communicate with various circuit modules. This made the design and test of these modules very difficult because there was no way to easily verify that the physical layout of any circuit module meshed with the context. The only thing we managed were some crude visual inspections where we created a background design that we used to make sure that the physical spacings of various ports, etc., were correct. Even so, this was a constant source of headache as we would load a newly-designed circuit module and it would sometimes not operate correctly because of a physical mismatch between ports.

What is needed for this type of system are CAD tools that allow a set of circuit modules to be associated with a physical context. At the most basic level, the CAD tool should at least provide a visual overlay of the context so that the designer can ensure that all parts of the circuit module interconnect correctly with the context. Beyond this, however, the tools should actually detect whether or not the physical layout is actually correct and warn the designer of any errors. This could be achieved relatively simply through a named association of the context with physical ports on the circuit modules. For example, if a port on a circuit element were named "addr31", it would be relatively simple to detect whether or not that port were physically located such that it would correctly connect with the physical context when loaded. In more advanced CAD tools, the tool might actually place and route the circuit module –provided that performance goals could be met– such that all ports were automatically lined up with their respective connections on the context. Implementing these features in a CAD tool would not be particularly difficult and would greatly improve the overall design process by removing a significant source of error.

5.3 Debugging

DISC turned out to be a very complex system, much more complex than first intended. This complexity was a direct result of the many different circuit modules used to implement a single application. Debugging DISC was a nightmare primarily because of all of the configurations that were used to implement a single application. General configuration (bitstream) management was problematic. When the system quit working for some reason, it was difficult to verify that we were using the correct version of some of the bitstreams (there were 30-40 of these things at any given moment). This problem could be easily avoided with simple database management tools that carefully document exactly where a bitstream came from.

The worst problems arose as a result of bugs that caused latent problems that were not evidenced until after the actual cause of the bug (some circuit element) had been removed. One of the most difficult bugs to find in DISC was when a buggy circuit module caused a piece of one of the wires that formed the static context to be removed when the module was swapped out of DISC. This caused apparently random behavior of some DISC programs because they depended upon the value of the disconnected wire and the detected value depended on the physical placement of the circuit module (which side of the wire it was on) and the previous value stored on the wire (held there by capacitance). To detect these kinds of problems, it is very important to have a kind of Layout-Versus-Schematic (LVS) tool that would enable a designer to readback the internal configuration data from the FPGA and convert it into a graphical representation that shows all physical connections so that the designer can determine what has been corrupted and correct the cause.

6 Conclusions

Domain-specific systems represent an important opportunity for FPGA-based systems. They provide a way to exploit the low-level flexibility of the device through the use of domain-specific libraries of specialized circuit elements. Moreover, reconfigurability of the FPGAs is also exploited by reusing these circuit elements to implement a wide variety of applications within the specific domain. Application areas such as image processing, signal processing, graphics, DNA analysis, and other areas are important areas that can use this approach to achieve high levels of performance at reasonable cost. Unfortunately, current CAD tools are a relatively poor match for the design needs of such systems. However, some of the additions listed in this paper should not be too difficult to implement and they would greatly ease the design of such systems.

References

1. D. A. Clark and B. L. Hutchings. Supporting FPGA microprocessors through retargetable software tools. In J. Arnold and K. L. Pocek, editors, *Proceedings of IEEE Workshop on FPGAs for Custom Computing Machines*, pages 195–203, Napa, CA, April 1996.
2. C. W. Fraser and D. Hansom. *A Retargetable C Compiler*. Benjamin/Cummings Company, 1995. ISBN 0-8053-1670-1.
3. J. D. Hadley and B. L. Hutchings. Design methodologies for partially reconfigured systems. In P. Athanas and K. L. Pocek, editors, *Proceedings of IEEE Workshop on FPGAs for Custom Computing Machines*, pages 78–84, Napa, CA, April 1995.
4. M. J. Wirthlin and B. L. Hutchings. A dynamic instruction set computer. In P. Athanas and K. L. Pocek, editors, *Proceedings of IEEE Workshop on FPGAs for Custom Computing Machines*, pages 99–107, Napa, CA, April 1995.

Technology Mapping by Binate Covering

Michal Z. SERVÍT [1] and Kang YI [2]

[1] Dept. of Computer Science and Engineering, Czech Technical University, Karlovo nám. 13, Prague, CZ 121 35 Czech Republic, e-mail: servit@cs.felk.cvut.cz

[2] Dept. of Computer Engineering, Seoul National University, Shillim-dong Kwanak-Gu, Seoul, 151-742 Korea, e-mail: yk@riact.snu.ac.kr

Abstract. Technology mapping can be viewed as the optimization problem of finding a minimum cost cover of the given Boolean network by choosing from given library of logic cells. The core of this problem in turn can be formulated as the *binate covering problem* that is NP-hard. A number of heuristics solving the binate covering problem has been proposed. However, no experimental comparison of efficiency of such techniques with respect to this specific domain has been published according to our knowledge. The aim of this paper is to analyze specific features of the binate covering formulation of the technology mapping, to propose and test a collection of heuristics using MCNC benchmarks and to select the most efficient heuristic algorithm.

1 Introduction

Logic synthesis takes the circuit specification at the functionality level and generates an implementation in terms of interconnection of logic cells from a given library. Since synthesis is a difficult task, it is often separated into two phases [3], [4]: *technology–independent optimization phase* (logic optimization), followed by a *technology mapping phase*. The optimization phase attempts to generate an optimum abstract representation of the circuit. The technology mapping phase inputs an interconnection of abstract operators - *the subject Boolean network* - and generates an interconnection of logic cells selected from a given library. Further we will assume that any abstract operator can be implemented by a single logic cell (i.e. we assume that the given network is *feasible*), and that no modifications of the given subject network are allowed while it may improve mapping results [1].

Under the above assumption we can easily decompose the technology mapping into two subsequent phases without loss of generality:

i. Construction of all (feasible) clusters: a cluster (also called match [4] or supernode [3]) is a subnetwork of the given subject network implementable by a single logic cell from the given library.

ii. The selection of clusters for a functionally correct implementation of the subject network while optimizing area, delay, or power.

The enumeration of all clusters is relatively simple problem; the hard part is selecting the optimum subset that satisfies all constraints. This is why we will focus our attention to the second phase. It is well known [3], [4] that it can be formulated as the *binate covering problem* where all constraints related to functional correctness are expressed as a product-of-sums (POS) Boolean formula while the optimization criteria are expressed as a cost function.

Because the binate covering problem is NP-hard (for formal proof see e.g. [8]) we resort to heuristic methods. A number of such methods has been proposed recently [2], [6], [13] (for review see [5]) with the aim of general applicability. On the other hand, domain specific procedures are used in many technology mappers [1], [4], [10] (for review see [3]) to solve essentially the same task. Upon to our knowledge, no exhaustive comparison of methods of both types using a technology mapping experimental data has been published so far while such an experiment has been reported in [13] for state minimization of incomplete FSMs. This is why we decided to propose a collection of heuristics and compare their efficiency using MCNC benchmarks.

The rest of the paper is organized as follows: Section 2 gives the problem formulation, Section 3 describes the proposed collection of heuristics. Section 4 shows some experimental results. In Section 5 we finally draw conclusions.

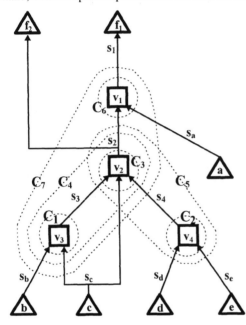

Fig. 1 Feasible subject network

2 Binate Covering Problem Formulation

Let a feasible subject network and a set of corresponding clusters be given. The subject network consists of primary input nodes, primary output nodes and internal nodes representing abstract operators (abstract gates/blocks) interconnected via (generally multiterminal) nets (Fig. 1). *Clusters* are subnetworks of the given network that can be mapped into a single-output library cell. We will say that a node or cluster *produces* a signal: for example node v_4 produces the signal s_4 and cluster C_3 produces the signal s_2. By $C(s)$ we will denote the set of all clusters producing the signal s : for example $C(s_2) = \{C_3, C_4, C_5\}$. Similarly we will say that a node or cluster *consumes* a signal: for example node v_2 consumes signals s_3, s_4, s_c and cluster C_5 consumes signals s_3, s_c, s_d, s_e.

Our task is to select an optimum subset S from the set of all clusters C that satisfies the following constraints [3, p. 114]:

• all signals consumed by primary output nodes have to be produced by clusters from S (*output constraints*), and

- if a cluster C_i is in S, each signal consumed by C_i should be produced either by a primary input node or by some cluster(s) in S *(implication constraints)*.

Output constraints ensure that each primary output is realized by a selected cluster. Implication constraints ensure that the network obtained in terms of the chosen clusters is a physically realizable circuit, i.e. no cluster selected has a dangling input.

It is well known that the above constraints can be expressed in terms of a product-of-sums (POS) Boolean formula. For each cluster C_i, a Boolean variable x_i is introduced: $x_i = TRUE \Leftrightarrow C_i \in S$, $x_i = FALSE \Leftrightarrow C_i \notin S$. The constraints produce clauses (sum terms) as follows:

- Given a primary output node v consuming a signal s, let $C(s) = \{C_{s1}, C_{s2} \dots C_{sk}\}$ be the set of clusters generating the signal s. Than at least one of them has to be selected, i.e. the *output constraint* for v can be written as follows:

$$(x_{s1} + x_{s2} + \dots + x_{sk})$$

- Given a signal s consumed by a selected cluster C_0, let $C(s) = \{C_{s1}, C_{s2} \dots C_{sk}\}$ be the set of clusters producing the signal s. Than at least one of then has to be selected, i.e. the implication constraint with respect to s and C_0 can be written as follows:

$$x_0 \Rightarrow (x_{s1} + x_{s2} + \dots + x_{sk}) \Leftrightarrow (\overline{x}_0 + x_{s1} + x_{s2} + \dots + x_{sk})$$

Take the product of all the sum clauses generated above to form a product-of-sums Boolean expression T. Any satisfying assignment to T is a solution to the binate covering problem, and, consequentially, to the original technology mapping problem.

For our example (Fig. 1), we get $T = T_1 \cdot T_2 \cdot \dots \cdot T_8$ where: $\qquad$ (1)

$$
\begin{aligned}
T_1 &= (x_6 + x_7) &&\cdots \text{ output constraint for } f_1 \\
T_2 &= (x_3 + x_4 + x_5) &&\cdots \text{ output constraint for } f_2 \\
T_3 &= (\overline{x}_3 + x_1) &&\cdots \text{ implication constraint for } C_3, s_3 \\
T_4 &= (\overline{x}_3 + x_2) &&\cdots \text{ implication constraint for } C_3, s_4 \\
T_5 &= (\overline{x}_4 + x_2) &&\cdots \text{ implication constraint for } C_4, s_4 \\
T_6 &= (\overline{x}_5 + x_1) &&\cdots \text{ implication constraint for } C_5, s_3 \\
T_7 &= (\overline{x}_6 + x_3 + x_4 + x_5) &&\cdots \text{ implication constraint for } C_6, s_2 \\
T_8 &= (\overline{x}_7 + x_2) &&\cdots \text{ implication constraint for } C_7, s_4
\end{aligned}
$$

The set $S = \{C_6, C_5, C_1\}$ is a solution to our technology mapping problem because the assignment $x_6, x_5, x_1 = TRUE$ and $x_7, x_4, x_3, x_2 = FALSE$ satisfies T.

Let a positive cost be associated with each cluster expressing area needed for the implementation of the cluster. Finding a satisfying assignment with the least sum of costs of all selected clusters - the least *total cost* - is the same as finding an optimum solution to the original technology mapping problem when the total area of cells is to be minimized. To simplify the argumentation we will assume further that all library cells have the same cost, i.e. just the number of selected clusters is to be minimized. Let us note that generalizations of this cost measure are quite straightforward [2], [5].

For our example, the solution set $S = \{C_6, C_5, C_1\}$ has the total cost equal to 3 while the solution set $S_1 = \{C_6, C_5, C_1, C_4\}$ has the total cost equal to 4. Notice that S_1 is a *redundant* solution because there exists a proper subset of S_1 that is a solution. On the other hand, S is an *irredundant* solution because no cluster can be deleted from S. It is convenient to distinguish between two types of redundancy: either S contains one or more clusters producing a signal that is not consumed by any other cluster in S *(ALPHA redundancy)* or S contains two or more clusters producing the same signal (*BETA redundancy*). Notice that C_4 is BETA redundant with respect to S_1.

A set of all output and implication constraints will be called the *basic set of constraints* because it expresses *all* conditions that any solution must obey. However, as Murgai et.al. [3, p.115] pointed out, a basic set can be *enhanced* by adding covering constraints in order to help heuristic procedures in obtaining somewhat better approximate solutions. Of course, their presence does not affect the optimum solution.

Covering constraints [1], [3] are constructed as follows. If an internal node v is covered by clusters $C_{v1}, C_{v2}...C_{vk}$, write a clause $(x_{v1} + x_{v2} +...+ x_{vk})$ indicating that at least one of the clusters $C_{v1}, C_{v2}...C_{vk}$ must be selected. Repeat it for each internal node of the network. Four covering constraints are added in our example: $(x_6 + x_7), (x_3 + x_4 + x_5 + x_7), (x_1 + x_4 + x_7), (x_2 + x_5)$.

Binate covering problems that originate in technology mapping have some domain specific features that can easily be identified (for formal proofs see [12]):

- Let PL_j denote a set of *positive* literals from T_j where T_j is from the *basic* set of constraints, i.e. PL_j represents a set of clusters producing the same signal. It holds: if $PL_j \cap PL_k \neq \Phi$ then $PL_j = PL_k$. See for example $PL_2 = PL_7 = \{x_3, x_4, x_5\}$. Notice that clusters C_3, C_4 and C_5 produce s_2.

- Let $x_i = TRUE$, $x_i \in PL_j$ and the formula T is satisfied, then all variables from $PL_j \setminus \{x_i\}$ can be set *FALSE* and T remains satisfied. Notice that it is sufficient to select just one of the clusters producing the same signal. The application of this rule allows to avoid generation of BETA redundant solutions (see Section 3.3).

- An irredundant solution can always be found by traversing a subject network from primary outputs to primary inputs provided that the network is acyclic. The application of this rule allows to avoid generation of redundant solutions (see Section 3.1).

To conclude this section: we will investigate both basic and enhanced set of constraints while minimizing the number of clusters selected.

3 Heuristic Algorithm Proposal

According to our experience [3], [5], [6], [7], a greedy algorithm combined with backtracking is appropriate to solve the binate covering problem. In this section, we will propose the core of such a procedure with several options that seems to be reasonable to investigate experimentally bearing in mind domain-specific knowledge.

The basic operation of our algorithm is reduction. The *reduction* consists of two steps: the value *TRUE* or *FALSE* is assigned to a variable x_i, and the formula T is simplified accordingly. For our example, we get by assignment $x_7 = FALSE$:

$$T = (x_6) \cdot T_2 \cdot T_3 \cdot T_4 \cdot T_5 \cdot T_6 \cdot T_7 \tag{2}$$

Notice that the above formula (2) can be satisfied only if $x_6 = TRUE$ because it contains the clause (x_6). This is why we denote the reduction that is based on existence of a clause in T that contains only one literal as *exact*. A reduction that is not an exact one is denoted as a *heuristic* reduction. For example:

$x_6 = TRUE$, $T = T_2 \cdot T_3 \cdot T_4 \cdot T_5 \cdot T_6 \cdot (x_3 + x_4 + x_5)$ is an exact reduction with respect to (2) and $x_7 = FALSE$, $T = (x_6) \cdot T_2 \cdot T_3 \cdot T_4 \cdot T_5 \cdot T_6 \cdot T_7$ is a heuristic reduction with respect to (1).

Using the notion of exact and heuristic reduction, we can formulate the core of the proposed algorithm as follows:

Step1: Perform all possible exact reductions.

Step2: If a solution was found – remove redundant clusters and STOP. If the reduced formula cannot be satisfied – backtrack to the last heuristic decision and assign the opposite value to the variable involved.

Step3: Perform a heuristic reduction and repeat Step 1.

The crucial part of this algorithm is Step 3: select a variable x_i from T (select a cluster C_i) and assign it a value (accept or reject cluster C_i). There are several possibilities: how to define a *candidate set* for variable selection, how to *select a variable* from the candidate set, and how to *assign a value* to the variable selected.

3.1 Candidate set

Two options are to be investigated: either to use the whole set of variables from T as the candidate set, or to consider only such variables that appear in positively unate clauses (i.e. clauses that do not contain an inverted literal) in the basic set of constraints (i.e. covering clauses are not taken into account!). We will call these variants as *complete* and *restricted* candidate set respectively. In our example (1), the complete candidate set is $\{x_1, x_2, ..., x_7\}$,   and the restricted candidate set is $\{x_3, x_4, ..., x_7\}$.

Notice that the restriction of the candidate set has the same effect as selecting candidate clusters by a traversal from primary outputs to primary inputs. The traversal has been used in [1] successfully. This motivated us to explore this option in our experiments.

When the restricted candidate set is used, the solution found is always irredundant provided that the subject network is acyclic [12]. Moreover, it ensures that a solution will be found without backtracking [12].

3.2 Variable selection

For variable selection, we propose to use slight modifications of score functions as defined in [2]. The score functions used are based on the following simple observation: the probability that a cluster C_i will be included in at least one minimal solution is directly proportional to the number of terms containing literal x_i, while the contribution from clause T_j containing literal x_i is inversely proportional to the number of literals contained in it. We define *weight* w_j of T_j as inversely proportional to the number of literals in T_j. It holds for our example (1): $w_1 = 1/2$, $w_2 = 1/3$, $w_3 = 1/2$ ... etc.

Direct score DS_i approximates the probability that a cluster C_i is a part of at least one minimal solution:

$$DS_i = \sum_{j=1}^{m} dw_j$$

where $dw_j = w_j$ if T_j contains literal x_i, otherwise $dw_j = 0$, and m is the number of variables in T. It holds for our example: $DS_1 = w_3 + w_6 = 1/2 + 1/2$, $DS_2 = w_4 + w_5 + w_8 = 1/2 + 1/2 + 1/2$, $DS_3 = w_2 + w_7 = 1/3 + 1/4$, ... etc.

Similar statement as above can be drawn for the probability that a cluster C_i will not be included in at least one minimal solution. *Indirect score* IS_i approximates this probability:

$$IS_i = \sum_{j=1}^{m} iw_j$$

where $iw_j = w_j$ if T_j contains literal $\bar{x}_i$, otherwise $iw_j = 0$. It holds for our example: $IS_1 = 0$, $IS_2 = 0$, $IS_3 = w_3 + w_4 = 1/2 + 1/2 = 1$, $IS_4 = w_5 = 1/2$, ... etc.

Obviously, a variable x_i with the highest value of DS_i or IS_i is a candidate for the value assignment. The difference $DIF_i = DS_i - IS_i$ can be used for the same purpose as well.

3.3 Value assignment

Three strategies of value assignment are to be investigated: Accept [2], Reject [2] and Accept-or-Reject.

Accept strategy assigns $x_i = TRUE$ to the variable x_i having the highest value of DS_i. Ties are resolved with respect to IS_i. For our example (1), the first assignment made is: $x_2 = TRUE$.

Reject strategy assigns $x_i = FALSE$ to the variable x_i having the highest value of IS_i. Ties are resolved with respect to DS_i. For our example (1), the first assignment made is: $x_3 = FALSE$.

Accept-or-Reject strategy assigns a value to the variable having the highest value of $|DIF_i|$: $x_i = TRUE$ if $DIF_i \geq 0$, otherwise $x_i = FALSE$. For our example (1), the first assignment made is: $x_2 = TRUE$.

Notice that the reject strategy provides irredundant solutions only, while the remaining ones may produce a redundant solution [2]. However, the notion of (heuristic) reduction can be *enhanced* by using the domain-specific features of technology mapping. If $x_i \in PL_j$ is set $TRUE$ (i.e. C_i is accepted), then all the remaining variables from PL_j are set $FALSE$ and the constraint formula is simplified accordingly. For our example (1), we get by assignment $x_3 = TRUE$:

$$x_4 = FALSE \text{ and } x_5 = FALSE \Rightarrow T = T_1 \cdot T_3 \cdot T_4 \cdot (x_2) \cdot (x_1) \cdot T_7 \cdot T_8$$

This allows us to avoid generation of BETA redundant solutions and to speed-up the whole process.

3.4 Orthogonal set of heuristics

It is possible to derive an orthogonal set of heuristics by combining three types of options: *basic* or *enhanced* set of constraints, *complete* or *restricted* candidate set, and *accept* or *reject* or *accept-or-reject* strategy. For obvious reasons, the *enhanced definition of reduction* is applied in all cases.

4 Experiments

MCNC benchmarks listed in Table 1 were used for experimentation. A feasible acyclic subject network was generated for each benchmark using the same technique as in [1]: a modified *misII* provided an optimized network that was decomposed by *ite* in order to obtain a feasible network. The set of clusters was derived for ACTEL ACT1 [9] library cell using the front-end of the mapper described in [1]. All computations were done on HP9000/735 (99 MHz, 41 MFLOPS, 80 Mbyte memory).

4.1 Problem size

Table 1 lists the number of nodes and clusters (*#clu*) for each benchmark as well as

| | # of nodes | | | | basic set of constr. | | | enh. set of constr. | | |
| | | | | | | after exact red. | | | after exact red. | |
name	input	output	internal	#clu	#cons	#clu	#cons	#cons	#clu	#cons
5xp1	7	10	40	60	119	69	43	159	43	89
alu2	10	6	171	208	429	118	117	600	117	198
alu4	14	8	88	92	156	5	8	244	8	9
apex2	39	3	104	117	241	28	25	345	25	40
apex6	135	99	310	450	646	372	317	956	317	512
apex7	49	37	95	113	191	42	40	286	40	60
b9	41	21	59	80	117	61	47	176	47	83
bw	5	28	65	141	380	309	128	445	128	359
c1908	33	25	91	263	526	379	186	700	186	472
c499	41	32	166	180	320	24	24	484	24	180
c5315	178	123	539	624	1247	239	177	1786	177	624
c880	60	26	173	235	441	329	176	614	176	443
clip	9	5	48	62	103	55	43	151	43	80
count	35	16	39	39	52	0	0	91	0	0
des	256	245	1308	1908	5275	2386	1136	6583	1136	2919
duke2	22	29	164	254	535	268	184	669	184	358
e64	65	65	95	125	184	31	61	279	61	125
f51m	8	8	40	43	74	5	7	114	7	9
misex1	8	7	18	31	65	31	25	83	25	43
misex2	25	18	38	47	70	13	17	108	17	21
rd73	7	3	28	37	64	14	14	92	14	19
rd84	8	4	48	53	118	29	25	166	25	49
rot	135	107	248	289	516	97	87	764	87	289
sao2	10	4	47	50	68	21	19	115	19	37
vg2	25	8	33	34	47	8	8	80	8	15
z4ml	7	4	12	14	24	6	5	36	5	14

Table 1 Sizes and complexity of benchmarks

numbers of constraints (#*cons*) for both basic and enhanced set of constraints. Additionally, problem sizes after exact reduction (after the first application of the Step 1) are listed as well.

4.2 Efficiency of heuristics

set of constr. cand. set strategy	enhanced restricted accept		enhanced restricted acc-or-rej.		enhanced restricted reject		enhanced complete accept		enhanced complete acc-or-rej.		enhanced complete reject	
	cost	time	cost	time	cost	time	cost	time	cost	time	cost	time
SUM	3918	2.29	3921	2.12	3953	4.29	3929	14.46	3927	2.60	3933	4.14

set of constr. cand. set strategy	basic restricted accept		basic restricted acc-or-rej.		basic restricted reject		basic complete accept		basic complete acc-or-rej.		basic complete reject	
	cost	time	cost	time	cost	time	cost	time	cost	time	cost	time
SUM	3933	1.83	3941	1.94	3971	3.90	3930	2.69	3932	2.94	3955	3.88

Table 2 Comparison of heuristics

Table 2 lists summarized costs of solutions (in the number of clusters used) and run times (in seconds) for all cluster selection heuristics proposed in this paper. Results of mis-pga(new)[1] (*mis-pga*) [3, p. 59], [10] and of the algorithm of Kang Yi and Chu-Shik Jhon (*ky*) [1] are presented in Table 3 in order to enable comparison between our algorithms and the best representatives of domain-specific mappers. Table 3 also contains costs of minimal solutions (*min cost*) provided by the commercial ILP solver CPLEX (for the formulation of our problem in terms of 0-1 LP see [11], [12]).

Mis-pga(new) is based on a different philosophy than the other mappers under comparison: it does not contain explicitly expressed phases of cluster enumeration and selection and it allows to modify the given subject network. This is why it is not possible to measure mis-pga cluster selection time. However, the comparison of run times of the whole technology mapping procedure is possible (see Table 3, columns *full time*). Additionally, results provided by mis-pga might be better than the minimum results provided under our assumption that no modifications of the given subject network are allowed. In reality, this happened for *vg2* and *duke2* only.

5 Conclusions and Future Work

Thorough analysis of the data collected can be summarized as follows:

- In average, the *enhanced* set of constraints provides better results than the *basic* one.
- In average, the *restricted* candidate set provides better results than the *complete* one.

---

[1] Standard options: ite_map -cn 1 -F 50 -d 4 -f 7 -rNU ; ite_map -cn 1 -d 4 -f 8 -r

- In average, the *accept* strategy provides slightly better results than the *accept-or-reject* strategy that, in turn, provides better results than the *reject* one.
- ALPHA type of redundancy appeared only if the *complete* candidate set and *accept* or *accept-or-reject* strategy is used.
- BETA type of redundancy was not observed.
- Backtracking was not employed.

The majority of above observations is in agreement with the results of theoretical analysis [12]. However, two phenomena seems to be rather confusing: we expected, based on results from [2], that the *reject* strategy will provide better quality solutions than the remaining strategies, and that a backtracking will be needed when the complete candidate set is employed. The explanation of both phenomena lies probably in specific properties of technology mapping binate covering problems.

The main achievement of our work is that the best heuristic from the orthogonal collection of heuristics under test was identified: *Enhanced* set of constraints / *Restricted* candidate set / *Accept* strategy. This heuristic provides better results (by 4% in average) than those provided by *mis-pga(new)* [10] and practically the same results as *ky* [1] in much shorter time. It should be noted that our approach is better

	enh. / res. / accept			mis-pga		ky			min
name	cost	time	full time	cost	full time	cost	time	full time	cost
5xp1	34	0.00	3.20	36	23.70	33	0.15	3.35	33
alu2	162	0.04	13.64	169	140.10	162	0.66	14.26	162
alu4	88	0.01	3.41	109	19.70	88	0.11	3.51	88
apex2	103	0.03	5.63	103	81.20	103	0.12	5.72	103
apex6	230	0.11	15.11	275	145.80	230	1.68	16.68	229
apex7	90	0.03	2.53	92	41.30	90	0.10	2.60	90
b9	56	0.01	1.21	62	36.20	55	0.06	1.26	55
bw	56	0.04	45.34	55	12.60	55	0.56	45.86	54
c1908	154	0.05	19.25	160	141.30	154	2.47	21.67	151
c499	164	0.02	8.62	164	60.90	164	1.94	10.54	164
c5315	520	0.19	49.09	558	448.40	521	4.54	53.44	519
c880	154	0.05	8.45	161	94.50	153	0.29	8.69	153
clip	42	0.02	4.32	43	46.00	41	0.04	4.34	41
count	39	0.02	0.42	39	8.60	39	0.01	0.41	39
des	1278	1.47	227.47	1285	1520.30	1280	38.76	264.76	1278
duke2	152	0.07	13.57	148	65.00	154	0.89	14.39	151
e64	94	0.03	0.63	94	1.20	94	0.31	0.91	94
f51m	39	0.01	3.11	40	31.30	39	0.05	3.15	39
misex1	16	0.01	1.11	16	8.40	16	0.04	1.14	16
misex2	38	0.00	0.40	38	3.30	38	0.06	0.46	38
rd73	28	0.01	1.51	29	13.50	28	0.03	1.53	28
rd84	47	0.01	4.31	49	59.90	47	0.04	4.34	47
rot	243	0.05	8.75	251	163.20	244	1.26	9.96	243
sao2	46	0.00	1.10	46	24.50	46	0.03	1.13	46
vg2	33	0.00	0.40	32	9.10	33	0.02	0.42	33
z4ml	12	0.01	0.52	14	13.10	12	0.01	0.52	12
SUM	3918	2.29	443.10	4068	3213.10	3919	54.23	495.04	3906

Table 3 Comparison of our best heuristic to *mis-pga(new)* and *ky*

especially in case of large and complex circuits. Moreover, no substantial improvement in the cost of solution can be expected in future because the results of the best of our heuristics are just about 3.7% above the minimum in the worst case (*bw*) and 0.3 % in average!

We are studying enhancements of the proposed techniques for the case when a collection of library cells with *different* costs is given (see e.g. Xilinx XC4000 family). Finally, we are analyzing the possibility to adopt the proposed techniques when *delay* or *power* minimization is the goal, i.e. to use them for performance oriented mapping. Preliminary results can be found in [12].

Acknowledgment

The authors would like to thank Dr. Rajeev Murgai for providing the code of mis-pga(new) and hints how to run it in the most efficient way.

The research of one of the authors (Michal Servít) was supported by COPERNICUS Coprodes CP 940453 grant.

References

[1]   Kang Yi, Chu-Shik Jhon: A New FPGA Technology Mapping Approach by Cluster Merging. In: R.W. Hartenstein, M. Glesner (Eds.): Field-Programmable Logic. Springer 1996, pp 366-370.

[2]   M. Servít, J. Zamazal: Heuristic Approach to Binate Covering Problem. EDAC'92 Proc., 1992, pp. 123-129.

[3]   R. Murgai, R. Brayton, A. Sangiovanni-Vincentelli: Logic Synthesis for Field-Programmable Gate Arrays. Kluver, 1995.

[4]   L Stok *et al.*: BooleDozer - Logic Synthesis for ASICs. IBM J. of Res. and Dev., 1995, Vol. 40, No. 4, pp. 407-430.

[5]   J. Zamazal: Boolean Satisfiability and Covering Problems - Design and Evaluation of Efficient Algorithms. PhD Dissertation, Czech Technical University, 1995.

[6]   O. Coudert, J. Madre: New Ideas for Solving Covering Problems. 31st DAC Proc., 1995.

[7]   M. Servít, J Zamazal: Decomposition and Reduction - General Problem-Solving Paradigms. VLSI Design J., 1995, Vol. 3, Nos. 3-4, pp. 359-371.

[8]   C. Papadimitriou, K. Steglitz: Combinatorial Optimization - Algorithms and Complexity. Prentice-Hall, 1982, pp. 406-409.

[9]   The Actel FPGA Data Book, Actel Inc., 1993.

[10] R. Murgai, K. Brayton, A. Sangiovanni-Vincentelli: An Improved Synthesis Algorithm for Multiplexor-based PGA's. 28th DAC Proc., 1992, pp. 380-386.

[11] Kang Yi, Soeng-Yong Ohm, Chu-Shik Jhon: An Efficient FPGA Technology Mapping Tightly Coupled with Logic Minimization. IEICE Trans. on Fundamentals of Electronics, Communications and Computer Sciences, to appear in September 1997.

[12] M. Servít, Kang Yi: Binate Covering Approach to FPGA Technology Mapping Problem. CTU Research Report under preparation.

[13] T. Kam *et al.*: Synthesis of Finite State Machines - Functional Optimization. Kluver, 1997.

VPR: A New Packing, Placement and Routing Tool for FPGA Research[1]

Vaughn Betz and Jonathan Rose

Department of Electrical and Computer Engineering, University of Toronto
Toronto, ON, Canada M5S 3G4 {vaughn, jayar}@eecg.toronto.edu

Abstract

We describe the capabilities of and algorithms used in a new FPGA CAD tool, Versatile Place and Route (VPR). In terms of minimizing routing area, VPR outperforms all published FPGA place and route tools to which we can compare. Although the algorithms used are based on previously known approaches, we present several enhancements that improve run-time and quality. We present placement and routing results on a new set of large circuits to allow future benchmark comparisons of FPGA place and route tools on circuit sizes more typical of today's industrial designs.

VPR is capable of targeting a broad range of FPGA architectures, and the source code is publicly available. It and the associated netlist translation / clustering tool VPACK have already been used in a number of research projects worldwide, and should be useful in many areas of FPGA architecture research.

1 Introduction

In FPGA research, one must typically evaluate the utility of new architectural features experimentally. That is, benchmark circuits are technology mapped, placed and routed onto the FPGA architectures of interest, and measures of the architecture's quality, such as speed or area, can then readily be extracted. Accordingly, there is considerable need for *flexible* CAD tools that can target a wide variety of FPGA architectures efficiently, and hence allow fair comparisons of the architectures.

This paper describes the Versatile Place and Route (VPR) tool, which has been designed to be flexible enough to allow comparison of many different FPGA architectures. VPR can perform placement and either global routing or combined global and detailed routing. It is publicly available from http://www.eecg.toronto.edu/~jayar/software.html.

In order to make meaningful FPGA architecture comparisons, it is essential that the CAD tools used to map circuits into each architecture are of high quality. The routing phase of VPR outperforms all previously published FPGA routers for which standard benchmarks results are available, and that the combination of VPR's placer and router outperforms all published combinations of FPGA placement and routing tools.[2]

The organization of this paper is as follows. In Section 2 we describe some of the features of VPR and the range of FPGA architectures with which it may be used. In Sections 3 and 4 we describe the placement and routing algorithms. In Section 5, we compare the number of tracks required by VPR to successfully route circuits with that required by other published tools. In Section 6 we conclude and outline some future enhancements which will be made to VPR.

---

1. This work was supported by a Walter C. Sumner Memorial Fellowship, an NSERC 1967 Scholarship, and the Information Technology Centre of Ontario.

2. Again, for those tools which have standard benchmark results to which we can compare.

2 Overview of VPR

Figure 1 outlines the VPR CAD flow. The inputs to VPR consist of a technology-mapped netlist and a text file describing the FPGA architecture. VPR can place the circuit, or a pre-existing placement can be read in. VPR can then perform either a global route or a combined global/detailed route of the placement. VPR's output consists of the placement and routing, as well as statistics useful in assessing the utility of an FPGA architecture, such as routed wirelength, track count, and maximum net length.

Some of the architectural parameters that can be specified in the architecture description file are:

- the number of logic block inputs and outputs,
- the side(s) of the logic block from which each input and output is accessible,
- the logical equivalence between various input and output pins (e.g. all LUT inputs are functionally equivalent),
- the number of I/O pads that fit into one row or one column of the FPGA, and
- the dimensions of the logic block array (e.g. 23 x 30 logic blocks).

In addition, if global routing is to be performed, one can also specify:

- the relative widths of horizontal and vertical channels, and
- the relative widths of the channels in different regions of the FPGA.

Finally, if combined global and detailed routing is to be performed, one also specifies:

- the switch block [1] architecture (i.e. how the routing tracks are interconnected),
- the number of tracks to which each logic block input pin connects (F_c [1]),
- the F_c value for logic block outputs, and
- the F_c value for I/O pads.

The current architecture description format does not allow segments that span more than one logic block to be included in the routing architecture, but we are presently adding this feature. Adding new routing architecture features to VPR is relatively easy, since VPR uses the architecture description to create a routing resource graph. Every routing track and every pin in the architecture becomes a node in this graph, and the graph edges represent the allowable connections. The router, graphics visualiza-

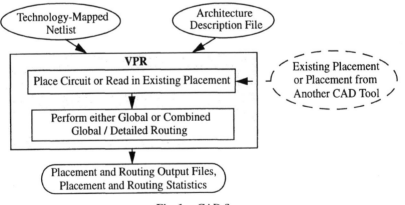

Fig. 1. CAD flow.

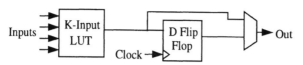

Fig. 2. Basic FPGA logic block.

tion and statistics computation routines all work only with this routing resource graph, so adding new routing architecture features only involves changing the subroutines that build this graph.

Although VPR was initially developed for island-style FPGAs [2, 3], it can also be used with row-based FPGAs [4]. VPR is not currently capable of targeting hierarchical FPGAs [5], although adding an appropriate placement cost function and the required routing resource graph building routines would allow it to target them.

Finally, VPR's built-in graphics allow interactive visualization of the placement, the routing, the available routing resources and the possible ways of interconnecting the routing resources.

2.1 The VPACK Logic Block Packer / Netlist Translator

VPACK reads in a blif format netlist of a circuit that has been technology-mapped to LUTs and flip-flops, packs the LUTs and flip flops into the desired FPGA logic block, and outputs a netlist in VPR's netlist format. VPACK can target a logic block consisting of one LUT and one FF, as shown in Figure 2, as this is a common FPGA logic element. VPACK is also capable of targeting logic blocks that contain several LUTs and several flip flops, with or without shared LUT inputs [6]. These "cluster-based" logic blocks are similar to those employed in recent FPGAs by Altera, Xilinx and Lucent Technologies.

3 Placement Algorithm

VPR uses the simulated annealing algorithm [7] for placement. We have experimented with several different cost functions, and found that what we call a *linear congestion* cost function provides the best results in a reasonable computation time [8]. The functional form of this cost function is

$$Cost = \sum_{n=1}^{N_{nets}} q(n) \left[\frac{bb_x(n)}{C_{av,x}(n)} + \frac{bb_y(n)}{C_{av,y}(n)} \right]$$

where the summation is over all the nets in the circuit. For each net, bb_x and bb_y denote the horizontal and vertical spans of its bounding box, respectively. The $q(n)$ factor compensates for the fact that the bounding box wire length model underestimates the wiring necessary to connect nets with more than three terminals, as suggested in [10]. Its value depends on the number of terminals of net n; q is 1 for nets with 3 or fewer terminals, and slowly increases to 2.79 for nets with 50 terminals. $C_{av,x}(n)$ and $C_{av,y}(n)$ are the average channel capacities (in tracks) in the x and y directions, respectively, over the bounding box of net n.

This cost function penalizes placements which require more routing in areas of the FPGA that have narrower channels. All the results in this paper, however, are obtained with FPGAs in which all channels have the same capacity. In this case C_{av} is a constant and the linear congestion cost function reduces to a bounding box cost function.

A good annealing schedule is essential to obtain high-quality solutions in a reasonable computation time with simulated annealing. We have developed a new annealing schedule which leads to very high-quality placements, and in which the annealing parameters automatically adjust to different cost functions and circuit sizes. We compute the initial temperature in a manner similar to [11]. Let N_{blocks} be the total number of logic blocks plus the number of I/O pads in a circuit. We first create a random placement of the circuit. Next we perform N_{blocks} moves (pairwise swaps) of logic blocks or I/O pads, and compute the standard deviation of the cost of these N_{blocks} different configurations. The initial temperature is set to 20 times this standard deviation, ensuring that initially virtually any move is accepted at the start of the anneal.

As in [12], the default number of moves evaluated at each temperature is $10 \cdot (N_{blocks})^{1.33}$. This default number can be overridden on the command line, however, to allow different CPU time / placement quality tradeoffs. Reducing the number of moves per temperature by a factor of 10, for example, speeds up placement by a factor of 10 and reduces final placement quality by only about 10%.

When the temperature is so high that almost any move is accepted, we are essentially moving randomly from one placement to another and little improvement in cost is obtained. Conversely, if very few moves are being accepted (due to the temperature being low and the current placement being of fairly high quality), there is also little improvement in cost. With this motivation in mind, we propose a new temperature update schedule which increases the amount of time spent at temperatures where a significant fraction of, but not all, moves are being accepted. A new temperature is computed as $T_{new} = \alpha \, T_{old}$, where the value of α depends on the fraction of attempted moves that were accepted (R_{accept}) at T_{old}, as shown in Table 1.

Table 1. Temperature update schedule.

Fraction of moves accepted (R_{accept})	α
$R_{accept} > 0.96$	0.5
$0.8 < R_{accept} \le 0.96$	0.9
$0.15 < R_{accept} \le 0.8$	0.95
$R_{accept} \le 0.15$	0.8

Finally, it was shown in [12, 13] that it is desirable to keep R_{accept} near 0.44 for as long as possible. We accomplish this by using the value of R_{accept} to control a range limiter -- only interchanges of blocks that are less than or equal to D_{limit} units apart in the x and y directions are attempted. A small value of D_{limit} increases R_{accept} by ensuring that only blocks which are close together are considered for swapping. These "local swaps" tend to result in relatively small changes in the placement cost, increasing their likelihood of acceptance. Initially, D_{limit} is set to the entire chip. Whenever the temperature is reduced, the value of D_{limit} is updated according to $D_{limit}^{new} = D_{limit}^{old} \cdot (1 - 0.44 + R_{accept}^{old})$, and then clamped to the range $1 \le D_{limit} \le$

maximum FPGA dimension. This results in D_{limit} being the size of the entire chip for the first part of the anneal, shrinking gradually during the middle stages of the anneal, and being 1 for the low-temperature part of the anneal.

Finally, the anneal is terminated when $T < 0.005 * Cost / N_{nets}$. The movement of a logic block will always affect at least one net. When the temperature is less than a small fraction of the average cost of a net, it is unlikely that any move that results in a cost increase will be accepted, so we terminate the anneal.

4 Routing Algorithm

VPR's router is based on the Pathfinder negotiated congestion algorithm [14, 8]. Basically, this algorithm initially routes each net by the shortest path it can find, regardless of any overuse of wiring segments or logic block pins that may result. One iteration of the router consists of sequentially ripping-up and re-routing (by the lowest cost path found) every net in the circuit. The cost of using a routing resource is a function of the current overuse of that resource and any overuse that occurred in prior routing iterations. By gradually increasing the cost of oversubscribed routing resources, the algorithm forces nets with alternative routes to avoid using oversubscribed resources, leaving only the net that most needs a given resource behind.

For the experimental results in this paper we set the maximum number of router iterations to 45; if a circuit has not successfully routed in a given number of tracks in 45 iterations it is assumed to be unroutable with channels of that width. To avoid overly circuitous routes and to save CPU time, we allow the routing of a net to go at most 3 channels outside the bounding box of the net terminals.

One important implementation detail deserves mention. Both the original Pathfinder algorithm and VPR's router use Dijkstra's algorithm (i.e. a maze router [15]) to connect each net. For a k terminal net, the maze router is invoked k-1 times to perform all the required connections. In the first invocation, the maze routing wavefront expands out from the net source until it reaches any one of the k-1 net sinks. The path from source to sink is now the first part of this net's routing. The maze routing wavefront is emptied, and a new wavefront expansion is started from the entire net routing found thus far. After k-1 invocations of the maze router all k terminals of the net will be connected.

Unfortunately, this approach requires considerable CPU time for high-fanout nets. High-fanout nets usually span most or all of the FPGA. Therefore, in the latter invocations of the maze router the partial routing used as the net source will be very large, and it will take a long time to expand the maze router wavefront out to the next sink. Fortunately there is a more efficient method. When a net sink is reached, add all the routing resource segments required to connect the sink and the current partial routing to the wavefront (i.e. the expansion list) with a cost of 0. Do not empty the current maze routing wavefront; just continue expanding normally. Since the new path added to the partial routing has a cost of zero, the maze router will expand around it at first. Since this new path is typically fairly small, it will take relatively little time to add this new wavefront, and the next sink will be reached much more quickly than if the entire wavefront expansion had been started from scratch. Figure 3 illustrates the difference graphically.

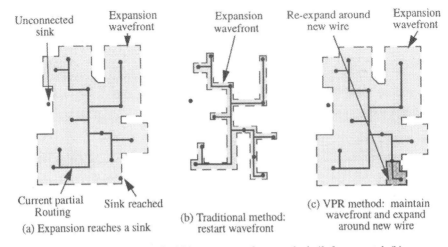

Fig. 3. When a sink is reached (a), a new wavefront can be built from scratch (b), or incrementally (c).

5 Experimental Results

The various FPGA parameters used in this section were always chosen to allow a direct comparison with previously published results. All the results in this section were obtained with a logic block consisting of a 4-input LUT plus a flip flop, as shown in Figure 2. The clock net was not routed in sequential circuits, as it is usually routed via a dedicated routing network in commercial FPGAs. Each LUT input appears on one side of the logic block, while the logic block output is accessible from both the bottom and right sides, as shown in Figure 4. Each logic block input or output can connect to any track in the adjacent channel(s) (i.e. $F_c = W$). Each wire segment can connect to three other wiring segments at channel intersections (i.e $F_s = 3$) and the switch box topology is "disjoint" -- that is, a wiring segment in track 0 connects only to other wiring segments in track 0 and so on.

5.1 Experimental Results with Input Pin Doglegs

Most previous FPGA routing results have assumed that "input pin doglegs" are possible. If the connection box between an input pin and the tracks to which it connects consists of F_c independent pass transistors controlled by F_c SRAM bits, it is possible to turn on two of these switches in order to electrically connect two tracks via the input pin. We will refer to this as an input pin dogleg. Commercial FPGAs, however,

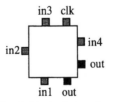

Fig. 4. Logic block pin locations.

implement the connection box from an input pin to a channel via a multiplexer, so only one track may be connected to the input pin. Using a multiplexer rather than independent pass transistors saves considerable area in the FPGA layout. As well, normally there is a buffer between a track and the connection block multiplexers to which it connects in order to improve speed; this buffer also means that input pin doglegs can not be used. Therefore, while we allow input pin doglegs in this section in order to make a fair comparison with past results, it would be best if in the future FPGA routers were tested without input pin doglegs.

In this section we compare the minimum number of tracks per channel required for a successful routing by various CAD tools on a set of 9 benchmark circuits.[1] All the results in Table 2 are obtained by routing a placement produced by Altor [16], a min-cut based placement tool. Three of the columns consist of two-step (global then detailed) routing, while the other routers perform combined global and detailed routing. VPR requires 10% fewer tracks than the second best router, and the third best router consists of VPR's global route phase plus SEGA for detailed routing.

Table 2. Tracks required to route placements generated by Altor.

Global R.	LocusRoute [17]		GBP [20]	OGC [21]	IKMB [22]	VPR	TRACER [24]	VPR
Detail R.	CGE [18]	SEGA [19]				SEGA [23]		
9symml	9	9	9	9	8	7	6	6
alu2	12	10	11	9	9	8	9	8
alu4	15	13	14	12	11	10	11	9
apex7	13	13	11	10	10	10	8	8
example2	18	17	13	12	11	10	10	9
k2	19	16	17	16	15	14	14	12
term1	10	9	10	9	8	8	7	7
too_large	13	11	12	11	10	10	9	8
vda	14	14	13	11	12	12	11	10
Total	123	112	110	99	94	89	85	77

Table 3 lists the number of tracks required to implement these benchmarks when new CAD tools are allowed to both place and route the circuits. The size column lists the number of logic blocks in each circuit. VPR uses 13% fewer tracks when it performs combined global and detailed routing than it does when SEGA is used to perform detailed routing on a a VPR-generated global route. FPR, which performs placement and global routing simultaneously in an attempt to improve routability, requires 87% more total tracks than VPR. Finally, allowing VPR to place the circuits instead of forcing it to use the Altor placements reduces the number of tracks VPR requires to route them by 40%, indicating that VPR's simulated annealing based placer is considerably better than the Altor min-cut placer.

5.2 Experimental Results Without Input Pin Doglegs

Table 4 compares the performance of VPR with that of the SPLACE/SROUTE tool

---

1. These benchmarks are available for download at http://www.eecg.toronto.edu/~lemieux/sega.

Table 3. Tracks required to place and route circuits.

Placement Global Routing Detailed Routing	Number of Logic Blocks in Circuit	FPR [25]	VPR SEGA	**VPR**
9symml	70	9	6	5
alu2	143	10	7	6
alu4	242	13	8	7
apex7	77	9	5	4
example2	120	13	5	5
k2	358	17	10	9
term1	54	8	5	5
too_large	148	11	7	6
vda	208	13	9	8
Total	--	103	62	**55**

set, which does not allow input pin doglegs. When both tools are only allowed to route an Altor-generated placement VPR requires 13% fewer tracks than SROUTE. When the tools are allowed to both place and route the circuits, VPR requires 29% fewer tracks than the SPLACE/SROUTE combination. Both VPR and SPLACE are based on simulated annealing. We believe the VPR placer outperforms SPLACE partially because it handles high-fanout nets more efficiently, allowing more moves to be evaluated in a given time, and partially because of its more efficient annealing schedule.

Table 4. Tracks required to place and route circuits with no input doglegs.

Placement Global + Detailed Route	Altor SROUTE [26]	**VPR**	SPLACE [26] SROUTE	**VPR**
9symml	7	**6**	7	5
alu2	9	**8**	8	6
alu4	12	**10**	9	7
apex7	9	**9**	6	4
example2	11	**10**	7	5
k2	15	**14**	11	9
term1	8	7	5	4
too_large	11	**9**	8	7
vda	12	**10**	10	8
Total	94	**83**	71	**55**

5.3 Experimental Results on Large Circuits

The benchmarks used in Sections 5.1 and 5.2 range in size from 54 to 358 logic blocks, and accordingly are too small to be very representative of today's FPGAs. Therefore, in this section we present experimental results for the 20 largest MCNC benchmark circuits [27], which range in size from 1047 to 8383 logic blocks. We use Flowmap [28] to technology map each circuit to 4-LUTs and flip flops, and VPACK to combine flip flops and LUTs into our basic logic block. The number of I/O pads that fit per row or column is set to 2, in line with current commercial FPGAs. Each circuit is placed and routed in the smallest square FPGA which can contain it. Input pin doglegs are not allowed. Note that three of the benchmarks, bigkey, des, and dsip, are pad-limited in the FPGA architecture assumed.

Table 5. Channel widths required to place and route 20 large benchmark circuits.

Circuit	# LBs	SEGA	VPR	Circuit	# LBs	SEGA	VPR	Circuit	# LBs	SEGA	VPR
alu4	1522	16	10	dsip	1370	9	7	s298	1931	18	7
apex2	1878	20	11	elliptic	3604	16	10	s38417	6406	10	8
apex4	1262	19	12	ex1010	4598	22	10	s38584.1	6447	12	9
bigkey	1707	9	7	ex5p	1064	16	13	seq	1750	18	11
clma	8383	≥ 24	12	frisc	3556	18	11	spla	3690	26	13
des	1591	11	7	misex3	1397	17	10	tseng	1047	9	6
diffeq	1497	10	7	pdc	4575	≥ 31	16	Total	--	≥ 331	197

Table 5 compares the number of tracks required to place and completely route circuits with VPR with the number required to place and globally route the circuits with VPR and then perform detailed routing with SEGA [23]. Table 5 also gives the size of each circuit, in terms of the number of logic blocks. The entries in the SEGA column with a ≥ sign could not be successfully routed because SEGA ran out of memory. Using SEGA to perform detailed routing on a global route generated by VPR increases the total number of tracks required to route the circuits by over 68% vs. having VPR perform the routing completely. Clearly SEGA has difficulty routing large circuits when input pin doglegs are not allowed.

To encourage other FPGA researchers to publish routing results using these larger benchmarks, we issue the following "FPGA challenge." Each time verified results which beat the previously best verified results on these benchmarks are announced, we will pay the authors $1 (sorry, $1 Cdn., not $1 U.S.) for each track by which they reduce the total number of tracks required from that of the previously best results. The technology-mapped netlists, the placements generated by VPR and the currently best routing track total are available at http://www.eecg.toronto.edu/~jayar/software.html.

6 Conclusions and Future Work

We have presented a new FPGA placement and routing tool that outperforms all such tools to which we can make direct comparisons. In addition we have presented benchmark results on much larger circuits than have typically been used to characterize academic FPGA place and route tools. We hope the next generation of FPGA CAD tools will be compared on the basis of these larger benchmarks, as they are a closer approximation of the kind of problems being mapped into today's FPGAs.

One of the main design goals for VPR was to keep the tool flexible enough to allow its use in many FPGA architectural studies. We are currently working on several improvements to VPR to further increase its utility in FPGA architecture research. In the near future VPR will support buffered and segmented routing structures, and soon after that we plan to add a timing analyzer and timing-driven routing.

References

[1] S. Brown, R. Francis, J. Rose, and Z. Vranesic, *Field-Programmable Gate Arrays*, Kluwer Academic Publishers, 1992.

[2] Xilinx Inc., *The Programmable Logic Data Book*, 1994.

[3] AT & T Inc., *ORCA Datasheet*, 1994.

[4] Actel Inc., *FPGA Data Book*, 1994.

[5] Altera Inc., *Data Book*, 1996.

[6] V. Betz and J. Rose, "Cluster-Based Logic Blocks for FPGAs: Area-Efficiency vs. Input Sharing and Size," *CICC*, 1997, pp. 551 - 554.

[7] S. Kirkpatrick, C. D. Gelatt, Jr., and M. P. Vecchi, "Optimization by Simulated Annealing," *Science*, May 13, 1983, pp. 671 - 680.

[8] V. Betz and J. Rose, "Directional Bias and Non-Uniformity in FPGA Global Routing Architectures," *ICCAD*, 1996, pp. 652 - 659.

[9] V. Betz and J. Rose, "On Biased and Non-Uniform Global Routing Architectures and CAD Tools for FPGAs," *CSRI Tech. Rep. #358*, Dept. of ECE, University of Toronto, 1996.

[10] C. E. Cheng, "RISA: Accurate and Efficient Placement Routability Modeling," *DAC*, 1994, pp. 690 - 695.

[11] M. Huang, F. Romeo, and A. Sangiovanni-Vincentelli, "An Efficient General Cooling Schedule for Simulated Annealing," *ICCAD*, 1986, pp. 381 - 384.

[12] W. Swartz and C. Sechen, "New Algorithms for the Placement and Routing of Macro Cells," *ICCAD*, 1990, pp. 336 - 339.

[13] J. Lam and J. Delosme, "Performance of a New Annealing Schedule," *DAC*, 1988, pp. 306 - 311.

[14] C. Ebeling, L. McMurchie, S. A. Hauck and S. Burns, "Placement and Routing Tools for the Triptych FPGA," *IEEE Trans. on VLSI*, Dec. 1995, pp. 473 - 482.

[15] C. Y. Lee, "An Algorithm for Path Connections and its Applications, "*IRE Trans. Electron. Comput.*, Vol. EC=10, 1961, pp. 346 - 365.

[16] J. S. Rose, W. M. Snelgrove, Z. G. Vranesic, "ALTOR: An Automatic Standard Cell Layout Program," *Canadian Conf. on VLSI*, 1985, pp. 169 - 173.

[17] J. S. Rose, "Parallel Global Routing for Standard Cells," *IEEE Trans. on CAD*, Oct. 1990, pp. 1085 - 1095.

[18] S. Brown, J. Rose, Z. G. Vranesic, "A Detailed Router for Field-Programmable Gate Arrays," *IEEE Trans. on CAD*, May 1992, pp. 620 - 628.

[19] G. Lemieux, S. Brown, "A Detailed Router for Allocating Wire Segments in FPGAs," *ACM/SIGDA Physical Design Workshop,* 1993, pp. 215 - 226.

[20] Y.-L. Wu, M. Marek-Sadowska, "An Efficient Router for 2-D Field-Programmable Gate Arrays," *EDAC*, 1994, pp. 412 - 416.

[21] Y.-L. Wu, M. Marek-Sadowska, "Orthogonal Greedy Coupling -- A New Optimization Approach to 2-D FPGA Routing," *DAC*, 1995, pp. 568 - 573.

[22] M. J. Alexander, G. Robins, "New Performance-Driven FPGA Routing Algorithms," *DAC*, 1995, pp. 562 - 567.

[23] G. Lemieux, S. Brown, D. Vranesic, "On Two-Step Routing for FPGAs," *Int. Symp. on Physical Design*, 1997, pp. 60 - 66.

[24] Y.-S. Lee, A. Wu, "A Performance and Routability Driven Router for FPGAs Considering Path Delays," *DAC*, 1995, pp. 557 - 561.

[25] M. J. Alexander, J. P. Cohoon, J. L. Ganley, G. Robins, "Performance-Oriented Placement and Routing for Field-Programmable Gate Arrays," *EDAC*, 1995, pp. 80 - 85.

[26] S. Wilton, "Architectures and Algorithms for Field-Programmable Gate Arrays with Embedded Memories," *Ph.D. Dissertation,* University of Toronto, 1997.

[27] S. Yang, "Logic Synthesis and Optimization Benchmarks, Version 3.0," *Tech. Report*, Microelectronics Centre of North Carolina, 1991.

[28] J. Cong and Y. Ding, "Flowmap: An Optimal Technology Mapping Algorithm for Delay Optimization in Lookup-Table Based FPGA Designs," *IEEE Trans. on CAD*, Jan. 1994, pp. 1 - 12.

Technology Mapping of Heterogeneous LUT-Based FPGAs

Maurice Kilavuka Inuani and Jonathan Saul

Programming Research Group,
Oxford University Computing Laboratory,
Wolfson Building, Parks Road,
Oxford OX1 3QD, UK.
Email: [Maurice.Inuani, Jon.Saul]@comlab.ox.ac.uk

Abstract. New techniques have been developed for the technology mapping of FPGAs containing more than one size of look-up table. The Xilinx 4000 series is one such family of devices. These have a very large share of the FPGA market, and yet the associated technology mapping problem has hardly been addressed in the literature. Our method extends the standard techniques of functional decomposition and network covering. For the decomposition, we have extended the conventional bin-packing (cube-packing) algorithms so that it produces two sizes of bins. We have also enhanced it to explore several packing possibilities, and include cube division and cascading of nodes. The covering step is based on the concept of flow networks and cut-computation. We devised a theory that reduces the flow network sizes so that a dynamic programming approach can be used to compute the feasible cuts in the network. An iterative selection algorithm can then be used to compute the set cover of the network. Experimental results show good performances for the Xilinx 4K devices (about 25% improvement over MOFL and 10% over comparable algorithms in SIS in terms of CLBs).

1 Introduction

Over the last decade, *field programmable gate arrays* (FPGAs) have become an important part of Integrated Circuit (IC) design. In particular, FPGAs have opened the way for the use of programmable devices in a variety of applications, from "glue-logic" to actual processors. The shorter turn-around times and greater flexibilities offered by programmable devices have been the major advantages of these devices over semi-custom ones. Their versatility makes them attractive for rapid prototyping and for low-volume production systems.

Technology mapping binds an optimized representation of the circuit to a set of logic elements, taking into account the two constraints of circuit area and delay. These measures are often conflicting and thus cannot be optimised together. The technology mapping problem is an important one, since it can be computationally expensive, and has a large effect on the speed and size of the final implementation.

A large part of the FPGA market is for devices based on lookup tables (LUTs). The Xilinx 4000 series of devices contains a very large and growing portion of this market, and yet technology mapping for this class of *heterogeneous* architectures – which have more than one size of LUT– has hardly been considered.

This paper is concerned with the area optimisation of heterogeneous LUT-based FPGAs. Our contribution is to show an effective scheme for decomposing a general Boolean network into heterogeneous nodes and optimising the number of logic blocks that cover the given network. This paper begins with some background and definitions in Sect. 5 and Sect. 3. In Sect. 4, we describe two algorithms for the technology mapping for area (in terms of the number of logic blocks) optimisation. Our results are then presented in Sect. 5.

2 Previous Work

A good deal of work has been done by other researchers on the technology mapping of homogeneous FPGAs, but very little work has done on heterogeneous FPGAs. Most authors have concentrated on one of the objectives of technology mapping, that is: delay minimisation, area minimisation, or mapping for placement. Some have attempted to work out trade-offs between the competing objectives.

The *Chortle* algorithms were some of the earliest for LUT-based FPGAs[1], [2], [3]. They were soon followed by the *MIS-pga* algorithms [4], [5], [6]. These algorithms make extensive use of various decomposition schemes and graph covering techniques. Other algorithms include the *TechMap* algorithms [7], [8].

The *FlowMap*[9] and *DAGMap*[10] algorithms are delay-optimisation algorithms. These algorithms were instrumental in showing how a delay-optimal solution (at least in terms of network depths) could be computed for a general Boolean network in polynomial time. The method used in **FlowMap** led to various variants [11], [12], [13], [14], [15].

Algorithms targeted to heterogeneous FPGAs include those by Chung and Rose, **Tempt** [16] and by Lee and Shragowitz, **MOFL** [17]. For **Tempt** the architecture considered is such that each basic block's output in a hard-wired logic block (HLB) was accessible. This assumption does not hold true for the Xilinx 4000 series FPGAs. **MOFL** uses a fuzzy logic approach in which the decision function optimises for delay and area as measured by the maximal path length, the node levels, the fanin and fanout sizes of nodes, and the number of CLBs covering a node.

3 Preliminaries

The Xilinx 4000 series FPGAs have, in each logic block, two "F" LUTs and an "H" LUT, arranged in a cascade fashion[18]. The outputs from the F LUTs are fed into two inputs of the H LUT. A third input of the H LUT comes from an external source, as do all four inputs of each F LUT. The hard-wired connection between LUTs of the same block is much faster than programmable connections between blocks.

In this paper, we refer to *logic elements* as nodes. Because the logic elements are LUTs, we shall often use the words LUT and node interchangeably. The FPGA architecture will incorporate two types of LUTs. One type will be called the α-type, which has a maximum of κ_α inputs. The second type will be called the β-type, which differs from the α-type in that two of its inputs are the outputs of two α-type nodes. In addition to which there are $(\kappa_\beta - 2)$ other inputs (from some other α or β nodes). Thus an α node can implement any function of 1 up to κ_α inputs, while a β node can implement any function of 1 up to κ_β inputs. We assume that $\kappa_\beta < \kappa_\alpha$. It follows that an α node can implement any function that a β node can, but not the converse.

Therefore, two α nodes may be associated with a β node. This is represented by a triple $(\alpha_1, \alpha_2, \beta)$ and is referred to collectively as a *block*. Equally, some α node might not have any association with any β node. Such an α node can be thought of as a block composed of (α) alone. For the Xilinx 4000 series FPGAs an "F" LUT is equivalent to an α node while an "H" LUT is equivalent to a β node. The *configurable logic block (CLB)* is the same as our block. It will be assumed that $\kappa_\alpha = 4$ and $\kappa_\beta = 3$ unless otherwise stated.

The following terms are used throughout the paper. The fanin set of either a subgraph or node x is given by $\mathcal{F}_{\text{IN}}(x)$; its fanout set is $\mathcal{F}_{\text{OUT}}(x)$. A node v is *k-feasible* if it has no more than k inputs (fanin nodes), i.e., $|\mathcal{F}_{\text{IN}}(v)| \leq k$. By extension, a network is k-feasible if all its nodes are k-feasible. A *cube* is a product term of literals determined by variables of a function. The *transitive fanin* of a node v is the union of $\mathcal{F}_{\text{IN}}(v)$ and the transitive fanins for each $f \in \mathcal{F}_{\text{IN}}(v)$. Thus, the transitive fanin of a primary input is the null set. The *transitive fanout* is defined likewise. However, a subset of the primary outputs may have non-empty transitive fanout sets.

In discussing graphs, we assume that a graph $G = (V, E)$, where V is the set of vertices and E the set of edges in G. The *fanin cone* of t, $\mathcal{N}_t$, is a subgraph rooted at t that includes the transitive fanin of t in its vertex set. The edge set is composed of all edges in the original DAG that connect two vertices contained in the vertex set.

4 Our Approach

In this section, we discuss the new algorithms that map a general Boolean network into an FPGA with two types of LUTs. The algorithms seek to find a suitable mix of the two types of LUTs to minimise the total number of logic blocks in the mapping.

A common design flow for FPGAs first decomposes the optimised circuit to make it feasible, before covering it with logic blocks. There are usually other refinements that simplify the circuit. Here we consider only the new algorithms specific to heterogeneous architectures. The decomposition algorithm works on a general Boolean network a node at a time. For our purposes, it suffices that the

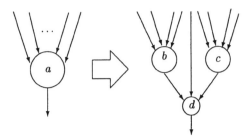

Fig. 1. Example of a decomposition of a 9-input node, a to two 4-input nodes (b and c) and one 3-input node, c.

Boolean network is κ_α-feasible. For example, Fig. 1 shows how a 9-input node can be decomposed to yield two nodes of 4 and one of 3 inputs. The covering algorithm seeks a minimal number of logic blocks from a mixture of κ_α- and κ_β-feasible clusters of nodes that covers the entire network. At the same time we seek to minimise the circuit delay by optimising the use of hardwired and programmable interconnections.

In Sect. 4.1 we discuss the cube-packing decomposition algorithm. Then in Sect. 4.2 we discuss the covering algorithm. Finally, in Sect. 4.3 we describe some network simplification techniques.

4.1 Cube-Packing Decomposition

There are several methods for decomposing a node, such as cube-packing, functional decomposition and co-factoring[5]. *Cube-packing* is a decomposition method that models this problem as a *bin-packing* problem [5], [1]. Very good results can be achieved with cube-packing under certain conditions. Equally, cube-packing is much simpler to implement and much quicker than many other schemes. To overcome some of the deficiencies of cube-packing, a modified bin-packing scheme is proposed. Furthermore, forming disjoint cubes and cascading derived nodes improves the performance of the cube-packing scheme.

In the following sections we discuss the bin-packing paradigm and our extension to the BFD algorithm for cube-packing. These are then applied to heterogeneous cube-packing node decomposition. We intend to use this decomposition method in concert with other methods although it can also be used on its own.

Bin-Packing: Francis et al. [1] and Murgai et al. [5] used the bin-packing method for node decomposition, where the cubes are considered as the items to be packed. Francis et al. use the *first-fit decreasing (FFD)* strategy, whereas Murgai et al. use the *best-fit decreasing (BFD)* strategy.

Our algorithm closely follows the one implemented by Murgai et al. since the BFD strategy gives better results than the FFD one. The criterion applied in our approach minimises the *increase* in the weight of the bin at each step, i.e., if an item can fit into more than one bin, the packing that minimises the increase in bin weight is chosen. The difference between our problem and the one solved by Murgai et al. is that we have two types of "bins" with different capacities.

Extensions to Bin-Packing: The crucial departure of our approach from others is that we maintain up to τ solutions, $S_1, \ldots, S_\tau$, at any one time. As τ tends to infinity, the algorithm implements an exhaustive search over the bin-packing solution space. Each time an item, x_i, is packed, the current solutions are each expanded separately. Thus, a solution S_j may have m_j bins available to x_i. By packing x_i into one of those bins we derive a new solution from S_j. Each such packing has an associated cost of 0. A final solution is found by using a new bin for x_i; this packing has a cost of 1. Therefore, there are $(m_j + 1)$ new solutions from S_j. After packing x_i into all current solutions $S_1, \ldots, S_n$, we may discard some so that there are no more than τ new solutions. When all items have been packed, only the best solution will be chosen for implementation.

A key aspect to our approach is the selection of the τ best solutions at each step. We use a greedy strategy which selects τ of the least costly solutions found so far. We define one packing option to be better than another if it uses fewer bins. In many instances, however, two packings may have the same number of occupied bins. To resolve this tie, we define the option with more total free space in its bins as the better one. The space in a bin is simply the difference between its capacity and its current level. In using the total free space as a criterion to break ties, we project that the more space there is available, the likelier it will be to fit the remaining (smaller) items into the currently occupied bins. However, this is not always the case since the space may be fragmented over many bins, so that a single item cannot fit into one bin. To counter this, we prefer options where the latest item is placed into its own bin since this gives some contiguous space within a single bin.

Compared to a conventional BFD approach, our scheme explores τ times as many options at each stage. Since τ is a constant, the increase in the complexity of our approach in time and space requirements is bounded by a constant factor, i.e., it has τ times the space and time requirements of the conventional BFD algorithm. Consequently, we able to vary τ depending on the size of the problem being solved. In general, we would expect that a larger τ gives better results.

Decomposing Functions: The decomposition of a cube can be reduced to the problem of forming a k-ary tree for a cube. If a cube is part of a multiple-cube function, then we would like the literals left in the decomposed cubes to be those that either occur frequently in other cubes or have the highest level. The reason for the first requirement is that cubes with many literals in common are far easier to pack than orthogonal cubes. The second requirement helps limit the circuit depth. The following example illustrates the first point.

Example 1. Let $f = x_1x_2x_3 + x_2x_4x_5x_6x_7 + x_5x_8x_9$. We need to extract 4 literals from the second term and represent them by a single literal. This will reduce the cube's size to 2 literals. There are 5 possible ways of selecting four literals from a set of 5. Consider two such groups: $y_1 = x_4x_5x_6x_7$ and $y_2 = x_2x_4x_5x_6$.

(1) $f_1 = x_1x_2x_3 + y_1x_2 + x_5x_8x_9$ can be packed into two bins: (x_1, x_2, x_3, y_1) and (x_5, x_8, x_9).

(2) $f_2 = x_1x_2x_3 + y_2x_7 + x_5x_8x_9$ needs at least three bins: (x_1, x_2, x_3), (x_5, x_8, x_9) and (y_2, x_7).

We see that f_1 is better than f_2 because of the choice of x_2 in f_1 which also occurs in the first cube. On the other hand, x_7 is unique to the second cube, making it more difficult to pack it with other cubes.

Given a multi-cube functions $f = d_1 + \cdots + d_m$, where d_i is a cube and $m > 1$, we first express f as $f = c \cdot (d_1^* + \cdots + d_m^*) = c \cdot D$. The disjunction D is cube-free and c the the largest cube divisor of f that divides it evenly. If c is not a constant, then c and D are decomposed separately; the results are amalgamated later. Considering D, if two cubes d_i and d_j intersect such that $e = d_i \cap d_j$ has at least κ_β literals, we introduce a new node corresponding to e and substitute it into D. In fact, it is more economical to find an e that occurs in the most cubes and which has the most literals. This can be done iteratively until no pairs of cubes have an intersection of the required size.

Then we partition the cubes of D into two sets: the nodes in S_A each have more than $(\kappa_\beta - 2)$ variables; while set $S_B = D - S_A$. We find a minimal set of κ_α-sized bins required for S_A and fill these, where possible, with nodes from S_B. The free slots in the inputs of the resulting nodes are filled by cascading the nodes. The residual nodes in S_B are then packed into $(\kappa_\beta\text{-}2)$-sized bins. These nodes are then matched with the previous ones to give logic blocks. It may be necessary to transfer some nodes from S_B to S_A, filling the nodes to size κ_α at the same time, in order to achieve the right balance between S_A and S_B.

4.2 Covering a Network

The network covering problem is solved by employing flow network techniques on the Boolean DAGs. Let $(X, \overline{X})$ be a *cut*, where the *source* $s \in X$, *sink* $t \in \overline{X}$ and $\overline{X} = V - X$. The sum of the capacities of the arcs originating in X and ending in $\overline{X}$ defines the capacity of the cut, $C(X, \overline{X})$ [19], [20].

Murgai et al. suggested a scheme where flows are used to enumerate the set of *super-nodes* in the graph [5]. More recently, Cong and Ding used a flow network method to compute depth-optimal mappings for Boolean graphs with homogeneous nodes [9]. Our approach is very similar to both these works except that we seek two types of super-nodes or equivalently, LUTs, and we seek an area-optimal solution.

In order to apply flows to a given Boolean graph, certain transformations on the graph are required. Given a node t we find the fanin cone, $\mathcal{N}_t$, which incorporates the transitive fanin of t. $\mathcal{N}_t$ has the following properties: (a) it is a directed graph, T, with a root t such that all the arcs are directed towards t; and (b) if v is a vertex of T, then $\mathcal{F}_{\text{IN}}(v)$ is a subset of the vertices of T.

Clearly, the leaf nodes of T are the primary inputs in the transitive fanin of t. To transform T into a flow network, we introduce a sink s and connect it to all the primary inputs in T. Then we assign the sink to t. We also transform the flow network by splitting to the each vertex $u \notin \{s, t\}$ to give two vertices u' and u'' and introducing a *bridging arc* (u', u'') [9]. We shall denote this transformed network as $\tilde{\mathcal{N}}_t = (\tilde{V}_t, \tilde{E}_t)$.

A consequence of this transformation is that the flow through any node u of T is limited to 1 in $\tilde{\mathcal{N}}_t$, due to the arc (u', u''). Therefore, a minimum cut is induced across a set of bridging arcs. This fact allows us to define the *node cutset* as the set of vertices whose bridging arcs the cut bisects.

Finding a Cutset: To compute the cuts, we use Dinic's augmenting path algorithm [20]. We only need to find $(k + 1)$ augmenting paths before deducing that the minimum cut is greater than k. The total running time for Dinic's algorithm is $\Theta(|A| + \sum_{i=1}^{m} l_i)$, where $|A|$ is the number of arcs, l_i is the length of the ith augmenting path, and m is the number of augmenting paths. In the worst case, it takes $\Theta(|V|^2 \cdot |A|)$, since l_i is bounded by $|V|$ and m by $|A|$. In our modification, we only examine k paths. Thus the running time for this new algorithm is $\Theta(k \cdot |A|)$. By slightly modifying the flow network, we can achieve a faster computation of a feasible, minimum cut in a network.

Our approach is to traverse the Boolean graph from primary inputs in a topological manner. For each non-primary input node, t, we shall define two types of cuts sets within its fanin cone: (a) $Cut_B(t)$ is a cutset of size no more than κ_β. At least one node within $Cut_B(t)$ has only a single fanout and not all its nodes are primary inputs. (b) $Cut_A(t)$ is any other cutset of size no more than κ_α.

Thus, given t we can construct $\tilde{\mathcal{N}}_t$ and run the max-flow algorithm on it to find a maximum flow f^*. Depending on the nature of the cutset found, we assign it to one of the two types of cutsets defined above.

Further feasible cuts, greater than the minimum cut, can be identified in $\tilde{\mathcal{N}}_t$ by first connecting each node in the minimum cutset to t. We then increase some the capacities of the bridging arcs of these nodes so as to admit more flow. By trying various combinations, we can find different cuts of larger volume than the min-cut.

The time complexity of the max-flow algorithm depends on the number of edges in the flow network. If $\tilde{\mathcal{N}}_t$ is constructed so that all nodes not in the fanin set of t have arcs from nodes in their min-cutset (rather than their fanin sets), then the resulting network could be smaller. We hypothosise that the omission of certain arcs and vertices in the graph from the network does not affect the flow within it. In particular, the min-cut is unaffected by the transformation.

Our approach to covering the graph is an iterative one that evaluates the number of logic blocks required for a subgraph rooted at a given node. This value, $Mes(v)$, is the minimal one for a node v and is determined by the available cuts. The algorithm works on the nodes in a topological fashion starting from the primary outputs. Only nodes that are roots of LUTs are processed. For these nodes the measure $Mes(v)$, where v is a given node, is re-evaluated by considering the mappings in other external subgraphs. The iteration ends when no further change occurs. We also able to limit the number of iterations.

4.3 Reducing a Network

We introduce two techniques that reduce the number of nodes in a network. The first collapses a node into all its fanouts if the fanouts remain k-feasible (k would vary with the type of node). The problem of selecting a maximal subset of all the possible collapses to implement is similar to the set covering problem. Our implementation uses a greedy strategy to solve this problem. This process can be carried out at any point within the design flow.

The second routine extends node simplification. We found that simplification usually makes some nodes infeasible while reducing the overall literal count. However, the actual number of nodes is unaffected. To improve it, we altered it to satisfy two criteria. (1) If the node was feasible, then it must not become infeasible. (2) The simplification is accepted if the complexity measures (number of literals and cubes) decrease and rejected if they increase. Our simplification routine is similar to the method used in SIS in all other respects [21].

5 Experimental Results

The algorithm described above have been implemented as: `binpack` (cube-packing decomposition of Sect. 4.1); `mcmap` (network covering of Sect. 4.2); `remove` (collapsing into fanouts of Sect. 4.3); and `gsimpify` (a modification of network simplification). The tables below show comparison between two given results. We indicate the raw difference in the usage of one resource as well as the percentage difference, in that order (except for CPU times).

We compared the new algorithms with equivalent ones in SIS [21]. The scripts are shown in the following table.

New	SIS	Remarks
remove	xl_partition	Node elimination
binpack	xl_ao	Cube-packing
gsimplify	simplify	Network simplification
	xl_ao	simplify may cause infeasibility
remove	xl_partition	
mcmap	xl_cover	Covering routines

Both scripts were preceded by the optimisations used in Mis-pga$_{new}$ [5]. We labelled each node as either an α or β LUT. In this way the number of CLBs can be estimated by placing as many pairs of α LUTs into the same CLB. In the following table each column contains the difference of our method to an equivalent one in SIS. Altogether, 103 circuits were analysed. We only show the circuits that produced different results for the two methods but the averages are taken over the entire set. The CPU times were measured on a SUN SPARCstation 5.

Table 1. Comparison with sis

Circuit	LUTs #	%	CLBs %	Time (ns)	Circuit	LUTs #	%	CLBs %	Time (ns)
i2	-41	-33.88	-52.46	-27.2	spla	-22	-9.61	-9.82	-0.5
C2670	-103	-34.80	-42.03	-2.5	apex1	0	0.00	-9.12	3.1
cm152a	-4	-36.36	-40.00	-0.1	t481	1	4.76	-9.09	-0.5
dalu	-196	-40.08	-38.79	-7.5	pair	-20	-4.07	-8.44	8.6
C3540	-176	-30.77	-38.77	-14.1	apex4	-4	-0.45	-8.37	102.9
C7552	-231	-32.49	-34.39	-19.5	bw	-4	-8.16	-8.00	-2.0
i10	-291	-28.06	-29.58	-16.6	apex3	8	1.40	-7.55	6.3
ex4	-56	-23.33	-29.46	3.1	table5	1	0.28	-7.51	4.6
C6288	-210	-28.57	-28.80	42.8	table3	-15	-4.67	-6.96	22.9
mux	-2	-11.76	-28.57	-1.0	alu2	-2	-1.63	-6.56	-14.8
C1355	-28	-25.45	-27.78	-13.1	cps	-17	-3.49	-5.68	-0.8
C499	-28	-25.45	-27.78	-14.1	x3	-13	-6.02	-5.66	0.7
rd84	-9	-15.00	-26.67	-3.6	i6	-6	-5.26	-5.26	-52.7
i3	-24	-22.64	-26.42	-21.9	misex2	-1	-2.70	-5.26	-1.8
C432	-25	-23.58	-25.49	-15.7	5xp1	-1	-2.50	-5.26	-1.0
frg2	-76	-23.03	-25.16	0.8	apex6	-11	-5.26	-4.90	0.7
term1	-13	-22.41	-25.00	-5.5	duke2	-1	-0.62	-4.00	-28.1
frg1	-13	-20.31	-25.00	-5.1	ttt2	-1	-1.75	-3.57	-4.3
i7	-3	-1.71	-22.73	-80.4	k2	0	0.00	-3.50	4.8
o64	7	16.28	-22.73	-6.3	apex7	-2	-2.94	-3.03	-7.7
apex2	-19	-17.59	-21.15	-14.8	vda	2	0.92	-2.86	0.6
clip	-6	-15.79	-21.05	-1.9	i8	-6	-1.84	-2.68	7.8
cordic	-4	-19.05	-20.00	-0.9	i9	-2	-1.02	-2.67	-57.7
misex3c	-21	-11.54	-19.32	10.5	des	-11	-1.05	-2.44	16.7
b9	-9	-16.36	-19.23	-4.4	alu4	1	0.48	-1.94	8.2
comp	-7	-16.67	-19.05	-2.4	C880	1	0.86	-1.89	-19.5
9sym	-6	-10.17	-17.24	-2.3	pdc	-3	-2.52	-1.72	-15.5
pcle	-3	-13.04	-16.67	-0.8	inc	-5	-13.89	0.00	-1.3
Z9sym	-7	-9.33	-16.22	-4.8	rd53	-1	-9.09	0.00	-0.4
9symml	-5	-6.33	-15.38	-4.8	b12	-2	-7.41	0.00	-0.8
i4	-4	-4.88	-15.38	-23.0	cm163a	-1	-7.14	0.00	-0.4
C5315	-39	-7.63	-14.49	5.7	alupla	-3	-6.12	0.00	-2.7
cmb	-2	-11.76	-14.29	-0.5	x2	-1	-5.88	0.00	-0.5
too_large	-17	-10.43	-14.10	4.0	x4	-1	-0.83	0.00	-23.6
rot	-35	-14.06	-14.05	2.1	pcler8	1	3.23	0.00	-1.4
misex3	-32	-11.90	-13.85	3.7	apex5	11	4.74	0.00	4.0
sao2	-4	-9.30	-13.64	-2.0	rd73	1	5.26	0.00	-0.4
Z5xp1	-2	-6.45	-13.33	-0.9	pm1	1	6.25	0.00	-0.5
lal	-1	-3.45	-13.33	-1.6	ex5	-1	-0.84	1.75	-14.6
cht	-6	-13.33	-13.04	-4.7	e64	1	1.16	2.33	-9.3
cm162a	-3	-17.65	-12.50	-0.5	f51m	3	14.29	11.11	-0.2
x1	-8	-6.30	-11.86	-21.7	f2	2	50.00	50.00	-0.6
c8	0	0.00	-10.53	-2.5	**Total**	-1886	-744.61	-1065.88	-348.6
cc	-1	-5.00	-10.00	-0.9	**Aver.**	-18.3	-7.2	-10.4	-3.4

Table 2. Comparison with MOFL

Circuits	CLBs	%	Levels	%	Delays (ns)	%
t481	-180	-92.78	-3	-37.50	-79.2	-57.94
c499	-71	-55.91	-4	-40.00	-35.2	-29.53
c1355	-62	-52.54	-7	-53.85	-25.7	-22.86
term1	-26	-45.61	1	20.00	0.9	1.35
c880	-40	-36.70	4	50.00	5.0	4.41
c1908	-47	-36.43	1	7.14	-19.6	-12.52
9symml	-21	-35.00	3	60.00	3.9	4.73
x2	-4	-33.33	0	0.00	1.7	4.34
apex7	-20	-29.85	3	75.00	6.4	8.31
z4ml	-2	-28.57	0	0.00	-5.8	-13.46
c432	-17	-27.87	4	33.33	20.3	13.52
alu2	-30	-26.32	3	23.08	-21.6	-12.03
x1	-24	-25.26	1	33.33	1.1	1.76
i9	-26	-25.00	-2	-28.57	-11.1	-11.94
lal	-4	-16.67	-1	-25.00	-7.3	-12.92
sct	-3	-16.67	2	50.00	20.3	36.31
alu4	-30	-15.96	5	31.25	11.0	5.58
example2	-11	-11.22	-1	-20.00	-5.9	-8.54
pcle	-1	-7.69	-1	-33.33	-16.4	-25.99
frg1	-2	-5.88	3	60.00	11.1	15.35
pcler8	-1	-4.35	0	0.00	-14.5	-17.70
c8	0	0.00	1	33.33	-15.5	-24.11
count	0	0.00	-1	-14.29	-38.6	-36.97
vda	3	2.03	0	0.00	-6.6	-6.17
b9	2	6.45	1	25.00	12.4	22.96
Averages	-24	-24.85	12	9.96	-8.36	-6.96
Std. Dev.	–	21.62	–	34.83	–	19.81

Table 1 shows that our script outperforms an equivalent one in SIS by 7.2% in terms of LUTs and 10.4% in terms of estimated CLBs. Moreover, it also takes less time on average.

The only published results on Xilinx 4K FPGAs we have are those of Lee and Shragowitz's MOFL [17]. Our comparison is based is based on the script given above followed by Xilinx's ppr version 5.1.0. We also incorporated the good_decomp [22] into our decomposition so that a choice is made from two decomposition methods for each infeasible node: (a) bin-packing alone and (b) good_decomp followed by bin-packing . In Table 2 we report principally the number of CLBs used for the benchmarks used in [17]. The columns contain the raw difference as well as the percentage difference. We used the XC4013PQ240 device, but have not been able to find out which device Lee and Shragowitz used. It should be noted that ppr has a significant influence on the final mapping, especially since it favours routing over placement, and therefore disperses LUTs over the FPGA. We achieve an average of 24% improvement in CLB utilisation over MOFL. We note, however, that MOFL optimises for delay as well as area. The levels reported the maximum number of LUTs any signal goes through. This figure does not distinguish between "H" and "F" LUTs and is obtained via XDelay in the XACT tools. The delay in our method is 7% better. We treat the delay results with caution since place and routing has the greater influence over them. Moreover, the devices used may be different.

6 Conclusion

We have addressed the problem of technology mapping for heterogeneous LUT-based FPGAs. Such FPGAs make up a large part of the market, and yet the technology mapping problem has not been fully addressed in the literature.

We have developed a node decomposition scheme based on bin-packing that improves on some of the deficiency of the conventional BFD and FFD cube-packing methods. We extended this to deal with heterogeneous FPGAs. Moreover the decomposition not only tries to optimise for area, but also takes delay considerations into account. The results from this are good.

Network covering has been extended to deal with heterogeneous FPGAs, and significantly enhanced by reducing the time need to compute clusters in a graph. Moreover, the set-covering step works globally to give very good results.

Future work will involve a further investigation on network simplification and its application to heterogeneous networks. We will also be examining BDD based methods for functional decomposition and logic re-synthesis. Cube-packing cannot always compete with a fully functional decomposition method, and so there is a need to incorporate some functional decomposition methods into the decomposition. We are currently implementing such a scheme in our overall mapping strategy and are looking at ways in which logic synthesis, decomposition and covering can be performed in concord with each other.

References

1. R. J. Francis, J. Rose, and K. Chung, "Chortle: A technology mapping program for lookup table-based field programmable gate arrays," in *27th Design Automation Conference*, pp. 613 – 619, 1990.
2. R. J. Francis, J. Rose, and Z. Vranesic, "Chortle-crf: Fast technology mapping for lookup table-based FPGAs," in *28th Design Automation Conference*, pp. 227 – 233, 1991.
3. R. J. Francis, J. Rose, and Z. Vranesic, "Technology mapping of lookup table-based FPGAs for performance," in *IEEE/ACM International Conference on Computer-Aided Design*, pp. 568 – 571, 1991.
4. R. Murgai, Y. Nishizaki, N. Shenoy, R. K. Brayton, and A. Sangiovanni-Vincentelli, "Logic synthesis for programmable gate arrays," in *27th Design Automation Conference*, pp. 620 – 625, 1990.
5. R. Murgai, N. Shenoy, R. K. Brayton, and A. Sangiovanni-Vincentelli, "Improved logic synthesis algorithms for table look-up architectures," in *IEEE/ACM International Conference on Computer-Aided Design*, pp. 564 – 567, 1991.

6. R. Murgai, N. Shenoy, R. K. Brayton, and A. Sangiovanni-Vincentelli, "Performance directed synthesis for table look up programmable gate arrays," in *IEEE/ACM International Conference on Computer-Aided Design*, pp. 572 – 575, 1991.

7. P. Sawkar and D. Thomas, "Area and delay mapping for table-look-up based field programmable gate arrays," in *29th Design Automation Conference*, pp. 368 – 373, 1992.

8. P. Sawkar and D. Thomas, "Performance directed technology mapping for look-up table based FPGAs," in *30th Design Automation Conference*, pp. 208 – 212, 1993.

9. J. Cong and Y. Ding, ""flowmap": An optimal technology mapping algorithm for delay optimization in lookup table based FPGA designs," *IEEE Trans. Computer-Aided Design*, vol. 13, pp. 1 – 12, Jan. 1994.

10. K.-C. Chen, J. Cong, Y. Ding, A. Kahng, and P. Trajmar, "DAG-Map: Graph based FPGA technology mapping for delay optimization," *IEEE Design and Test of Computers*, pp. 7 – 20, Sept. 1992.

11. J. Cong and Y. Ding, "On nominal delay minimization in LUT-based FPGA technology mapping," *Integration - The VLSI Journal*, vol. 18, pp. 73 – 94, 1994.

12. J. Cong, Y. Ding, T. Gao, and K.-C. Chen, "An optimal performance-driven technology mapping algorithm for LUT-based FPGAs under arbitrary net-delay models," *Computers & Graphics*, vol. 18, pp. 507 – 516, July 1994.

13. J. Cong and Y. Ding, "On area/depth trade-off in LUT-based FPGA technology mapping," *IEEE Transactions on Very Large Scale Integration (VLSI) Systems*, vol. 2, pp. 137 – 148, June 1994.

14. J. Cong and Y. Ding, "Beyond the combinatorial limit in depth minimization for LUT-based FPGA designs," in *IEEE/ACM International Conference on Computer-Aided Design*, pp. 110 – 114, 1993.

15. J. Cong and Y.-Y. Hwang, "Simultaneous depth and area minimization in LUT-based FPGA mapping," in *ACM/SIGDA International Symposium on Field-Programmable Gate Arrays*, (Monterey, California, USA), pp. 68 – 74, ACM, Feb. 1995.

16. K. Chung and J. Rose, "Tempt: Technology mapping for the exploration of FPGA architectures with hard-wired connections," in *29th Design Automation Conference*, (Anaheim, CA, USA), pp. 361 – 367, IEEE Computer Society Press, June 1992.

17. J.-Y. Lee and E. Shragowitz, "Technology mapping for FPGAs with complex block architectures by fuzzy logic technique," in *1st Asia and South Pacific Design Automation Conference*, (Japan), 1995.

18. Xilinx, *The Programmable Logic Data Book*. Xilinx Inc., San Jose, California, USA, 1993.

19. B. M. E. Moret and H. D. Shapiro, *Algorithms from P to NP: Design and Efficiency*, vol. 1. California, USA: Benjamin/Cummings Publishing Co., 1991.

20. T. H. Cormen, C. E. Leiserson, and R. L. Rivest, *Introduction to Algorithms*. MIT Press, 1993.

21. E. M. Sentovich, K. J. Singh, L. Lavagno, C. Moon, R. Murgai, A. Saldanha, H. Savoj, P. R. Stephan, R. K. Brayton, and A. Sangiovanni-Vincentelli, "SIS: A system for sequential circuit systems." Memorandum No. UCB/ERL M92/41, May 1992.

22. R. K. Brayton, R. Rudell, A. Sangiovanni-Vincentelli, and A. R. Wang, "MIS: A multiple-level logic optimization system," *IEEE Trans. Computer-Aided Design*, vol. 6, pp. 1062 – 1081, Nov. 1987.

Technology-Driven FSM Partitioning for Synthesis of Large Sequential Circuits Targeting Lookup-Table Based FPGAs

Klaus Feske, Sven Mulka, Manfred Koegst and Günter Elst

FhG IIS Erlangen - Department EAS Dresden
Zeunerstr. 38, D-01069 Dresden, Germany, e-mail: feske@eas.iis.fhg.de

Abstract. Different to common approaches we propose a novel Finite State Machine (FSM) partitioning procedure which takes technology-specific features into consideration. Moreover, partitioning and state encoding are performed simultaneously. We discuss the method utilizing the technology of Lookup-Table (LUT)-based FPGAs in detail. Our implementation can be used as an add-on for usual FPGA synthesis systems. We inserted this procedure into the FPGA design flow and achieved average area reductions by 38% and saved circuit depth by 29% for large FSM benchmarks.

1 Introduction

The top-down synthesis process of large sequential circuits is very complex. Design decisions at upper process steps like Finite State Machine (FSM) partitioning and state encoding directly influence low level stages e.g. logic optimization and technology mapping, and therefore they affect the design quality drastically. The question is: Can we open an additional optimization potential when taking the target technology into account already in earlier design steps? Technology driven synthesis onto FPGAs has so far addressed mainly the logic optimization and technology mapping design steps [MuBS93, BDPS94, Schl93, LeWE96]. Different to common approaches [Sauc93, AlBe94, Belh94, Syn96] we propose a novel FSM partitioning approach which takes technology-specific knowledge into consideration and exploit it for the simultaneous state assignment process.

In the paper, we discuss our method utilizing the technology of Lookup-Table (LUT)-based FPGAs. A LUT with n inputs implements any Boolean function of up to n variables. Such a technology is available e.g. for the Xilinx XC4000 family. Fig. 1 shows a simplified configurable logic block (CLB) containing two 4-input function generators (LUT) and a pair of flip-flops [Xili96].

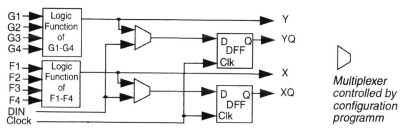

Fig. 1. Simplified configurable logic block of the Xilinx 4000 architecture [Xili96]

In order to enable the practical application of our method we implemented it as an experimental system „XPART". This system can be used as an add-on for usual FPGA synthesis systems. Inserting this procedure into the FPGA design flow of our FSM synthesis workbench [FFFK96], we achieved average area reductions by 38% and saved circuit depth by 29% for large FSM benchmarks.

The paper is organized as follows. First we introduce the basic idea of our approach. Section 3 describes the FSM partitioning method and outlines the adapted state encoding procedure illustrated by a small example. Finally, we show the advantages of this approach by experimental results.

2 Which is the Best State Encoding Targeting Lookup-Table Based FPGAs?

Aiming at programmable logic, it is said in [Alle93]: „Because of the wide input gates available in PLDs, fully encoded state machines are usually used. However, in register rich FPGAs with narrower gates, one-hot state machines are usually preferred". Our paper takes a critical look at this common application. For encoding a FSM with n states there is a wide range between the maximum code width of n

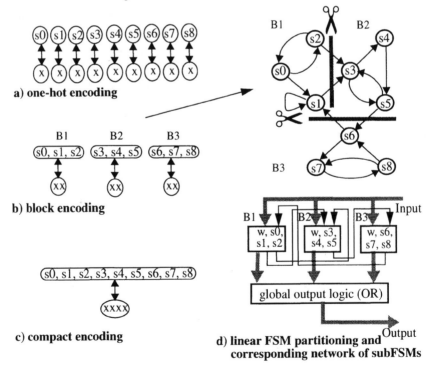

a) one-hot encoding

b) block encoding

c) compact encoding

d) linear FSM partitioning and corresponding network of subFSMs

Fig. 2. Relations between encoding, linear FSM partitioning, and corresponding LUT-based network of subFSMs B1, B2, B3 (‚x' symbolizes a bit of state code corresponding to the related set of states)

(one-hot encoding, Fig. 2a) and the minimal width of $\lceil log_2 \; n \rceil$ (fully encoded, Fig. 2c). We exploit this range by generating so-called block codes. Such a block encoded FSM corresponds to a network of cooperating subFSMs (visualized in Fig. 2d). To produce such a network, we evolve a related FSM synthesis procedure which is characterized by a technology driven FSM partitioning combined with a simultaneous state encoding.

3 An Approach Based on Linear FSM Partitioning

3.1 Preliminary Definitions

First we recall some basic definitions which are helpful for the subsequent problem formulations. A 5-tuple $A = (X, Y, S, f, g)$ is called Finite State Machine (FSM) where X, Y, and S is a finite set of inputs, outputs, and states, respectively. The unique mapping $f: X \times S \rightarrow S$, and $g: X \times S \rightarrow Y$ is the transition function, and the output function, respectively. To specify a FSM in a easy readable form we utilize the State Transition Table (STT) and the State Transition Graph (STG). The STT is a tabular representation of A. Each row of it can be described as a 4-tupel, $(in_i, ps_i, ns_i, out_i)$ | $i \in \{1,2,...,k\}$, where k is the number of rows, in_i and out_i is an assignment of the input and output, respectively, ps_i and ns_i is a symbolic present state (ps) and next state (ns), respectively. The directed graph G(S,E,W), called STG, is a graphical representation of A, where S is the set of vertices, E with $E \subseteq S \times S$ is the set of edges $e_i = (ps_i, ns_i)$ and $W = \{(in_i, out_i)\}$ is the set of weights of the edges. If $(s_i, s_j) \in E$ it is said that s_i is a predecessor of s_j, $s_i = Pre(s_j)$, and vice versa s_j is a successor of s_i, $s_j = Post(s_i)$.

3.2 Linear FSM Partitioning

A lot of FSM partitioning methods have been published. They are based on different communication structures and operational principles (summarized in [AsDN92]). Based on the so-called linear partitioning [GrMu89] approaches to design a FSM $A = (X, Y, S, f, g)$ as a network of cooperating subFSMs B_r are proposed in [Puttk91, BarB93, KGFF94]. This principle is defined by a partition P of S, where each block of P defines a subFSM B_i. Such a subFSM is characterized by an additional wait state w (representing all states outside of the corresponding block) as well as by additional conditions (for remaining in the wait state or leaving it). As an example for a given FSM A and the partition $P = \{(s0,s1,s2),(s3,s4,s5),(s6,s7,s8)\}$ Fig. 3 shows the STTs of the resulting subFSMs $B1$, $B2$ and $B3$ and the STG for $B1$ including the wait state w.

We customize this principle for the synthesis of large FSMs targeting LUT-based FPGAs. For this purpose we use a communication structure based on additional inputs ic (conditions for remaining in the wait state or leaving it) and on additional outputs oc (control for activating the next subFSM) so that the STT of a subFSM is defined by $B_r = (in, is, ps, ns, oc, out)_r$ as shown in Fig. 3. Our goal is to synthesize a network $B = \{B_r\}$ which realizes FSM A and minimizes the cost function

$$Cost(A) = Cost_{GO} + \left(\sum_{(B_r \in B)} Cost(B_r) \right)$$

where Cost_{GO} is the cost of the global output logic (Fig. 2d) and

$$Cost(B_r) = \lceil ld|S_r| \rceil + |oc_r| + |out_r|$$

is an estimation of the number of outputs of subFSM B_r. This value correlates to the number of required LUTs

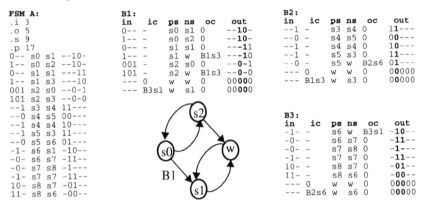

FSM A:

```
.i 3
.o 5
.s 9
.p 17
0-- s0 s1 --10-
1-- s0 s2 --10-
0-- s1 s1 ---11
1-- s1 s3 ---10
001 s2 s0 --0-1
101 s2 s3 --0-0
--1 s3 s4 11---
--0 s4 s5 00---
--1 s4 s4 10---
--1 s5 s3 11---
--0 s5 s6 01---
-1- s6 s1 -10--
-0- s6 s7 -11--
-0- s7 s8 -1---
-1- s7 s7 -11--
10- s8 s7 -01--
11- s8 s6 -00--
```

B1:

in	ic	ps	ns	oc	out
0--	-	s0	s1	0	--10-
1--	-	s0	s2	0	--10-
0--	-	s1	s1	0	---11
1--	-	s1	w	B1s3	---10
001	-	s2	s0	0	--0-1
101	-	s2	w	B1s3	--0-0
---	0	w	w	0	00000
---	B3s1	w	s1	0	00000

B2:

in	ic	ps	ns	oc	out
--1	-	s3	s4	0	11---
--0	-	s4	s5	0	00---
--1	-	s4	s4	0	10---
--1	-	s5	s3	0	11---
--0	-	s5	w	B2s6	01---
---	0	w	w	0	00000
---	B1s3	w	s3	0	00000

B3:

in	ic	ps	ns	oc	out
-1-	-	s6	w	B3s1	-10--
-0-	-	s6	s7	0	-11--
-0-	-	s7	s8	0	-1---
-1-	-	s7	s7	0	-11--
10-	-	s8	s7	0	-01--
11-	-	s8	s6	0	-00--
---	0	w	w	0	00000
---	B2s6	w	s6	0	00000

Fig. 3. Principle and STT representation of linear FSM partitioning

3.3 Algorithm

Our approach starts with an initial n-block partition, $n = |S|$. This corresponds to an one-hot encoding of the FSM and, in fact, to a network of n cooperating subFSMs.

Algorithm **optimizeFSM(A):**

```
List<Block> optimizeFSM( A )              // returns optimized
{                                          //      linear FSM partition
        Block B_r ;                        // SubFSM
        List<Block> B = ∅ ;                // set of all SubFSM
        forAll( s ∈ S )
        {
                B_r = buildBlock( s );     // performs initial partition
                B = B ∪ { B_r };           //      corresponding to
        }                                  //      one hot encoded FSM
        newCost = getCost( A );
        do                                 // global iteration
        {
                oldCost = newCost ;
                forAll( B_r ∈ B )
                {
                        while ( optimizeBlock( B_r ));   // take up adjoining states
                }                                        // while gain
                newCost = getCost( FSM );
        } while( newCost < oldCost );
        return B ;
}
```

To optimize network $B = \{B_r\}$ we apply algorithm **optimizeBlock**. Let S_r, $S_r \subseteq S$ be the set of states of the current subFSM B_r. Then this algorithm improves the local solution for subFSM B_r by moving a further state s, $s \in S - S_r$, into S_r.

Algorithm **optimizeBlock:**

```
size_t optimizeBlock( B_r )                          // returns gain
{
        Block parent ;                               // SubFSM
        gain  = 0;                                   // Cost
        forAll (( ( s ∈ S - S_r ) ∧ ∃s' ( s' ∈ S_r ∧ ( s' = Pre(s) ∨ s' = Post(s) ) ) )
                                                     // outside, adjoining B_r
        {
                parent      = getParent( s );        // provide Block containing s
                oldCost     = getCost( B_r )+getCost( parent );
                moveState( s, parent,B_r );
                newCost     = getCost( B_r )+getCost( parent );
                if(  newCost ≤ oldCost )
                        gain = gain + newCost - oldCost ;
                else
                        moveState( s, parent,B_r );  // s moving back
        }
        return gain;
}
```

b0	in	ic	ps	ns	oc	out
	0--	-	s0	w	b0s1	--10-
	1--	-	s0	w	b0s2	--10-
	---	0	w	w	0	00000
	---	b2s0	w	s0	0	00000

b1	in	ic	ps	ns	oc	out
	0--	-	s1	s1	0	---11
	1--	-	s1	w	b1s3	---10
	---	0	w	w	0	00000
	---	b0s1	w	s1	0	00000
	---	b6s1	w	s1	0	00000

b2	in	ic	ps	ns	oc	out
	001	-	s2	w	b2s0	--0-1
	101	-	s2	w	b2s3	--0-0
	---	0	w	w	0	00000
	---	b0s2	w	s2	0	00000

B1	in	ic	ps	ns	oc	out
	0--	-	s0	s1	0	--10-
	1--	-	s0	s2	0	--10-
	0--	-	s1	s1	0	---11
	1--	-	s1	w	B1s3	---10
	001	-	s2	s0	0	--0-1
	101	-	s2	w	B1s3	--0-0
	---	0	w	w	0	00000
	---	B3s1	w	s1	0	00000

Fig. 4. Merging of adjoining subFSMs

Moving a state s to a subFSM B_r is only feasible, if for all outputs o of oc_r and out_r and for all next state functions holds:

$$\left| in_r(o) \right| + \left| ic_r(o) \right| + \left\lceil ld \left| S_r \cup \{s\} \right| \right\rceil \le max .$$

Whereby *max* is the maximum number of LUT-inputs and $in_r(o)$, $ic_r(o)$ is the set of relevant primary and communication inputs associated to the output o, and $\lceil ld|S_r \cup \{s\}|\rceil$ is the number of next state functions which corresponds to the minimal code width for B_r. Fig. 4 visualizes the merging of the adjoining initial sub-FSMs $b0$, $b1,b2$ to $B1$ of our example according to this process. The technology-driven merging of adjoining subFSMs is terminated if they are embedded in a minimized number of CLBs according to the adapted encoding of their states.

3.4 Adapted State Encoding

To reduce the total number of CLBs we further combine FSM partitioning and state assignment of the subFSMs. We have to solve the following problem: For a given sub-FSM described by the STT $B_r = (in_r, ic_r, ps_r, ns_r, oc_r, out_r)$ find a state assignment for S_r, such that the realized combinational circuit uses a minimum number of CLB's of the target LUT- architecture.

The state encoding procedure is characterized by the following three features:
- the subFSMs are encoded independently,
- it is easy to find the local optimum of state assignments for each subFSM since the number of its states is relatively small ($|S|$ =4),
- additionally, outputs and code bits will be shared analogous to [KoFF93].

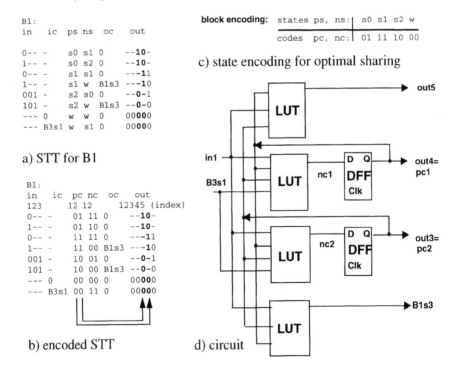

Fig. 5. Adaptive State Encoding for Sharing Output and Code Bits of subFSMs

This procedure [KoFF93, Mulk94] is used for calculating the value of *getCost* in the Algorithm *optimizeBlock* as well as for finding local state assignments for sub-FSMs according to a minimized CLB count. Fig. 5 illustrates the principle of column minimization for subFSM B1: Obviously, in relation to B1 the inputs in2, in3 and the outputs out1, out2 are redundant and can be deleted. Using the state assignment (s0,s1,s2,w) $\rightarrow$ (01,11,10,00) we can share (out4, pc1) and (out3, pc2). Thus the number of required LUTs is reduced from six to four.

4 Implementation and Experimental Results

Our approach is implemented as the experimental program XPART. XPART itself can be embedded in our workbench for FSM synthesis targeting LUT-based FPGAs as shown in Fig. 6. This experimental environment includes six alternatively applicable state encoding and/or FSM-partitioning strategies: group a) using the encoding procedures jedi (1), nova (2), one-hot (3) of the UCB [SSLM92] and group b) using XPART procedures for FSM partitioning and state encoding applying standard approach (4), using exact cost (5) and partitioning and one-hot encoding (6). For logic optimization and technology mapping we have a choice between the tools TOS-TUM [LeWE96] and SIS -xl.

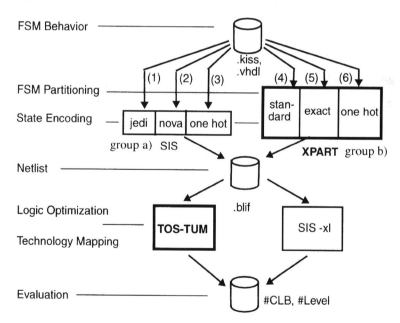

Fig. 6. Flow-chart of the Workbench for FSM-synthesis targeting LUT-based FPGAs

Our test FSMs are large industrially used Moore sequencers [KoWa93] having about 100 states (#s), approximately 100 outputs (#o), and more then 100 state transitions (#p) as shown in the first column of Table 1.

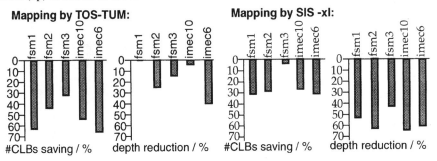

Fig. 7. Design improvement by integration of XPART
(reduction of #CLB and circuit depth in %)

| Design | Tool | mapping by TOS-TUM | | | mapping by SIS | |
		#level	#CLB	CLB-delay	#level	#CLB
fsm1	(1)jedi	8	252	30,5	56	212
	(2)nova	9	347	34,5	52	225
#s=74,	(3)SIS -oh	4	258	16,5	13	291
#p=165,	**(4)xpart -s**	4	121	16,5	6	160
#o=67	**(5)xpart -e**	4	93	16,5	6	145
	(6)xpart -oh	4	94	16,5	6	157
fsm2	(1)jedi	6	245	24.0	72	207
	(2)nova	7	267	26,5	55	205
#s=77,	(3)SIS -oh	4	152	16,5	11	180
#p=130,	**(4)xpart -s**	4	112	16,5	6	143
#o=76	**(5)xpart -e**	3	88	12,5	4	128
	(6)xpart -oh	3	88	12,5	4	129
fsm3	(1)jedi	7	369	26,0	50	254
	(2)nova	9	394	35,0	47	274
#s=79,	(3)sis -oh	4	221	16,5	9	407
#p=166,	**(4)xpart -s**	4	154	14,0	5	245
#o=138	**(5)xpart -e**	4	150	16,5	6	244
	(6)xpart -oh	4	150	16,0	6	245
imec10	(1)jedi	7	329	26,0	55	297
	(2)nova	7	340	26,0	40	284
#s=96,	(3)SIS -oh	4	308	16,5	9	289
#p=130,	**(4)xpart -s**	4	157	16,0	4	208
#o=120	**(5)xpart -e**	4	136	16,0	3	208
	(6)xpart -oh	4	137	16,5	4	207
imec6	(1)jedi	8	264	30,0	82	199
	(2)nova	8	308	31,0	76	205
#s=75,	(3)SIS -oh	4	337	16,5	13	304
#p=152,	**(4)xpart -s**	3	102	12,0	5	141
#o=150	**(5)xpart -e**	3	91	10,0	5	135
	(6)xpart -oh	3	91	10,0	5	136

Table 1. Results

Fig. 7 summarizes our results depicting the relative improvements achieved by integrating our add-on tool XPART into the FPGA design flow and applying technology mapping by SIS -xl and TOS-TUM alternatively. For five benchmarks the area reduction (in terms of CLBs) and the saved circuit depth (in terms of logic levels) are represented. Note, that in the diagrams the best results of group a) have been checked in comparison with the best results from group b). The average savings are 38% for area (#CLB) and 29% for the circuit depth (#level). Table 1 demonstrates the results achieved in detail.

Moreover our experiences show that:

- In contrast to using encoding procedures nova or jedi the computation time for mapping of circuits partitioned and encoded by XPART is very low.
- Concerning the logic depth TOS-TUM mapping produces better results then SIS.
- But overall, using XPART as add-on improves all results in terms of circuit depth and CLB count.

5 Conclusions

We have presented a new approach for FSM partitioning and state encoding targeting a wide spread LUT architecture. Our algorithm starts with an initial n-block partitioning and improves it by moving and encoding states simultaneously according to technology based constraints. We inserted the implemented procedure XPART into the design flow of our FSM synthesis workbench and achieved significant improvements (#CLB, #levels, CPU-time) of the synthesized designs especially for large Moore sequencers.

Currently we extent the experimental environment including the synthesis tool SYNOPSYS and we round off our approach by taking different technology based constraints into account.

Acknowledgment

This work has been partially supported by DFG Project SFB 358. The authors thank Prof. Kurt J. Antreich and Ch. Legl, Technical University of Munich, for their support, fruitful discussions and for the preparation of mapping results using TOS-TUM.

References

[Alle93]   Allen, D.: Automatic One-hot Re-encoding for FPGAs. Proc. of FPL'92, Vienna, Austria, Springer-Verlag Berlin Heidelberg 1993, pp.71-77.

[AlBe94]   Alfke, P.; New, B.: Implementing state machines in LCA devices. In The Programmable Logic Data Book, pp. 8-169 - 8-172. Xilinx, Inc., San Jose, CA, 2nd edition, 1994. XAPP 027.001.

[AsDN92]   Ashar, P.; Devadas, S.; Newton, A.R.: Sequential Logic Synthesis. Kluwer Academic Publishers, Boston/Dortrecht/London, 1992.

[BaBr93]   Baranov, S.; Bregman, L.: Automata decomposition and synthesis with PLAM. Microprocessing and Microprogramming 38 (1993) pp. 759-766.

[Belh94]   Belhadj, H.: State Assignment Selection for FPGAs and CPLDs. Proceedings JESSI AC-8. 5th Workshop, Munich, April 11-13, 1994.

[BDPS94]   Burgun,L.; Dictus, N.; Prado Lopes, E.; Sarwary, C.: A Unified Approach For FSM Synthesis On FPGA Architectures. Proceedings Euromicro '94, September 5-8, 1994, Liverpool, England. pp. 660-669.

[FFFK96]   Fehlauer, E. et. al.: Workbench for Design Improvement. Baltic Electronics Conference, October 7-11, 1996 Tallinn, Estonia.

[GrMu89]   Grass, W.; Mutz,M.: Modular Implementation of FSM Observing Topological Constraints, IFIP-Congress Computer Hardware Description Languages and there Applications, 1989, pp.183-196.

[KoFF93]   Koegst, M.; Feske, K.; Franke, G.: FSM State Assignment by Sharing Inputs and Outputs. 3rd International Design Automation Workshop (Russian Workshop'93), July 1993, Moskow, Russia,pp.12-19.

KGFF94]   Koegst, M.; Grass, W.; Feske, K.; Franke, G.: Simultaneous State Encoding and Communication Cost Optimization for FSM Net Design. Proceedings Euromicro '94, September 5-8, 1994, Liverpool, England. pp. 653-658.

[KoWa93]   Kottsieper, J.; Waldschmidt, K.: Application of the novel associative programmable array- structure Multi- Match PLA in synthesis of decomposed finite state machines. Microprocessing and Microprogramming 38, 455-465, North-Holland, 93.

[LeWE96]   Legl, Ch.; Wurth, B.; Eckl, K.: A Boolean Approach to Performance-Directed Technology Mapping for LUT-Based FPGA Designs. Proc. 33rn IEEE/ACM Design Automation Conference, DAC 1996, Las Vegas, Nevada, U.S.A., June 3-7, 1996.

[Putk91]   Puttkammer, A.: Entwicklung, Implementierung und Bewertung von Methoden zur Modularisierung von Steuerwerken. Diplomarbeit, Universität Passau, Fachbereich Mathematik und Informatik, 1991.

[MuBS93]   Murgai, R.; Brayton, R.K.; Sangiovanni- Vincentelli, A.: Sequential Synthesis for Table Look Up Programmable Gate Arrays. 30-th ACM/IEEE DAC, 1993, pp. 224-229.

[Mulk94]   Mulka, S.: Partitionierung und Synthese endlicher Automaten für CPLD- und FPGA-Implementierung. Diplomarbeit, Technische Universität Dresden, Fakultät Elektrotechnik, Dresden 1994.

[Sauc93]   Saucier, G.: Synthesis of a Finite State Machine on any target. EDAC'93, Paris.

[Schl93]   Schlichtmann, U.: Boolean Matching and Disjoint Decomposition for FPGA Technology Mapping. Proc. of IFIP Intern. Workshop on Logic and Architecture Synthesis, Grenoble, France, Nov. 6-8, 1993.

[SSLM92]   Sentovich, E. M.; Singh, K. J.; Lavagno, L. et al.: "SIS: A System for Sequential Circuit Synthesis", Memorandum No. UCB/ERL M92/41, Electronics Research Laboratory, Department of Electrical Engineering and Computer Science, University of California, Berkeley, CA, May 1992.

[Syn96]   Synopsys:." FPGA Express User's Guide." Synopsys Inc., Mountain View, CA, September 1996.

[Xili96]   XILINX: "The Programmable Logic Data Book 1996." Xilinx, Inc., San Jose, CA, August 1996.

Technology Mapping of LUT based FPGAs for Delay Optimisation

Xiaochun Lin[1], Erik Dagless[1], and Aiguo Lu[2]

[1] Dept. of Electrical & Electronic Engineering
University of Bristol, UK
[2] Institute of Electronic Design Automation
Technical University of Munich
D-80290 Munich, Germany

Abstract. This paper presents a new LUT based technology mapping approach for delay optimisation. To optimise the circuit delay after layout, the wire delays are taken into account in our delay model. In addition, an effective approach is proposed to trade-off the CLB delays and the wire delays so as to minimise the whole circuit delay. The trade-off is achieved in two phases, mapping for area optimisation followed by new delay reduction techniques. Based on a standard set of benchmark examples, experimental results after PPR layout have shown that the proposed approach outperforms state-of-the-art approaches.

1 Introduction

Most of the previous LUT based mapping approaches for delay optimisation apply the unit delay model, which assumes that the wire delay is zero. This is an acceptable approximation for small FPGA chips. With the increasing size of FPGAs, it becomes a necessity to involve the wire delays in logic synthesis. For this reason, the proposed approach considers both cell delays and wire delays. The cell delays are determined by the logic blocks, whereas the wire delays are caused by the wire lengths. To reduce both delays, we divide our mapping approach into two phases. In the first phase, the technology mapping for area optimisation is performed. This minimises the circuit size, thus the wire lengths and wire delays. Based on the area-optimally mapped results, the critical nodes are determined using the proposed delay model. To involve the wire delays in the delay model, the delay contributions of fanout load are considered in deciding the critical nodes. Then the second phase restructures the critical nodes using the proposed techniques. Doing so the critical nodes are mapped for delay optimisation whereas the non-critical nodes are mapped for area optimisation.

2 Previous Work

There have been a lot of research achievements in logic synthesis for LUT based FPGAs. Previous work can be roughly divided into three areas: area optimisation

[15, 19, 12], delay optimisation [2, 15, 3, 11, 9, 5, 1, 7, 18] and routability driven synthesis [11, 16].

According to the different techniques applied, technology mapping for delay optimisation can be further divided into 3 categories: decomposition followed by covering [2, 3, 6, 9, 1], mapping for delay optimisation followed by area reduction [4], and mapping for area optimisation followed by level reduction [15, 11, 7]. Comparisons of the previous achievements have shown that the strategy of *mapping for area optimisation followed by level reduction* provides satisfactory results. Among them, TechMap [15] used clique partitioning to perform level reduction, ABLO-d [10] applied three operations to perform level reduction, and ALTO [7] utilised delay oriented area optimisation and delay oriented Roth-Karp decomposition to achieve level minimisation. Our approach will follow this strategy but with more effective delay reduction techniques. Unlike above approaches that determine critical nodes based only on the node levels, we are based on the arrival time of each node. The arrival time of a node contains both the CLB delays and also the delays of fanout loads[3].

3   Technology Mapping for Delay Optimisation

In the following, our two-phased mapping approach will be described, which is the area-optimal mapping phase followed by the delay reduction phase.

3.1   Mapping for area optimisation

In this phase, we apply ABLO-o [10] to perform technology mapping for area optimisation, which uses the LUT directed kernel extraction and bin-packing for area optimisation [12]. After this, all nodes in the network are feasible and they are area-optimally mapped. Then we can determine the critical nodes in the mapped network.

The novel feature in determining the critical nodes is that we consider both the cell delays and the wire delays. The CLB delay is a constant for each logic block and so the level count can reflect the cell delays. The wire delays are related to the interconnect length and also the number of switches the net passes through. This is certainly unavailable at the stage of technology mapping. However, it is generally true that a multiple-pin net will have a longer interconnect length than a 2-pin net. For this reason, we use the fanout count to estimate the wire delays. In our delay model, the arrival time d_n of a node n is defined as

$$d_n = D + MAX\{d_{xi}|_{i=\{1,2,...,k\}}\} + \alpha \times F_n,$$

where D is the CLB delay, d_{xi} is the arrival time of the ith fanin to node n, F_n is the number of fanouts of node n, and α is a user-specified parameter.

---

[3] FlowMap-d [3] used a delay model for LUT based technology mapping, in which the wire delay is considered to be proportional to the fanout size of a net.

After the arrival time of each node in the mapped network is decided, the primary output nodes that have the maximum arrival time are defined as the critical primary outputs. Starting from the critical primary outputs, a backward tracing is performed towards primary inputs. Let the required time of the critical primary outputs be equal to the maximum arrival time. The backward tracing determines the required time of the transitive fanin nodes of the critical primary outputs. Then the critical nodes are found under the condition that the arrival time is equal to the required time.

After critical nodes are determined, they are sorted according to their sensitivity to the circuit delay. For example, F and G in Figure 1 are two critical primary outputs, whose transitive critical fanins are A, B, C, D, and E. If the delay of node D is reduced, the circuit delay will not be changed as the delay of node C remain unchanged. The same case occurs for node E. However, if the delay of node C is reduced, the circuit delay will then be reduced. This means that the critical node C is more sensible to the circuit delay than nodes D and E. So node C will have more priority to be restructured in the following delay reduction phase.

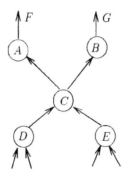

Fig. 1. Sensitivity of critical nodes to the circuit delay.

3.2 Reducing circuit delays

The proposed delay reduction phase contains 3 steps: partial collapsing followed by remapping, collapsing followed by remapping, and fanout optimisation.

A. Partial collapsing and remapping. In this step, the critical fanins of a critical node will be collapsed into this node, and then the node will be remapped using the minimum delay.

For example, consider a subnetwork with two nodes F and X: $F = Xab + D + Xc$ and $X = Yi + hej'$. Suppose F is the critical node under consideration, X is its critical fanin whose arrival time is 4.8, variables Y and D both have the arrival time of 3.2, and variables a, b, c, e, h, i, and j are primary inputs. After X is collapsed to F, F becomes

$F = Yabi + hej'ab + D + Yci + hcej'$.

Remapping it to XC3000 gives $F = YR + D + T + S$, $R = abi + ci$, $T = hej'ab$, and $S = hcej'$. It can be seen that the arrival time of F has been reduced with the penalty of two more LUTs. The detail of the remapping techniques will be discussed in the next section.

B. Collapsing and remapping. In this step, a critical node will be collapsed into the two-level form so as to eliminate the influence of multi-level logic optimisation and technology mapping for area optimisation. This will certainly provide more chance for this node to be mapped with fewer delays. After a node is collapsed, the remapping techniques discussed below will be used to perform mapping for delay optimisation.

C. Fanout optimisation. As the fanout count of a node contributes to the circuit delay, the fanout optimisation technique is proposed as the third step. This step acts as an additional effort to reduce the interconnect delay. Two methods, selective collapsing and trading off area for delay, are used.

(1) For a multiple fanout critical node, it will be collapsed to its fanouts based on the following condition. Let node n be the critical node being considered, and nodes A, B, and C are its fanouts. If collapsing node n to node A means A remains feasible, n will be collapsed to A, and the fanout count of n is reduced by one. Otherwise, n will not be collapsed. Similarly, if node n can be collapsed to B or C, it will be collapsed. In this way, the fanout load of critical nodes will be reduced.

(2) For those nodes which have a big fanout count, node duplication will be performed to further reduce the fanout loads of critical nodes. Assume that node n is under consideration, which has M fanouts ($M > 6$). Node n is duplicated to form a new node n_1 so that n fanouts to $M/2$ nodes and n_1 fanouts to other $M - M/2$ nodes. This increases the size by one LUT but reduces the fanout count of node n. To avoid using bigger FPGA devices to implement the circuits, the area increase of duplication is controlled. Suppose the circuit is mapped to N CLBs. According to a rule of 80% cell usage, the device that is most suitable to implement the circuit can be determined, and say it has C_N cells. Then the maximum area increase allowed for node duplication is $C_N - N$ CLBs.

4 Remapping Techniques

In the remapping process, we use an enhanced bin-packing algorithm to perform delay optimisation. This is quite effective for those node functions with orthogonal cubes. For other node functions, we first apply a set of decomposition methods, then use bin-packing.

4.1 Bin-packing for delay optimisation

To map a node function F into k-input LUTs, the cubes of function F are considered as boxs and the LUTs are treated as bins. Firstly, the cubes of function F are sorted in increasing order of their delays. The cube with the minimal delay will be packed first, the cube with the second minimal delay will be packed second, and so on. If cubes have the same delay, the cube with more literals will be packed first. At each packing step, the minimal delay cube is selected and fitted into existing bins. If it can be fitted into several bins, the best bin for this cube is determined. The best bin is the one which has the maximum delay. If the cube cannot be fitted into any existing bin, a new bin is created and the cube is placed in it. When all the cubes have been assigned to the bins, the bin of cubes with minimal delay is closed. This generates a single literal cube. This new cube is then placed in a bin as before under the condition that it cannot increase the delay of that bin. Otherwise, it is not packed. The process is repeated until only the bins with the maximum delay are left.

For those bins with the maximum delay, they are processed so that they have the priority to be packed in the output LUT. Let the number of placed single literal cubes be g. The condition that bins with the maximum delay can be placed in the output LUT is

$$\sum |sup(LUT_m)| + LUT_s - g - 1 <= k,$$

where $\sum |sup(LUT_m)|$ is the number of variables in the bins with the maximum delay, LUT_s is the number of LUTs used to implement F, and k is the maximum number of inputs to a LUT. Otherwise, a new bin is created and those unpacked single literal cubes will be placed in it.

4.2 Different decomposition methods

In our approach, three decomposition methods have been implemented to decompose those node functions with non-orthogonal cubes, which are modified Shannon cofactoring, best decomposition, and LUT directed kernel extraction.

A. Modified Shannon cofactoring. Shannon cofactoring is a classical approach to break big nodes into a set of small ones. It is mainly based on the formula

$$F = xF_x + x'F_{x'}.$$

For example, $F = abcd + ah + a'mn + e$. If a is selected, then $F = aF_a + a'F_{a'}$, $F_a = F|_{a=1} = bcd + h + e$, $F_{a'} = F|_{a=0} = mn + e$.

Realising that if cofactoring is performed recursively, a node function will usually be decomposed to a set of sub-functions and each sub-function is feasible. Because we use bin-packing to perform LUT implementation, it will do nothing for feasible nodes. This makes the decomposed network dominate the final results which are not delay-optimal. For this reason, In our modified Shannon cofactoring approach, we selects the variable with the maximum arrival time as the variable to be decomposed so that the late arrival variable is implemented in the output LUT of the decomposed network. In addition, we only perform cofactoring for the variables with the maximum arrival time.

For example, a function is $F = Xabd + Xabc + X'efg + hi$. All variables except X are primary inputs and so X has the maximum arrival time. After cofactoring, we have $F = XK_1 + X'K_2$, $K_1 = abd + abc + hi$, and $K_2 = efg + hi$. All nodes except K_1 in the decomposed network are feasible. After performing bin-packing, we have $F = XK_4 + X'K_2$, $K_3 = abc + abd$, $K_4 = K_3 + hi$, and $K_2 = efg + hi$.

Collapsing K_4 gives the final mapped results: $F = XK_3 + Xhi + X'K_2$, $K_3 = abc + abd$, $K_2 = efg + hi$.

The above modified cofactoring method is quite effective for delay reduction in the step *partial collapsing and remapping*. If *collapsing and remapping* is applied, other decomposition approaches have to be applied because of the following two reasons.

1. All the variables in the node function to be decomposed are primary inputs. There is no advantage to use the modified cofactoring.

2. While the above method is useful for delay reduction, it has the disadvantage of requiring more areas because of the logic duplication.

Obviously, the above result can also become: $F = XK_3 + hi + X'K_2$, $K_3 = abc + abd$, and $K_2 = efg$ if logic duplication is avoided. This is especially necessary for those node functions collapsed to the two-level form. Therefore, other decomposition approaches are required for area minimisation.

B. Best decomposition. This decomposition method applies kernel extraction approach proposed in [14] to perform decomposition. Here the *co-kernel cube matrix* is formed for each node function. All kernels are created and the best kernels in reducing the number of literals in Sum-of-Products (SOPs) form are selected as the decomposition functions (new divisors). This, of course, helps to break big nodes into a set of small ones.

For example, suppose a function F is
$F = abce + abde + ace'i + ade'i + a'fgh$.
After selecting one best kernel, it becomes $F = a'fgh + Kbe + Ke'i$, $K = ac + ad$.

C. LUT directed kernel extraction. The main advantage of this method is to reduce the support count of the function to be decomposed. In addition to the literal count, the support count is another important factor affecting the area cost and delay cost for LUT based FPGAs [12].

Consider the above example again.
$F = abce + abde + ace'i + ade'i + a'fgh$.
After performing LUT directed kernel extraction [12], it gives $F = a'fgh + aK$ and $K = bec + bde + ce'i + de'i$, in which the variables available in the decomposition function K are not available in the composition function F. So the support count of two functions is minimised.

5 Experimental results

The proposed approach has been implemented in 'C' and added into SIS [17]. To evaluate it, a standard set of benchmark examples has been synthesised. Table 1 gives our results and those provided by state-of-the-art approaches. To compare our results with others, the level count and the LUT count of our results are listed. Columns labelled *levs* give the maximum number of levels of mapped circuits, and columns labelled *luts* give the number of nodes (LUTs). The inputs to our approach are the optimised circuits using ABLO [12], which are the same as those used by ABLO-d [10]. As our approach is an enhancement of ABLO, we call it **ABLO-new**. Comparing columns 2 and 3 with columns 4 and 5 of Table 1 indicates that our approach requires 10% fewer levels and 2% fewer LUTs than ABLO-d [10]. Although the achievement of 10% fewer levels sounds not a lot, it should be pointed out that some of ABLO-d results have reached the minimal levels (e.g., z4ml, rd73, f51m, and misex1). As for 2% fewer LUTs, the LUT count of our approach is diversely affected in order to further reduce the circuit level. Take *count* as an example. When its level is 3, our approach only requires 36 LUTs (ABLO-d requires 42 LUTs for the same level count). When our approach reduces its level to 2, the area increases to 52 LUTs. The similar case happens for examples *rot* and *C880*. Nevertheless, if both the level count and the LUT count are considered, it can be seen that the proposed approach outperforms ABLO-d on 17 examples, equal on 6 and worse only on 2.

While different optimised circuits will affect the final mapped results, we still list the reported results provided by some state-of-the-art approaches. If final mapped results are concerned, it is seen that our approach requires 22% fewer levels and 36% fewer LUTs than MIS-pga [13], 27% fewer levels and 40% fewer LUTs than Flowmap-r [4]. If compared with recently published algorithms, ABLO-new also requires 8% fewer levels and 18% fewer LUTs, 15% fewer levels and 30% fewer LUTs than ALTO [7] and BDD [1], respectively.

To further evaluate the proposed approach, the mapped designs are placed and routed on real FPGA chips using a vendor's layout tool PPR [8] so that the final maximum delay of mapped circuits can be obtained. As we only have the mapped results of ABLO-d, we then compare with ABLO-d here. The inputs to PPR are mapped circuits of both ABLO-new and ABLO-d. The maximum delay from primary inputs to primary outputs (in terms of nanoseconds) after running PPR are shown in Table 2, in which all our mapped results reported in Table 1 that can be fitted into one XC3000 chip are listed. Columns labelled with *Device* are the types of FPGA chips used, which is based on a rule of 80% chip usage. Columns labelled with *Delay* are the maximum circuit delays after PPR. For easy comparisons, the CLB count (columns labelled with *CLB*) and the level count (columns labelled with *lev*) are also reported in the table. It is noted that ABLO-new requires 8.3% fewer circuit delays when compared with ABLO-d, whereas their average level difference among these examples fitted into one FPGA chip is 9%. This can be explained by investigating the example *count*. For this example, ABLO-d requires 3 levels while ABLO-new only requires 2 levels (33% fewer levels). However, because ABLO-new used more extra areas

Name	ABLO-new		ABLO-d		MIS-pga (delay)		FlowMap-r		ALTO		BDD	
	levs	luts	levs	luts	levs	luts	levs	luts	levs	luts	levs	luts
z4ml	2	5	2	5	2	10	3	13	2	5	2	6
misex1	2	15	2	16	2	17	2	15	2	14	2	18
vg2	3	25	3	23	4	39	4	38	3	26	4	33
5xp1	2	17	2	21	2	21	3	23	2	19	2	19
count	2	52	3	42	4	81	4	73	3	47	4	80
9sym	3	7	3	7	3	7	5	61	3	7	3	6
9symml	3	7	3	7	3	7	5	58	3	7	3	6
apex7	3	78	4	54	4	95	4	80	4	77	4	80
rd84	3	10	3	10	3	13	4	43	3	13	3	13
e64	4	113	4	146	5	212	-	-	3	144	-	-
alu2	5	76	5	87	6	122	8	148	6	61	6	95
duke2	4	111	4	114	6	164	4	187	4	156	5	218
apex6	4	181	4	183	5	274	4	232	4	229	5	236
sao2	3	32	3	29	5	45	-	-	3	38	4	26
rd73	2	7	2	7	2	8	-	-	2	8	2	8
misex2	2	31	2	32	3	37	-	-	2	37	2	41
f51m	3	14	3	24	4	23	-	-	3	15	3	81
clip	3	23	3	28	4	54	-	-	3	33	-	-
b9	3	35	3	37	3	47	-	-	3	36	3	46
apex2	6	85	6	99	6	116	-	-	-	-	-	-
rot	6	215	8	183	7	322	6	243	7	214	-	-
alu4	6	91	8	96	11	155	10	245	8	259	-	-
C880	6	114	8	89	9	259	8	211	8	96	8	147
C5315	8	401	10	438	10	636	-	-	-	-	-	-
misex3	6	216	-	-	-	-	-	-	6	251	-	-
Total	88	1745	98	1777								
Total	88	1745			113	2764						
Total	54	1004					74	1670				
Total	80	1475							87	1792		
Total	55	817									65	1160

Table 1. Comparison of experimental results.

(52 LUTs, 24% more than ABLO-d) to implement it, its final delay after PPR is only 15% better than ABLO-d's. Therefore, this justifies our basic consideration that it is important to reduce both the level count and the LUT count (i.e., cell delays and wire delays) for achieving final delay optimisation.

The improvement in performance produced by our approach can be attributed to the following three factors: (1) minimising area while minimising circuit delay so as to reduce both block delay and wiring delay, and also determining critical nodes based on the arrival time; (2) enhanced bin-packing algorithm to remap the critical nodes for delay reduction after partial collapsing and collapsing; (3) better decomposition strategies.

Name	ABLO-new				ABLO-d			
	Device	CLB	levs	Delay	Device	CLB	levs	Delay
z4ml	XC3020LPC84	5	2	24.6	XC3020LPC84	5	2	24.6
misex1	XC3020LPC84	15	2	34.1	XC3020LPC84	16	2	36.2
vg2	XC3020LPC84	25	3	42.5	XC3020LPC84	23	3	41.1
5xp1	XC3020LPC84	17	2	31.0	XC3020LPC84	21	2	31.9
count	XC3020LPC84	52	2	43.8	XC3020LPC84	42	3	51.5
9sym	XC3020LPC84	7	3	35.2	XC3020LPC84	7	3	35.2
9symml	XC3020LPC84	7	3	42.4	XC3020LPC84	7	3	42.4
apex7	XC3042LTQ144	78	3	56.5	XC3042LTQ144	54	4	61.9
rd84	XC3020LPC84	10	3	35.8	XC3020LPC84	10	3	35.8
e64	XC3090LTQ176	113	4	97.3	XC3090LTQ176	146	4	119.0
alu2	XC3030LPC84	76	5	74.8	XC3030LPC84	87	5	88.1
duke2	XC3042LPC84	111	4	95.1	XC3042LPC84	114	4	107.7
sao2	XC3020LPC84	32	3	49.7	XC3020LPC84	29	3	49.1
rd73	XC3020LPC84	7	2	26.1	XC3020LPC84	7	2	26.1
misex2	XC3020LPC84	31	3	36.9	XC3020LPC84	32	3	37.8
f51m	XC3020LPC84	14	3	37.9	XC3020LPC84	24	3	44.2
clip	XC3020LPC84	23	3	44.6	XC3020LPC84	28	3	45.2
b9	XC3020LPC84	35	3	47.8	XC3020LPC84	37	3	47.9
apex2	XC3042LPC84	85	6	116.4	XC3042LPC84	99	6	123.1
alu4	XC3042LPC84	91	6	116.6	XC3042LPC84	96	8	131.4
C880	XC3042LTQ144	114	6	101.5	XC3042LTQ144	89	8	118.2
Total			71	1190.6			77	1298.4

Table 2. The final delays (in terms of nanoseconds) after PPR routing.

6 Conclusions

A mapping approach for performance optimisation was presented in this paper. Unlike previous approaches which determine the critical nodes under the unit delay model, the critical nodes are determined based on the arrival time of each node, which contains both the cell delays and delay contributions of fanout load. An effective delay reduction process was proposed in a three-stepped algorithm, which possesses the advantage of minimising the area increase when the delay is reduced. The delay reduction techniques (or remapping techniques) are mainly based on modified cofactoring and bin-packing. Experimental results have shown the benefits of the proposed approach.

References

1. Chang, S. C., Sadowska, M., Hwang, T. T.: Technology Mapping for TLU FPGAs Based on Decomposition of Binary Decision Diagrams. IEEE Trans. on Computer-Aided Design of Integrated Circuits and Systems, **15(10)** (1996) 1226-1235

2. Cong, J., Ding, Y.: FlowMap: An Optimal Technology Mapping Algorithm for Delay Optimization in Lookup-Table Based FPGA Design. IEEE Transactions on Computer-Aided Design of Integrated Circuits and Systems, **13(1)** (1994) 1-12

3. Cong, J., Ding, Y., Chen, K.: An Optimal Performance-driven Technology Mapping Algorithm for LUT based FPGAs under arbitrary net-delay models. Int. Conf. on Computer-Aided Design and Computer Graphics, (1993) 599-604

4. Cong, J., Ding, Y.: On Area/Depth Trade-off in LUT-Based FPGA Technology Mapping. 30th Design Automation Conference (DAC), (1993) 213-218

5. Cong, J., Hwang, Y. Y.: Structural Gate Decomposition for Depth-Optimal Technology Mapping in LUT-based FPGA Design. 33th Design Automation Conference, (1996) 726-729

6. Francis, R. J., Rose, J., Vranesic, Z.: Technology Mapping of Lookup Table Based FPGAs for Performance. IEEE International Conference on Computer-Aided Design, (1991) 568-571

7. Huang, J. D., Jou, J. Y., Shen, W. Z.: An Iterative Area/Performance Trade-Off Algorithm for LUT-Based FPGA Technology Mapping. IEEE International Conference on Computer-Aided Design, (1996) 13-17

8. Xilinx Inc.: The Programmable Gate Array Data Book, (1994)

9. Legl, C., Wurth, B., Eckl, K.: A Boolean Approach to Performance-Directed Technology Mapping for LUT-based FPGA Designs. 33th Design Automation Conference, (1996) 730-733

10. Lu, A.: Logic Synthesis for Field Programmable Gate Arrays. PhD Thesis, University of Bristol (1995)

11. Lu, A., Dagless, E., Saul, J.: DART: Delay and Routability Driven Technology Mapping for LUT Based FPGAs. International Conference on Computer Design (ICCD), (1995) 409-414

12. Lu, A., Dagless, E., Saul, J.: Tradeoff literals against support for Logic Synthesis of LUT based FPGAs. IEE Proceedings on Computers and Digital Techniques, **143(2)** (1996) 111-119

13. Murgai, R., Shenoy, N., Brayton, R. K., Sangiovanni-Vincentelli, A.: Performance Directed Synthesis for Table Look Up Programmable Gate Arrays. IEEE International Conference on Computer-Aided Design, (1991) 572-575

14. Rudell, R.: Logic synthesis for VLSI design. Ph.D thesis, UC Berkeley (1989)

15. Sawkar, P., Thomas, D.: Performance Directed Technology Mapping for Look-Up Table Based FPGAs. 30th Design Automation Conference, (1993) 208-212

16. Schlag, M., Kong, J., Chan, P. K.: Routability-Driven Technology Mapping for Lookup Table-Based FPGAs. International Conference on Computer Design: VLSI in Computer and Processors, (1992) 86-90

17. Sentovich, E., Singh, K. J., Lavagno, L., Moon, C., Murgai, R., Saldanha, A., Savoj, H., Stephan, P. R., Brayton, R. K., Sangiovanni-Vincentelli, A.: SIS: A System for Sequential Circuit Synthesis. Memorandum No. UCB/ERL M/92/41, Electronics Research Laboratory, University of California, Berkeley, (1992)

18. Togawa, N., Sato, M., Ohtsuki, T.: Maple: A simultaneous technology mapping, placement, and global routing algorithm for Field-Programmable Gate Arrays. IEEE International Conference on Computer-Aided Design, (1994) 156-163

19. Wurth, B., Eckl, K., Antreich, K.: Functional Multiple-Output Decomposition: Theory and an Implicit Algorithm. 32th Design Automation Conference, (1995) 54-59

Automatic Mapping of Algorithms onto Multiple FPGA-SRAM Modules

S.J.B.Acock, K.R.Dimond

Abstract

This paper describes the processes that have been developed to allow the automatic mapping of technology independent algorithms onto a network of reconfigurable hardware modules. The VHDL language is used at the behavioural level of abstraction to describe the algorithm. Each reconfigurable hardware module comprises of a single FPGA and SRAM. A mesh configuration of these modules provides the resources for data manipulation and data storage required by the algorithm. By pre-processing the algorithm prior to synthesis, and then performing network partitioning on the synthesised netlist, a hardware implementation is realised on multiple modules. This approach means that an algorithm may be mapped onto a versatile hardware system which is not constrained by the limitations of the target technology.

Introduction

With the introduction of Hardware Description Languages such as VHDL (VHSIC Hardware Description Language) [LRM93], either a part or a complete system may be specified as an algorithm, independent of the target technology. At the highest level of abstraction, this specification allows the designer to create a behavioural description of the systems functionality. Synthesis tools such as the Synopsys Design Analyzer, then translate this VHDL code into an intermediate format which is optimised and mapped on to the target technology using a specific device library. The output from this tool can be VHDL for subsequent simulation, in addition to a netlist format required by the place and route tools supporting the target device. In the case of an FPGA target device, the design will be constrained by gate capacity, data storage, and routing resources.

A requirement of many current designs is some form of data storage heap. This may be required for caching, pipelining processes, queuing data, etc. In implementing anything more than relatively small banks of memory, the architecture of FPGA devices are poorly equipped.

In this paper we address the inherent constraints and limitations of FPGA technology outlined above, by introducing a network of hybrid FPGA-SRAM processing modules as the target hardware. Each hybrid module contains a single FPGA device and a single SRAM device. By interconnecting these modules according to the design requirements, gate capacity and storage limitations are overcome. In this approach, the utilisation of each FPGA may also be set so that the routing of each device is feasible.

To facilitate the mapping of an algorithm on to the hybrid FPGA-SRAM modules, we have developed processes and supporting CAD tools to aid the designer. We have assumed that the algorithm will be specified in VHDL, and that a commercial synthesis tool (in our case we used the Synopsys design tools) is used to map the design to a module FPGA. Conceptually, our work research proposes two processes that aid the designer in implementing a system onto a network of hybrid FPGA-SRAM modules. These are illustrated by the design flow diagram of Fig.1.

The first process is the optimisation of the source VHDL code to translate data arrays into references for external memory residing on the hardware module. This *intermediate* VHDL code is then synthesised to the target FPGA technology and converted into a netlist format. The second process we have developed is to partition this netlist using multi-way partitioning techniques. The objective functions here are to partition for minimum cutsize of nets between the modules, and to maintain references to external memory in the correct modules. The partitioning enables large data sets which are larger than the capacity of the memory associated with a single FPGA to be mapped to two or more modules in a distributed memory system.

This paper introduces the hybrid FPGA-SRAM module architecture, and the processes used in the CAD tools that have been developed to automate the mapping of a design onto this hardware.

Hybrid FPGA-SRAM Module Architecture

We have designed a hybrid FPGA-SRAM module which is expandable depending upon the requirements of the design algorithm. By providing each module with four ports, they may be interconnected in a mesh configuration. The interconnected modules realise a distributed logic and memory system.

The design requirements for the FPGA-SRAM module are:

- In-circuit reconfigurable logic;
- A standalone system;
- The FPGA will have dedicated pins for SRAM connection;
- An expandable interconnection of modules
- Support for varying FPGA capacities

The architecture of a single module is shown in Fig.1. The modules are designed to be accommodated into a fixed interconnection structure. In designing such a structure, optimum use has been made in utilising the I/O capability of the FPGA. The modules are interconnected to maximise the number and visibility of nets between modules. The buses of nets connecting the modules are made manually.

A photograph of the hybrid FPGA-SRAM module is shown in Fig.2.

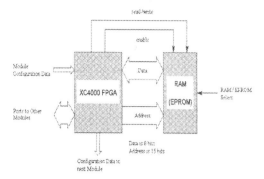

Fig. 1. FPGA-SRAM Module Architecture.

Comparison with other Virtual Platforms

A number of virtual hardware platforms currently exist, and our hardware has been designed to take account of the advantages and disadvantages of each. Some platforms such as the *BORG* [BOR94], *Anyboard* [ANY92] and *Transmogrifier* [TRA94] are based on a PC card architecture. Our hardware provides unlimited expansion as a standalone system and can be independent of a PC. The Duke University *GERM* [GER94] is also a standalone system, but does not have the advantage of our on module SRAM.

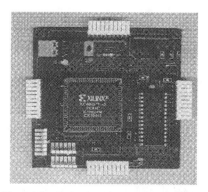

Fig. 2. The Hybrid FPGA-SRAM Module.

A VHDL Pre-Processor for Distributed Memory

An algorithm that describes a typical digital system may be decomposed into two distinct parts, those of data processing and data storage. In mapping such an algorithm onto a system of hardware, resources must be made available for these distinct requirements. This is realised by the FPGA-SRAM hardware. To perform the actual mapping of the algorithm, a pre-processing tool has been developed so that references to data array variables are directed at memory held by the SRAM hardware in the synthesised design. Data manipulation functions are retained within the FPGA.

Algorithm Pre-Processing

We have developed a pre-processor which parses VHDL code prior to synthesis. This code is first simulated for design validation, and then submitted to the pre-processor. This identifies variable assignments in the architectural body and substitutes references to the external SRAM on the module. At the same time, the interface description in the VHDL is modified to include ports for external memory, address and control signals. In this way an *intermediate* VHDL model is created for synthesis.

Data Types and Array Assignments

Temporary storage of data makes use of data *variables* in the architectural body. As data *array variable* types use a greater amount of memory, it is these types that are transformed for storage in SRAM. In this way, the SRAM memory of the modules is divided into memory mapped blocks for each array that is stored. An example of a variable assignment statement that would be transformed by the parser is shown:

```
trace_array (index) := new_data; - (1)
new_data := trace_array (index); - (2)
```

The statements in (1) and (2) are equivalent to writing and reading memory locations. After pre-processing they become external memory assignment statements for *intermediate* VHDL:

```
sram_data1 <= int_to_std_data (new_data); - (3)
new_data := std_data_to_int (sram_data1); - (4)
```

Where the functions *int_to_std_data* and *std_data_to_int* perform the task of converting data types in the algorithm to a subtype of the resolved enumeration type std_logic. This is required to create the correct logic in the synthesised design. The signal port *sram_data1* interfaces to the memory data bus, and references module 1. In the present version of the pre-processor, the array data types are limited to constrained integers of 8 bits, which is the width of the SRAM data bus.

Memory Addressing

External memory addressing is implemented by converting the index of the data array in the VHDL code to an 8-bit std_logic data type. This is synthesised as 8 lines which connect to the SRAM on a module. Addressing multiple memory devices in each module is achieved by a series of conditional IF-THEN-ELSE statements which are dependent upon the value of a memory address variable derived from the data array index in the source VHDL. Data is directed to a specific SRAM according to the range of addresses specified by the conditional statement. This is illustrated below:

```
IF mem_address < 32768 AND mem_address < 65536 THEN
    sram_addr2 <= mem_address;
    sram_data2 <= int_to_std_data(filtered_data);
    sram_rw2 <= '0';
    sram_en2 <= '0';
END IF;
```

In this example segment of pre-processed code the SRAM in module 2 is directed data in a write cycle by the conditional statement. The variable *mem_address*, determined by the original array index must be conditional upon the 32 K module memory size.

Distributed Memory Arrays

For a network of FPGA-SRAM modules the algorithm will not be constrained by the capacity of a single module. In implementing a distributed memory system, the array variable limits are monitored during pre-processing so that when the capacity of one SRAM is exceeded, an SRAM in a connecting module is utilised. This assumes that control logic is also mapped to the new module and that furhter SRAM devices that are used do not reside in modules with no logic functions in the FPGAs.

Concurrent Processes

Within a behavioural VHDL architecture each assignment statement is executed concurrently, unless it appears within a PROCESS construct, in which case all assignments are made sequentially. There could be several PROCESS constructs, each one executed concurrently depending upon the events in their respective sensitivity lists. This concurrency has implications for a data segmented system that stores multiple data arrays. Multiple memory accesses may occur during data assignments within each PROCESS at the same time. Memory clashes can be avoided by ensuring that external memory accesses can only occur concurrently for different modules i.e. Data array assignments in different processes never access the same memory at the same time.

In our work we employ different clock signals in the sensitivity lists of separate PROCESS statements. By using out of phase clocks the memory of a single module is never accessed at the same time by concurrent processes. A clocking scheme is also necessary to ensure data accesses are synchronous for external memory.

Synthesis

The intermediate VHDL code generated is in a form suitable for synthesis, observing correctly placed library statements and any packages which are required. The pre-processed intermediate VHDL code is then submitted to a synthesis tool. We used the Synopsys Design Analyzer. The technology libraries used were for the Xilinx XC4000 series FPGA devices. The optimised design is saved in the Xilinx Netlist Format (XNF) and then submitted to the Syn2xnf tool for checking and merging to produce a single XFF file.

Multi-Way Network Partitioning

To transform a design into a form suitable for mapping onto multiple FPGA-SRAM modules, a network partitioning tool is required. Our approach has been to apply an iterative improvement lookahead algorithm to the Xilinx XFF netlist generated by the Syn2xnf tool. We have based our network partitioning algorithm on the work of Krishnamurthy [KRI84] and Sanchis [SAN89], with important extensions to take account of the references to external memory. We also take into consideration distributed memory during partitioning, the constraints imposed by the physical limitations of the FPGA devices, and the interconnectivity of the modules. Objective Functions for Cost Minimisation. To achieve a realisable partition of the source network, a number of objective functions have been identified:

(i) The maximum number of I/O pins used by each FPGA must be minimised and pin constraints must not be violated.

(ii) For a multiple FPGA-SRAM module system, the number of crossing nets must be minimised

(iii) The nets of a cell v_i should be restricted to a single port P, crossing a partition from a subset V_{b1} to a subset V_{b2} where $V_{b1} \in V$ and $V_{b2} \in V$. The total number of ports $P = 1,2,3...m$ is dependent upon the network topology.

(iv) For a given block b, if a net e_u is directly connected to the memory device, then any cell v_i that is incident on net e_u should be locked in block b for all passes of the algorithm.

(v) The size constraint of each FPGA device must not be exceeded. More specifically, the size should be restricted to upper and lower bounds, these are imposed to accommodate routing limitations within the devices in addition to maintaining a well balanced partition.

(vi) The capacity of each module's memory device must not be exceeded. For the network there will be memory of total size M bytes. Each module contains a memory device $Mb \in M$ which will have a size constraint $S(M_b)$. If a block b has all elements of M_b full, then another memory device in the network must be used.

Our network partitioning algorithm has been designed to achieve these objective functions and network constraints, completely automating the mapping of a netlist onto multiple FPGA-SRAM modules.

Overview of the Core Iterative Improvement Algorithm

The source netlist describing the network is written into a cell database, describing the attributes of each logic cell in a an array of bucket linked list data structures as described by Fiducci and Matthyeses, but extended to K-way partitions. Initially, the algorithm calculates the number of blocks (each block equalling one module) required to accommodate all of the cells in the source network. This includes checking the FPGA device type specified in the source Xilinx netlist and referencing a list of supported devices with their logic specification for the number of CLBs/IOBs they contain. A parameter is set which ensures that each FPGA does not reach 100% utilisation.

There is an initial random partition of the cells across all blocks. An iterative improvement algorithm is then employed to move cells between blocks according to their gain vectors (as described in the work of Krishnamurthy), at the same time observing the physical and interconnectivity constraints of the blocks. This stage consists of a number of passes, which are assessed upon the overall improvement in cutsizes of all the partitions. If there is no improvement after three consecutive passes then the cell database is restored to the best pass.

Clustering Highly Connected Cells

The efficiency of the algorithm is enhanced through exploiting the concept of level gains introduced by Krishnamurthy. We use this concept to ensure that highly connected cells are retained as clusters during a move to a new block, creating a more optimum partition than the standard approach of Fiducci and Matthyeses. this has the additional advantage in that timing critical logic can be bound together in this scheme. Clustering is only permitted if the network constraints are not violated.

Constraining the Partitioned Network

An even distribution of the network cells amongst all blocks is desirable to facilitate efficient FPGA usage, and to prevent an asymmetric cell migration to a single block. For subsets $Vb \in V$, where $b = 1,2,3...B$, this will be subjected to the block limitation : $Cg \leq |Vb| \leq Ch$ where C_g gives the minimum size constraint, and C_h gives the maximum size constraint and $S(V)$ the total size, for $0 < C_g \leq C_h < S(V)$.

Network Topology

The network topology is dependent upon the number of FPGA-SRAM modules required by the design and ensures maximum visibility between modules. The channel width of each module port is limited to a number of I/O pins. By evaluating the number of nets that will be contributed to a partition during each cell move, if this exceeds the channel width the move is rejected.

Memory Interfacing

Memory references from the pre-processed intermediate VHDL code produced by the VDSPP tool are retained in the source netlist submitted to the network partitioning algorithm. Through the parsing of the memory signal port names in the netlist, nets connected to the signal ports are identified. This then allows the algorithm to determine which cells are directly connected to the memory port nets. A module number for such a cell is stored in the cell database and has the effect of fixing the cell in the specified module. In this way logic cells are present in the FPGA to interface directly to its associated memory SRAM.

Netlist Output

The partitioned cell database is written to multiple Xilinx netlist files for submission to the Xilinx suite of place and route tools. A sophisticated set of pin allocation functions have been developed to correctly assign pin numbers to each netlist in the network.

A Design Application for Multiple FPGA-SRAM Modules

A VHDL Model of an image contrast enhancement algorithm is used to illustrate the mapping of a design that requires separate logic and data storage over multiple modules. The design is essentially an ALU coupled to an image store, ideally suited to our network of FPGA-SRAM modules. The image store is 64Kbytes, so it will require two modules. The algorithm processes individual 8-bit pixels as they are written to the image store increasing their value by a fixed number if they are above 128, and decreasing their value if they are below 128. This provides the contrast enhanced image in the image store. A block diagram of the pre-processed VHDL model is shown in Fig.3.

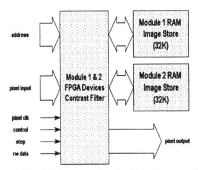

Fig. 3. Functional Block Diagram for the Pre-Processed VHDL Model.

A code segment of pre-processed VHDL is shown in Fig.4. The interface description contains the new signal ports for the external SRAM memory in the network. The architectural specification contains data assignments to write and read from the image store. Note the two different SRAM references automatically inserted by the pre-processor. After synthesising this design using the Synopsys Design Analyzer, it is submitted to the network partitioning algorithm. This reports 305 logic cells, which will require four FPGA-SRAM modules, assuming an XC4003 FPGA.

The logic cells incident on the nets connecting to the SRAM signal ports are locked into the respective modules. Partitioning the design results in a logic cell distribution of 92,91,61and 61cells. The four individual netlists for each module are then written to files for routing by the Xilinx place and route tools.

Conclusions

We have designed and demonstrated an automated approach to the mapping of high level abstract algorithms described in VHDL onto a network of FPGA-SRAM modules. By devising a hybrid FPGA-SRAM architecture, the constraints of limited logic and data storage cells in a conventional FPGA are overcome. Multiple FPGA and SRAM devices allow large algorithms to be considered for implementation in versatile programmable logic environment for design prototyping and implementation. By pre-processing the design prior to synthesis, external memory references may be incorporated into the code, removing array variables which would use up the CLB resources in an FPGA. This pre-processing stage could additionally be adapted to format a source VHDL design into formatted code suitable for synthesis.

```vhdl
use work.filter_utils.all;
use work.vdspp_utils.all;

library ieee;
use ieee.std_logic_1164.all;


ENTITY contrast_filter IS
    PORT ( pixel_input : IN word;
        address : IN addr_word;
        pixel_clk : IN bit;
        control : IN bit;
        step : IN adjust;
        rw_data : IN bit;
        pixel_output : OUT word ;
        sram_addr1 : out std_addr;
        sram_data1 : inout std_data;
        sram_rw1 : out bit;
        sram_en1 : out bit;
        sram_addr2 : out std_addr;
        sram_data2 : inout std_data;
        sram_rw2 : out bit;
        sram_en2 : out bit);
END contrast_filter;

ARCHITECTURE behavioural OF contrast_filter IS
BEGIN
    variable mem_address : mem_addr;
    start : PROCESS ( pixel_clk )
        VARIABLE filtered_data : word;
    BEGIN
        IF pixel_clk'EVENT and pixel_clk = '1' THEN
            IF rw_data = '0' THEN
                -- Write data to framestore
                filtered_data := ctr_enh(step,control,pixel_input);
                mem_address := address;
                IF mem_address < 32768 THEN
                    sram_addr1 <= mem_address;
                    sram_data1 <= int_to_std_data( filtered_data);
                    sram_rw1 <= '0';
                    sram_en1 <= '0';
                END IF;
                IF mem_address > 32768 AND mem_address < 65536 THEN
                    sram_addr2 <= mem_address;
                    sram_data2 <= int_to_std_data( filtered_data);
                    sram_rw2 <= '0';
                    sram_en2 <= '0';
                END IF;
            ELSE
                -- Read data from framestore
                mem_address := address;
                IF mem_address < 32768 THEN
                    sram_addr1 <= mem_address;
                    filtered_data := std_data_to_int(sram_data1);
                    sram_rw1 <= '1';
                    sram_en1 <= '0';
                END IF;
                IF mem_address > 32768 AND mem_address < 65536 THEN
                    sram_addr2 <= mem_address;
                    filtered_data := std_data_to_int(sram_data2);
                    sram_rw2 <= '1';
                    sram_en2 <= '0';
                END IF;
                pixel_output <= filtered_data;
            END IF;
        END IF;
    END PROCESS start;
END behavioural;
```

Fig.4. Pre-Processed VHDL Model for Image Contrast Enhancement.

We assume that the designer will use a conventional synthesis tool such as the Synopsys Design Analyzer. This is used to generate a Xilinx netlist format netlist that is then partitioned into multiple netlists targetted at each FPGA-SRAM module.

Multi-way network partitioning techniques are used to create the netlists. The algorithms we employ here also determine precisely the number of modules required in the final design. The core algorithms used include the enhancement to accommodate a distributed memory network, and additionally have a specific knowledge of the physical network topology onto which the design will be mapped.

References

[ACO96] S.J.B.Acock, K.R.Dimond, "Hardware Implementations of Algorithms on Networks of FPGA Processors", *IEE Professional Group C2 Colloquium* Digest No. 96/029: "Digital System Design using Synthesis Techniques", pp.3/1-3/6, Feb. 1996.

[ANY92] D.E.Van den Bout, J.H.Morris, D.A.Thomae, S.Labrozzi, S.Wingo, D.Hallman, "Anyboard: An FPGA-Based, Reconfigurable System", *IEEE Design and Test of Computers*, pp.21-30, Sept. 1992.

[BOR94] P.K.Chan, "A Field Programmable Prototyping Board: XC4000 BORG User's Guide", *Computer Engineering, University of California*, Santa Cruz, 1994.

[FID82] C.M.Fiducci, R.M.Mattheyses, "A Linear Time Heuristic for Improving Network Partitions", *19th Design Automation Conference Proceedings*, pp.175-181, 1982.

[GER94] J.D.Sterling Babcock, A. Dollas, B.Ward, "Generic Reusable Module (GERM)", *Dept. Electrical Engineering, Duke University*, Durham, North Carolina, Sept. 1994.

[KRI84] B.Krishnamurthy, "An Improved Min-Cut Algorithm for Partitioning VLSI Networks", *IEEE Transactions on Computers*, Vol. C-33, No. 5, pp.438-446, May 1984.

[LRM93] "IEEE Standard VHDL Language Reference Manual", *IEEE Standards Board, The Institute of Electrical and Electronic Engineers Inc.*, 1993.

[SAN89] L.A.Sanchis, "Multi-Way Network Partitioning", *IEEE Transactions on Computers*, Vol. 38, No. 1, pp.62-81, Jan. 1989.

[TRA94] D.Galloway, D.Karchmer, P.Chow, D.Lewis, J.Rose, "The Transmogrifier: The University of Toronto Field-Programmable System", *Technical Report CSRI-306*, June 1994.

[XIL96] Xilinx, "The Programmable Logic Data Book", *Xilinx Inc.*, 1996.

FPLD HDL Synthesis Employing High-Level Evolutionary Algorithm Optimisation

R. Bruce Maunder, Zoran A. Salcic and George G. Coghill

Department of Electrical and Electronic Engineering
University of Auckland
20 Symonds St., Auckland, New Zealand
r.maunder@auckland.ac.nz

Abstract. This paper presents a novel approach to optimising high level designs for circuits to be implemented on FPLDs. The aim is to search the design space using an evolutionary algorithm to find solutions that optimise circuit speed and circuit size under given constraints. To ensure correct circuit operation, a library of synchronous functional modules are used and interchanged in the circuit, altering only each module's data type (not its functionality). After modifying a circuit, modules for data synchronising and type conversions are added automatically. It is these extra modules that cause the search to become non-linear, indicating that the combination of optimised sub-circuits does not necessarily give an overall optimised circuit. The input to the synthesiser and optimiser is a netlist of modules, while the output is a completely specified Altera Hardware Description Language (AHDL) listing ready to be compiled. The main advantages of the method are that existing sub-circuits can be utilised, circuits can often be fit into available hardware without being redesigned, advances in algorithms and sub-circuit designs can be utilised, and low-level compilers and optimisers are left to their speciality.

1 Introduction

New high density field programmable devices (which exceed 100 000 usable gates) are of a capacity that does not always require very low level mapping of a design onto the target technology. Tradeoffs are acceptable and can be made between design effort and final circuit size and speed. Assuming some inefficiency is allowable then tools such as those presented in this paper can bridge the gap between an abstract design and a final circuit solution; thus, removing some of the onus from the designer. It is especially useful for partitioning a design into limited hardware. One of the main foreseeable applications is in the development of application specific execution units, perhaps spread over multiple devices, for custom configurable processor cores [1]. A useful subset of these is neural networks that take advantage of stochastic modules.

A circuit to be optimised and then synthesised is described by a netlist of interconnected synchronous modules. This netlist is altered through having modules swapped with other modules from the module library. The aim of this is to find modules that perform exactly the same function, but on different data types, such as serial, parallel or stochastic representations. By changing the data type of the module, different size and timing requirements will be used, enabling a search of the design space. Also, modules that have different pipelining structures, such as those created using Altera's Library of Parameterized Modules (LPM) provide the same tradeoffs with different parameters. Modifying the modules' types provides non-linear changes in the circuits size and time requirements. Also, the data flow analysis is a recursive procedure. Lastly, the design space is very large for even moderately sized circuits. It is therefore infeasible to use an analytic search method to find an optimal solution.

The intent of our method is to provide high level solutions of a circuit within given constraints. Changing the data types of parts of a circuit or deciding on the acceptable latencies of subcircuits is otherwise a design consideration aided by a designer's experience. Rather than replacing conventional synthesis from a high-level input, this method forms a preprocessor that removes that abstraction from the netlist and optimises it in novel ways using an evolutionary algorithm. The output of the synthesiser is an AHDL listing that fully describes the designed circuit, including a state machine controller to control the timing of the circuit. It includes functional prototypes so that the synchronous library modules may be linked in at compile time. The advantage of having the output in AHDL is that existing tools, specialising in the conversion of AHDL to the configuration bit-stream of the target device, can be utilised.

2 Design-Space Search

Conventional design relies upon the experience and knowledge of a designer to find a solution that optimally complies with all the requirements. The parameters that are considered by this research are circuit size, the number of clock cycles for operation, and the accuracy of the results. These may be weighted to control the optimisation process. There are also hard constraints that must be met — for example, there may be only a limited number of available FPLD logic cells. The design space search is effected by modifying the data types of various module components of the circuit. The search is non-linear; it not sufficient to optimise parts of the circuit and then combine these optimised sub-circuits. Even individual modules that are both smaller and faster may not be the most efficient when delays and converters are introduced. The size of the design space of m modules and d types (including different versions of the same type) is d^m assuming there is a module of every type for each module present in the circuit. Even in considering parallel, serial and stochastic types for a 25 module circuit, there are 8.5×10^{11} combinations. An evolutionary algorithm was chosen as the search method for finding a near-optimum solution, although if trivially small then an exhaustive search is feasible.

The design space is dramatically altered by even small modifications to module interconnections. Figure 2.1 shows the entire 'serial/parallel' design space of a circuit containing nine adders in series and the smearing effect of adding three feedforward connections. Note that there are many configurations that give the same point in the design space.

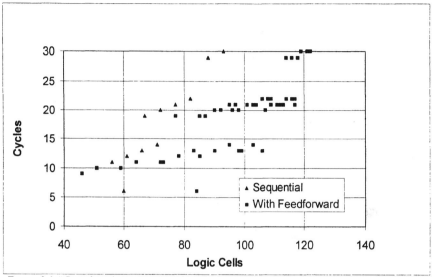

Figure 2.1 "Complete design spaces of two six-adder circuits. Smearing is shown through the addition of three feedforward module interconnections."

2.1 Synchronous Modules

Each synchronous module is fully described in AHDL (Altera Hardware Description Language) and has an associated file that lists its requirements and attributes. These modules and their information files form a library. The library currently contains 78 modules including arithmetic, logic and conditional functions as well as a few relating to neural network components. With minor modifications to the interface and the creation of an information file, most existing AHDL circuits can be used by the synthesiser and optimiser. There is no limitation on the complexity of a module, and it may contain loops and feedback within itself. One constraint, however, is that the number of cycles for its operation must be known a priori — it may be necessary to consider the worst case. The two files that describe each module are:

Module library information files (*.inf): Each textual information file describe the various parameters of a particular synchronous module. Described module parameters include function, number of cycles, propagation delay, number of cells (for a specific device architecture), the type of data it processes, whether it is clocked, number of inputs and outputs, and required control signals. The required number of cells is based on a single implementation and the sum of these is only an approximation to a complete circuit's requirements.

Module library text design files (*.tdf): Each design file contains the AHDL code for a synchronous module that will be linked in at compile-time. Each is encapsulated in a standard interface shell which conforms to the given information file specification.

2.2 Module Data Types

The data types (DTs) currently considered are unipolar 8 bit parallel, 8 bit serial and 1000 bit stochastic (which has 6 bits of accuracy). Stochastic numbers consist of a probabilistic serial bit stream where the proportion of '1's dictates the represented value. Stochastic DTs allow result accuracy and processing time to be modified to obtain the optimum balance. The length s of a stochastic bit stream that is v bits accurate is given by $s = 2^{(2v-2)}$ [2]. The outcome of using different DTs is the different routability and calculation processing requirements in terms of silicon resource usage and function processing time.

A more subtle advantage is the suitability of some data types for specific functions. For example, the multiplication of two stochastic numbers requires only a single two-input logic gate. Similarly, stochastic numbers are very useful for artificial neural networks. A circuit to perform input summation and implement a sigmoidal activation function of n streams requires an n-input lookup table or a collection of cascaded smaller lookup tables [3].

3 Hardware Description Language Synthesis

The task of the synthesiser is to take a circuit description and by removing any abstraction, create an unambiguous HDL listing. The synthesiser uses textual input files as discussed below, and provides textual output files with an AHDL listing that is always further synthesisable by Altera's synthesiser.

3.1 Synthesiser Input Files

The synthesiser utilises the library files described in section 2.1 as well as a netlist file *(*.net)* that describes which modules are to be included in the circuit, their data type and the interconnection between modules. A 32-bit Microsoft Windows application has been developed in C++ that provides a graphical utility for creating and editing the netlist files. Additional information is included in the netlist files created by this utility to dictate relative visual layout positions of the synchronous modules.

The netlist does not require information about clocks, functional prototypes, global clear signals, delays, converters and numerical types, or latencies — as this is handled by the synthesiser.

3.2 Synthesiser Operation and Algorithm

The synthesiser module takes a netlist of interconnected synchronous modules as its input. It also utilises a list of available modules and each module's parameters. From these it generates AHDL code that when compiled along with the AHDL descriptions of the synchronous modules, produces the configuration bit stream for an appropriately targeted FPLD. The synthesis algorithm is outlined in figure 3.1. Special care is taken in step 4 if an output is required to be converted to the same data type for two different modules. In this case, only one converter is added and the connection is moved such that the input to the second module is taken from the output of the converter. This eliminates the insertion of redundant converters.

1.	Load netlist.
2.	Check netlist integrity.
3.*	Include random number generator and address generator if stochastic numbers are used.
4.*	Add converters where necessary.
5.*	Determine data flow paths by conversion to an action/event network.
6.*	Add delays so that all input data arrives at each module in unison.
7.*	Calculate critical path and *number of cycles* for circuit operation.
8.	Synthesise main AHDL module.
	8.1 Functional prototypes.
	8.2 Main input and output connections.
	8.3 Module interconnections.
	8.4 Control lines connections to the scheduler.
	8.5 Clock and global clear signals where necessary.
	8.6 Sharing and uncorrelation of stochastic address lines.
	8.7 Commented code so it is possible for a user to make modifications for special needs.
9.*	Estimate *number of cells* (including delays and converters).
10.	Synthesise the scheduler AHDL module.
	10.1 State machine for resetting, initiation and control of each module.
	(* Functions are used for every individual in every generation of the EA)

figure 3.1 "AHDL synthesis algorithm overview."

The synthesiser also performs the following error checking:

- All module inputs are present and each is from only one source (to prevent conflicts).
- All module outputs are connected to at least one node.
- Overall circuit inputs and outputs are connected.
- All referenced modules are available.
- Appropriate converters are available to be added.
- Feedback is not present.

Data type converters are automatically inserted where necessary to make sure that the data that is presented at the inputs of each module is of the correct type. Delay modules of the appropriate data type are also automatically inserted into the netlist and given control signals at the correct times to ensure that data arrives at each module's inputs in phase and on the correct cycle. This means that a designer does not necessarily have to consider the times taken for each path through the circuit. A module that takes two cycles can be placed in parallel with a module that takes five cycles, for example, and data integrity will be maintained. If stochastic numbers are used then a pseudo-random number generator is automatically synthesised that is required for the creation of stochastic bit streams [4].

A state machine 'scheduler' is also synthesised to give the required internal control signals to each module at the correct times. Relative times for control signals are given for each module in its information file. These relative times are offset from the clock cycle at which all the input data arrives at a module to give the actual clock cycles for the pulses from the scheduler.

The analysis of the interconnected modules is performed by creating an action/event network representation. From this network, the critical path is deduced. This path dictates the shortest time that data will take from entering the circuit to leaving it. Once the duration of the critical path is known, delays can be inserted to ensure data appears at all module inputs in phase. The method used to accomplish this is to determine where *floats* are present and eliminate them. A float, in this case, is the number of clock cycles data is ready before it is required. They are removed by increasing the amount of time taken for a module to perform its function, which in practice is achieved by the addition of one or more delay modules. By definition, the critical path has a float of zero and therefore needs no extra delays [5].

The synthesiser has enough information to both estimate the circuit size (in logic cells) as well as the number of cycles taken for circuit operation. This information is used in the cost function calculation by the evolutionary algorithm.

4 Evolutionary Algorithm (EA) Optimisation

The task of the evolutionary algorithm is to search through the design space. An initially random population of possible solutions is modified by each individual undergoing the genetic operations of mutation and crossover. An 'individual' that represents a possible solution is a list of the synchronous modules that make up the circuit. The genetic operations that are performed on the individual have the effect of modifying some of the modules' data types but not their functionality. This means that the circuit is still guaranteed to perform the correct function. The *mutation* operator modifies the type of data that a module operates on as long as the functional module of the new type exists and there is provision for converting the data to the new data type. The probability of a module being mutated is given to the algorithm as a parameter. The *crossover* operator used is a single point crossover that creates two new individuals by breaking two individuals at a point and connecting the first portion of each to the last portion of the other. The crossover probability is given to the EA when it is initialised.

Reproduction of individuals into the subsequent generation is based on *roulette wheel selection*. An individual's fitness is calculated by how well it matches the constraints of size and accuracy. Parameters are given to the EA when initialising it which state how important size and time are and whether there are any hard constraints that *must* be met. Before the fitness of an individual is calculated, the necessary data-type conversion modules are added and delay modules are inserted. These additions will affect the circuit size and may or may not affect the critical path of circuit operation, hence affecting the circuit's processing time. The fitnesses are scaled before the selection process to reduce the rate of convergence to a particular solution. Without suitable scaling in place, "super" individuals can dominate in the early stages of the algorithm. Conversely, scaling is useful near the end of an evolutionary algorithm to prevent the population from becoming too similar and from ceasing to converge further [6].

4.1 Evolutionary Algorithm Configuration File

If an evolutionary algorithm configuration file (*.ga) is specified on the command line then optimisation is performed. The '*.ga' file is a text file containing configuration parameters for the evolutionary algorithm. The four fundamental parameters it contains are the number of generations, populations size, crossover probability and mutation probability. There are also two parameters which dictate how focused (*f*) the algorithm should be on optimising the circuit for size, time or a linear combination of the two. The linear combination utilises two positive fixed point numbers fs and ft which comprise part of the fitness function as $f = fs \times size + ft \times time$. These parameters must be tuned for a specific circuit. Parameter tuning is usually performed by monitoring the results of a small EA run and making adjustments. Lastly, there are two parameters; the first gives a *size limit* over which the fitness of a circuit is penalised by the second *size penalty* parameter. There are two similar parameters for time. The aim of these four parameters is to guide the evolution of the circuit to keep within desired constraints through penalties. It is important not to entirely dismiss individuals that do not fully comply because they could still be very close to a near-optimum solution that *does* comply with the requirements.

5 Results

The example circuit used consists of nine modules; five multipliers and four adders. It has three parallel inputs *a*, *b* and *c*, and three parallel outputs *d* and *e*. The circuit is an arbitrary implementation of the functions $d=2b(a+c)$ and $e=2c(a+b)$. Table 5.1 shows the EA results for runs with different guiding weights for the relative importance of size and time. An exhaustive search of the design space is given in figure 5.2, and the three solutions described in table 5.1 can be seen. Figure 5.3 shows the average population fitnesses of three evolutionary runs optimising for a combination of size and time.

Parameter↓ \ Optimised for→	Time	Size	Time and Size
Focus on cycles (*ft*)	1.00	0.00	4.95
Focus on cells (*fs*)	0.00	1.00	0.16
Number of converters	0	5	7
Number of logic cells	961	495	709
Number of cycles	8	33	25
Parallel adders	4	0	2
Serial adders	0	4	2
Parallel multipliers	5	0	2
Serial multipliers	0	5	3

Table 5.1 "Results for trials with different fitness weightings guiding the evolutionary algorithm toward three different goals."

Results from trials that allowed stochastic modules showed that if loss in accuracy and increase in circuit processing time could be tolerated then overall circuit size could be dramatically reduced. The overhead for the single required random number generator is quite high as shown in table 5.4, but the savings in logic cell usage for multipliers are also quite large. It should be noted that the length of a stochastic bit stream is only evident in the conversion back to a parallel representation due to the summation operation. The increase in latency for parallel to stochastic conversion and stochastic multiplication is one or zero cycles each depending on their configuration.

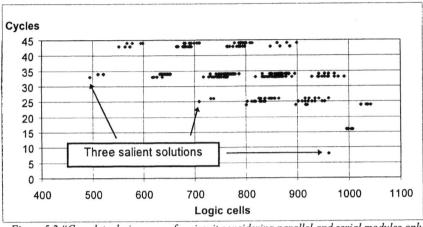

Figure 5.2 *"Complete design space for circuit considering parallel and serial modules only. Note the three salient solutions that were found by the EA"*

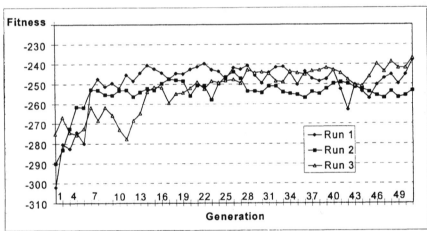

Figure 5.3 *"Average population fitness of three EA runs seeking to find a compromise of smallest circuit size and least number of cycles for circuit operation. fs=0.16, ft=4.95"*

10k Cells	Module Function
87	Parallel to stochastic converter
73	Random number generator (1 per circuit)
10	Stochastic to parallel converter
181	Parallel multiplier
1	Stochastic multiplier (scaled output)

Table 5.4 *"Logic cell usage for stochastic related modules implemented on '10k' devices."*

The time taken for a complete analysis of a circuit including adding delays and converters and determining the critical path takes $O(n^2)$ time where n is the number of modules in the circuit. This analysis takes most of the computing resource during an EA run. For a nine module circuit, the analysis of an individual netlist takes up to 0.6 seconds on a 166MHz Pentium (running 32 bit C++ code), while the same analysis for a fifteen module circuit takes approximately 2.0 seconds. For a circuit consisting of 50 modules, processing a generation of 20 individuals takes about 5 minutes. The required number of generations to converge to a solution in trials has been between 50 and 200.

6 Conclusions

This paper has presented a method that optimises circuits by modifying the types of data that is processed in various parts of the circuit. The optimisation is performed by an evolutionary algorithm that carries out a robust search of the design-space. Parameters can be weighted so that circuits can be altered to fit within available hardware without redesign. The output from the synthesiser contains no abstraction — it is able to be compiled for implementation in an FPLD. The method of optimisation and synthesis also provides the following advantages:

- Existing synchronous modules can be easily converted for use with it.
- It benefits from other areas of optimisation and research, such as better AHDL compilation.
- New modules created from better algorithms or designs are employable.
- It can optimise a circuit given multiple weighted constraints. This is especially useful in FPLD designs where size is often a hard constraint.
- The interface to a human designer is at a high level, and it removes many of the timing details of synchronous design.

More data types will be considered in future work, and the advantages and disadvantages analysed. Further data types will include representations with greater and fewer numbers of bits. Modifications may be made to the evolutionary algorithm to test the search abilities of different genetic operators. In addition, the advantage of different mappings of solutions to their genetic representations could be analysed along with automatic EA parameter tuning. Larger and smaller parallel and serial representations will be considered as well as hybrid types that have both parallel and serial components. Finally, a C-like language interface has yet to be developed and this will simplify the conversion to netlists of algorithms in software form.

References

[1] Salcic, Z., and Maunder, B.: CCSimP - An Instruction-Level Custom-Configurable Processor for FPLDs. *Proceedings 6th International Workshop on Field-Programmable Logic and Applications, Darmstadt, Germany*, pages 280-289, September 1996

[2] van Daalen, M., Jeavons, P., and Shawe-Taylor, J.: A Stochastic Neural Architecture That Exploits Dynamically Reconfigurable FPGAs. *Proceedings IEEE Workshop on FPGAs for Custom Computing Machines*, pages 202-211, April 1993.

[3] Bade, S.L., and Hutchings, B.L.: FPGA-Based Stochastic Neural Networks - Implementation. In *Proceedings of IEEE Workshop on FPGAs for Custom Computing Machines, Napa, CA*, pages 189-198, April 1994.

[4] van Daalen, M., Jeavons, P., Shawe-Taylor, J., and Cohen, D.: Device For Generating Binary Sequences for Stochastic Computing. *Electronic Letters*, 29(1):80-81, Jan 1993.

[5] Battersby, A.: *Network Analysis for Planning and Scheduling.* 2nd Ed. MacMillan and Company Ltd., London, 1967.

[6] Goldberg, D.E.: *Genetic Algorithms: in Search, Optimization and Machine Learning.* Addison-Wesley, 1989.

An Hardware/Software Partitioning Algorithm for Custom Computing Machines

Anton Velinov Chichkov
avc@dalton.inesc.pt

Carlos Beltrán Almeida
cfb@dalton.inesc.pt

INESC/IST, R. Alves Redol 9 1000 Lisbon Portugal
Fax: +351 1 3145843

Abstract

In this paper an Hardware/Software partitioning algorithm is presented. Appropriate cost and performance estimation functions were developed, as well, as techniques for their automated calculation. The partitioning algorithm that explores the parallelism in acyclic code regions is part of a larger tool kit specific for custom computing machines. The tool kit includes a parallelising compiler, an hardware/software partitioning program, as well as, a set of programs for performance estimation and system implementation. It speeds up the computationally intensive tasks using a FPGA based processing platform to augment the functionality of the processor with new operations and parallel capacities. An example was used to demonstrate the proposed partitioning techniques.

1.   Introduction

In this paper an hardware/software (HW/SW) partitioning algorithm, that divides one system in a way allowing for concurrent execution, is described. The objective of the undergoing research is the development of a *configuration compiler* [1]. The configuration compiler is a software tool having as input a C program and producing as output an hardware description and a software program that, together, will perform the same function as the input C program with a significant improvement of performance.

Computationally intensive applications typically spend most of their execution time within a small fragment of the executable code [2]. A general-purpose machine can substantially improve its performance by adapting the processor's configuration to these frequently accessed pieces of code. The processing platform, with FPGA based segments added to it, can be reconfigured to customise the architecture to some individual tasks [3]. Such an architecture retains its general-purpose nature, while having the performance benefits of application-specific architectures.

Partitioning the software program between the microprocessor and the reconfigurable resources is an hardware/software co-design problem. The HW/SW partitioner determines which portion of the C code should be implemented in hardware. The remaining code is implemented in software. Typically, this decision is based on the communication required between the partitions and the timing and area constraints of the system. A variety of algorithms and tools for Hardware/Software

divided into basic blocks by the compiler. Each basic block can include only one statement or a sequence of them.

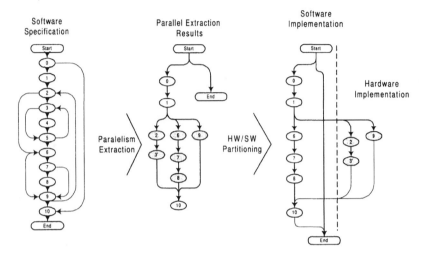

Figure 2 - HW/SW Partitioning with parallelism extraction

This initial division of the source by the compiler is used to build the dependence graph. The basic blocks are vertices of the graph and the edges represent control or data dependencies between them. The DCDG is an hierarchical model. The hierarchy is represented by the use of *link vertices*. There are two types of links: *call* and *cycle*. Call link represents the hierarchy of the source program, for example, functional calls. The call cycle is used to model cycles. The body of a cycle is represented by a DCDG. The back jump is eliminated and the number of cycles is controlled by the *cycle vertex*.

3.   System Partitioning

The task of the system partitioning is accomplished in two steps. In the first step the identification of the implicit parallelism is performed creating the DCDG model of the system which represents the local parallelism. In order to obtain the DCDG, the following analysis are performed: control flow *Figure 3-a*, post dominance *Figure 3-b*, control dependence *Figure 3-c* and data dependence analysis *Figure 3-d*. The resulting flow graph is scheduled *Figure 3-e* in order to solve the additional data dependencies caused by the parallelisation. All the dependencies conflicts not solved by this algorithm need an explicit synchronisation mechanism. All the analysis and transformations performed in this step are made at basic block level. The term basic block is used as it is defined in compiler design texts [2]. The methodology and the algorithms developed for this analysis are described in [7].

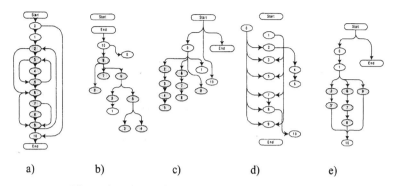

Figure 3 - Process of parallelisation of the C function.
a) Control flow graph. b) Post dominance tree. c) Control dependence graph. d) Data
dependence graph. e) Data and control dependence graph.

The second step of the algorithm determines which independent parallel branches of
the DCDG are implemented in hardware or in software. First, it analyses the
independent parallel branches in order to decide if they are suitable for hardware
implementation. Not all graph components can be transferred to hardware, some are
better implemented in software. For example, all operations dealing with standard
input and output are more suitable to a software implementation. The C language
library functions linked to the program in assembly code remain also in software.
This analysis is performed during the parsing of the input RTL file [7]. The basic
blocks are marked as "SW" if they contain some of the above components. After that
the performance of the independent parallel branches containing no blocks marked
"SW" is evaluated. The final decision for the implementation of the branches in
hardware is based on the value of its cost function. The cost function balance the
goals of maximising the performance and minimising the hardware size. This is
achieved by the use of two terms, one indicating the performance improvement, the
other the hardware size [6]. The performance term is weighed heavily to ensure that
a solution with increased performance is found, i.e., minimising the hardware size is
secondary. The cost function is:

$$Cost = Kp \times (\eta_{sw} - \eta_{hw}) - Ks \times N_{CLB}$$

Where the η_{sw} expresses the latency of the software implementation of the
branch, η_{hw} expresses the latency of the hardware implementation of the branch, Kp
and Ks are weight coefficients for the performance and the hardware size. The N_{CLB}
represents the hardware size as a number of *CLBs* used for the implementation. The
algorithm searches for the maximum positive value of the function. For the hardware
implementation only branches with *positive* cost function are scheduled. The
examples presented are made with $Ks = 0$ and $Kp = 1$ therefore considering only the
performance.

3.1   Performance Estimation

In order to make effective trade-offs during partitioning, it is necessary to make good estimates about software and hardware performance. One approach to the estimation would be to synthesise the hardware and compile the software for a given graph model and evaluate the resulting system. However, the process of compilation and synthesis is time intensive and may not be suitable when evaluating trade-offs among software and hardware implementations. In this work the time estimation is done using flow graph models for both hardware and software before the implementation.

Software Performance Estimation

An estimation of the software performance is characterised by assignment of delays to the operation vertices of the DCDG model. Each basic block in the flow graph is characterised by a number of memory accesses, m_{bb}, and a number of assembly-level operations $n_o(bb)$. The software delay time of the basic block, δ_{bb}, can be computed as follows [5]:

$$\delta_{bb} = \sum_{i=1}^{n_o(bb)} \delta_{op_i} + m_v \times \delta_m$$

where the δ_{op} is the time of the execution of the assembly-level operation represented by the number of clock cycles and δ_m is the memory access time represented by the number of clock cycles. In the expression of the memory access delay, the time of the address computation is not included, because in *RTL* it is represented by an assembly-level operation.

The latency of each operation depends on the processor instruction set and on the delays associated with these instructions. However, the estimation tool is independent from the choice of the CPU. It uses the machine description file to configure itself for the specific CPU architecture. The machine description file includes a table defining the delays of the RTL operations for this architecture. The machine model assumed in this work is the DLX [9] processor, which provides a good architectural model for study, not only because of the popularity of this type of machine, but also because it has an easy to understand architecture. After the calculation of the delays of each operation the latency of the basic block δ_{bb} can be calculated. The computation of the delays of the basic blocks is performed during the depth-first traversal of the graph. The delays are represented by a number of clock cycles needed to perform the operation.

The next step is the calculation of the branch delay. This is accomplished by the calculation of the corresponding graph length. For the hierarchical graphs the latency of the call vertex is equal to the length of the sub-graph to which it is linked. The latency of the cycle link vertex is equal to the length of the graph representing the cycle body multiplied by the number of executions of the cycle. The latency of the cycle link vertex can be unbounded if the number of executions of the cycle is data dependent. To avoid unbound delays in this approach is assumed that all cycles are executed once. The methodology for graph length calculation used in this work is based on [5] adapted for the system model. If the graph includes conditional vertices the length of the graph is a vector. The maximum length of the graph, i.e. the biggest component of the vector, is used as the estimation. As an example, the length of the graph represented in *Figure 4-a* is calculated.

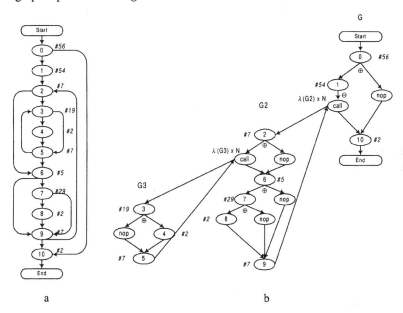

Figure 4 – Performance estimation of the software implementation of the function.

The length computation should be done using the hierarchical representation of the graph *Figure 4-b*. First, the lengths of the graphs G3 and G2 are calculated. Finally the length of the graph l(G) is computed.

$$l(G3) = \delta_3 \Theta(\delta_4 \oplus \delta_{nop})\Theta\delta_5$$

$$l(G3) = 19\Theta(2 \oplus 0)\Theta7$$

$$\underline{l}(G3) = (28,26)$$

Length of the sub graph G3:

$$l(G2) = \delta_2\Theta(\delta_{call3} \oplus \delta_{nop})\Theta\delta_6\Theta((\delta_7\Theta(\delta_8 \oplus \delta_{nop})) \oplus \delta_{nop})\Theta\delta_9$$

$$l(G2) = 7\Theta(\delta_{call3} \oplus 0)\Theta5\Theta((29\Theta(2 \oplus 0)) \oplus 0)\Theta7$$

$$\underline{l}(G2) = (78,76,57,50,48,29)$$

$$l(G) = \delta_0 \Theta((\delta_1 \Theta \delta_{call2}) \oplus \delta_{nop}) \Theta \delta_{10}$$

$$l(G) = 56\Theta((54\Theta \delta_{call2}) \oplus 0)\Theta 2$$

$$\underline{l}(G) = (190,188,196,162,160,141,58)$$

The maximum length of the software implementation of this graph is 196 cycles.

Hardware performance estimation

The hardware performance estimation gives the execution time of the hardware implementation of the branch η_{HW}:

$$\eta_{hw} = \begin{cases} (t_{hw} + t_{ovh}) - t_{swb} & for \quad t_{swb} < (t_{hw} + t_{ovh}) \\ \\ 0 & for \quad t_{swb} \geq (t_{hw} + t_{ovh}) \end{cases}$$

where t_{hw} is the execution time of the operations implemented in hardware, t_{ovh} is the overhead time caused by the interface circuits over the partitioning edge, and t_{swb} is the execution time of the software parallel branch. The overhead time depends on the interface implementation. In this work the interface through shared memory is assumed, and t_{ovh} is equal to the memory access time. The model used here assumes 10 software cycles. The time t_{swb} is obtained from the graph length of the branch and the software clock cycle.

The t_{hw} is estimated from the implementation by prediction. The system is implemented in reprogrammable FPGAs based hardware, where the delay time includes: Configurable Logic Blocks (CLBs) delay, wire routing segments delay and Programmable Interconnection Point (PIP) delay. The hardware performance estimation method is based on [8], adapted for the flow graph model.

Component delay estimation

Given a flow graph, the corresponding Lookup Table (LUT) mapping and CLB construction can be predicted. To predict LUT mapping, the algorithm groups the operations in K-input LUT, checking whether each visited statement can be included into the current LUT. If the total number of inputs is less than K+1 then the operation is included, otherwise, it is rejected.

In the XC4000 architecture, there are several possible CLB configurations. CLBs with different configurations have different LUT patterns and delays. Priority is given to patterns that can make good use of the H LUT and therefore would result in a more compact CLB netlist [8]. The algorithm starts with the LUT netlist and produces the CLB netlist by grouping LUTs into CLBs. At the end a CLB tree is build. Once the number of CLBs and their configuration has been obtained, their delay is known [10] *Figure 5*.

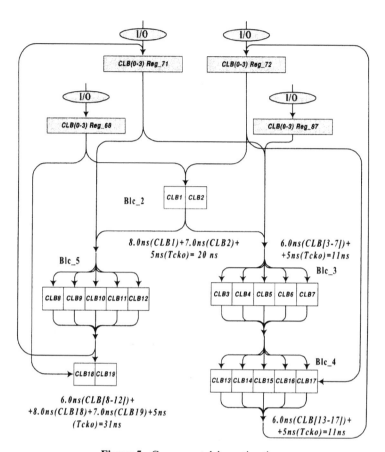

Figure 5 - Component delay estimation

Predicting placement

The prediction of the wire delay is carried out using the distances between blocks, which are known only after the placement.

Xilinx 4000 series employs a particular Manchester-carry chains implemented with pass transistors as part of the logic in configurable logic block. This results in a natural order for data flow, being the least significant bit of one word at the bottom of a column of *CLBs* and the most significant bit at the top. The placement is obtained by slicing top down the graph of *CLBs*, placing the first *CLB* in the bottom left corner of the *FPGA*.

Wiring delay estimation

To predict the delay between points A and B, D(A,B), the orthogonal distances (x, y) are first calculated. Then, wire type *single-length* lines, *double-length* lines and *long* lines are assigned to the wire. This assignment also decides the number of

PIPs and the number of segments between points A and B, and subsequently, the point to point delay $D(A,B)$ [8] see *Figure 6*.

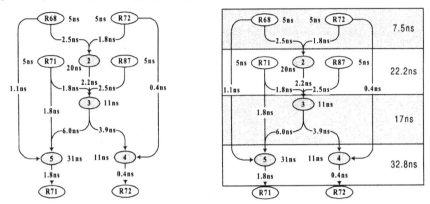

Figure 6 - The performance estimation process for branch composed by vertices 2,3,4,and 5.

Example:

Performance evaluation of the branch composed by vertices 2, 3, 4 and 5:

Software performance evaluation - the software execution of the branch takes 35 clock cycles or 1400ns *Figure 4*.

Hardware performance evaluation – Figure 6 shows the calculated values for the vertex and edge delays. The calculation is much too long to be presented in detail. The clock cycle for the example of the *Figure 6* will be 33ns and overall execution time will be 132ns.

Interface overhead evaluation - the data transfer takes 10 SW cycles resulting in 800ns. Total HW execution time 932ns.

4.   Conclusions and Future Work

In this paper is presented an HW/SW partitioning algorithm for custom computing machines. Appropriate cost function and performance estimation functions were developed, as well as, techniques for their automated calculation. An example was used to demonstrate the functionality of the proposed techniques. The partitioner is a module in a software tool having as input a C program and producing as output an hardware description and a software program that, together, will perform the same function as the input C specification. A significant improvement of the performance is obtained due to the implementation of specific segments of the program in HW and to the concurrent execution of the resulting HW and SW.

Some additional work is needed in order to increase the accuracy of the placement prediction.

5. References

[1] P. Athanas and H.Silverman, "Processor reconfiguration trough instruction-set metamorphosis: architecture and compiler", Computer, vol. 28, no. 3,pp. 11-18, March 1993.

[2] Alfred V. Aho, Ravi Sethi and Jefferey D. Ullman, "Compilers: Principles, Techniques and Tools", Addison Wesley, 1986.

[3] Peter M. Athanas, "An Adaptive Machine Architecture and Compiler for Dynamic Processor Reconfiguration", Technical Report LEMS-101, Brown University, February, 1992.

[4] Rolf Ernst Jorg Henkel Thomas Benner, "Hardware-Software Co-synthesis for Microcontrollers", IEEE Design & Test of Computers, December 1993, page 64.

[5] Rajesh Kumar Gupta, "Co-Synthesis of Hardware and Software for Digital Embedded Systems", Ph.D. dissertation Stanford University, December 10, 1993.

[6] D. D. Gajski, Frank Vahid Sanjiv Narayan Jie "Specification and design of embedded systems", Gong University of California at Irvine. PTR Prentice Hall 1994.

[7] Anton Chichkov, C. Beltrán Almeida, "Identification and Optimisation of Parallelism in Hardware/Software Partitioning", International Workshop on Logic and Architecture Synthesis, Grenoble France, December 1996.

[8] F. Kurdahi, Ms. Min Xu, "Area & Timing Estimation Techniques for Lookup Table-Based FPGA with Application to High-Level Synthesis", International Workshop on Logic and Architecture Synthesis, Grenoble France, December 1996.

[9] John L. Hennessy, David A. Patterson, "Computer Architecture a Quantitative Approach", Morgan Kaufmann Publishers, INC. San Mateo, California.

[10] Xilinx, "The Programmable Logic Data Book", 1993.

The Java Environment for Reconfigurable Computing

Eric Lechner and Steven A. Guccione

Xilinx Inc.
2100 Logic Drive
San Jose, CA 95124 (USA)
Eric.Lechner@xilinx.com
Steven.Guccione@xilinx.com

Abstract. The *Java Environment for Reconfigurable Computing (JERC)* is a software environment for reconfigurable coprocessor applications. This environment consisting of only a standard Java compiler and a set of libraries. Using JERC, configuration, reconfiguration and host run-time operation is supported. JERC also features design compile times on the order seconds and built-in support for parameterized macros.

1 Introduction

In recent years, there has been an increasing interest in reconfigurable logic based processing. These systems attempt to use reconfigurable logic to implement algorithms directly in hardware, thus increasing performance.

By one count, at least 50 different hardware platforms have been built to investigate this novel approach to computation [5]. Unfortunately, software seems to lag behind hardware in this area. Most systems today employ traditional circuit design techniques, then interface these circuits to a host computer using standard programming languages.

Work done in high-level language support for reconfigurable logic based computing currently falls into two major approaches. The first approach is to use a traditional programming language in place of a hardware description language [2] [6]. This still requires software support on the host processor.

The second major approach is compilation of standard programming languages to reconfigurable logic coprocessors. These typically attempt to detect computationally intensive portions of code and map them to the coprocessor [3] [4] [7] [9] [10] [8]. These compilation tools, however, are usually tied to traditional placement and routing back-ends and have relatively slow compilation times. They also provide little or no run-time support for dynamic reconfiguration.

The *Java Environment for Reconfigurable Computing (JERC)* represents a novel approach to hardware / software codesign for reconfigurable logic based coprocessors. Using the *JERC* libraries and standard Java, configuration, reconfiguration and host interface software for coprocessing applications is supported in a single piece of code. Additionally, since this tool does not make use of the traditional placement and routing approach to circuit synthesis, compilation times are on the order of seconds. This combines to produce a development environment which very closely resembles those used for modern software development.

2 The Design Flow

Design of an application using a reconfigurable logic coprocessor currently requires a combination of two distinct design paths. The first, and perhaps most significant portion of the effort involves circuit design using traditional CAD tools. This design path for these CAD tools typically consists of entering a design using a schematic editor or hardware design language, generating a netlist for this design, importing this netlist into an FPGA placement and routing tool, which finally generates a file used to configure the FPGA logic.

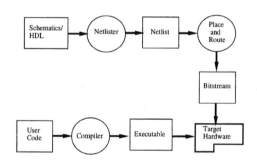

Fig. 1. Traditional design flow.

Once the configuration data has been produced, the next task is to provide software to interface the host system to the reconfigurable logic coprocessor. This task is usually completely decoupled from the task of designing the circuit, and hence is often difficult and error-prone. This dual path design flow is shown in Figure 1.

In addition to the problems of interfacing the hardware and software in this environment, there is also the problem of design cycle time. Any change to the circuit design requires a complete pass through the hardware design tool chain. This process is time consuming, with the place and route portion of the chain typically taking several hours to complete.

Finally, this approach provides no support for reconfiguration. The traditional hardware design tools provide support almost exclusively for static design. It is even difficult to imagine constructs to support run-time reconfiguration in environments based on schematic or HDL design entry.

In contrast, the *JERC* environment consists of a library of functions which permit logic and routing to be specified and configured in a reconfigurable logic device. By making calls to these library functions, circuits may be configured and reconfigured. Additionally, host code may be written to interact with the reconfigurable hardware. This permits all design data to reside in a single system, often in a single *Java* source code file.

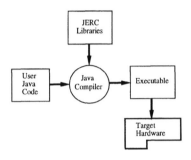

Fig. 2. JERC design flow.

In addition to greatly simplifying the design flow, as shown in Figure 2, the *JERC* approach also tightly couples the hardware and software design processes. Design parameters for both the reconfigurable hardware and the host software are shared. This coupling provides better support for the task of interfacing the logic circuits to the software.

3 The JERC Abstraction

JERC takes a layered approach to abstracting the reconfigurable logic. At the lowest layer, *Level 0*, *JERC* supports all accessible hardware resources in the reconfigurable logic. Extensive use of constants and other symbolic data makes Level 0 usable, in spite of the necessarily low level of abstraction.

The current platform for the *JERC* environment is the *XC6200DS* Development System [12]. This system consists of a a PCI board containing a Xilinx *XC6216* FPGA [11]. In the *XC6200*, Level 0 support consists of abstractions for the reconfigurable logic cells and all routing switches, including the clock routing. The code for Level 0 is essentially the bit-level information in the *XC6200 Data Sheet* (cite data sheet) coded into Java.

While Level 0 provides complete support for programming all aspects of the device, it is very low level and may be too tedious and require too much specialized knowledge of the architecture for most users. Although this layer is always available to the programmer, it is expected that level 0 support will function primarily as the basis for the higher layers of abstraction. In this sense, Level 0 is the "assembly language" of the JERC system.

Above the Level 0 abstractions is the Level 1 abstraction. This abstraction permits simpler access to logic definition, clock and clear routing and the host interface.

The most significant portion of the level 1 abstraction is the logic cell definition. This permits cells in the XC6200 to be configured as standard logic operators. Currently, *AND, NAND, OR, NOR, XOR, XNOR, BUFFER* and *INVERTER* combinational logic elements are supported. These may take an optional registered output. Additionally, a *D flip-flop, toggle flip-flop* and a *register* logic cell is defined. All of these logic operators are defined exclusively using *JERC* level 0 operations, and hence are easily extended. Figure 3 gives a diagram of the Level 1 cell abstraction.

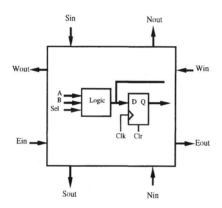

Fig. 3. The JERC Level 1 cell abstraction.

A second portion of the JERC Level 1 abstraction is the *Register* interface. In the XC6200, columns of cells may be read or written via the bus interface. The *Register* interface allows registers to be constructed and accessed symbolically.

In addition to the logic cell and register abstractions, the clock routing is abstracted. Various global and local clock signals may be defined and associated with a given logic cell.

4   A Counter Example

This section describes a simple counter based on toggle flip flops using the Level 1 abstraction. In less than 30 lines of code, the circuit is described and configured and clocking and reading of the counter value is performed. In addition, the structure of this circuit permits it to be easily packaged as a parameterized object. Such and object based approach would permit counters of any size to be specified and placed at any location in the XC6200.

The implementation process is fairly simple. First, the logic elements required by the circuit are defined. These circuit element definitions are abstractions and are not associated with any particular hardware implementation.

Once these logic elements are defined, they may be written to the hardware, configuring the circuit. Once the circuit is configured, run time interfacing of the circuit, usually in the form of reading and writing registers and clocking the circuit, is performed. If the application demands it, the process may be repeated, with the hardware being reconfigured as necessary.

The counter example contains 9 basic logic elements. Two of these are *Registers* which simply interface the circuit to the host software. These two registers are used to read the value of the counter and to toggle a single flip flop, producing the *local clock*.

To support the flip flops in the XC6200, clock and clear inputs must also be defined. The *global clock* is the system clock for the device and must be used as the input to any writable register. In this circuit, the flip flop which provides the software controlled local clock must use the global clock.

The *local clock* is the output of the software controlled clock, and must be routed to the toggle flip flops which make up the counter. Finally, all flip flops in the XC6200 need a *clear* input. In this circuit, the *clear* input to all flip flops is simply set to logic zero.

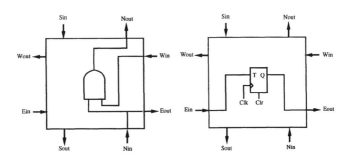

Fig. 4. The carry and toggle flip flop cell definitions.

These four logic elements provide all of the necessary support circuitry to read, write, clock and clear the hardware. The remaining logic elements are used to define the counter circuit itself.

The first logic element in the circuit is the *clock*. This is just a single bit *Register* which is writable by the software. Toggling this register via software control produces the clock for the counter circuit.

The next element is a toggle flip flop, *tff*. This flip flop is defined as having an input coming from the west. This element provides the state storage for the counter. Next, the *carry* logic for the counter is simply an AND gate with inputs from the previous stage and the output of the current stage. This generates the "toggle" signal for the next stage of the counter. Figure 4 gives a graphical representation of these two logic cells.

Finally, a logic *one* cell is implemented for the carry input to the first stage of the counter. Figure 5 give the *JERC* code for describing the basic logic elements. The *pci6200* object passed to each of the logic definitions is the hardware interface to the *XC6200DS* PCI board.

```
Pci6200 pci6200 = new Pci6200N(null); // Hardware interface
pci6200.connect();
Register counterReg =   new Register(COLUMN, counterMap, pci6200);
Register clockReg =     new Register(COLUMN, clockMap, pci6200);
ClockMux localClock =   new ClockMux(ClockMux.CLOCK_IN);
ClockMux globalClock = new ClockMux(ClockMux.GLOBAL_CLOCK);
ClearMux clear =        new ClearMux(ClearMux.ZERO);
Logic tff =             new Logic(Logic.T_FLIP_FLOP, Logic.EAST);
Logic clock =           new Logic(Logic.REGISTER);
Logic one =             new Logic(Logic.ONE);
Logic carry =           new Logic(Logic.AND, Logic.NORTH, Logic.WEST);
carry.setEastOutput(Logic.NORTH); // Set carry output
```

Fig. 5. The logic element definition code.

Once this collection of abstract logic elements is defined, they may be instantiated anywhere in the XC6200 cell array. This is accomplished by making a call to the *write()* function associated with each object. This function takes a *column* and *row* parameter which define the cell in the XC6200 to be configured. Additionally, the hardware interface object is passed as a parameter. In this case, all configuration is done to *pci6200*, a single *XC6200DS* PCI board.

The code in Figure 7 performs all configuration. In the *for()* loop, the *carry* cells go in one column with the *tff* toggle flip flops in the next column. A *local clock* and a *clear* is attached to each *tff* toggle flip flop. A graphical representation of the location of these cells is shown in Figure 6.

Below the *for()* loop, a constant "1" is set as the input to the carry chain. Next the software controlled clock is configured. This is the *clock* object, with its *local Clock* routing attached to the toggle flip flops of the counter. Finally, the *global clock* is used to clock this software controlled clock.

Once the circuit is configured, it is a simple matter to read and write the *Register* objects via the *set()* and *get()* functions. In Figure 8, the clock is toggled by alternatively writing "0" and "1" to the *clock register*. The *counter register* is used to read the value of the counter. Next to the code is an actual trace of the execution of this code running on the *XC6200DS* development system.

While this is a simple example for demonstration purposes, it makes use of all of the features of *JERC*. This includes register reads and writes, as well as features such as software driven local clocking. Other more complex circuits have been developed using JERC, but differ primarily only in the size of the code, not the number of features needed to provide system level support for reconfigurable processing.

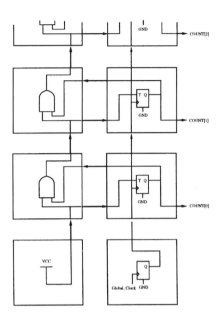

Fig. 6. The relative locations of cells in the counter.

```
/* Configure cells */
for (i=ROW_START; i<ROW_END; i++) { // The counter
    carry.write((COLUMN-1), i, pci6200);
    tff.write(COLUMN, i, pci6200);
    localClock.write(COLUMN, i, pci6200);
    clear.write(COLUMN, i, pci6200);
    } /* end for() */
one.write((COLUMN-1), (ROW_START-1), pci6200); // Carry in
clock.write(COLUMN, (ROW_START-1), pci6200); // Clock
localClock.set(ClockMux.NORTH_OUT);
localClock.write(COLUMN, ROW_START, pci6200);
globalClock.write(COLUMN, (ROW_START-1), pci6200);
```

Fig. 7. The configuration code.

```
for (i=0; i<5; i++) {
    clockReg.set(0); // Toggle clock
    clockReg.set(1);
    System.out.println("Count: " +
        counterReg.get());
    } /* end for() */
```

```
C:\java\JERC>java Counter
Count: 0
Count: 1
Count: 2
Count: 3
Count: 4
C:\java\JERC>
```

Fig. 8. The run time code and the execution trace.

5 Drawbacks of JERC

While *JERC* provides a simple, fast, integrated tool for reconfigurable logic based processing, there are still several drawbacks. First, *JERC* is currently a manual tool. Since it is possible to perform reconfiguration, it is necessary for the programmer to exercise tight control over the placement and routing of circuits. For highly repetitive designs, this is not a problem, but using *JERC* for large, unstructured designs is not recommended.

Also, *JERC* relys on abstractions to simplify the design process. Unfortunately, with this abstraction and simplification comes a loss of flexibility. Many hardware resources are ignored or abstracted away at the higher *JERC* levels. Of course, Level 0 operation is always an option, but this requires detailed knowledge of the underlying architecture, and may be quite tedious.

Finally, *JERC* currently provides no timing analysis of the underlying circuits. This is perhaps a more fundamental problem with reconfigurable systems in general. If true dynamic reconfiguration is possible, analyzing the possible circuits is likely to prove to be a difficult problem. But it should always be possible for the programmer to do simple critical path analysis, which should give a reasonable upper bound on the clock speed.

6 Future Plans

Work on *JERC* is continuing. Three particular directions are being investigated. First, the object oriented nature of *Java* permits libraries of parameterized macrocell-like objects to be built. This could significantly increase the productivity of users of *JERC*.

The second area of investigation is using *JERC* as a basis for a traditional graphical CAD tool. While this would be useful for producing static circuits, it is not clear how temporal reconfiguration would be managed. It would, however, trade very fast compilation times in exchange or the manual design style of *JERC*.

The last area of investigation is higher levels of abstraction. One possibility is to add some limited automatic placement and routing capability.

7 Conclusions

JERC represents a novel approach to reconfigurable computing. Using a single inexpensive, off the shelf development tool, circuits can be constructed, reconfigured and interfaced to host systems.

Perhaps more importantly, the compile times necessary to produce these circuits and run-time support code is on the order of seconds. This is many orders of magnitude faster than the design cycle time of traditional CAD tools.

This permits development in an environment that in nearly all ways operates like a modern intergrated software development environment.

Early experiences has shown *JERC* to be a fast and friendly alternative to existing approaches to algorithm development for reconfigurable computing.

Acknowledgements

Thanks to the engineering staff of the Xilinx Development Corporation for XC6200 and XC6200DS support. And thanks to Eric Dellinger for support of the JERC concept.

References

1. Peter M. Athanas and Harvey F. Silverman. Processor reconfiguration through instruction-set metamorphosis. *IEEE Computer*, 26(3):11–18, March 1993.
2. David Galloway. The transmogrifier C hardware description language and compiler for FPGAs. In Kenneth L. Pocek and Jeffrey Arnold, editors, *IEEE Symposium on FPGAs for Custom Computing Machines*, pages 136–144, Los Alamitos, CA, April 1995. IEEE Computer Society Press.
3. Maya Gokhale and Aaron Marks. Automatic synthesis of parallel programs targeted to dynamically reconfigurable logic arrays. In Will Moore and Wayne Luk, editors, *Field-Programmable Logic and Applications*, pages 399–408, 1996. Proceedings of the 5th International Workshop on Field-Programmable Logic and Applications, FPL 95. Lecture Notes in Computer Science 972.
4. Steven A. Guccione. *Programming Fine-Grained Reconfigurable Architectures.* PhD thesis, The University of Texas at Austin, May 1995.
5. Steven A. Guccione. List of FPGA-based computing machines. World Wide Web page http://www.io.com/~guccione/HW_list.html, 1997.
6. Shaori Guo and Wayne Luk. Compiling ruby into FPGAs. In Will Moore and Wayne Luk, editors, *Field-Programmable Logic and Applications*, pages 188–197, 1996. Proceedings of the 5th International Workshop on Field-Programmable Logic and Applications, FPL 95. Lecture Notes in Computer Science 972.
7. Reiner W. Hartenstein, Alexander G. Hirschbiel, Michael Reidmüller, Karin Schmidt, and Michael Weber. A novel ASIC design approach based on a new machine paradigm. *IEEE Journal of Solid-State Circuits*, 26(7):975–989, July 1991.
8. Christian Iseli and Eduardo Sanchez. A C++ compiler for FPGA custom execution unit synthesis. In Kenneth L. Pocek and Jeffrey Arnold, editors, *IEEE Symposium on FPGAs for Custom Computing Machines*, pages 173–179, Los Alamitos, CA, April 1995. IEEE Computer Society Press.
9. James B. Peterson, R. Brendan O'Connor, and Peter M. Athanas. Scheduling and partitioning ANSI-C programs onto multi-FPGA CCM architectures. In Kenneth L. Pocek and Jeffrey Arnold, editors, *IEEE Symposium on FPGAs for Custom Computing Machines*, pages 178–187, Los Alamitos, CA, April 1996. IEEE Computer Society Press.
10. Markus Weinhardt. Portable pipeline synthesis for FCCMs. In Reiner W. Hartenstein and Manfred Glesner, editors, *Field-Programmable Logic: Smart Applications,*

New Paradigms and Compilers, pages 1–13, 1996. Proceedings of the 6th International Workshop on Field-Programmable Logic and Applications, FPL 96. Lecture Notes in Computer Science 1142.

11. Xilinx, Inc. *The Programmable Logic Data Book*, 1996.
12. Xilinx, Inc. *XC6200 Development System*, 1997.

Data Scheduling to Increase Performance of Parallel Accelerators

R. W. Hartenstein, J. Becker, M. Herz, U. Nageldinger
University of Kaiserslautern
Erwin-Schrödinger-Straße, D-67663 Kaiserslautern, Germany
Fax: ++49 631 205 2640, email: abakus@informatik.uni-kl.de
www: http://xputers.informatik.uni-kl.de/

Abstract

The paper presents a general data and task scheduling technique for parallel accelerators with one or more processing modules and the capability for local and shared memory access. Multiple tasks and their data are mapped onto the processing modules and the host, providing a high degree of parallelism. The system performance is enhanced by avoiding unnecessary data transfers. It is shown, how overall system performance can be increased by appropriate data distribution.

1. Introduction

Due to the increasing demands in computation power, it has shown, that there are certain applications, which take too much time to compute on a standard von Neumann Computer, despite of remarkable developments in system speed. Such algorithms can be found e.g. in image or digital signal processing, multimedia applications and others.

CCMs try to boost the performance by adding an external accelerator to the host computer. They are connected to the host via a coprocessor bus or an I/O bus. A subclass of CCMs uses field-programmable logic for the accelerator. Therefore, these CCMs are called F-CCMs. Field-programmable logic provides the advantage of a manifold of possible configurations of the accelerator, which is therefore not restricted to execution of one dedicated application. In the following, we will consider only this class of accelerators. F-CCMs can be classified by the way the data memory is attached and accessed. Typically, four classes of memory accessing-methods can be distinguished:

a) Remote memory access

b) Shared memory access

c) Local memory

d) Local memory and shared memory access

Hereby, it has been seen, that the shared and local memory solution involve the potential to gain a remarkable increase of performance. Further details and example architectures for F-CCMs concerning these four memory accessing modes as well as a complete classification scheme can be found in [HBK96].

Our own approach follows the Xputer paradigm [HHS91] [Hirs91]. The current prototype is called MoM-3 (Map oriented Machine 3) [BHK95] . The MoM-3 uses multiple Xputer modules, which have local memories as well as access to the host memory, therefore following the accessing mode d) above. Data, which is accessed frequently, is copied to the local memory. Data, which is accessed only once remains on the host. Therefore there is the necessity for an effective distribution of the data onto the different memory resources at compile time, what we call data scheduling. This paper will focus on the problems and appropriate solution strategies relevant to these issues. In our current Application Development Environment, the Code-X framework [HBH96a], an input application is distributed into parts for both the host and the Xputer accelerator to get a high performance. To exploit more parallelism, data scheduling has to take place, which is shown in the following. The paper is organized as follows:

Section 2 gives an overview about the target hardware and the Xputer architecture. In section 3, the function of the Code-X framework is described briefly. A technique to schedule the data inside Code-X is presented in section 4.

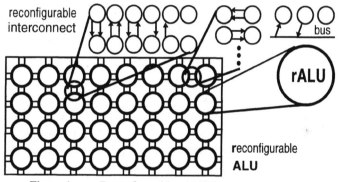

Figure 1.   Reconfigurable rALU Array.

2.   The Xputer hardware

A new kind of structurally programmable technology platform is needed. To support highly computing-intensive applications we need structurally programmable platforms providing word level parallelism instead of the bit level parallelism of FPGAs. We need FPAAs (field-programmable ALU arrays) instead of FPGAs. For flexibility to implement applications with highly irregular data dependencies, each PE (processing element) should be programmable individually. Each PE should be a reconfigurable ALU (rALU). Also the local interconnect between particular PEs or rALUs should be programmable individually. A good solution is the Kress Array (figure 1) [Hart95], [Kres96]. It permits to map highly irregular applications onto a regularly structured hardware platform. To realize the integration of such *soft ALUs* into a computing machine, a deterministic data sequencing mechanism is also needed, because the von Neumann paradigm does not efficiently support "soft" hardware [HBH96b]. As soon as a data path is changed by structural programming, a von Neumann architecture would require a new tightly coupled instruction sequencer. A good backbone paradigm for implementing such a deterministic reconfigurable hardware architecture based on data sequencing is the procedurally data-driven Xputer paradigm (figure 2), already published many times in the past [HHS91]. Like shown in figure 2, an Xputer consists of a 2-D Data Memory, a reconfigurable ALU and one or more Data Sequencers. The Xputer paradigm conveniently supports *soft ALUs* like in the *rALU array* concept (reconfigurable ALU array), and it bridges the modelling gap and also the education gap.

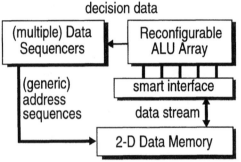

Figure 2.   Xputers: basic machine principles

The task of the data sequencer is to feed a data stream into the rALU. For many applications, the access sequence for the data words follows regular patterns. The data sequencer exploits this situation by providing a hardware capable of generating a wide variety of access sequences, which can be described by only a few parameters. We call such an access sequence a scan pattern (see figure 3). The data sequencer calculates a new position of the scan pattern in the two-dimensional memory. This handle position is not used directly to fetch a data word from the resulting location. Instead, the whole area, which contains the data words for one calculation of the rALU, called scan window, is moved to the handle position, allowing access to all data words inside the scan window in an application dependent order.

So, an Xputer execution works as follows, considering the example in figure 3: First, the scan window is moved to the next position according to the scan pattern. In the example, the scan pattern is a simple sequence of linear movements, each 3 steps down

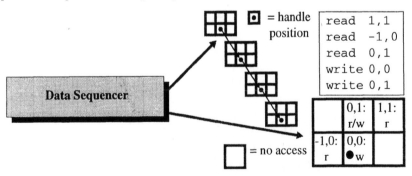

Figure 3.   Illustration of Xputer Execution Mechanism

and 2 steps to the right. Then, the data words at positions (1,1), (-1,0) and (0,1), relative to the scan window handle, are read and passed to the rALU in this sequence and the rALU starts the calculation. After the rALU is finished, two data words are written back to positions (0,0) and (0,1). Then, the scan window is moved again one step of the scan pattern. This process repeats until the scan pattern is completed.

Note, that the scan pattern describes explicitly the sequence of accesses to the data words, which can be realized using only a few parameters. As the parameters include the step widths and boundaries of a scan pattern, it is easy to calculate the number of steps of a scan pattern. Together with the access sequence of the data inside the scan window, the number of memory accesses for one scan pattern can be calculated. This statistical data is used later on in the data scheduling strategy.

3. Code-X - Parallelizing Accelerator Compilation

Structural software being really worth such a term would require a source notation like the C language and a compiler which automatically generates structural code from it. For such a new class of accelerator hardware platforms a completely new class of compilers is needed, which generate both, sequential and structural code: partitioning compilers, which separate a source into two types of cooperating code segments:

- *structural software* for the accelerator(s), and
- *sequential software* for the host.

In such an environment parallelizing compilers require two levels of partitioning:

- host/accelerator (or sequential/structural software) partitioning for optimizing performance, and
- a structural/sequential partitioning of structural software (second level) for optimizing the hardware/software trade-off of the Xputer resources.

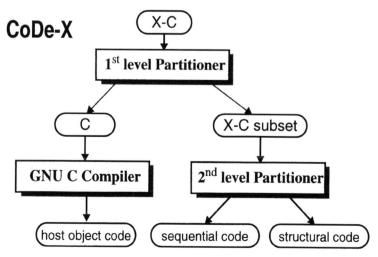

Figure 4.   Overview on CoDe-X Accelerator Compilation

For Xputer-based accelerators the partitioning application development framework CoDe-X is being implemented, based on two-level hardware/software co-design strategies [Beck97]. CoDe-X accepts X-C source programs (Xputer-C, figure 4), which represents a C dialect. CoDe-X consists of a 1st level partitioner, a GNU C compiler, and a 2nd level partitioner. The X-C source input is partitioned in a first level into a part for execution on the host (host tasks, also permitting dynamic structures and operating system calls) and a part for execution on the accelerator (Xputer tasks). Parts for accelerator execution are expressed in a C subset, which lacks only dynamic structures, [Schm94] (*structural software*). At second level this input is partitioned by an X-C subset compiler in a sequential part for the data sequncer, and a structural part for the rALU array. By using C extensions within X-C experienced users may hand-hone their source code by including directly data-procedural MoPL code (Map-oriented Programming Language [ABHK94], [Beck97]) into the C description of an application. Also less experienced users may use generic MoPL library functions similar to C function calls to take full advantage of the high acceleration factors possible by the Xputer paradigm.

The profiling-driven first level of the dual CoDe-X partitioning process is responsible for the decision which task should be evaluated on Xputer-based accelerators and which one on the host. Generally, four kind of tasks can be determined:

- *host tasks* (containing dynamic structures,
- *Xputer tasks* (candidates for Xputer migration),
- *MoPL-code segments* included in X-C source, and
- generic *Xputer library function* calls.

The *host tasks* have to be evaluated on the host, since they cannot be performed on Xputer-based accelerators. This is due to the lack of an operating system for Xputers for handling dynamic structures like pointers, recursive functions etc., which can be done more more efficiently by the host. The generic *Xputer library functions* and included *MoPL-code segments* are executed in any case on the accelerator. All other tasks are *Xputer tasks*, which are the candidates for the iterative first level partitioning step determining their final allocation based on profiling-driven simulated annealing [Beck97]. The granularity of *Xputer tasks* depends on the hardware parameters of the current accelerator prototype, e.g. the maximal nesting depth of for-loops to be handled by data sequencers, the number of available PEs within one rALU array, etc. For further details and examples about CoDe-X' accelerator compilation techniques and necessary scheduling, as well as run time synchronization and reconfiguration strategies see [Beck97].

4.   Data scheduling in the Code-X framework

After the principles of the Code-X framework have been illustrated briefly, we will take a closer look at its data distribution, called data scheduling. We will introduce a new extension to Code-X, which distributes the tasks and data of the application onto single Xputer modules and the host. As the host is handled the same way as the Xputer modules, we will not distinguish the host from Xputer modules, except for the time to access data memory. Our strategy tries to avoid unnecessary copy operations between different data memories. The Xputer architecture supports our approach, as the execution time of a Xputer task can be estimated quite well (see section 2).

Code-X creates a so-called *total execution graph (TEG)*, which represents the total execution order of all tasks, containing all their sequential / parallel execution dependencies. The TEG contains synchronization points called concurrency update points, which divide the graph in discrete sequential timesteps with (one or more) parallel tasks in each step [Beck97].

As each Xputer module (and the host, of course) has its own memory, application data transfers occur between different memories. Additionally, there are different kinds of memory accesses, with different memory access times. An access to the local memory of the same module is faster than inter-module memory access.

To increase application performance, the memory transfers, i.e. copy operations between modules should be minimized. For our approach, a performance analysis of the tasks is needed. We exploit the fact, that it is possible to calculate the number of memory accesses of a scan pattern, like explained in section 2. We use a greedy strategy, which tries to reduce copy operations between modules. As the modifications in the task graph are always inside a step between two concurrency update points, the total execution time can only decrease.

We introduce the following values, which are used by our strategy:

For a given task T and a data block D, let $scansteps(T,D)$ denote the number of steps of the scan pattern and $windowsteps(T,D)$ the number of read and write operations inside the scan window of this task on the data block. Both values can be calculated and are generated as statistical data by the X-C compiler and the Datapath Synthesis System, which is part of the Xputer compiler section. Then we can express the total number of memory accesses $accesses(T,D)$ as:

$$accesses(T, D) = scansteps(T, D) \cdot windowsteps(T, D)$$

Now, let t_{mem} be the time to access a data word in memory. Let $t_{MoM\text{-}bus}$ and $t_{hostbus}$ be the additional delay of the MoM-bus and the host bus respectively. We can write t_{local}, the time for a local memory access, t_{board} the time for an access to a different module and t_{host}, the time for an access from or to the host memory as:

$$t_{local} = t_{mem}$$

$$t_{board} = t_{mem} + t_{MoM\text{-}bus}$$

$$t_{host} = t_{mem} + t_{MoM\text{-}bus} + t_{hostbus}$$

The values t_{mem}, $t_{MoM\text{-}bus}$ and $t_{hostbus}$ are hardware-dependent. We will use the unified term t_{remote} for the time to access non-local memory. Depending, if the remote module is the host or an Xputer module, t_{remote} is either t_{board} or t_{host}.

Together with the size of a data block D, which we denote by $size(D)$, we can express the time to copy D from one module to another as:

$$copytime(D) = size(D) \cdot (t_{local} + t_{remote})$$

Let *penalty(T,D)* be the additional time needed by a task *T*, if it has to access data block *D* remotely. Let *remotes* be the number of remote accesses already scheduled in the timestep. Hereby, the potential remote access of *T* is counted, too, so that *remotes* is equal to or greater 1. Then, *penalty(T,D)* is estimated by:

$$penalty(T, D) = (t_{remote} - t_{local}) \cdot accesses(T, D) \cdot remotes$$

For a set of datablocks $DS = \{D_1,...,D_n\}$, the penalties sum up to the accumulated penalty *acpenalty(T,DS)*:

$$acpenalty(T, DS) = \sum_{Di \in DS} penalty(T, D_i)$$

Using the above definitions, we will now describe our mapping-strategy.

The TEG from Code-X is processed timestep by timestep from the start to the end. In the first step, all tasks, which can be executed in parallel, are assigned to different Xputer modules. As Code-X performs a resource constraint scheduling, it is provided, that at any time, there are not more tasks scheduled than modules are available.

For all other timesteps, only the tasks, which start in that step are considered one by one. Let T_0 be the current task. There are several possible cases:

a) There is no predecessor to T_0. In this case, the task is assigned to a free module.

b) Task T_0 has one predecessor T_p, which has itself no other successors besides T_0. Here, T_0 is mapped onto the same module as T_p. Therefore, the application data can stay in the same module and needs not to be copied.

c) There are multiple predecessors to T_0, which have no additional successors. So, T_0 obviously needs data from several sources. In this case the *penalty(T_0,D_i)* is checked for every data block D_i needed by T_0, to find out, which data is most important. The task is then preferably mapped onto the module, which holds this data. Data blocks D_j from other modules are copied to the module of T_0, if *penalty(T_0,D_j)* is greater than *copytime(D_j)*. Otherwise, the data is accessed remotely.

d) Task T_0 has one or more predecessors $T_{p1},...,T_{pm}$, which have themselves successors $T_1,...,T_n$ besides T_0. Let *DS* be the set of all data blocks, which are needed by the $T_i \in \{T_0,T_1,...,T_n\}$. For each T_i and each $T_{pj} \in \{T_{p1},...,T_{pm}\}$ the accumulated penalty *acpenalty(T_i,DSP_{ij})* is calculated, where $DSP_{ij} \subseteq DS$ is the set of data blocks T_i would have to access remotely, if it was not mapped onto the module of T_{pj}. All penalties are sorted and the task T_i with the highest value is mapped onto its prefered module. Then, the data blocks on this module are removed from the set *DS*. The penalties are calculated again and the process repeats, until all tasks are mapped.

After the mapping of a task, it can be expected, that some data blocks are missing. Therefore, for the task T_i the set DM_i is determined, consisting of the data blocks needed by T_i, which are not available locally. Like in case c), a data block $D \in DM_i$ is copied to the local memory, if *penalty(T_i,D)* is greater then *copytime(D)*.

To illustrate our strategy, we use a demonstration example given in [Beck97]. This example consists of seven tasks. The simplified TEG is shown in the left part of figure 5. Each task performs a two dimensional cellular automaton on one or more

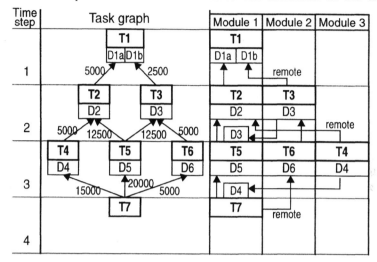

Figure 5.   Example mapping

input arrays of the same size of 50 by 50 words. Thus, *size(D)* is 2500 for each data block. The example is mapped on three Xputer modules, needing no host tasks. The hardware parameters measured for the Xputer prototype are t_{local} = 170ns and t_{remote} = t_{board} = 280 ns. The numbers next to the data dependencies in figure 5 denote *accesses(T,D)*. The resulting mapping is shown in the right part of figure 5. We will discuss the steps of the algorithm in the following:

In step 1, T1 is mapped onto module 1. In step 2, situation d) of the algorithm applies. The accumulated penalties result to:

acpenalty(T2,{D1a}) = *penalty(T2,D1a)* = 110 ns * 5000 * 1 = 0.55 ms

acpenalty(T3,{D1b}) = *penalty(T3,D1b)* = 110 ns * 2500 * 1 = 0.275 ms

So, T2 is mapped to module 1. For T3, which is mapped onto the next free module 2, *copytime(D1b)* = 450 ns * 2500 = 1.125 ms. As this is longer than the estimated penalty, T3 will do remote access to D1b.

We will examine the next step in more detail. At first, the set of all needed data blocks is determined to DS = {D2,D3}. There are six sets DSP_{ij}. $DSP_{T4T2} = DSP_{T5T2} =$ {D2}, $DSP_{T5T3} = DSP_{T6T3}$ = {D3} and $DSP_{T4T3} = DSP_{T6T2}$ = {}. The penalties are:

acpenalty(T4,{D2}) = *acpenalty(T6,{D3})* = 0.55 ms

acpenalty(T5,{D2}) = *acpenalty(T5,{D3})* = 1.375 ms

So, T5 is mapped first. As *penalty(T5,D2)* = *penalty(T5,D3)*, T5 is mapped arbitrarily onto module 1. The *copytime(D3)* is again 1.125 ms. As it takes less time to make a local copy of D3 than to access it remotely, D3 is copied to module 1. Now the calculation is repeated for T4 and T6. There is no remote access from T5, so *remotes* is still 1.

D2 is removed from *DS*, so *DS* = {D3}. We get: DSP_{T6T3} = {D3} and DSP_{T4T3} = { }. So, we get zero penalty, if T4 is not mapped onto module 2. For T6 *acpenalty(T6,{D3})* = 0.55 ms. Therefore, T6 is mapped onto module 2. Finally, T4 is put to the next free module, number 3. As the calculated penalty is shorter than the copytime, D2 is accessed remotely.

In the last step, situation c) of the algorithm applies. The three penalties calculate to *penalty(T7,D4)* = 1.65 ms, *penalty(T7,D5)* = 2.2 ms and *penalty(T7,D6)* = 0.55 ms. Obviously, T7 is best mapped after T5. As *penalty(T7,D4)* is also greater than the copytime, D4 is copied to module 2, while D6 is accessed remotely.

5.　Conclusions

A data and task scheduling technique for Xputers has been presented. One or more processing modules with local memory can be handled. Multiple tasks and their data are mapped onto the processing resources, providing a high degree of parallelism. The mapping avoids unnecessary data transfers between modules.

Therefore, overall system performance can be increased by using the introduced data scheduling technique. It was shown, that the Xputer prototype MoM-3 suits best for this method. The MoM-3 hardware allows the calculation of exact performance data, which is used to control the scheduling.

The Code-X framework has been presented briefly. It allows the simultaneous programming of both the accelerator and the host by a two-level partitioning approach and a parallelizing compiler. The scheduling technique introduced is an extension to Code-X to optimize the task distribution and to enhance the overall system performance.

6.　References

[ABHK94]　A. Ast, J. Becker, R. Hartenstein, R. Kress, H. Reinig, K. Schmidt: Data-procedural Languages for FPL-based Machines; 4th Int. Workshop on Field Programmable Logic and Appl., FPL'94, Prague, Sept. 7-10, 1994, Springer, 1994

[AHR94]　A. Ast, R. Hartenstein, H. Reinig, K. Schmidt, M. Weber: A General Purpose Xputer Architecture derived from DSP and Image Processing. In Bayoumi, M.A. (Ed.): VLSI Design Methodologies for Digital Signal Processing Architectures, Kluwer Academic Publishers 1994.

[BHK95]　J. Becker, R. W. Hartenstein, R. Kress, H. Reinig: High-Performance Computing Using a Reconfigurable Accelerator; Proc. of Workshop on High Performance Computing, Montreal, Canada, July 1995

[Beck97]　J.Becker: A Partitioning Compiler for Computers with Xputer-based Accelerators, Ph.D. dissertation, Kaiserslautern University, 1997

[HaRe95]　Reiner W. Hartenstein, Helmut Reinig: Novel Sequencer Hardware for High-Speed Signal Processing; Workshop on Design Methodologies for Microelectronics, Smolenice Castle, Slovakia, September 1995

[Hart95]　R. Hartenstein, et al.: A Datapath Synthesis System for the Reconfigurable Datapath Architecture; ASP-DAC'95, Makuhari, Chiba, Japan, Aug. 29 - Sept. 1, 1995

[HBH96a]   Reiner W. Hartenstein, Jürgen Becker, Michael Herz, Rainer Kress, Ulrich Nageldinger: A Parallelizing Programming Environment for Embedded Xputer-based Accelerators; High Performance Computing Symposium '96, Ottawa, Canada, June 1996

[HBH96b]   R. Hartenstein, J. Becker, M. Herz, U. Nageldinger: A General Approach in System Design Integrating Reconfigurable Accelerators; Proc. of IEEE 1996 Int'l. Conference on Innovative Systems in Silicon; Austin, Texas, USA, Oct. 9-11, 1996

[HBK95]   R. W. Hartenstein, J. Becker, R. Kress, H. Reinig, K. Schmidt: A Reconfigurable Machine for Applications in Image and Video Compression; Conf. on Compression Techniques & Standards for Image & Video Compression, Amsterdam, Netherlands, March 1995

[HBK96]   Reiner W. Hartenstein, Jürgen Becker, Rainer Kress: Custom Computing Machines vs. Hardware/Software Co-Design: from a globalized point of view; 6th International Workshop On Field Programmable Logic And Applications, FPL'96, Darmstadt, Germany, September 23-25, 1996, Lecture Notes in Computer Science, Springer Press, 1996

[HHS91]   R. Hartenstein, A. Hirschbiel, K. Schmidt, M. Weber: A novel Paradigm of Parallel Computation and its Use to implement Simple High-Performance Hardware; Future Generation Computing Systems 7 (1991/92), p. 181 - 198

[Hirs91]   A. Hirschbiel: A Novel Processor Architecture Based on Auto Data Sequencing and Low Level Parallelism; Ph.D. Thesis, University of Kaiserslautern, 1991

[Kres96]   R. Kress: A fast reconfigurable ALU for Xputers; Ph. D. dissertation, Kaiserslautern University, 1996

[Schm94]   K. Schmidt: A Program Partitioning, Restructuring, and Mapping Method for Xputers; Ph.D. Thesis, University of Kaiserslautern, 1994

An Operating System for Custom Computing Machines based on the Xputer Paradigm

Rainer Kress

Siemens AG, Corporate Technology
D-81730 Munich, Germany

Reiner W. Hartenstein, Ulrich Nageldinger

University of Kaiserslautern
D-67663 Kaiserslautern, Germany

Abstract. The paper presents an operating system (OS) for custom computing machines (CCMs) based on the Xputer paradigm. Custom computing tries to combine traditional computing with programmable hardware, attempting to gain from the benefits of both adaptive software and optimized hardware. The OS running as an extension to the actual host OS allows a greater flexibility in deciding what parts of the application should run on the configurable hardware with structural code and what on the host-hardware with conventional software. This decision can be taken late - at run-time - and dynamically, in contrast to early partitioning and deciding at compile-time as used currently on CCMs. Thus the CCM can be used concurrently by multiple users or applications without knowledge of each other. This raises programming and using CCMs to levels close to modern OSes for sequential von Neumann processors.

1 Introduction

With the availability of FPGAs a new class of computing machines has been made possible, the custom computing machines (CCMs). Custom computing tries to combine traditional computing (microprocessor and memory) with programmable hardware, attempting to gain from the benefits of both adaptive software and optimized hardware. A number of different application areas have been proposed for these configurable computing machines. A key characteristic of the applications which reach high performance is the availability of some amount of *semi-static* information. Semi-static means it changes frequently enough to require programmable hardware, but slowly enough to provide the opportunity to improve performance through customized hardware. Three main application areas are [MaSm97]: *situation based* configuration, i.e. configuration changes only at a relatively slow rate; *time-sharing* of hardware resources; and *dynamic circuit generation*, i.e. in evolutionary systems. The latter application uses the full advantage of reconfigurable hardware.

Custom computing machines (CCMs) can be distinguished by their operation mode. Either CCMs run stand-alone or with the aid of a host computer. The use of a host computer starts from simple I/O tasks up to a sophisticated partitioning between tasks scheduled to run on the configurable hardware and the microprocessor of the host. In this paper, we consider host based CCMs. Their programming requires techniques from the hardware/software codesign area [Buch94].

Programming of CCMs can be distinguished into two major groups. The first group contains CCMs, where hardware and software is programmed separately. In most cases the host is programmed using a procedural language like C and the configurable hardware is programmed using schematic entry or synthesized using a hardware description language (HDL) like VHDL. The synchronization is then completely in the responsibility of the user. In the second group, common description languages are used. Either this is a hardware description language (HDL) [GHK90] or a hardware-oriented programming language [SWA93]. These systems require a partitioning step to separate code to be executed on the configurable hardware and on the host.

The new approach proposed in this paper uses also a common description language. But it compiles and synthesizes as much functions as possible for both the programmable hardware and the host. This allows a greater flexibility in deciding what runs on hardware and what on software. The decision can be taken late - at run-time - and dynamically, in contrast to early partitioning and deciding at compile-time. Thus the CCM can be used concurrently by multiple users or applications without knowledge of each other, providing a virtual processor for each user as known from modern operating systems (OS) for von Neumann-processors. The operating system required for such a novel approach is explained using an exemplary CCM based on the Xputer paradigm [HHW90].

The paper is organized as follows: section 2 gives a brief overview on the target architecture. Details can be found in [HaRe95]. Section 3 describes the programming environment CoDeX [HBH96]. The novel approach using an operating system to control the programmable hardware is explained in section 4. Finally the paper is concluded.

2 Overview on the Xputer architecture

Xputer-based CCMs may consist of several Xputer modules. The modules are connected to a host computer. Making use of the host simplifies disk access and all other I/O operations. After setup, each Xputer module runs independently from the others and the host computer until a complete task is processed. Each module generates an interrupt to the host when the task is finished. So, the host is free to concurrently execute other tasks in-between. This allows the use of the Xputer modules as a general purpose acceleration board for time critical parts in an application.

The basic structure of an Xputer module consists of three major parts (figure 1):

- the data memory
- the reconfigurable arithmetic and logic unit (rALU) including several rALU subnets with multiple scan windows (SW)
- the data sequencer (DS)

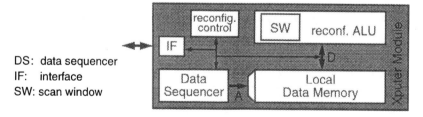

DS: data sequencer
IF: interface
SW: scan window

Fig. 1. Xputer module

The data memory contains the data which has to be accessed or modified during an application. The data is arranged in a special order to optimize the data access sequences. This arrangement is called datamap. The data sequencer (DS) computes generic address sequences described by few parameters to access the data. These sequences are called scan patterns. Data sequencing in general means, that the data is addressed in the correct sequence by generic scan patterns and loaded into the rALU. All data manipulations are done in the rALU. It consists of several rALU subnets. After finishing, the rALU signals the end of the computation to the data sequencer.

For the evaluation of standard C programs, word-oriented operators such as addition, subtraction, or multiplication are required. This realization of a reconfigurable architecture for word-oriented operators needs a more coarse grained approach for the logic block architecture of the rALU than supplied by commercially available FPGAs. Therefore, in the current Xputer prototype, a parallel ALU array is used. The ALU array features logic blocks, called DPUs (Datapath Units), which can be loaded with a complete arithmetic or logic operator. The DPUs are capable of executing all operators of the C-language, and additional operators can be added by microprogramming. In the prototype the rALU consists of 96 DPUs. As the regular structure of the parallel ALU array reaches across chip boundaries, the array consisting of several chips can be seen as a large parallel ALU array. Thus, it can be easily adapted to the size of the problem. A more detailed description can be found in [Kres96].

3 The CoDe-X Framework

This section gives an overview on the hardware/software co-design framework CoDe-X (figure 2) [HBH96]. Input language to the framework is the programming language X-C (Xputer C). X-C provides the complete functionality of C with optional extensions (Xputer specific functions) for experienced users. The input is partitioned in a first level into a part for execution on the host and a part for execution on the accelerator. The part for the Xputer-based accelerator is then partitioned in a second level into a sequential part for programming the data sequencer and a structural part for configuring the rALU. Special Xputer-specific functions in the original description of the application allow to take full usage of the high speed-up of the Xputer paradigm. The experienced user may add new functions to a library such that every user can access them.

3.1 Profiling-driven Host/Accelerator Partitioning

The first level of the partitioning process is responsible for the decision which task should be evaluated on the Xputer and which one on the host. Generally three kinds of tasks can be determined:

- tasks which contain dynamic structures, called host tasks,
- Xputer tasks, and
- Xputer-library functions.

The host tasks have to be evaluated on the host, since they cannot be performed on the Xputer accelerator. The Xputer tasks are the candidates for the performance analysis on the host and on the Xputer. The Xputer-library functions are used to get the highest performance out of the Xputer. An experienced user has developed a function library with Xputer-library functions together with their performance values. These functions can be called directly in the input C-programs. The usage of a programmable accelerator arise two problems: the accelerator needs reconfiguration at run-time and in a few cases, the datamap has to be reorganized also at run-time.

3.2 Experienced User Features

The experienced user can use and build optionally a generic function library with a large repertory of Xputer-library functions described in the language MoPL [ABH93]. It fully

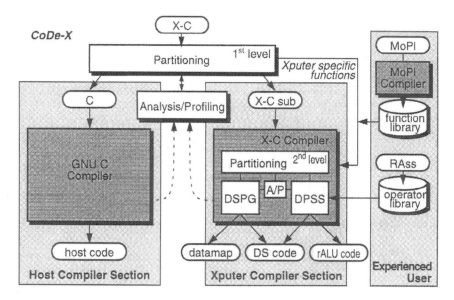

Fig. 2. Overview of the CoDe-X environment

supports all features of the Xputer modules. This optional data-procedural language features make highest acceleration factors provided by Xputer hardware possible.

3.3 Resource-driven second level Partitioning

The X-C compiler [Schm94] translates a program written in a subset of the X-C language into code which can be executed on the Xputer without further user interaction. It comprises four major parts: the data sequencer parameter generator (DSPG), the datapath synthesis system (DPSS), and the analysis/profiling tool (A/P) together with the partitioner. For each part of code delivered from the first partitioning level, the X-C compiler produces a trade-off between area, that means number of DPUs in the rALU and performance.

4 Execution Mechanisms

To enable efficient access to the Xputer hardware, the software interface must provide a way of coupling the Xputer-Tasks to host tasks, which should be conform to the host OS. Thus, the Xputer hardware, namely the rALU, the local memory of the Xputer, the configuration memory of the data sequencer appear as resources to be managed by a software interface.

The current approach for such an interface is the MoM-Runtime-System (MRTS). MoM (Map oriented Machine) is the name of the Xputer prototype. In the MRTS approach, all resources are given temporally to one task. The host-application has to start up the runtime-system, which will take over the Xputer with its modules for the time of its existence in the system, load the configuration and the input data, start an Xputer module, output the results of the calculations and finish (figure 4). This method enables only one task to use the accelerator module at a time, as the runtime-system has to establish a temporary, exclusive control connection to the Xputer, which lasts only during the lifetime of the MRTS. Therefore, there can be only one single process (controlled by a single user) using the Xputer. Thus, a global, static scheduling of all Xputer-tasks of this process is necessary. Furthermore, the temporary control connection mentioned above contains direct accesses to the hardware, including the reception of interrupts from the Xputer.

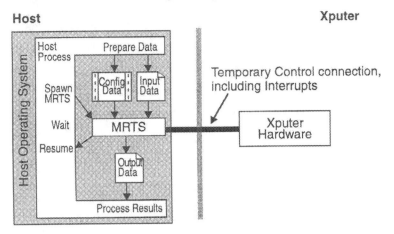

Fig. 3.   Primitive Xputer-task execution

4.1   The Xputer Operating System and its executing mechanisms

A new approach has been evolved, which tries to avoid the above restrictions. This approach tries to establish similar powerful resource management, flexibility and efficiency for the Xputer hardware like modern Operating Systems show for standard system resources. Therefore, we call our approach XOS, the Xputer operating system.

The XOS is started only once. It provides access to the Xputer for multiple host tasks. The basic structure of the XOS embedding can be seen in figure 4.

The capabilities of the XOS include dynamic scheduling of tasks and dynamic management of the Xputer memory. Due to the organization of the rALU, the DPUs can also be managed as a resource similar to memory. This enables multiple tasks (configurations) to be present at a time inside the ALU array. The same is true for the configuration memory of the data sequencer.

As the XOS resembles a kind of server in the host-OS, processes can easily communicate and therefore use the Xputer capabilities, making the Xputer a powerful computation resource to traditional applications. The possibilities for a host process to use the Xputer hardware are described in the following.

4.2   Library functions

Library function execution on Xputer modules only. First, a library can contain Xputer specific functions (figure 4 top). This is the most comfortable way for programmers to make use of the Xputer, as there is just one call to the appropriate function, and there is no knowledge about Xputer programming needed. The functions in the libraries have to be generic, that means, that e.g. array boundaries have to be provided by parameters from the side of the host process. These parameters must then be replaced in the

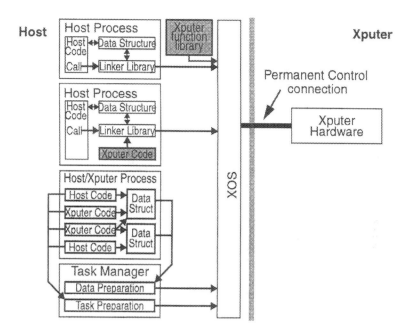

Fig. 4. The XOS approach showing three modes of utilising the Xputer hardware

Xputer function while it is loaded. Therefore, a linker concept is necessary for this mode of operation. This type of Xputer-access is already provided in the CoDe-X framework.

Library function on Xputer modules or the host. Furthermore, a library function may exist not only as an Xputer implementation but also as code executable by the host (figure 4 bottom). Then, the XOS can consider the current load of the Xputer and dynamically select either the host code or the Xputer code to gain the best performance.

Such a scenario is depicted in detail in figure 5. It is assumed, that a host process has an array of integers named „image" with the dimensions 10 by 10. A filter operation is to be executed on the data in „image".

To achieve this, the programmer has linked the program with a special linker library. This library contains a function named „XpCall" in this paper, which initiates the execution of a library function. The parameters for XpCall are:

- The identifier of the requested function („Filter")
- Pointers to the arrays („image") containing the data.
- Additional parameters, like array boundaries, filter coefficients etc. For the sake of simplicity, we assume only the array boundaries (10,10) to be passed.

The function XpCall has three main tasks:

1) Contacting the XOS as a client and passing the request, including parameters and array pointers. This data exchange between the XOS and the host process can be established using the interprocess communication (IPC) facilities of the host operating system.

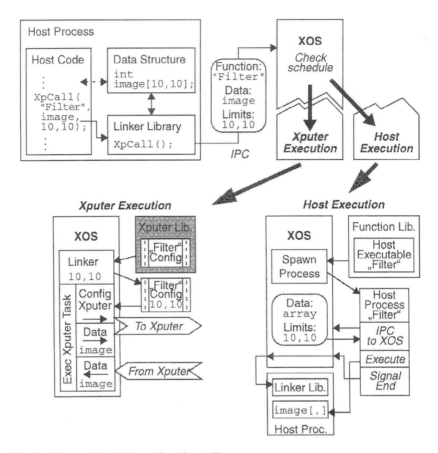

Fig. 5. Example for Library function call

2) Make the data areas (these will normally be arrays) accessible to the XOS or, generally spoken, to a foreign process. On some operating systems, this is a non-trivial task due to sophisticated memory management or security issues. It might be necessary to establish a shared memory segment for the data areas. The Xputer modules directly support shared memories.

3) An obvious requirement for the above two tasks is the knowledge of XpCall about the parameter requirements of the library function. In the example, XpCall must interpret its first argument as an array pointer, but the other two arguments as plain integers. These two variable types need different handling, as the required memory areas to have to be made accessible by foreign processes. Therefore, XpCall knows, that the function „Filter" needs one array pointer and two plain integers. If the argument list does not match these requirements, an error should be issued. Note that it is still in the responsibility of the programmer, that the first argument is in fact a valid pointer, as XpCall sees only the value of the variable.

There are several ways to let XpCall know what the function „Filter" needs:

- XpCall simply consults the function library. This would make the linker library, where XpCall resides, unnecessary big, as there must be code included to scan the library and extract the argument list.
- Obtain the information from the XOS, which has to look up the function in the library anyway. This would require a total of three messages between the XOS and the host process: One to pass the function ID to XOS, one to pass the argument list to the process and one to transfer the parameters.
- Encoding the arguments in the function ID, similar to C++ compilers. This is the preferred possibility. In our example, if the function ID is supposed to be a character string, the function „Filter" could be called „Filter_P1I2" indicating, that there are one pointer and two integer parameters to be passed. It provides the possibility for generic functions, which have the same name but different argument lists. Consider e.g. a function „Filter_P1I11", which resembles a filter function taking a pointer, two array boundaries and nine more integers resembling additional coefficients. Then the function „Filter_P1I2" would be derived from „Filter_P1I11" by simply providing default values for the nine missing parameters. The default values would be stored with the library function and applied by the XOS.

When the XOS receives the request, it looks up the function in the function library, testing if it exists. The current scheduling decides whether to execute the function on the Xputer or on the host (see also section 5). If the host is chosen, the XOS would create a host process for the function. The lower part of figure 6 shows, how the execution on Xputer or host respectively works.

Xputer. If the function is to be run on the Xputer, the XOS looks up the Xputer configuration in the function library. Normally, this configuration has to be completed with certain parameters. In our example, this would be the array limits (10,10). This procedure of satisfying extern references in the Xputer configuration is similar to a „linking" process, which is why we call this part of the XOS the linker. After the linking, the configuration is ready to be executed and the XOS will schedule an Xputer task for the function.

Host. If the function is to be run on the host, the XOS looks up the according host executable in the function library. This executable is then spawned as a host process. The new process then contacts the XOS via IPC, receiving the parameters of the function. Having direct access to the data areas of the calling process through the passed pointers, the function can execute and work on the original arrays. When finished, the function signals its termination to the XOS, which passes the signal on to the caller.

4.3 Host processes with Xputer code

Besides using a function provided by a library, an experienced programmer may also create Xputer code directly. To generate the Xputer code, the experienced programmer can use the language MoPL [ABH93].

To execute this code, the program calls a function of a linker library, which handles the communication to the XOS. The resulting configuration code may then be converted to a C data object and linked to the application. As an example, we assume an application, which executes a filter operation onto an internal array, like the example shown in figure 5. Again, this filter should run on the accelerator. Instead of using a library function, the programmer has written an own function in MoPL. The MoPL compiler generates an Xputer configuration for the filter function. In order to make the application consist of one file, the configuration can easily be transformed into a C source file by creating an array containing the configuration binary. The source file can then be compiled and linked together with the application source code. To enable Xputer access for the application, the programmer has again linked the code with the linker library, which coordinates communication to the XOS.

At run time, the application contains the Xputer code as a normal data object. When the filter is to be executed, a function from the linker library named „XpLoad" is called. The arguments to this function are a pointer to the array with the Xputer configuration and a pointer to the data area. Note, that the array limits are included in the Xputer code, which has been designed by the programmer to match the application requirements. The function „XpLoad" does basically the same as „XpCall", which is used for functions from the Xputer library in the example in section 4.2. The difference lies in the simpler handling of the accelerator code, which is transferred directly from the C-array instead of the library. Furthermore, the function will be executed on the Xputer in any way. There is no option for the XOS to use a traditional implementation.

5 Dynamic task scheduling

To provide an efficient use of the Xputer hardware, a good scheduling of the Xputer tasks has to be performed. As the accelerator should be a global resource to the system, the ability of multitasking is desirable. In this section, the requirements for a task scheduling are discussed.

For the processing of interrupts, the XOS should provide preemptiveness. That means the XOS should be capable to suspend any running task at any time. To preserve the status of the old task, a context switch must be performed. Therefore, the current status of the hardware must be saveable and loadable. In modern von Neumann processors, the whole processor status is contained in a set of registers. Normally, there is a special instruction, which saves or loads the complete register set to or from the system stack, i.e. a location in the main memory. The status of an Xputer module consists of much more data:

- the current status of the data sequencer, consisting of the current configuration as well as the current working-location,
- the status of the rALU, which comprises the current configuration (ranges from 20 to 100 kbits for 96 DPUs), as well as the status of every used register in each DPU.

It is easily seen, that the status of the Xputer contains more data than that of a traditional processor. Therefore, both data sequencer and rALU provide the possibility to hold multiple configurations in their configuration memories. The context switch is performed as a switch between two configurations. We are aware, that the maximal number of configurations in the hardware limits the capabilities of the XOS, but a transfer of the configuration data to and from an external memory would need too much time.

For the dynamic scheduling, information from the static scheduling of CoDe-X is used. For each task, its predecessor, successor and mobility is stored. From the profiler of the CoDe-X framework, the acceleration factor, required configuration time, and the time for memory access (if not in a local memory) is saved. A cost function (1) using this information creates a priority for the currently available tasks ready for execution.

$$\text{cost function} = f(\text{acc. factor}, t_{config}, t_{memory}, \text{mobility}) \qquad (1)$$

Running only a single application concurrently on the Xputer results in the same schedule as provided by the CoDe-X environment.

6 Conclusions

An operating system (OS) for custom computing machines (CCMs) based on the Xputer paradigm has been presented. The OS running as extension to the actual host OS allows a greater flexibility in deciding what runs on the configurable hardware and what on the host hardware. This decision is taken late, at run-time and dynamically, in contrast to early partitioning and deciding at compile-time as used currently by CCMs. Thus a CCM can be used concurrently by multiple users or applications without knowledge of each other. Compared to a previous runtime-system, the XOS approach is its logical extension, narrowing the gap to capabilities of modern operating systems for von Neumann computers.

References

[ABH93]   A. Ast, J. Becker, R. Hartenstein, R. Kress, H. Reinig, K. Schmidt: MoPL-3: A New High Level Xputer Programming Language; 3rd Int´ Workshop On Field Programmable Logic And Applications, Oxford, 7. - 10. September 1993

[Buch94] Klaus Buchenrieder: Hardware/Software Co-Design, An Annotated Bibliography; IT Press Chicago, 1994

[GHK90]   M. Gokhale, B. Holmes, A. Kopser, D. Kunze, D. Lopresti, S. Lucas, R. Minich, P. Olsen: SPLASH: A Reconfigurable Linear Logic Array; International Conference on Parallel Processing, pp I 526 - I 532, 1990

[HBH96]   R. W. Hartenstein, J. Becker, M. Herz, R. Kress, U. Nageldinger: A Parallelizing Programming Environment for Embedded Xputer-based Accelerators; High Performance Computing Symposium '96, Ottawa, Canada, 1996

[HaRe95] R. W. Hartenstein, H. Reinig: Novel Sequencer Hardware for High-Speed Signal Processing; Workshop on Design Methodologies for Microelectronics, Smolenice Castle, Slovakia, Sept. 11-13, 1995

[HHW90]   R. W. Hartenstein, A. G. Hirschbiel, M. Weber: A Novel Paradigm of Parallel Computation and its Use to Implement Simple High Performance Hardware; InfoJapan'90 - International Conference memorizing the 30th Anniversary of the Computer Society of Japan, Tokyo, Japan, 1990

[Kres96] R. Kress: A fast reconfigurable ALU for Xputers; Ph. D. dissertation, Kaiserslautern University, 1996

[MaSm97] W. Mangione-Smith: Configurable Computing: Concepts and Issues; Task Force on Configurable Computing, Proceeding of the HICSS 97, pp. 710-712, 1997

[Schm94] K. Schmidt: A Program Partitioning, Restructuring, and Mapping Method for Xputers; Ph.D. Thesis, University of Kaiserslautern, 1994

[SWA93]   A. Smith, M. Wazlowski, L. Agarwal, T. Lee, E. Lam, P. Athanas, H. Silverman, S. Ghosh: PRISM-II: Compiler and Architecture; IEEE Workshop on FPGAs for Custom Computing Machines, FCCM'93, Napa, CA, April 1993

Fast Parallel Implementation of DFT Using Configurable Devices[*]

Andreas Dandalis and Viktor K. Prasanna

Department of Electrical Engineering-Systems
University of Southern California
Los Angeles, CA 90089-2562, USA
{dandalis, prasanna}@usc.edu
http://ceng.usc.edu/~prasanna
Tel: +1-213-740-4336, Fax:+1-213-740-4418

Abstract. In this paper we propose a fast parallel implementation of Discrete Fourier Transform (DFT) using FPGAs. Our design is based on the Arithmetic Fourier Transform (AFT) using zero-order interpolation. For a given problem of size N, AFT requires only $O(N^2)$ additions and $O(N)$ real multiplications with constant factors. Our design employs $2p + 1$ PEs ($1 \le p \le N$), $O(N)$ memory and fixed I/O with the host. It is scalable over p ($1 \le p \le N$) and can solve larger problems with the same hardware by increasing the memory. All the PEs have fixed architecture. Our implementation is faster than most standard DSP designs for FFT. It also outperforms other FPGA-based implementations for FFT, in terms of speed and adaptability to larger problems.

1 Introduction

The Discrete Fourier Transform (DFT) plays a fundamental role in digital signal processing. The complexity and computation time of algorithmic approaches for forward computation of DFT, are essential issues in algorithms where many forward DFTs are required while one inverse Fourier transform must be performed at the end. For a problem of size N, the sequential computation time of a straightforward approach is $O(N^2)$ and is characterized mainly by the large number of complex multiplications and additions. This fact limits the computational performance of the approach as well as the algorithmic efficiency of implementations using Field Programmable Gate Arrays (FPGAs).

The FPGA based implementations for computing the DFT, proposed in [12, 8, 11], use the Fast Fourier Transform (FFT) to reduce the computation time and complexity to $O(N \log_2 N)$. The basic computation unit is the butterfly. Butterfly is a repetitive structure that has 2 inputs and 2 outputs. It involves one complex multiplication, one complex addition and one complex subtraction.

---

[*] This research was performed as part of the MAARC project (Models, Algorithms and Architectures for Reconfigurable Computing, http://maarc.usc.edu). This work is supported by DARPA Adaptive Computing Systems program under contract no. DABT63-96-C-00049 monitored by Fort Hauchuca.

For a problem of size N, the algorithm requires $\log_2 N$ stages with $N/2$ butterflies in each stage. Even though these designs optimize the structure of the butterfly, the complexity still remains high. All these designs are solutions optimized for a particular problem size. For larger problems, re-design is required resulting in area penalty. Parallelism is not exploited and the designs are not scalable except the one proposed in [11]. In [8], the idea of FPGAs with an external multiplier is used to overcome the critical issue of complex multiplication. This solution has still problems since it adds extra control/complexity and requires a large number of I/O pins for interfacing the multiplier chip. In spite of this, the computation time is not attractive. The implementation in [11] uses the CORDIC approach for optimizing the butterfly by eliminating multiplications. Again, the resulting performance is not attractive.

In this paper we propose a novel parallel, scalable, partitioned solution for computing the DFT using FPGAs, based on the Arithmetic Fourier Transform (AFT). Using this approach, we can solve larger problems with fixed hardware, simply by increasing the memory size. We can linearly speed-up the computation proportionally to the number of PEs employed and achieve superior performance compared with previous FPGA-based solutions. Also, it offers faster solution compared with most standard DSP designs for computing the DFT. The key idea of our design is the use of an algorithmic approach to the problem. Contrary to traditional approaches, we perform an algorithmic design for reconfigurable devices, based upon the architecture/features of the device. While known techniques map an algorithm for DFT onto the device and perform device dependent optimizations, our methodology employees algorithm synthesis techniques instead of logic synthesis. This alleviates the FPGA's restriction of fast/compact adders vs slow/area-consuming multipliers. Complex multiplication is a critical issue in DSP applications and can lead to poor performance of FPGA-based solutions. AFT turns to be a suitable algorithmic approach for FPGAs since it is less complex than the FFT and performs real multiplications with constant factors instead of complex multiplications.

The Arithmetic Fourier Transform is based on the Möbius inversion formula of series and has been shown to be competitive with the conventional FFT in terms of accuracy, complexity and speed [9]. It needs $O(N^2)$ additions and $O(N)$ real multiplications by constant coefficients. It reduces the computation time of DFT to $O(N)$. In our design, two sets of p PEs $(1 \leq p \leq N)$ and an additional PE are used for computing $2N+1$ Fourier coefficients [7]. Our design is scalable over p $(1 \leq p \leq N)$, thus it can achieve $O(p)$ speed-up. It is also a partitioned solution since it can solve larger problems by increasing the memory size in proportion to the N the size of the problem. In each set, all PEs have the same architecture and perform additions and zero-order interpolation. The additional PE performs the scaling of the intermediate values by constant factors. All the PEs are cascaded using pipelining. The data as well as the control signals move from left to right. The complete design requires $O(N)$ memory and has fixed I/O bandwidth. External memory is used for storing the scaling factors as well as intermediate Fourier coefficients. Constant coefficients multiplier (KCM) [2], is used for performing the scaling operation. KCMs use the Distributed Arithmetic approach (DA) and turn out to be a very efficient choice for digital signal processing in terms of speed and area. The compact size and high performance of the KCMs compared with standard full multipliers, are promising features that make the AFT algorithm an efficient solution for computing the DFT using FPGAs. In addition, the parallel/modular structure, the regular architecture as well as the fixed, independent of the problem size I/O bandwidth, make our approach an attractive solution for implementation in FPGAs.

Preliminary estimations shows that our design achieves speed-up of 2-10 over most standard DSP designs for 256-FFT. Compared with the Fastest FFT in the West [12], the CORDIC approach [11] and the implementation in [8], our design outperforms these solutions in terms of speed and adaptability to larger size problems. Our preliminary implementation reported here using Xilinx devices, can be further optimized resulting in higher speed and less area.

This paper is organized as follows. In Section 2 we describe the Arithmetic Fourier Transform while in Section 3 we introduce our scalable architecture for AFT. In Section 4 the computation time and area estimations are shown. Finally in Section 5 comparisons are discussed and concluding remarks are made.

2   Arithmetic Fourier Transform

The Arithmetic Fourier Transform (AFT) is based on the Möbius inversion formula of series. Since it involves only additions and real multiplications by constant factors, it is computationally less complex than FFT while it achieves $O(\log_2 N)$ speed-up over it. An introduction to AFT is given below and detailed descriptions of it can be found in [7, 9].

Given $2N$ input samples $A(m), m = 0, 1, ..., 2N - 1$, we compute an average and $2N$ alternating averages over them. All these averages are scaled by constant factors and then the Möbius inversion formula is applied for the computation of $2N + 1$ Fourier coefficients. The Möbius inversion formula theorem [5] and the definition of the alternating average, are the key mathematical tools for the AFT algorithm.

Theorem (The Möbius inversion formula) Let $f(n)$ be a non vanishing function in the interval $1 \le n \le N$ and $f(n) = 0$ for $n > N$, where n, N are positive integers. If

$$g(n) = \sum_{m=1}^{\lfloor N/n \rfloor} f(mn)$$

then

$$f(n) = \sum_{k=1}^{\lfloor N/n \rfloor} \mu(k)g(kn)$$

where $\lfloor ... \rfloor$ denotes the integer part of a real number and $\mu(k)$ is the Möbius function.

The Möbius function is defined as

$$\mu(n) = 1 \qquad \text{if } n = 1$$
$$\mu(n) = (-1)^r \quad \text{if } n = p_1 p_2 ... p_r \text{ where } p_i (\text{i=1,2,...,r}) \text{ distinct primes}$$
$$\mu(n) = 0 \qquad \text{if } p^2 \mid n \text{ for some prime } p$$

Definition (Alternating Average) The $2n$th alternating average $B(2n, \alpha)$ of the $2n$ values $A(mT/2n + \alpha T), 0 \leq m \leq 2n - 1$, is defined as:

$$B(2n, \alpha) = \frac{1}{2n} \sum_{m=0}^{2n-1} (-1)^m A(mT/2n + \alpha T)$$

where α is a shifting factor, $-1 < \alpha < 1$. Assuming now a finite Fourier series $A(t)$ with period T, we can represent it as:

$$A(t) = a_0 + \sum_{n=1}^{N} a_n \cos 2\pi n f_0 t + \sum_{n=1}^{N} b_n \sin 2\pi n f_0 t$$

where $f_0 = 1/T$, a_n and b_n are the real and imaginary parts of the Fourier coefficients of the non vanishing function in the interval $-N \leq n \leq N$ and a_0 is the mean of $A(t)$. Applying the Möbius inversion formula to $A(t)$, we can compute the $2N + 1$ Fourier coefficients in terms of the alternating averages as follows:

$$a_0 = \frac{1}{2n} \sum_{m=1}^{2N} A(m), \quad a_n = \sum_{l=1,3,...}^{\lfloor N/n \rfloor} \mu(l) B(2n, 0)$$

$$b_n = \sum_{l=1,3,...}^{\lfloor N/n \rfloor} \mu(l)(-1)^{\frac{l-1}{2}} B(2n, \frac{1}{4nl})$$

where $n = 1, 2, ..., N$ and $m = 0, 1, ..., 2N - 1$.

For computing the alternating averages $B(2n, 0), B(2n, \frac{1}{4nl})$ from the input samples $A(m)$, we use zero-order interpolation for computational efficiency [9] . In this method we interpolate an unknown value $A(mT/2n + \alpha T)$ to a known input sample $A(i)$, where i is the integer part of $mT/2n + \alpha T$. The resulting error due to this approximation is shown to be tolerable [9, 10]. The AFT computation method presented above, requires $(2N + 1)^2$ additions and $(2N + 1)$ multiplications with constant factors, for computing $2N + 1$ Fourier coefficients. The reduced complexity, the use of scaling by constant factors instead of complex multiplications and the amenability to parallel processing makes AFT more desirable computationally than FFT [10].

In Figure 1 we can show the structure of the AFT algorithm. In Part I the non-scaled alternating averages are computed while in Part II the scaling operation and the computation of a_0 take place. Finally, in Part III the $2N$ Fourier coefficients are computed. Multiplications with constany factors are performed only in Part II while in the other parts additions and zero-order interpolation are performed. The Möbius values in the last part define the sign of the alternating averages.

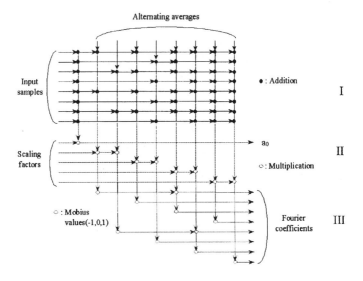

Fig. 1. Structure of AFT

3 A Scalable Architecture for AFT

In this section we show a scalable architecture to map the AFT algorithm (see Figure 2). Each part of the architecture corresponds to a part of the AFT structure in Figure 1. Data and control signals flow from left to right.

In Part I, p PEs are employed to compute the alternating averages. An input buffer B_I of size $2N \times w$ is used for storing the input window of $2N$ samples, where w denotes the number of bits in each input sample. Assuming that p divides N, each window is fed N/p times into the pipe. Let $PE_{1,i}$ denote the ith PE in Part I, $1 \leq p \leq N$ and $n = 1, 2, ..., N$. In $PE_{1,n\,mod\,p}$, the alternating averages $B(2n, 0)$ and $B(2n, \frac{1}{4nl})$ are computed during the $\lceil n/p \rceil th$ feeding of the input data window into the pipe. Each PE checks if the received data is needed for its computation based on the zero-order interpolation. The interpolation is implemented using local registers and a comparator for checking the index of the received sample. Every $2N$ units, p alternating averages are computed. Thus, the total computation time for $2N$ alternating averages is $\frac{2N^2}{p} + 2p - 1$ units. All the PEs in this part have the same architecture and consist of one adder, one comparator and local registers. The local registers are used for performing the interpolation as well as for interconnection with other PEs.

In Part II, one PE is employed for scaling the averages computed in the previous part and for computing the mean a_0. This PE is denoted as PE_{mul}. Totally $2p$ scaled averages are computed every $2N$ time units. PE_{mul} employes

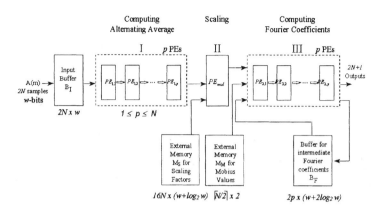

Fig. 2. Overall Architecture

one Constant Coefficient Multiplier (KCM) [2] for computing a_0 as well as for scaling $B(2n, 0)$, $B(2n, \frac{1}{4nl})$ by constant factors during each step. Using the hybrid technique described in [2], we have to precalculate 16 values for each scaling factor. The hybrid tehnique of multiplication is a hexamedical equivalent of the long hand method. Since a single hex digit represents four bits, the look-up table for each constant factor has entries for 0 to $15(F)$. Thus, the size of the external memory M_S would be $16N \times (w + \log_2 w)$. The set of precalculated values of constant factor $\frac{1}{2i}$ is stored in 16 consecutive locations of the memory, starting from the ith memory location where $1 \le i \le N$. PE_{mul} also employes local registers for interconnection with other stages.

In Part III, p PEs are employed to compute the Fourier coefficients. Similar to Part I, the PEs in this part are denoted as $PE_{2,i}$. All the PEs have the same architecture. Each of them consists of one adder, one comparator and local registers. As in the first group, each processing element checks if the received average is needed in its partial sum. This checking is performed using the index of the incoming average. The Möbius values required for the computation of the partial sums are provided by an external memory M_M. The memory size is $\lceil \frac{N}{2} \rceil$ and the stored values are $\{-1, 0, 1\}$. Few local registers are employed for controlling the data flow between consecutive PEs. A buffer B_f is employed for storing the intermediate results of the computation. When a new set of $2p$ alternating averages are available to Part III, the intermediate results of the computation and the Möbius function values are fed back from the rightmost to the leftmost PE. The size of B_f is $2p \times (w + 2\log_2 w)$ where w denotes the number of bits of each input sample. In this part, only the Fourier coefficients $a_1, b_1, a_2, b_2, ..., a_{N/3}, b_{N/3}$ are computed since for $n > N/3$, a_n and b_n are equal to alternating averages $B(2n, \alpha)$.

Since the computation time of $2N$ alternating averages is $\frac{2N^2}{p} + 2p - 1$ units, the total computation time for $2N+1$ Fourier coefficients becomes $\frac{2N^2}{p} + 3p + \lceil \frac{p}{2} \rceil$ units. The throughput rate of KCM critically affects the overall performance since it determines the minimum time unit. The architecture employes $2p$ adders, 1 KCM, few local registers and external memories. The total size of the external memories is $O(N)$. A key advantage of the design is that it is scalable over p, $1 \leq p \leq N$, thus it can linearly speed-up the computation by increasing p. It can also solve larger size problems (at a lower throughput rate) by simply increasing the memory but still using the same number and structure of PEs.

4 Performance Estimates

Table 1 lists the estimated area for various components of our architecture. We estimated the area of each of the functional blocks and the total area for each PE was then derived. All the PEs in a group (I or II) have the same architecture. Thus, each of them occupies the same constant area. We have assumed 16-bit input data and that our design is mapped onto Xilinx XC4000 series of FPGAs.

CLBs per Function	$PE_{1,i}$	$PE_{2,i}$	PE_{mul}
Registers	110	140	50
16-bit comparator	2	2	-
16-bitadder	16	16	-
16-bit KCM	-	-	230
Control	2	2	2
Total CLBs	130	160	282

Table 1. PE area requirements in terms of number of CLBs to realize the function

Our design consists of p $PE_{1,i}$, p $PE_{2,i}$ and one PE_{mul}. Assuming $2N$ input samples, the total time for computing the $2N+1$ Fourier coefficients is $\frac{2N^2}{p} + 3p + \lceil \frac{p}{2} \rceil$ time units. The time unit is determined by the performance of the KCM since Part II is the most time-consuming stage of our design. The pipelined performance of a 16-bit operand KCM (after place and route) is $50MHz$ [2] using the -3 speed grade components. The performance of a 16-bit adder is in the range of $100MHz$ using -3 speed grade components [6]. Thus, there is enough time for the PEs in parts I and III to complete their operations using $50MHz$

clock rate. Currently, -2 speed grade devices are available. Thus, $50MHz$ system clock rate is an achievable goal for the entire design. Table 2 shows the area requirements and estimate of the computation time for two designs. We assume 256 input samples and $50MHz$ system clock rate.

Hardware	Area Requirements	Computation Time
$p = 8$	2602 CLBs, 3 XC4025	4124 time units (82.48 μsec)
$p = 16$	4922 CLBs, 5 XC4025	2104 time units (42.08 μsec)

Table 2. Area and performance estimates

5 Comparisons and Conclusions

In this paper, we have proposed a novel parallel, scalable, partitioned solution for computing the DFT using FPGAs. Our solution based on the AFT turns out to be more efficient than the FFT based approach in terms of area and speed. Our design is scalable over $1 \leq p \leq N$, where p is the number of PEs employed. We can also linearly speed-up the computation proportionally to p. The architecture of the PEs and the I/O bandwidth are fixed and independent of the problem size. The required memory is $O(N)$. Our design can solve larger problems (with reduced throughput) with fixed hardware.

In Figure 3 and in Table 3 the execution times of various designs for 256-FFT are shown. The input samples are 16-bit data for all the designs. Figure 3 shows the results of a benchmark evaluation [12] of DSP-based and Xilinx FPGA-based designs. Our design achieves speed-up of $2 - 10$ over most single chip DSP designs for 256-FFT.

Implementation	Area Requirements	Computation Time
Xilinx 3 nodes [12]	1 XC4025	102.4 μsec
Xilinx 70MHz[12]	1 XC4025	223 μsec
Xilinx 60MHz [12]	1 XC4025	312.5 μsec
PDSP16116/A [8]	2 Chips	61.4 μsec
CORDIC [11]	10 XC4010	5000 μsec

Table 3. Performance of FPGA-based designs for 256-FFT

---

[1] 1000-FFT

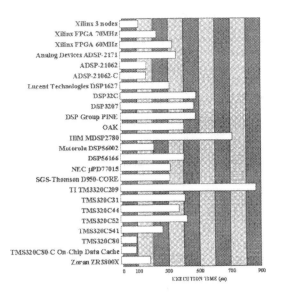

Fig. 3. Performance of DSP-based and FPGA-based designs for 256-FFT (from[12])

In Table 3 the performance of five FPGA-based implementations are shown. Three of them are from "The Fastest FFT in the West" [12]. In that work a radix-2 butterfly FFT design was used. The implementation in [8] makes use of one Altera FPGA and a PDS P16116/A 16-bit complex multiplier to overcome the critical problem of performing complex multiplication in FPGAs. A radix-4 design and necessary control were mapped onto the Altera FPGA. The implementation in [11] uses the CORDIC approach to eliminate the complex multiplications. Even though it computes a 1000-point FFT, the performance is not attractive compared with our approach. Table 3 also shows the area requirements of these implementations.

Our design is faster than the earlier FPGA-based implementations. The implementations in Table 3 are designs optimized for a particular problem size and device features and need to be redesigned for larger problems. Our design can also handle larger problems with the same fixed hardware by increasing the memory. Known implementations exploit the power of a single chip while we have developed a scalable and partitioned solution with high performance and adaptability to larger problems. Our performance estimation has been based on a preliminary implementation and no optimizations have been performed. Further improvement of the performance of our design is possible. Parallelism can be exploited; for example parallel I/O can improve the performance significantly. Many registers can be eliminated by efficiently performing zero-order interpolation. Other interpolation approaches (such as first-order) can also be exploited.

The work reported here is part of the USC MAARC project. This project is developing algorithmic techniques for realising scalable and portable applications using configurable computing devices and architectures. Contrary to traditional approaches to configurable computing, in our approach the user "sees" the architecture/device features and uses algorithm synthesis techniques instead of logic synthesis. We are developing computational models and algorithmic techniques based on these models to exploit dynamic reconfiguration. In addition, compilation onto reconfigurable hardware is also addressed. Some related results can be found in [1], [3], [4].

References

1. K. Bondalapati and V. K. Prasanna, "Reconfigurable Meshes: Theory and Practice", *Reconfigurable Architectures Workshop, Int. Parallel Processing Symposium (IPPS)*, April 1997.
2. K. Chapman, "Constant Coefficient Multipliers for the XC4000C", XILINX Application Note 054, Dec. 1996.
3. S. Choi, Y. Chung and V. K. Prasanna, "Configurable Hardware for Symbolic Search Operations", submitted to *Int. Conf. Parallel and Distributed Systems*, Dec. 1997.
4. Y. Chung S. Choi and V. K. Prasanna, "Parallel Object Recognition on an FPGA-based Configurable Computing Platform", submitted to *Int. Workshop on Computer Architecture for Machine Perception*, Oct. 1997.
5. L. K. Hua, "Introduction to Number Theory", New York: Springer-Verlag, 1982.
6. B. New, "Estimating the Performance of XC4000E Adders and Counters", XILINX Application Note 018, July 1996.
7. H. Park and V. K. Prasanna, "Modular VLSI Architectures for Computing the Arithmetic Fourier Transform", *IEEE Trans. Signal Processing*, vol. 41, no. 6, pp. 2236-2246, June 1993.
8. R. J. Petersen and B. L. Hutchings, "An Assessment of the Suitability of FPGA-Based Systems for use in Digital Signal Processing", in *Proc. Int. Workshop on Field-Programmable Logic and Applications*, Aug. 1995.
9. I. S. Reed, D. W. Tufts, X. Yu, T. K. Truong, M. Shih and X. Yin, "Fourier analysis and signal processing by use of the Möbius inversion formula", *IEEE Trans. Acoust., Signal Processing*, vol.38, no. 3, pp. 458-470, Mar. 1990.
10. I. S. Reed, M. T. Shih, T. K. Truong, E. Hendon and D. W. Tufts, "A VLSI architecture for simplified arithmetic Fourier transform algorithm", in *Proc. Int. Conf. Application Specific Array Processor*, 1990.
11. R. Wilson, "Reprogrammable FPGA-based techniques provide prototyping aid", *Electronic Engineering Times*, Mar. 11, 1996.
12. XILINX DSP Application notes, "The Fastest FFT in the West", http://www.xilinx.com/apps/displit.htm

Enhancing Fixed Point DSP Processor Performance by adding CPLD's as Coprocessing Elements

David Greenfield
Target Applications Manager

Caleb Crome
MegaCore Engineer

Martin S. Won
Applications Supervisor

Altera Corporation
3 West Plumeria
San Jose, CA
USA

Doug Amos
European Applications Manager

Altera Europe
Holmers Farm Way
High Wycombe, Bucks
UK

Introduction

DSP processors can easily implement complete algorithms with impressive performance; however, one function within the system implementation often takes up an inordinate amount of processing bandwidth, which effectively minimizes the bandwidth of the entire system. These high-bandwidth functions are often low in complexity but high in throughput demands. Programmable logic devices can be utilized as DSP coprocessors to off load these functions thereby freeing up the DSP processor to implement the more complex functions with greater speed, dramatically improving overall system performance. System-level functions that are enhanced with the DSP coprocessor approach include spread-spectrum modems, fast Fourier transform acceleration, and machine vision.

The purpose of this paper is to discuss a means of enhancing the overall performance of fixed-point DSP processor based systems by off loading low-complexity, high-throughput functions onto programmable logic devices (PLD's) acting as DSP coprocessors. This method utilizes a low-cost PLD that significantly improves overall system performance without adding significantly to the overall system cost or severely impacting system board space requirements. The paper will begin by examining existing DSP design options. Next, the arithmetic function capabilities of programmable logic devices (these functions being the foundation of most DSP functions) will be examined. Afterwards, the paper will explore programmable logic's capacity to act as the most common type of DSP function: the finite-impulse response (FIR) filter. Finally, a few specific application examples where programmable logic has been used to supplement a DSP processor will be presented.

DSP Design Options

There are several options available for designers to build DSP functions. The most commonly used ones are: DSP processors, ASIC's, and Application-Specific Standard Products (ASSP's). Each of these options have their own set of advantages and disadvantages. Fixed-point DSP processors are a typical low-cost option, but are too slow to address real-time applications. Floating-point DSP processors may be fast enough for these applications but are too expensive. ASIC solutions are typically high-performance and have two options: a "build your own" (e.g. multiply accumulate engine) or a core approach. DSP cores are typically limited to high-volume applications but all ASIC solutions have the disadvantage of lengthening time-to-market owing to factory fabrication cycles. Factory fabrication also renders ASIC much less flexible than DSP Processors but at least ASIC's can be altered. ASSP's on the other hand, which are fixed dedicated functions such as FIR's and Multipliers, allow the least user alteration of any of the options.

The Application of Programmable Logic

Programmable logic fills the gap where both flexibility and high-speed real-time performance are required for specific DSP applications. The graph below in Figure 1 shows conceptually where programmable logic (specifically, FLEX programmable logic from Altera Corporation) compares in speed and flexibility to the traditional DSP solutions:

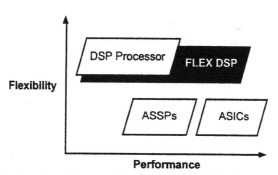

Fig. 1   Area of best application of various DSP Solutions

For low-throughput designs, any solution will adequately support DSP computational needs; a low-cost microcontroller or microprocessor provides an excellent solution. As performance requirements increase to the 10KSPS to 1MSPS range, DSP processors provide an ideal solution that addresses both performance and flexibility.

Between 1 and 10 MSPS a DSP processor reaches limitations and alternate solutions must be examined. These solutions include multiple DSP processors or DSP cores and programmable logic (both as primary processor and as coprocessor). In the range of 10 MSPS to 150 MSPS, PLD's provide ideal performance. The number of functions also impacts overall system bandwidth needs - the more functions performed, the earlier the solution reaches bandwidth limitations. This idea is portrayed in Figure 2 below:

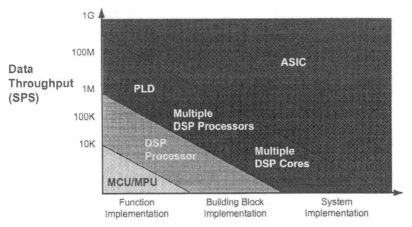

Fig. 2 Complexity/Performance Domain for DSP Solutions

Above 150 MSPS, an ASIC is the only single-chip solution that provides adequate performance. An ASIC will also provide an attractive option at lower performance level if the volume is high enough.

Arithmetic Capability of Programmable Logic

DSP functions are composed largely of arithmetic operations. The speed of programmable logic devices in performing DSP-type functions is therefore dependent on their ability to perform arithmetic operations. In order to understand programmable logic in this capacity, the structure and composition of programmable logic devices will be examined. Specifically, this section will focus on look-up table-based PLD's, which are the PLD's that are best suited for DSP functions.

Look-Up Tables, or LUTs can implement any function of N inputs, where N is the number of inputs to the LUT (see Figure 3 below). Functions that require more than 4 inputs are split between multiple LUTs. In FLEX devices, the LUTs are supplemented by carry chains, which are used to build fast adders, comparators, and counters. For example, an 8-bit adder can run at a data rate of 172MHz.

Beyond addition, LUT-based PLD's also provide high speed multiplication. The following table shows performance and utilization data for different-sized pipelined multipliers in FLEX 8000 PLD's:

Multiplier Size	Fmax (MHz)	Logic Cells	Latency
8x8	103.09	145	4
10x10	86.95	251	5
12x12	81.99	337	5
16x16	68.46	561	5

Note: The pipelined multipliers used in this benchmark were built using the parameterisable multiplier LPM_MULT available from Altera Corporation.

FIR Filters in PLD's

Since it is apparent that programmable logic can perform well in the arithmetic functions that compose most DSP-type functions, the next step is to study the implementation of an actual DSP function in a PLD. In this section, a FIR filter design is placed into the FLEX architecture and its characteristics are examined.

A conventional 8-tap FIR filter structure has eight 8-bit registers arranged in a shift register configuration. The output of each register is called a tap and is represented by $x(n)$, where n is the tap number. Each tap is multiplied by a coefficient $h(n)$ and then all products are summed (see figure 3).

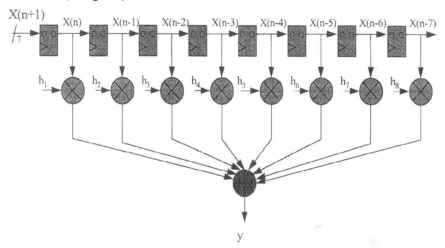

Fig. 3 Typical Multiply-Accumulate-based FIR Filter

The equation for the filter in Figure 3 is:

$$y(n) = x(n)h_1 + x(n-1)h_2 + x(n-2)h_3 + x(n-3)h_4 +$$
$$x(n-4)h_5 + x(n-5)h_6 + x(n-6)h_7 + x(n-7)h_8$$

In the FIR filter implementations described in this paper, the multiplications take place in parallel, which means that only one clock cycle is required to calculate each result. A common term used to describe this function is a "multiply and accumulate" or MAC.

Consider a linear phase FIR filter with symmetric coefficients (the coefficients are symmetric about the center taps). Linear phase means that the phase of signals going through the filter varies linearly with frequency. If the filter depicted in Figure 3 had symmetric coefficients, then the following would be true:

$$h(1) = h(8), \ h(2) = h(7), \ h(3) = h(6), \ h(4) = h(5)$$

Which means that the equation for the output could be converted into the following:

$$y(n) = h_1 \times [x(n) + x(n-7)] +$$
$$h_2 \times [x(n-1) + x(n-6)] +$$
$$h_3 \times [x(n-2) + x(n-5)] +$$
$$h_4 \times [x(n-3) + x(n-4)]$$

By factoring the coefficients out, the function now only requires 4 multiplication operations, instead of 8. This conversion reduces the multiply hardware required by 50%.

The vector multiplier multiplies four constants, h1, h2, h3, h4, by four variables, s1, s2, s3, s4. The fact that the coefficients are constant can be used to build a more efficient LUT-based multiplier than the standard multiplier approach. Specifically, this approach takes advantage of the fact that there are a limited number of total possible products for a given multiplicand. To understand this approach, consider the case where the multiplicands are 2-bit numbers:

If the impulse response h_i is:

$$h_1 = 01, \quad h_2 = 11, \quad h_3 = 10, \quad h_4 = 11$$

and s_i is:

$$s_1 = 11, \quad s_2 = 00, \quad s_3 = 10, \quad s_4 = 01$$

Writing out the multiplication in long form results in:

```
hi = 01      11      10      11
si = 11      00      10      01
     ----    ----    ----    ----
     01      00      00      11 =   100 = p0
     01      00      10      00 =   011 = p1
     ----    ----    ----    ---- ------
     011     000     100     011 = 1010 = yi
```

where p_0 and p_1 are partial products.

Each partial product (p_i) is uniquely determined by the four bits $s_i^{(1-4)}$. Since all h_i are constant, there are only 16 possible partial products (p_i) for each value of $s_i^{(1-4)}$. These 16 values can be stored in a LUT of 4-bit inputs and outputs in a programmable logic device. To calculate the final result yi, a shift-add operation is used to accumulate each pi. The diagram below shows the contents of the LUT for the given h_i value for our example.

```
si0            p0 -- LUT Value              si0            p0
0000   =>   00+00+00+00 = 0000     1000   =>   01+00+00+00 = 0010
0001   =>   00+00+00+11 = 0011     1001   =>   01+00+00+11 = 0100
0010   =>   00+00+10+00 = 0010     1010   =>   01+00+10+00 = 0011
0011   =>   00+00+10+11 = 0101     1011   =>   01+00+10+11 = 0110
0100   =>   00+11+00+00 = 0011     1100   =>   01+11+00+00 = 0100
0101   =>   00+11+00+11 = 0110     1101   =>   01+11+00+11 = 0111
0110   =>   00+11+10+00 = 0101     1110   =>   01+11+10+00 = 0110
0111   =>   00+11+10+11 = 1000     1111   =>   01+11+10+11 = 1001
```

Figure 4 displays a visual conception of a 4-input, 2-bit vector multiplier. The LSB (bit 0) of s_i goes to the least significant LUT. The MSB (bit 1) of si goes to the most significant LUT. The outputs of each LUT (the corresponding p_i) is then added to obtain the result

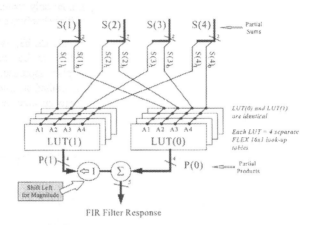

Fig. 4 Visual Conception of LUT-based 2-bit FIR

Applying this vector multiplier concept to FIR filters allows programmable logic to achieve high performance and low resource utilization shown in the following table.

FIR Filter	Utilization (Logic Cells)	Performance (MSPS)	
		A-2	A-3
8-Tap Parallel	296	101	74
16-Tap Parallel	468	101	75
24-Tap Parallel	653	100	74
32-Tap Parallel	862	101	75
64-Tap Serial	920	7	5

The resource utilization of a FIR filter in a PLD grows as both the input value width increases and as the number of taps increases.

Case Studies of PLD's used as DSP Coprocessors

Having established the efficiency of implementing DSP-type functions in programmable logic in terms of performance and resource utilization, we can examine specific case studies where PLD's have been used effectively as DSP coprocessors. The first case involves a design for a spread spectrum modem for a wireless LAN. Two different designs were considered for the modem; in the first, both correlation and data demodulation were performed by the DSP processor, a TMS320C25-50.

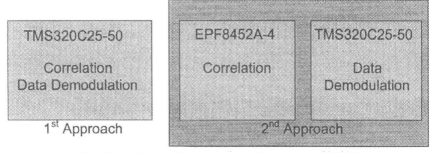

Fig. 5 Repartition of System between Processor and CPLD's

In the second approach, a PLD (FLEX 8000) is used to perform the correlation (see Figure 10 above).

The efficiency of each system was measured, given pseudo-random number (PN) sequences as input (filter size). The results are outlined in the table below.

Filter Size (PN)	System Chip Rate	Required LCELL's	System Configuration
15-Bit Sequence	10 MHz	-	DSP Processor
15-Bit Sequence	100 MHz	266	DSP Processor & EPF8452A-2
15-Bit Sequence	66 MHz	266	DSP Processor & EPF8452A-4
31-Bit Sequence	66 MHz	553	DSP Processor & EPF8636A-4

The PLD coprocessor implementation increases performance by a factor of 6 with a low-cost PLD and by a factor of 10 with a high-performance PLD. Increasing the PN size with the single-chip DSP processor implementation would have significantly degraded performance. In PLD's, increased PN is handled by adding parallel resources, while providing the same performance.

Fast Fourier Transform

In the next case study, we examine a system that requires a Fast Fourier Transform (FFT). FFT's are often used to calculate the spectrum of a signal. An N-point FFT produces N/2 bins of spectral information spanning from zero to the Nyquist frequency. The frequency resolution of the spectrum is Fs/N Hz per bin, where Fs is the sample rate of the data. The number of computations required is approximately N(logN).

Many applications though only require a narrow band of the entire signal spectrum. The FFT calculates the entire spectrum of the signal and discards the unwanted frequency components. Multirate filtering techniques let you translate a frequency band to the baseband and reduce the sample rate to 2x the width of the narrow band. An FFT performed on reduced-sample-rate data allows either greater resolution for same amount of computations or equivalent resolution for a reduced amount of computations.

Fixed-point DSP processors can perform both the FFT algorithm and the pre-processing frequency translation. One method of translation takes advantage of the aliasing properties of the rate compressor. This rate compression in the frequency domain results in images that are spaced at harmonics of the sampling frequency. The modulation and the sample rate reduction can be done simultaneously; the signal is rate-compressed by a factor of M to get the decimated and frequency-translated output. An (N/M)-point FFT is then performed on the resulting signal, producing the spectrum of the signal, which contains only the narrow band.

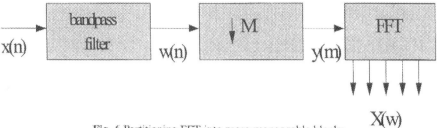

Fig. 6 Partitioning FFT into more manageable blocks

Since the filtering tasks can be separated in to different processes, different devices can be used to performs the tasks. Specifically, a PLD can be used to perform the preprocessing, and the DSP processor can be used to perform the FFT operation. To understand the benefits a PLD would bring, we must first examine the performance of a single DSP processor when performing the whole operation.

The following table contains the results for an ADSP2101 processor.

	Execution Time DSP Processor	Decimation Factor	Max Sampling Rate
1024-point FFT[1]	5.3 msec		193 KHz
128-point FFT[1]	0.49 msec		
128-tap FIR for 1024 Samples	0.22 msec		
128-point FFT 128-tap FIR	0.71 msec	8	1.4 MHz
128-point FFT 128-tap FIR	0.49 msec coprocessor	8	2.1 MHz

[1]Radix-2 DIT FFT with conditional Block Floating Point

The ADSP2101 can sustain a maximum sampling rate with a 1024-point FFT of 193 KHz. Maximum frequency is determined by dividing the execution time (5.3 msec) by 1024 points to get 5.2 µsec per point which corresponds to frequency of 193 KHz. A 128-point FFT can support a higher sample rate but also involves a decimation factor of 8 to elevate the maximum sampling rate further. The input decimation filtering can be done optimally with a 128-tap FIR filter; this FIR filter in the ADSP2101 takes 0.22 msec for the 1024 points that are downsampled to provide the 128 points for the FFT. This 0.22 msec comes from 13.3 µsec for a 128-tap FIR filter divided by decimation factor (8) times the 128 points. The DSP processor must timeshare the two tasks and can support a maximum sampling rate of 1.4 MHz.

Offloading the decimating filter to a DSP coprocessor enables the maximum system frequency to go back to 2.1 MHz. A serial, 128-tap, 10-bit FIR filter can fit into a single programmable logic device (such as an EPF10K30). This serial implementation supports 5 MSPS throughput which easily supports the maximum sample rate. The DSP processor is now free to focus entire bandwidth on the FFT algorithm which elevates the maximum frequency substantially. Figure 7 below shows a block diagram of this implementation.

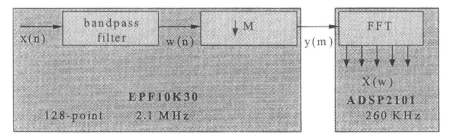

Fig. 7 Implementation of FFT Blocks in CPLD and Processor

Another option is to implement the FFT completely in a programmable logic device. Megafunction compilers are available that provide impressive performance for FFT processing with complete compilation flexibility. FFT's obviously requires larger PLD device than the pre-processing filter; resource utilization and performance statistics are outlined in the table below (the PLD's used for these measurements were FLEX 10K devices):

Length (points)	Precision	Memory	Size (LE's)	Speed
512	8 Data 8 Twiddle	Dual - Internal	1150	94 μsec
1024	16 Data 16 Twiddle	Dual - External	2993	207 μsec
32K	16 Data 16 Twiddle	Dual - Internal	3100	9.8 msec

System Implementation Recommendations

The first step in the process is to evaluate system bandwidth requirements. If the DSP processor is not operating at capacity, PLD Coprocessing will not add any benefit. If however, the DSP processor operates at full bandwidth capacity and critical functions/algorithms must wait for processor resources, PLD'S as a coprocessor may provide a significant performance benefit. The system should then be analyzed to determine which function/algorithm depletes the bandwidth. If a single function uses greater than 1/2 of available bandwidth this function may be offloaded efficiently; functions related to filtering (preprocessing, decimating, interpolating, convoluting, FIR, IIR, etc.) will most efficiently be implemented in programmable logic devices.

Summary

Programmable logic provides an ideal balance between the flexibility of a DSP processor and the performance of a DSP ASIC solution. Programmable logic also provides a strong complement to a DSP processor to offload computationally intensive functions/algorithms as a DSP coprocessor. In addition to improving system performance, this coprocessor methodology also acts to protect investments that have been made in DSP processor tools, code, and experience by extending the potential applications that could initially be done with a given DSP processor.

A Case Study of Algorithm Implementation in Reconfigurable Hardware and Software

Mark Shand

Digital Equipment Corporation, Systems Research Center
130 Lytton Ave, Palo Alto, CA 94301
shand@acm.org

Abstract. We present a case study of implementation of a combinatorial search problem in both reconfigurable hardware and software. The particular problem is the search for approximate solutions of overconstrained systems of equations over GF(2). The problem is of practical interest in cryptanalysis. We consider the efficient implementation of exhaustive search techniques to find the best solutions of sets of up to 1000 equations over 30 variables. *Best* is defined to be those variable assignments that leave the minimum number of equations unsatisfied.

As we apply various techniques to speed up this computation, we find that the techniques, whether inspired by software or reconfigurable hardware, are applicable to both implementation domains. While reconfigurable hardware offers greater raw compute power than software, new microprocessor with wide datapaths and far higher clock speeds do not lag far behind. Software also benefits from faster compilation times which prove important for some optimizations.

1 Introduction

We present a case study of implementing a combinatorial search problem in both reconfigurable hardware and software. Although we did not produce running hardware, the hardware design was taken to a point where both the circuit size and time required for the search execution could be accurately predicted. The software implementation was carried through to optimized C code running on a Digital 64-bit Alpha processor. The particular problem is the search for approximate solutions of overconstrained systems of equations over GF(2). We consider only brute force approaches to the search. To the best of our knowledge the problem is not amenable to non-brute force approaches. The problem size is held at 1000 equations in 30 variables. Thus the problem state space is of size 2^{30}.

The initial goal of this work was to implement this problem on FPGA-based coprocessors. However, we have found that careful coding on a modern microprocessor with good compiler support can also yield very respectable performance, such that an FPGA-based solution, albeit faster, may not be worth pursuing.

FPGA-based reconfigurable computing machines are often seen as having major advantages over traditional computer architectures on problems involving repetitive computation on small or variable-sized integer data because they can tailor their datapaths to the problem at hand. However, for many years now, techniques have been known that use conventional ALU datapaths to process several small integer variables in parallel, in a SIMD manner ([1], [2]). Certainly these techniques entail some overhead but the speedups gained more than compensate, particularly on modern CPUs with 64-bit wide integer ALUs [3].

We will show that it is useful to consider implementations in both SIMD-style software and reconfigurable hardware in parallel. Techniques that emerge more naturally in one implementation domain can often be fruitfully applied to the other. We will also point out some drawbacks in current reconfigurable hardware systems that limit their performance in this and similar problems.

The paper is organized as follows. Section 2 presents the search problem in mathematical terms. Section 3 describes a basic hardware search engine. Section 4 discusses opportunities for parallelism that speed-up the search with modest amounts of extra circuitry. Section 5 covers the software implementation. Section 6 compares the hardware and software solutions. Section 7 concludes the paper.

2 Linear Systems over GF(2)

Consider a system of M formulae over N variables:

$$r_j(X) = \left(\bigoplus_{i=0}^{N-1} w_{ij}x_i \right) \oplus v_j, \quad 0 \le j < M \tag{1}$$

Arithmetic is performed over the Galois Field GF(2). In less mathematical terms this means values are represented in one bit and all computations are modulo 2. Addition and multiplication are the familiar Boolean operations of *exclusive-or* and *and* respectively.

For a given set of w_{ij}'s and v_j's, if $M > N$ we cannot guarantee that an assignment to each x_i can be found such that each formula in (1) is equal to zero $\forall j : 0 \le j < M$. Nevertheless, we can define a figure of merit, R, which can be used to rank the different possible assignments.

$$R(X) = \sum_{j=0}^{M-1} r_j(X) \tag{2}$$

$R(X)$ counts the number of formulae that do not evaluate to zero for a given assignment to X. We seek those assignments of X that minimize $R(X)$. Note that the arithmetic used to compute $r_j(X)$ is over GF(2) whereas that used to combine the single bit $r_j(X)$'s to form $R(X)$ is over the integers.

3 Basic Hardware Search Engine

In this section, we consider the hardware implementation of search engines that can handle up to 1000 equations ($M \le 1000$) in 30 variables ($N \le 30$). This implies 2^{30} possible assignments of X that must be evaluated. We will consider techniques to find the B best ranked X, where B is a small integer, say 16.

For quantitative evaluation of our proposed implementation, we assume the use of configurable hardware based on Xilinx 3000 series FPGAs [4]. These contain a regular two-dimensional mesh of programmable cells called configurable logic blocks (CLBs) in Xilinx terminology. Each CLB contains two registers and two lookup tables (LUTs) that allow it to implement either two 4-input combinatorial logic functions or one 5-input combinatorial logic functions.

3.1 Principal Constraints

In this section we consider the search order. We find that the enumeration of X should be confined to the outer loop to prevent an excess of intermediate results, and the inner loop should be at least parallelized to the level of the M equations to allow the equation parameters to be wired into the processing elements.

The rank of a particular X, $R(X)$, is the sum of the results of each equation $r_j(X)$. We assume that the computation proceeds by enumerating each possible X and maintaining a sorted list of the best $R(X)$ so far encountered. It is impractical to proceed in a different order, for instance evaluating for successive j, $r_j(X)$ for all possible X since this would require the partially accumulated ranks to be stored (and repeatedly updated) of which there are one billion.

The computation can be performed most efficiently if we have enough hardware to hold locally to each processing element in a distributed store, the w_{ij} coefficients needed for computation of each $r_j(X)$. If enough local store is not available we are likely to be required to reload these coefficients from some central store. The bandwidth required for such reloading would create a bottleneck.

3.2 Gray Codes

In principle, a simple counter could be used to enumerate X. Considerable advantages accrue if instead we use a *Gray code*. A Gray code ensures that only one bit in the representation of X changes for each new X. A Gray-coded counter, G, may be derived from a conventional counter, C, by setting $g_i = c_i \oplus c_{i+1}$.

Suppose that the ith bit of X changes, $x_i' = \neg x_i$. There are two cases to consider:

$$\begin{aligned}
&\text{if } w_{ij} = 0 \text{ then } w_{ij}x_i' = w_{ij}x_i = 0 \\
&\qquad \Longrightarrow r_j(X') = r_j(X) \\
&\text{if } w_{ij} = 1 \text{ then } w_{ij}x_i' = x_i' = \neg x_i = \neg w_{ij}x_i \\
&\qquad \Longrightarrow r_j(X') = \neg r_j(X)
\end{aligned}$$

Note that the value of $r_j(X')$ is independent of the value of x_i and depends solely on the value of $r_j(X)$ and of w_{ij}. Thus the processing element that computes the value of $r_j(X')$ needs to know only the value of $r_j(X)$ and which bit i changes in going from X to X'.

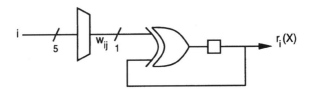

Fig. 1. Formula Evaluation Cell

3.3 Formula Evaluation Cell

The Figure 1 shows the hardware for evaluating successive values for one $r_j(X)$.

The five-bit index i is broadcast from a central Gray-coded enumerator which is shared amongst the evaluation circuits for all formulae. The Xilinx implementation of this circuit requires one and a half CLBs; one for the LUT that decodes i into w_{ij} and a half for an exclusive-or and a register that holds the current value of $r_j(X)$. Parallel evaluation of the 1000 formulae requires 1500 CLBs.

3.4 Population Count

The formula evaluations produce 1000 one-bit values which are added together to yield $R(X)$. The binary representation of this value needs 10 bits. The operation to produce it is a population count. We consider an implementation based on a tree of full-adders. A full-adder takes three values a, b and c of weight 2^k and produces outputs r and s of weights 2^{k+1} and 2^k respectively, where $a + b + c = 2r + s$. A portion of the required full-adder tree is shown in Figure 2.

The population count circuit requires approximately 1000 full-adders. The actual figure will be slightly higher than this due to boundary cases which cause some full-adder inputs to be tied to zero. In a Xilinx implementation this will consume one CLB per full-adder.

3.5 Implementation on DECPeRLe-1

DECPeRLe-1 [5] is an FPGA-based coprocessor built at Digital's Paris Research Laboratory (PRL) in 1992. Its central processing resource is a 4×4 matrix of Xilinx 3090-100 FPGAs containing a total of over 5000 CLBs. The formula evaluation and population count circuitry consumes roughly half of these. The remaining CLBs provide plenty of room to implement the comparators and registers to track the current B best ranked assignments of X.

Given the regularity and the ample opportunities for pipelining, a clock speed of 40MHz is not unreasonable. The proposed design evaluates one assignment of X per cycle and at 40MHz requires 25 seconds to enumerate the 2^{30} states.

We also considered implementation of this computation on the newer FPGA-based coprocessors (TURBOchannel Pamette and PCI Pamette) built by Digital and described in detail in [6] and [7] . These boards have a 2×2 matrix of

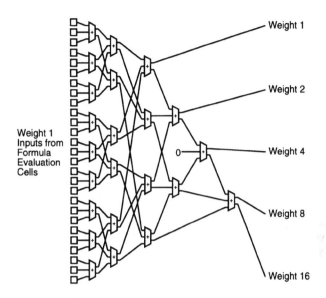

Fig. 2. Population Count Using a Tree of Full-Adders

Xilinx 4010 FPGAs. With only 1600 CLBs, these boards are not large enough to contain the formula evaluation and population count circuitry for the problem size considered.

Surprisingly, we could find no way to use these smaller boards to accelerate the computation over the pure software implementations discussed in section 5. Although the boards provide a powerful computational resource, they can only implement part of the computation and the cost of communicating between them and other parts of the system outweighs any processing advantages they can give.

4   Opportunities for Parallelism

We will consider two classes of methods to parallelize the enumeration of X: methods that make no particular assumptions about the formulae being treated; and methods that depend on w_{ij} and require preprocessing of the formulae. The distinction is important because in case two preprocessing may dominate the time taken for the search, particularly if it changes the wiring to be downloaded into the FPGA-based coprocessor. In the Xilinx technology, logic changes in LUTs can be trivially merged into a precompiled design, whereas wiring changes involve a circuit compilation process that takes, at best, several minutes to run. A software analogue is changing initialized data versus recompiling modified source code. Note also that some of our methods make assumptions about the statistical distribution of w_{ij} coefficients that may not hold in all contexts where solutions to these types of equations are sought.

4.1 Sharing Formula Evaluation Cells

As a first step, let us remove x_0 from the enumeration by computing in parallel the two cases $x_0 = 0$ and $x_0 = 1$. This effectively doubles the number of formulae because we must now evaluate formulae (3) and (4) in each cycle.

$$w_{0j} \cdot 0 \oplus \left(\bigoplus_{i=1}^{N-1} w_{ij} x_i \right) \oplus v_j \tag{3}$$

$$w_{0j} \cdot 1 \oplus \left(\bigoplus_{i=1}^{N-1} w_{ij} x_i \right) \oplus v_j \tag{4}$$

However, observe that (4) is either identical to or the inverse of (3), depending on the value of w_{0j}. This sharing is exploited by the circuit in Figure 3. In the Xilinx technology, the exclusive-or that computes $r_i(X + 1)$ may be merged into the logic in the population count for $x_0 = 1$. Thus, this circuit enumerates two states per cycle, double the speed of our previous circuit, at the cost of a second population count circuit, a 1.4 times increase in CLBs.

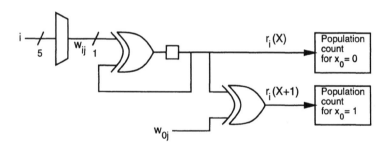

Fig. 3. Sharing Formula Evaluation Cells

We can apply this technique to x_1 and successive variables, but the relative gain decreases since the size of the total circuit quickly becomes dominated by the population count circuits. In any case, evaluating all possible values x_0 and x_1 in a single cycle by this method requires four population count circuits which with the formula evaluation circuit exceeds the CLB resources of DECPeRLe-1.

4.2 Partitioning on w_{ij} Values in Hardware

Much greater savings can be obtained by sharing population count circuits. This is possible if we are permitted to partition the formulae based on the w_{ij} values.

Suppose that in K of the M formulae $w_{0j} = 1$. Without loss of generality, let us sort the formulae so that these correspond to $j : 0 \leq j < K$

$$r_j(X \oplus 1) = w_{0j} \oplus r_j(X)$$
$$= \begin{cases} 1 - r_j(X) \text{ if } j < K \\ r_j(X) \quad \text{ if } j \geq K \end{cases} \tag{5}$$

$$R(X) = \left(\sum_{j=0}^{K-1} r_j(X) \right) + \left(\sum_{j=K}^{M-1} r_j(X) \right) \tag{6}$$

$$R(X \oplus 1) = K - \left(\sum_{j=0}^{K-1} r_j(X) \right) + \left(\sum_{j=K}^{M-1} r_j(X) \right) \tag{7}$$

Equations (6) and (7) show how the bulk of the population count circuitry may be shared. A circuit implementing this sharing appears in Figure 4.

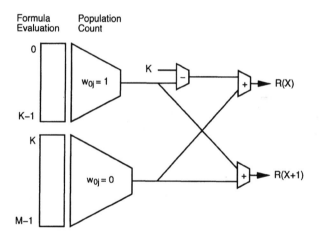

Fig. 4. Sharing Population Count Circuitry

This technique may be applied recursively to w_{1j}, w_{2j}, etc. The overhead of extra adders at the top of the population count tree roughly doubles for each successive application, but remains less than the cost of the population count tree until applied six or seven times. This corresponds to a speed-up of 64 or 128 over our original DECPeRLe-1 execution time estimate of 25 seconds. The chief difficulty posed by the technique is that the preprocessing of the w_{ij} values may require a rewiring of the circuit. The preprocessing overhead is far more manageable in software which is where we have put most implementation effort.

5 Software Implementation

Software implementation has concentrated on Digital's Alpha AXP 21064, with much attention given to maximally exploiting the 64-bit integer datapath of this processor[8]. Measurements on a number of Alpha platforms indicate that the critical parts of the algorithm fit entirely in on-chip caches so the only important parameter in determining speed is CPU clock rate.

5.1 Formula Evaluation

In section 3.1 we argued that in hardware the w_{ij} coefficients needed to be held locally to the processing elements to avoid a performance penalty due to their repeated reloading. Likewise in software these coefficients need to be rapidly loaded to the ALU. The coefficients of 1000 formulae in 30 variables occupy 4 kilobytes and fit comfortably in the 8 kilobyte first level data cache of the 21064.

The coefficients are packed such that adjacent bits share a common i index. A 1000-bit vector represents the current value of the formulae. A Gray-coded counter selects the index of the variable to toggle and the corresponding 1000-bit coefficient vector is exclusive-ored with the current value. This requires 32 loads, 16 xors and 16 stores which can execute in 48 cycles on the 21064.

5.2 Population Count

Population count is designed to exploit the full 64-bit wordlength of the Alpha ALU. Mask and shift operations extract odd and even bits to create 64-bit words containing 32 2-bit fields, the contents of which are either 0 or 1. Three such fields can be added together without danger of overflow into the next field. These adds are performed using the 64-bit register-to-register add, thus 32 fields are processed per instruction. 16 words are required to hold the 1000 1-bit inputs. When these are combined into 2-bit fields, of value 0 to 3, only 11 words are required since only 334 ($\lceil 1000/3 \rceil$) such fields are needed. A new sequence of mask and shift operations extracts the 2-bit fields into 4-bit zero-padded fields. Now 5 such fields can be added without fear of overflow—the maximum value in a 2-bit field is 3 and $5 \times 3 = 15$ which can be represented in 4 bits. There are 66 ($\lceil 1000/15 \rceil$) 4-bit fields, now requiring only 5 words[1]. The process continues until a single field containing the 10-bit population count remains. Portions of the code are reproduced in Figure 5. This code requires approximately 170 cycles on the 21064. On a 200MHz 21064 it takes 25 minutes to enumerate the 2^{30} possible assignments of X and record the 16 best.

5.3 Partioning on w_{ij} Values in Software

Naturally, the ideas of section 4.2 can be applied to software. Indeed refinements are possible. The subtraction from K can be replaced by a subtraction from $2^n - 1$ followed by a post correction. The subtraction from $2^n - 1$ can be implemented as a negation over n bits, and the post correction can be folded into

```
m0 = 0x5555555555555555L;     /* sum single bits to pairs */
d0 = (v[0] & m0) + ((v[0] >> 1) & m0) + (v[1] & m0);
d1 = ((v[1] >> 1) & m0) + (v[2] & m0) + ((v[2] >> 1) & m0);
d2 = (v[3] & m0) + ((v[3] >> 1) & m0) + (v[4] & m0);
...
m1 = 0x3333333333333333L;     /* sum double bits to nibbles */
n0 = (d0 & m1) + ((d0 >> 2) & m1)
   + (d1 & m1) + ((d1 >> 2) & m1) + (d2 & m1);
n1 = ((d2 >> 2) & m1) + (d3 & m1)
   + ((d3 >> 2) & m1) + (d4 & m1) + ((d4 >> 2) & m1);
...
m2 = 0x0f0f0f0f0f0f0f0fL;     /* sum nibbles to bytes */
B0 = (n0 & m2) + ((n0 >> 4) & m2) + (n1 & m2) + ((n1 >> 4) & m2)
   + (n2 & m2) + ((n2 >> 4) & m2) + (n3 & m2) + ((n3 >> 4) & m2)
   + (n4 & m2) + ((n4 >> 4) & m2);
m3 = 0x00ff00ff00ff00ffL;     /* sum bytes to shorts */
...
```

Fig. 5. C Code Fragment of Population Count Algorithm

the comparison against current best ranked assignments. These refinements, invented for the software implementation, can can in turn be applied back to the hardware implementation yielding simplifications in that domain too.

We have implemented six levels of partitioning, which implies that we evaluate 64 states on each iteration. Our current code assumes that w_{ij} coefficients are distributed such that the 64 partitions are of almost equal size[1]. The population count code depends on this. The assumption can be removed either by padding the 1000 formulae with extra dummy formulae to even out the distribution, or by generating and compiling population count code tailored to the particular distribution found in the input formulae.

Extending beyond six levels of partitioning proves difficult to implement and is unlikely to yield an appreciable performance improvement.

5.4 Comparision and Update

With six levels of partitioning we find that the comparison of new ranks against the current best encountered ranks needs attention. Otherwise, time spent in this code can dominate the computation. The 64 new ranks are returned from the population count routine packed into 16-bit fields. Each time the best encountered ranks are updated, we precompute 64 packed 16-bit fields that contain the value of the worst of the current best ranks combined with the post correction for the corresponding new rank. We can now perform 64-bit wordwise subtraction and extract the high bits of the 16-bit fields. Any non-zero high bit represents a borrow in the subtraction which triggers a detailed comparison of each of the 64 new ranks. The detailed comparison is slow but rare.

---

[1] In terms of Figure 4, $K \approx M/2$, and so on for each subpartition.

5.5 Software Performance

Test were performed under DEC OSF/1 V3.0. The C compiler switches -O2 -migrate were used, enabling the GEM compiler [9]. They use six levels of partitioning. Times are for a series of 21064-based Alpha systems. Despite widely differing memory systems the run time is almost a linear function of CPU clock speed, indicating that all critical parts of the algorithm fit in on chip caches. These are uniprocessor times. The computation can be parallelized trivially across multiple CPUs by dividing the search space across the processors.

Model	Clock rate	CPU time
DEC 3000 M400	133MHz	75.64s
DEC 3000 M300	150MHz	69.45s
DEC 2000 300	150MHz	68.52s
DEC 3000 M500	150MHz	67.33s
AlphaStation 200 4/166	166MHz	60.60s
AlphaServer 2100 4/190	190MHz	52.10s
DEC 3000 M800	200MHz	49.91s
DEC 3000 M900	275MHz	36.88s

6 Comparison of Hardware and Software

To make a technologically fair comparison of hardware and software we should concentrate on the 150MHz Alphas which are contempraneous with the DECPeRLe-1 hardware. We acknowledge that the silicon process used in the 150MHz alpha is more advanced than the process used in the FPGAs on DECPeRLe-1, however we argue that FPGA processes have consistently lagged those used on leading edge processors so the comparison is fair in terms of what can be expected to be commercially avaliable at any point in time. Current FPGAs can be clocked at probably two or three times the speed of DECPeRLe-1 and could form the basis of machines with larger logic matrices than DECPeRLe-1 thus allowing greater parallelism. On the other hand CPUs have gotten faster too: software, with six levels of partitioning, takes 17 seconds on a 400MHz Alpha 21164.

Platform	Technique	Running time
Alpha 21064 150MHz	No partitioning	2039s
DECPeRLe-1 (estimate)	No partitioning	25s
Alpha 21064 150MHz	Six levels of partitioning	67.33s

Using our simple algorithm of section 3 the reconfigurable machine is 60-100 times faster than a 150 MHz Alpha. However algorithmic techniques let us speed-up the software to within a factor of two of the initial hardware speed. These same algorithmic techniques can be applied in hardware provided that a data dependent circuit recompilation is not required. If so, the reconfigurable machine can maintain a healthy lead. If however, a data dependent circuit recompilation is required the advantage of the reconfiguarble machine is lost. Whereas software compiles in seconds, traditional FPGAs require tens of minutes if not hours. In the combinatorial problem we are considering here, once a particular input data set is solved, a circuit dedicated to that data set is of no further interest.

7 Conclusions

As we apply various techniques to speed up this computation, we find that the techniques, whether inspired by the SIMD-style software or the reconfigurable hardware, are applicable to both implementation domains. There is important synergy in considering both implementation domains simultaneously.

Many of the constraints of one domain have analogues in the other. However, the analogues are not exact: some constraints that are minor for software become major drawbacks in reconfigurable hardware systems. Reconfigurable hardware offers more raw compute power than current microprocessors but certain operations, such as circuit compilation, are much more expensive than software compilation, limiting the scope of some of our techniques. Both technologies exhibit sharp degradations in performance as certain problem size thresholds are crossed: for instance exceeding the total logic resources in reconfigurable hardware, or the size of the first level cache in software. However, the degradations are more catastrophic in reconfigurable hardware. These drawbacks suggest areas for future work in the design of reconfigurable hardware systems.

Acknowledgements. I thank F. Reblewski of MetaSystèmes for providing this problem and suggesting the use of Gray Codes for state enumeration.

References

1. M. Beeler, R. W. Gosper, R. Schroeppel, *HAKMEM* MIT AI Lab Memo 239, 29 Feb. 1972
2. Leslie Lamport, *Multiple byte Processing with Full-Word Instructions*, Communications of the ACM, August 1975, Volume 18, Number 8.
3. Mark Shand, Wang Wei, Göran B. Scharmer, *A 3.8ms latency correlation tracker for active mirror control based on a reconfigurable interface to a standard workstation*, Photonics East Symposium '95. SPIE, October 1995. SPIE Volume 2607.
4. Xilinx, Inc., *The Programmable Gate Array Data Book*, Xilinx, 2100 Logic Drive, San Jose, CA 95124, USA, 1994.
5. J. Vuillemin, P. Bertin, D. Roncin, M. Shand, H. Touati, P. Boucard, *Programmable Active Memories: Reconfigurable Systems Come of Age*, IEEE Transactions on VLSI Systems, March 1996.
6. M. Shand, *Flexible Image Acquisition using Reconfigurable Hardware*, IEEE Workshop on FPGAs for Custom Computing Machines, April 19-21 1995.
7. http://www.research.digital.com/SRC/pamette
8. Richard L. Sites (editor), *Alpha Architecture Reference Manual*, Digital Press, 1992.
9. David S. Blickstein, Peter W. Craig, Caroline S. Davidson, R. Neil Faiman Jr., Kent D. Glossop, Richard B. Grove, Steven O. Hobbs, William B. Noyce, *The GEM Optimizing Compiler System*, Digital Technical Journal, Volume 4, Number 4, 1992.

A Reconfigurable Data-Localised Array for Morphological Algorithms

Anjit Sekhar Chaudhuri & Peter Y. K. Cheung
Department of Electrical & Electronic Engineering
Imperial College of Science, Technology & Medicine
Exhibition Road, London SW7 2AZ, U.K.
e-mail: anjit.chaudhuri@ic.ac.uk / p.cheung@ic.ac.uk

Wayne Luk
Department of Computing
Imperial College of Science, Technology & Medicine
180 Queen's Gate, London SW7 2BZ, U.K.
e-mail: w.luk@ic.ac.uk

Abstract. This paper describes a parallel array architecture for performing morphological operations on images using dynamically reconfigurable Field Programmable Gate Arrays. The key feature of this architecture is the data-localised arrangement, which significantly reduces data flow and continuous reconfiguration, which leads to efficient utilisation of the area. The development of implementations of the various possible operations in morphological algorithms, using skeletonisation as an example, is discussed.

1 Introduction

Mathematical morphology, which involves the study of shapes [4], is now a fast growing area of image processing. Many parallel architectures have been proposed to perform morphological operations efficiently in hardware ([1], [11], [13]). However, the processing elements often include functional units that are not utilised for a significant proportion of the execution time of morphological algorithms.

Recently, various reconfigurable Field Programmable Gate Arrays (FPGAs) have been developed. In particular, the Xilinx XC6200 family allows rapid dynamic partial reconfigurations. Exploiting these properties led to the development of the new architecture for morphology proposed here. A data-localised approach, where data is loaded once, while the control is reconfigured between operations, reduces the data I/O bottleneck which is now the most significant limitation on performance. The trend towards increasingly rapid reconfiguration will further suit this approach.

This paper is organised as follows. The first section is an introduction to morphological theory, describing important formulations of the structuring element operator that are suited to hardware implementations, and how algorithms may be partitioned. Next, a few conventional architectures are described, problems in implementing morphological operations are explained, and the new architecture is proposed. Finally, we describe early performance results, as well as the future work required in developing this architecture.

2 Morphology

2.1 Binary Digital Morphology

Morphology is a non-linear space-domain filtering process. It may be performed on grey-scale images, but this study is being restricted to binary images. The filter operator is known as a *structuring element*. In morphology, an image (A) is 'probed' with the structuring element (B), according to the rules of the chosen morphological operation, to produce the desired result. The basic morphological operations are *dilation* and *erosion*, while *opening* and *closing* are compound operations. The following equations describe these operations using vector sets:

Binary dilation: $\mathcal{D}\,(A, B) = A \oplus B = \bigcup_{b \in B} A + b$

Binary erosion: $\mathcal{E}\,(A, B) = A \ominus (-B) = \bigcap_{b \in B} A - b$

Opening : $\mathcal{O}(A, B) = \mathcal{D}\,(\mathcal{E}\,(A, B), B)$

Closing : $\mathcal{C}(A, B) = \mathcal{E}\,(\mathcal{D}\,(A, B), B)$

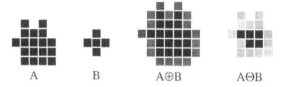

$$\begin{array}{cccc} A & B & A \oplus B & A \ominus B \end{array}$$

Fig. 1. Basic morphological operations

The set of points A and B where the image is defined (equal to 1) is called the *domain*.

The choice of structuring element is dependent on the information that is required from the filtering process; some knowledge of common structuring elements can be exploited to give more efficient architectures.

Re-writing the morphological operations in terms of Boolean operations gives:

$$[A \oplus B](c) = \bigvee_{w \in W} (A(c-w) \wedge B(w))$$

$$[A \ominus (-B)](c) = \bigwedge_{w \in W} (A(c+w) \vee B'(w))$$

where $\vee$ represents OR, and $\wedge$ represents AND, and B' denotes the complement of B. The window W is usually fixed, and is not the same as the domain of the structuring element ($B(w) = 0$ if $w \notin B$, and similarly for A). This enables the development of data-local architectures.

Other data-local operations required in morphological algorithms are min / max or image subtraction (usually between previous and current values for each pixel). Global image parameters are sometimes needed, usually as a test condition. These are all much simpler to implement than erosion and dilation.

From the definitions of erosion and dilation, rules for decomposition of structuring elements can be derived:

if $B = B1 \oplus B2$, then $A \oplus B = A \oplus B1 \oplus B2$

and $A \ominus B = A \ominus (B1 \oplus B2) = (A \ominus B1) \ominus B2$

Decomposition is guaranteed for convex structuring elements ([9] gives an optimal solution for this class), but is impossible with many non-convex ones, such as diagonal crosses. Instead, unions of structuring elements can be used:

for B1, B2 such that $B = (B1 \cup B2)$

$A \oplus B = A \oplus (B1 \cup B2) = (A \oplus B1) \cup (A \oplus B2)$

$A \ominus B = A \ominus (B1 \cup B2) = (A \ominus B1) \cap (A \ominus B2)$

B_1 and B_2 should be mutually exclusive to minimise the number of sub-elements required. Both rules can be extended to any number of sub-elements of the original. They can be combined into a general rule for any structuring element B [5], allowing morphological operations to be performed iteratively with small sub-elements.

2.2 Skeletonisation

An example of a typical morphological algorithm is skeletonisation, which can be used to achieve lossless compression. The aim is to find all points in an image which are equidistant from the two (or more) nearest points on the boundary, which can be achieved morphologically by:

$$SK(A) = \bigcup_{i=0}^{I} S_i(A); \qquad\qquad S_i(A) = (A \ominus n_iB) \setminus O(A \ominus n_iB, B)$$

where '\' represents set difference.

In the Euclidean, B is a circle and n_i is equal to i. This can only be approximated digitally using two sub-elements [10]. The algorithm requires a stopping condition test to be performed, as I cannot be known a priori. All $S_i(A)$ for $i > I$ are empty sets, and this can be tested for by taking the OR of all pixel values.

3 Hardware Issues

3.1 Decomposition for hardware

While the size of a structuring element is variable, it is necessary in any hardware implementation to have a fixed neighbourhood window size for the transition function by which it is being modelled, so that the interconnection and processor can be precisely defined. This could be achieved by having a window large enough to contain any desired structuring element. However, decomposing to sub-elements of a pre-determined maximum size can considerably reduce hardware requirements. The ideal domain for many architectures is a 3*3 area with the origin at the centre point. A morphological algorithm can perform automatic decomposition; a sequential algorithm in [9] has been parallelised for the proposed architecture.

347

3.2 Current Morphological Architectures

Many architectures have been proposed for performing morphological operations in hardware, both of a general nature ([1], [11]) and specific to certain limitations on structuring element or image type, e.g. [13]. The following briefly covers the main classes of architecture in use, and their relative merits.

Parallel array processors are direct realisations of tessellation automata [11]. This refers to architectures where there is a cellular system of processors, all performing the same transition function (a function of the neighbouring processors' current states) at any given time. This function may change with time. Cells must be identical, containing a state register and the neighbourhood processing unit, and the interconnections are defined by the neighbourhood relation.

Pipeline architectures consist of multiple stages, each of which performs a transition function on the image data as it is presented to the processor sequentially. The image is fed into a shift register in raster-scan order, and neighbourhood window values are read into a logic module from the appropriate register cells.

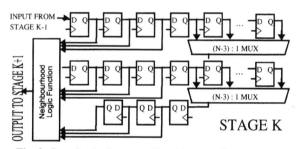

Fig. 2. Generic pipeline stage K architecture diagram

Transition functions may be modelled by a RAM or tree of gates. Reconfiguring for a new transition function is trivial in a RAM, but for a tree of gates, conventional hardware requires the inclusion of registers, which feed their contents into the tree, to store the structuring element values.

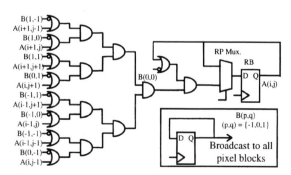

Fig. 3. Pixel layout for erosion with structuring element registers

A weakness of parallel arrays is that limited cell capacity requires the partitioning of images, which causes undesirable edge effects that have to be adjusted for. In addition, they have long input / output delays (the entire array must be filled before processing may start). For these reasons, many people have preferred to develop pipeline architectures. However, there are a few advantages with parallel arrays that can still make them worth considering if the weaknesses can be overcome:

(i) an array of parallel processors uses the same hardware to perform all the different operations, whereas pipeline systems need separate blocks for every stage (unless cyclical pipelines are used).
(ii) parallel arrays only have local data-flow reducing propagation delays.
(iii) parallel arrays provide considerably faster throughput. For a N*N image that can be processed without partitioning, the maximum possible rate (in pixels / cycle) at which the image is processed is $O(N^2)$, compared to $O(1)$ for pipelines.

4 Exploiting Reconfiguration

The aim of our research is to exploit reconfigurable hardware to perform a wide range of morphological algorithms, without limitations on structuring element types, such that a realisable combination of minimal real-estate and maximum speed is achieved. For any implementation, if individual operations are assigned an area (A(op)) and operation time parameter (T(op)) then the performance metrics may be evaluated.

Algorithms may be partitioned into single operations or groups so that performance is optimised. In a fixed hardware operation each pixel requires:

$$A_{fix} = A^s + \sum_{op \in allops} A^d(op)$$

where A_s represents the 'area' of hardware common to different operations. As only one operation, op(t), will be performed at any given time, represents the amount of redundant hardware. For reconfigurable systems, however, the total area is:

$$A_{fix}^{red}(op(t)) = \sum_{\substack{op \in allops \\ op \neq op(t)}} A^d(op)$$

and redundant real-estate for op = op(t) is only $\max(A_d(op)) - A_d(op(t))$.

The algorithm execution time, with M operations, is:

$$T(a \lg) = \sum_{m=1}^{M} T(op(m))$$

for both a fixed and a reconfigurable parallel array. Both array types will have a similar processing time of operations, $T_p(op)$, but there is a reconfiguration overhead time $T_c(op)$. The lower the ratio $T_c/T_p(alg)$, and the higher $A_{fix}/A_{rec}(pix)$, the more worthwhile dynamic reconfiguration becomes.

The trade-offs between a dynamically reconfigurable implementation and a fixed hardware implementation are:

• Reconfigurable implementations require less real-estate, or in the case of arrays, can fit a larger array on the same chip.

- Fixed implementations do not require any configuration time, and the total time taken to implement an algorithm is the processing time alone.

When using a fixed area IC, such as FPGAs, that requires the image to be partitioned, the aim is to minimise the time taken, as the available area will be used in sequence to process regions of the array. This is especially important if the algorithm is to be performed in real-time.

Reconfiguration can also be used to overcome the size restriction of a hardware system. A major drawback with array processors is that even the largest array processors cannot usually process the whole of a typical image simultaneously [1]. A mathematical formulation for partitioning algorithms in a hardware / software environment is given in [8]. With morphology, an added problem is that the area of overlap between partitions is variable, determined by the size of the structuring element. Various techniques for efficient partitioning are currently being evaluated.

A derivation of the relationship of the time taken for any implementation with respect to number of FPGA cells available, the size of the full structuring element, the cells needed per pixel and the image size, shows that if the number of cells per pixel is reduced, the corresponding drop in number of partitions required is more than proportional, because of the reduction in the proportion of the image that is in overlapping regions. This helps to offset the loss due to reconfiguration, hence leading to the conclusion that a dynamically reconfigurable parallel array will outperform its static equivalent, and can be considered as a viable alternative to pipelined architectures.

5 The Data-Localised Reconfigurable Parallel Array Architecture

The data-localised parallel array architecture processes individual pixels in a fixed region in the FPGA, so that the data-flow is minimised. The difference between this and conventional parallel arrays is that different operations are performed by reconfiguring the FPGA, *while the data remains held in internal registers*, rather than sending instructions to a fixed processor configuration. The architectural model is shown in Fig 4.

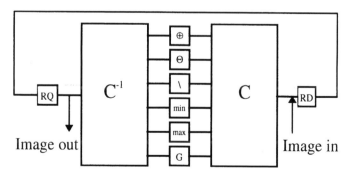

Fig. 4. The reconfigurable architecture

C and C^{-1} denote control blocks for configuring the current operation from the various possibilities shown using the model described in [6]. For most operations, the architecture resembles a tessellation automata, but the architecture allows any possible configuration, which is needed for global operations amongst others. Erosion and dilation blocks may be optimised for the given structuring element, an example for $B(w) = \begin{bmatrix} 1 & 0 & 0 \\ 1 & 0 & 1 \\ 0 & 0 & 0 \end{bmatrix}$ is given in Fig. 5.

(From neighbouring RBs)

A(i,j-1)

A(i+1,j-1)

A(i,j+1)

RB Mux.

RB

D Q

Fig. 5. Dilation for B(w) above

For reconfigurable implementations, it is important to derive the maximum differential area as given in section 4, and the exact nature of all common blocks. This determines placement on the FPGA. RD and RQ, the common blocks shown in Fig. 4, represent the inputs and outputs to a register bank for the results of operations. These registers must be protected during reconfigurations to preserve results for future operations. The analysis of a number of algorithms suggests that three registers will usually suffice: one to hold the original image (RA in the table below), another for the output of each operation (RB), and the third for holding intermediate results required later (RC). The exact use of these registers is flexible, and extra registers can be added if hardware allows.

The substitution $\aleph(A \ominus (n_i\text{-}1)B, B) = (A \ominus (n_i - 1)B) \ominus B \oplus B = (A \ominus n_i.B) \oplus B$ allows the skeletonisation algorithm to be performed as shown in Fig. 6.

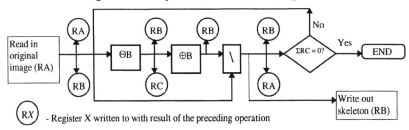

Fig. 6. Control flow diagram for skeletonisation

Various operations can be merged without increasing the maximum differential area, such as the feedback write and test for termination. This reduces both the number of reconfigurations and hardware redundancy.

The register values after each operation are shown in Table 1.

Config. no.	Operation	RA	RB	RC
1	Read in image, write to temp. reg.	A	A	X
2	ERODE	A	A⊖B	A
	Retain in temp. register	A	A⊖B	A⊖B
3	DILATE	A	A⊖B⊕B	A⊖B
4	Image Subtract (RA - RB)	A	$S_0(A)$	A⊖B
5	Write out first skeleton	A	$S_0(A)$	A⊖B
	Overwrite image	A⊖B	$S_0(A)$	A⊖B
	Test for termination	A⊖B	$S_0(A)$	A⊖B
6	Start of new cycle (ERODE)	A⊖B	A⊖B⊖B	A⊖B

Table 1. Register table for skeletonisation

Dynamic reconfiguration is controlled by a host processor, which downloads the appropriate configuration as the algorithm proceeds (like a control processor broadcasting instructions in a SIMD machine). This is usually achieved by having the FPGAs on special boards. However, the architecture is not board specific and can be adapted for different FPGAs, though the dependence on exploiting features specific to an FPGA such as the XC6200 may significantly affect performance. The other roles of the host are as controller of register accesses and to perform sequential parts of algorithms, much as any hardware / software co-designed system.

6 Prototype Implementation on the Riley-2 system

The architecture was implemented on a Xilinx XC6216 FPGA [3]. This has a 64*64 array of Configurable Logic Blocks, each of which contains a D flip-flop, and a series of multiplexers that can be configured to give any one or two input gate. The FPGA contains routing at the local level, as well as between 4*4 and 16*16 cell regions.

The XC6216, by allowing direct register accesses of up to 32-bits from an external interface, considerably eases the input/output problem, without requiring dedicated user i/o circuitry. The 'wildcard' configuration technique allows simultaneous configuration of multiple identically cells. This is useful with cellular automata, but global operations may take somewhat longer to reconfigure. The configuration time of the FPGA, which is essentially a setup cost, may be further reduced by partial reconfiguration techniques [6]. Despite this, in relation to the processing time, it is still far from insignificant. In addition, the architecture of FPGAs is essentially more suited to implementing parallel arrays than pipelines. For these reasons the XC6216 was considered a suitable chip for this implementation.

The Riley-2 board [2] includes:

- An Intel i960RP processor and shared memory.
- 4 XC6216-PQ240-2 FPGAs and 4 256KB RAMs (local to each FPGA).
- 33MHz PCI bus connection to a host.

- A XC4013 FPGA for interconnecting the PCI and reconfigurable resource busses. The FPGAs are arranged in a ring that allows designs to overflow, or alternatively larger arrays can be implemented.

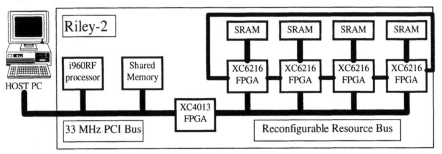

Fig. 7. Block diagram of Riley-2

Layouts have been produced and optimised for erosion and dilation phases for selected structuring elements. 1024 different configurations are required for all 3*3 structuring elements for both erosion and dilation. The first layouts were used as a template for a specially written VHDL source code generator. However, this still requires synthesis, place and route to be performed for each layout, and since they derive from a common template, the aim is to generate configuration files directly. All pixel blocks can fit in 4*4 CLBs so the circuit can be reconfigured in 20 writes (including control registers) with wildcarding. On the Riley-2 a 32*32 image can be processed at one instant in time, i.e. 1024 pixels in approximately 41 cycles. In comparison, a fixed processing element would require a block of 7*7 CLBs. Taking into account the partition overlap due to a 3*3 structuring element, a 1024 * 1024 image requires 1089 partitions with the reconfigurable architecture, while the fixed architecture needs 3481, and this degrades faster with larger structuring elements.

The Riley-2 has only recently become available for testing, and at the time of writing, no successful test runs had been performed. Full testing on Riley-2 will take place in the near future, to confirm theoretical performance metrics, and to compare performance with a pipeline architecture. However, a graphical user interface, incorporating routines based on theoretical requirements and tools for run-time reconfiguration in [7] is being developed to allow menu based implementation of algorithms on the Riley-2 system. This provides a framework that will make implementation of algorithms simple. It is already possible to do so, once a control flow such as in Fig. 6 has been derived, so a methodology is to be developed for efficient conversion of algorithms from mathematical description to the control flow description. Simulations of algorithms can already be performed on the PC. It is also envisaged that this application will produce optimised i960 code, to ensure that the maximum achievable reconfiguration and partition loading rate is not compromised by any bottlenecks elsewhere.

7 Conclusions and Future Development

Our work suggests that the drawbacks in implementing parallel array architectures for morphology can be overcome by the use of dynamic reconfiguration techniques, combined with a data-localised approach. The architecture is intended to be device independent, so generic code will be developed for any FPGA. In the near future, it is expected that context switching FPGAs [12] will be widely available, making it possible to reconfigure in one cycle, in which case this architecture will have the same processing time per partition as arrays of fixed processing elements. Finally, a version of this architecture that can deal with grey-scale images will be designed.

References

[1] L. Abbott, R. M. Haralick & X. Zhuang, *"Pipeline Architectures for Morphologic Image Analysis"*, Machine Vision & Applications, pp. 23-40, 1988.

[2] P. Y. K. Cheung, *"Riley-2: A flexible platform for codesign and dynamic reconfigurable computing research"*, this volume.

[3] S. Churcher, T. Kean & B. Wilkie, *"The 6200 Fastmap Processor Interface"*, in Field Programmable Logic & Applications, W. Moore & W. Luk (Eds.) , pp. 36-43, LNCS 975, Springer 1995.

[4] C. R. Giardina & E.R. Dougherty, *"Morphological Methods in Image and Signal Processing"*, Prentice Hall, 1988

[5] A. C. P. Loui, A. N. Venetsanopoulos & K. C. Smith, *"Flexible Architectures for Morphological Image Processing and Analysis"*, IEEE Transactions on Circuits & Systems for Video Technology, Vol. 2, No. 1, pp. 72-83, March 1992.

[6] W. Luk, N. Shirazi & P. Y. K. Cheung, *"Modelling and Optimising Run-Time Reconfigurable Systems"*, IEEE Symposium on FPGAs for Custom Computing Machines, 1996.

[7] W. Luk, N. Shirazi and P. Y. K. Cheung, *"Compilation Tools for Run-Time Reconfigurable Designs"*, IEEE Symposium on Field-Programmable Custom Computing Machines, 1997.

[8] W. Luk, T. Wu & I. Page, *"Hardware-Software Co-design of Multidimensional Programs"*, IEEE Workshop on FPGAs for Custom Computing Machines, 1994.

[9] H. Park & R. T. Chin, *"Optimal Decomposition of Convex Morphological Structuring Elements for 4-Connected Parallel Array Processors"*, IEEE Transactions on Pattern Analysis & Machine Intelligence, Vol. 16, No.3, pp. 304-313, March 1994.

[10] I. Pitas & A. N. Venetsanopoulos, *"Morphological Shape Decomposition"*, IEEE Transactions on Pattern Analysis & Machine Intelligence, Vol. 12, No.1, pp. 38-45, January 1990.

[11] S. R. Sternberg, *"Computer Architectures Specialized for Mathematical Morphology"*, Algorithmically Specialized Parallel Computers, pp. 169-176, 1985.

[12] S. Trimberger, D. Carberry, A. Johnson & J. Wong, *"A Time-Multiplexed FPGA"*, IEEE Symposium on Field-Programmable Custom Computing Machines, 1997.

[13] D. Wang & D.-C. He, *"A Fast Implementation of 1-D Grayscale Morphological Filters"*, IEEE Transactions on Circuits and Systems - II: Analog & Digital Signal Processing, Vol. 41, No. 9, pp. 634-636, September 1992.

Virtual Radix Array Processors (V-RaAP)

B Bramer, D Chauhan, M K Ibrahim and A Aggoun

DSP Systems Research Group
Faculty of Computing Sciences and Engineering
De Montfort University
The Gateway, Leicester LE1 9BH, UK

Abstract

The V-RaAP software (written in C++) enables the implementation of Radix Array Processors, expressed as iterative equations, into Field Programmable Gate Arrays (FPGA's). The V-RaAP software is not a compiler which translates C++ into VHDL. Rather, the V-RaAP C++ program is a high-level explicit description of structural VHDL that implements a RaAP algorithm in a hierarchical manner. When the C++ program is executed it automatically generates a number of VHDL files containing the entity and architecture specifications of the components which make up the final design. The code generated is equivalent to a hand crafted VHDL design. This is made possible because the RaAP iterative algorithm contains all the information specifying the functionality and interconnects of the architectures.

Introduction

Radix Array Processors (RaAP) [1-9] provide systems designers a wide range of implementation solutions, one for each radix, with different trade-offs with respect to speed and area. In recent years, the Radix approach has been used for the design and implementation of *sub-digit* pipelined digit serial architectures in a structured manner [7]. Sub-digit pipelined digit serial computation offers high speed at low cost. These structures have been shown to offer significant advantages over other existing realisations. RaAP has also been used for non-conventional arithmetic such as residue number systems and finite fields with similar improvements over existing structures [8-9].

When using ASICs to implement the RaAP, there are two possible solutions. The first is to evaluate the target architecture for different radices and then select the one that gives the best result for the final implementation into silicon. However, the resulting hardware is specific for a particular radix design and if the same architectures are to be implemented using a different radix, a new ASIC device has to be produced. The second approach, which is the ideal choice, is that a single RaAP device should allow the implementation of different radices to give the different trade-off for a particular architecture. However, this approach in realising the RaAP concept into silicon has proved difficult, since it requires the design of arrays whose basic cells can very in size (radix). Research work has shown that large radix cells can be composed from smaller radix cells, but this would put severe limitations on the type of internal architectures used for the radix cells.

FPGAs provide an ideal solution for exploiting the RaAP concept. The use of FPGAs will allow the virtual implementation of the RaAP engine without the need to actually produce a hardware device that can implement the different radix solutions. This new approach is called Virtual RaAP (V-RaAP). The V-RaAP concept is made possible because of one of the main advantages of the RaAP approach, which is that structures can be described completely in terms of iterative equations, even down to the bit level. It is this characteristic that is exploited in the current work, i.e. in the automatic generation of *structural* VHDL code from an *explicit description* using C++. This ability to explicitly specify the final architecture in terms of structural VHDL using C++ is one of the novel aspects of the approach. This way the use of synthesis tools is kept to a minimum.

Existing approaches of generating VHDL from C/C++ or any other high level sequential languages are based on developing either (i) a compiler/synthesiser which translates C++ into VHDL, or (ii) the inclusion of components from "ready built" VHDL routines stored in libraries. In these cases the final VHDL implementation is transparent to the users and they have little or no control over it. We term this the implicit approach.

The main difference of the V-RaAP approach from the conventional techniques is that the algorithmic nature of RaAP allows automatic generation of detailed structural VHDL code for RaAP using C++, without using (a) Compiler/Synthesis tools and/or (b) ready made VHDL routines.

This paper presents a novel technique to generate VHDL code for highly recursive algorithms such as radix multipliers. The software system consists of a library of generic 'C++' functions, which describe elements of the V-RaAp architecture, and can be called by an application program to generate radix array processors of any radix in any combination of optimised serial/parallel elements. When executed the resultant program generates a series of VHDL source files, which may then be compiled, simulated, synthesised and finally downloaded into the target FPGA family. Using the system not only increases the speed of development (the same C++ application program may be used to generate circuits for any given data word size) but more importantly the reliability of these structures is guaranteed, since the 'C++' functions based on the radix algorithm have been fully tested. In addition, many algorithm designers already have experience programming C or C++ for other targets such as DSP devices and would find the C++ based software easier to use than working directly in VHDL.

The parametric specification within most of the current tools is at the behavioural VHDL level and not the structural; the main exception being vendor specific tools. The V-RaAP approach allows the parametric specification of structural VHDL using C++.

Radix-2n multiplication algorithm

The equation, which has been implemented in the V-RaAP software is:

$$\{C_{ij}, S_{ij}\} = U_i V_j + S_{i+1, j-1} + C_{i, j-1}$$

The C++ code uses this algorithm to generate VHDL which instantiates the basic optimised arithmetic cells and the interconnects between them. The basic multiplier cell used consists of one n-bit multiplier, one n-bit adder and one $2n$-bit adder. Fig. 1 shows a diagram of such a cell where the s and c inputs to the adder are n-bits wide, therefore we only need an n-bit adder to add s and c. A $2n$-bit adder is needed for the second addition since the product UV is $2n$-bits.

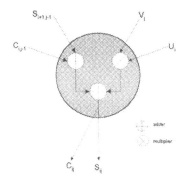

Fig. 1 Architecture of the radix -2n arithmetic cell

Radix -2n serial/parallel multiplier

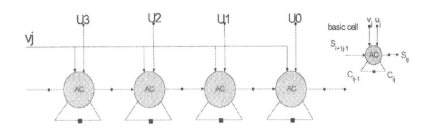

Fig. 2 Generalised 4 bit radix serial/parallel multiplier

The principle of the Serial Parallel Multiplier is shown in Fig. 2. The index j denotes the recursion number, while the index i represents the ith cell in the serial/parallel multiplier. It has been proven [1] that the function of the ith cell at the jth recursion will produce $u_i v_j$ and the result is added to the partial product $S_{i+1, j-1}$ accumulating from the left and the carry $C_{i, j-1}$ saved in the same cell from the previous recursion. The n least-significant bits of the result are the new accumulating partial product S_{ij} and are shifted to the right, while the n most-significant bit are the new carry C_{ij} and are saved in the cell for the next cycle. For example, if the data size were 16-bit U0 to U3 would consist of a serial stream of four 4-bit values with 4-bits of V being fed into each cell.

Using C++ to generate VHDL structures for Radix Multipliers

The V-RaAP software is *not* a compiler which translates C++ into VHDL. Rather, the C++ program is a high-level *explicit description* of structural VHDL, which implements the RaAP algorithm in a hierarchical manner. When the C++ program is executed it generates a number of VHDL files containing the entity and architecture specifications of the components which make up the final design. The code generated is equivalent to someone hand crafting VHDL.

The software contains a range of functions, which generate VHDL entities for various types of multipliers and associated components (serial to parallel converters, etc.). When executed, a typical application program would prompt the user for the type and size of the multiplier, thus allowing the examination of different trade-offs between speed and gate count for a range of serial/parallel combinations of different data word sizes. The generated VHDL code is then compiled and simulated with an appropriate test bench (which can also be generated by the C++ program). If results are satisfactory the design may be synthesised and downloaded into a target FPGA device.

What makes this approach novel is that the C++ code is based on the RaAp algorithm. The interconnects between the underlying entities are generated automatically and are therefore less prone to error than equivalent hand coding in VHDL. Another important issue is the speed in which these multipliers can be designed and tested. Using VHDL to design a multiplier in the traditional way would take several hours, whereas once the C++ program is written the VHDL may be generated in a few seconds. Fig. 3 shows the code of the C++ function which generates a digit serial/parallel multiplier (as shown in Fig. 2) of any size taking inputs u and v and generating the result sum (the line numbers down the left hand side are not part of the code but to enable identification in the following discussion). Fig. 4 shows the resultant VHDL code generated for a serial/parallel multiplier of data word size 16-bits using a multiplier digit size of 4-bits. The C++ function of Fig. 3 uses the size of parameter u (type std_logic_vector in line 2) to determine the data word size and the size of v (line 2) to determine the digit size. The appropriate number of multipliers of the required size can then be generated and interconnected (the FOR loop from lines 22 to 34). The program which generated the file of Fig. 4 prompts the user for the data word size and digit size and then generates (within a few seconds) ten VHDL files (including Fig. 4 and Fig. 5) which make up the design, i.e. basic arithmetic elements, adders and multipliers of various word sizes, serial/parallel and parallel/serial converters, test bench, etc. During simulation the test bench reads text input from a data file (using std.textio) and the results may be viewed with the appropriate tool.

Fig. 5 shows the VHDL code generated for a 4-bit version of the radix-2n arithmetic cell shown in Fig. 1. This instantiates three other entities, v4rad a 4-bit multiplier, a4rad a 4-bit adder, and a8rad an 8-bit adder, which form the elements of Fig. 1. In Fig. 5 the signal carry and output sum are latched for use in the digital serial/parallel multiplier shown in Figs. 2, 3 and 4.

```
1    // serial/parallel radix digit multiplier sum = u * v
2    void sp_digit_mult(std_logic_vector &u,  std_logic_vector &v,
3               std_logic reset, std_logic clk, std_logic_vector &sum)
4    {
5      int word_size = u.size, digit_size = v.size;
6      String name, title(String("Multiplier: ") + word_size + " bit ");
7      name = String("v") + word_size + "sp" + digit_size;
8      if (!entity_created(entity_name, name))
9        {
10         // define local signals, etc.
11         std_logic_vector u("U", word_size), v("V", digit_size),
12                  sum("SUM", digit_size), zero("ZERO", digit_size),
13                  s[word_size/digit_size];
14         std_logic carry("CARRY"), reset("RESET"), clk("CLK");
15         // set up description, open entity and set up inputs & signals
16         title = title +" radix " + digit_size + " digit serial/parallel";
17         entity sp(name, title);
18         sp.input(u, v, reset, clk);
19         signal(zero);
20         signal(s, "S", digit_size, word_size/digit_size);
21         // now generate a sequence of arithemetic cells
22         for (int i=0; i < word_size/digit_size; i++)
23           {
24           std_logic_vector temp("",digit_size);
25           signal(temp);
26           temp = u(i*digit_size, (i+1)*digit_size-1);
27           // S (sum) input is either zero or signal s[i+1]
28           if (i+1 == word_size/digit_size)
29               lvcell(temp, v, zero, reset, clk, s[i]);
30           else
31               // sum output is either output SUM or signal s[i]
32               if (i == 0) lvcell(temp, v, s[i+1], reset, clk, sum);
33               else        lvcell(temp, v, s[i+1], reset, clk, s[i]);
34           }
35         sp.output(sum);
36        }
37      // instansiate the multiplier in the calling entity
38      output(name + " port map(" + u.name + ", " + v.name + ", "
39          + reset.name + ", " + clk.name + ", " + sum.name + "); ");
40   }
```

Fig. 3 C++ code to generate a serial/parallel multiplier component

The code in Fig. 3 is (the numbers down the left hand side refer to line numbers):

2	input parameters u and v (type std_logic_vector)
3	input parameters reset and clk (the clock) and output parameter sum
5	determine the multiplier word_size and digit_size of the arithemetic cells
6-7	set up the entity name, in this case v16sp4 indicating the word and digit size
8	if the entity does not already exist execute lines 9 to 36
11-14	define and initialise local variables to hold signals, inputs, outputs, etc.
16	specify a title for the entity
17	open the entity - this starts to create the file of Fig. 4 outputting lines 1 to 3 and starts to create the entity and architecture parts (lines 4 and 12)
18	specify the inputs to the entity - this adds lines 5 to 8 to the entity in Fig. 4
19	signal zero, generates line 13 to the architecture of Fig. 4
20	signals to hold the interconnects between the cells, generates line 14 of Fig. 4

21-34   a FOR loop which generates the main body of the architecture of Fig. 4

24-25   define temp and make it a signal, generates lines 15, 24, 25 and 26 of Fig.4

26   assign temp 4-bits of parameter u, generates lines 28, 30, 32 and 34 of Fig. 4

28-33   instantiates the arithmetic cells to multiply two 4-bit numbers

29   instantiates the last cell in the structure (the left-hand cell in Fig. 2) generates line 35 of Fig. 4

32   instantiates the first cell in the structure (the right hand cell in Fig. 2) generates line 29 of Fig. 4 where the output of the cell is parameter sum

33   instantiates intermediate cells, generates lines 31 and 33 of Fig. 4

35   specify sum to be an output of the entity, generates line 9 of Fig. 4

36   closes the entity and writes Fig. 4 to file v16sp4.vhd, it also adds a v16sp4 component to the architecture of the component which called the function

38-39   instantiates the v16sp4 component in the architecture of the component which called the function

```
1     -- file v16sp4.vhd, Multiplier: 16 bit radix 4 digit serial/parallel
2     library IEEE; use IEEE.std_logic_1164.all;
3
4     entity v16sp4 is port(
5         U: in std_logic_vector (15 downto 0);
6         V: in std_logic_vector (3 downto 0);
7         RESET: in std_logic;
8         CLK: in std_logic;
9         SUM: out std_logic_vector (3 downto 0));
10    end v16sp4;
11
12    architecture RTL of v16sp4 is
13    signal ZERO: std_logic_vector (3 downto 0):="0000";
14    signal S0,S1,S2,S3,x: std_logic_vector (3 downto 0);
15    signal temp13: std_logic_vector (3 downto 0):="0000";
16    component lv4cell port(
17        U: in std_logic_vector (3 downto 0);
18        V: in std_logic_vector (3 downto 0);
19        S: in std_logic_vector (3 downto 0);
20        RESET: in std_logic;
21        CLK: in std_logic;
22        SUM: out std_logic_vector (3 downto 0));
23    end component;
24    signal temp30: std_logic_vector (3 downto 0):="0000";
25    signal temp31: std_logic_vector (3 downto 0):="0000";
26    signal temp32: std_logic_vector (3 downto 0):="0000";
27    begin
28    X0: temp13 <= U(3 downto 0);
29    X1: lv4cell port map(V, temp13, S1, RESET, CLK, SUM);
30    X2: temp30 <= U(7 downto 4);
31    X3: lv4cell port map(V, temp30, S2, RESET, CLK, S1);
32    X4: temp31 <= U(11 downto 8);
33    X5: lv4cell port map(V, temp31, S3, RESET, CLK, S2);
34    X6: temp32 <= U(15 downto 12);
35    v4cell port map(V, temp32, ZERO, RESET, CLK, S3);
36    end RTL;
```

Fig. 4 VHDL code for a 16-bit data word 4-bit digit serial/parallel multiplier

```
1       -- file lv4cell.vhd,  4 bit serial/parallel arithemetic cell
2       library IEEE; use IEEE.std_logic_1164.all;
3
4       entity lv4cell is port(
5           U: in std_logic_vector (3 downto 0);
6           V: in std_logic_vector (3 downto 0);
7           S: in std_logic_vector (3 downto 0);
8           RESET: in std_logic;
9           CLK: in std_logic;
10          SUM: out std_logic_vector (3 downto 0));
11      end lv4cell;
12
13      architecture RTL of lv4cell is
14      component v4rad port(
15          U: in std_logic_vector (3 downto 0);
16          V: in std_logic_vector (3 downto 0);
17          SUM: out std_logic_vector (7 downto 0));
18      end component;
19      signal temp15: std_logic_vector (7 downto 0):="00000000";
20      component a4rad port(
21          A: in std_logic_vector (3 downto 0);
22          B: in std_logic_vector (3 downto 0);
23          SUM: out std_logic_vector (4 downto 0));
24      end component;
25      signal temp16: std_logic_vector (4 downto 0):="00000";
26      component a8rad port(
27          A: in std_logic_vector (7 downto 0);
28          B: in std_logic_vector (7 downto 0);
29          SUM: out std_logic_vector (8 downto 0));
30      end component;
31      signal temp22: std_logic_vector (7 downto 0):="00000000";
32      signal temp21: std_logic_vector (8 downto 0):="000000000";
33      signal t_add: std_logic_vector (8 downto 0):="000000000";
34      signal TSUM: std_logic_vector (3 downto 0):="0000";
35      signal TCARRY: std_logic_vector (3 downto 0):="0000";
36      signal CARRY: std_logic_vector (3 downto 0):="0000";
37      begin
38      X0: v4rad port map(U, V, temp15);
39      X1: a4rad port map(S, CARRY, temp16);
40      X2: temp22 <= "000" & temp16;
41      X3: a8rad port map(temp22, temp15, temp21);
42      X4: t_add <= temp21;
43      X5: TSUM <= t_add(3 downto 0);
44      X6: TCARRY <= t_add(7 downto 4);
45      X7: process(RESET, CLK)
46          begin
47          if RESET = '1' then
48              SUM <= "0000";
49              CARRY <= "0000";
50          else
51              if CLK'event and CLK = '1' then
52                  SUM <= TSUM;
53                  CARRY <= TCARRY;
54              end if;
55          end if;
56          end process;
57      end RTL;
```

FIG 5 VHDL code for a 4-bit data word arithmetic cell

Consider lines 29, 32 and 33 of Fig. 3, which call the function lvcell, which is the radix-2n arithmetic cell shown in Fig. 1. The first call will create the file lv4cell.vhd, shown in Fig 5, which contains the VHDL code for a 4-bit arithmetic cell, add the component statement in lines 16 to 23 of Fig. 4 and instantiates the cell in line 35 of Fig. 4. Further calls to lvcell will only instantiate the cell (lines 29, 31 and 33 of Fig.4).

Preliminary results generating VHDL files for multipliers

To test the prototype V-RaAP software it was used to generate VHDL files for 8, 16 and 32-bit multipliers ranging from fully serial, through various digit size serial/parallel to fully parallel. Tables 1, 2 and 3 present the results for 8, 16 and 32-bit multipliers respectively when synthesised for the Xilinx XC3000 FPGA family. In each table the columns show the results for each multiplier type, first and second rows present the number of gates and registers synthesised and the last row shows the number of clock cycles which would be required to carry out the multiply and latch the result.

To test the multipliers a C++ program was implemented which generated VHDL files for a filter of a specified data word size and number of multiplier/adder stages, together with the digit size to be used for the arithmetic (allowing fully serial to fully parallel multiply/add). Various combinations of filters were tested using the ViewLogic simulator with the expected results.. Unfortunately it has not been possible to synthesise filters more complex than the serial arithmetic based due to problems with the synthesis tool (it stops with an error message and a request that the problem be reported to the vendor).

	Serial	2-bit digit	4-bit digit	parallel
gates	22	100	156	193
registers	16	16	16	0
Clock cycles	16	8	4	1

Table 1 8-bit multipliers

	serial	2-bit digit	4-bit digit	8-bit digit	parallel
gates	44	200	312	500	741
registers	32	32	32	32	0
Clock	32	16	8	4	1

Table 2 16-bit multipliers

	serial	2-bit digit	4-bit digit	8-bit digit	16-bit digit	parallel
Gates	88	400	624	1000	1716	2818
Registers	64	64	64	64	64	0
clock	64	32	16	8	4	1

Table 3 32-bit multipliers

Future Work

Fig. 6 shows the approach to be taken with the V-RaAP software. A library of basic cells optimised for the target device will provide fundamental arithmetic elements. A library of C++ routines builds on these to enable the designer to implement and rapidly prototype RaAP algorithms. The next stages of V-RaAp development will be:

1. to optimise basic arithmetic entities to improved the overall gate count on synthesis, e.g. a 1-bit arithmetic cell (Fig. 2) synthesised from a VHDL boolean specification generally takes three CLBs on a Xilinx XC3000 device whereas an equivalent generated schematically using a CLB component from the XC3000 library takes only one CLB

2. to synthesise a range of filters and download into a target FPGA device for testing

3. to implement more efficient RaAP algorithms for a variety of DSP functions

4. to assess the basic knowledge a writer of C++ V-RaAP would need of VHDL

5. to modify V-RaAP to minimise this knowledge of VHDL

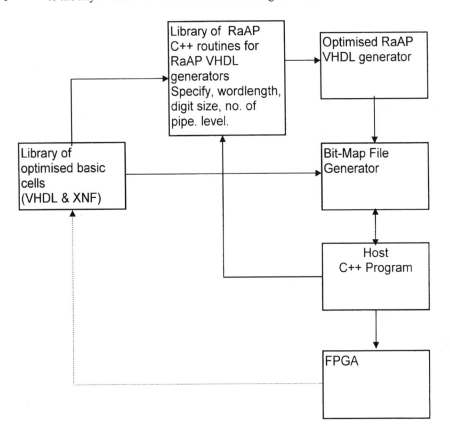

Fig. 6 The Virtual RaAP implementation (dotted line indicates pre-processing)

Conclusion

This paper presented a novel use of C++ programming which allows the generation of a regular array of radix multipliers described in VHDL. It is demonstrated that FPGA's provide an ideal solution for exploiting the Virtual Radix Array concept. The V-RaAP system allows the design of radix multipliers for different data word size in various combinations of digit serial/parallel elements. This enhances the speed at which different types of multipliers can be developed and, more importantly, less prone to human error than the equivalent hand crafted design. The implication this has is different types of radix multipliers can very quickly be reconfigured into an FPGA without any non-recurring engineering costs over ASIC solutions. In addition a wide range of implementation trade-offs are available to the programmer that require neither a dedicated RaAP device nor a general purpose DSP processor. The adaptability of RaAP generic C++ routines and the reconfigurability that FPGAs offer, makes this an ideal virtual RaAP engine.

Acknowledgements are due to Eithne Kane of Xilinx technical support for help with Xilinx devices and software and to DeMontfort |University for funding through the vice-chancellor's Research Initiative Program.

References

1. Ibrahim, M K, (1993) Radix Multiplier Structures: A structured design methodology. IEE Proceedings part E, 140, pp185-190.
2. Aggoun, A, Ibrahim, M K, and Ashur, A, (1995) Radix Multiplication Algorithms, International Journal of Electronics, Vol.79, No.3, pp329-345.
3. Bashagha, A E, Ibrahim, M K, (1995) A new high radix non-restoring divider architecture, International Journal of Electronics, Vol.79. No.4, pp.455-470.
4. Ashur, A, Ibrahim, M K, and Aggoun, A., (1996) Systolic digit serial multiplier, IEE Proceedings - Circuits, Devices and Systems, Vol.143, No.1, pp14-20.
5. Bashagha, A E, Ibrahim, M K, (1996) Non-restoring radix 2^k square rooting algorithm, the Journal of Circuits and Systems, and Computers, 1996, Vol.6, No.3, pp267-285.
6. Bashagha, A E, Ibrahim, M K (1996), High radix digit-serial division, IEE Proceedings on Circuits, Systems, and Devices, Vol 143,
7. Aggoun, A, Ibrahim, M K, and Ashur, A, Bit-level pipelined digit serial Processors, Accepted for publications in the IEEE Transactions on Circuits and Systems.
8. Mekhallalati, M, Ibrahim, M K, Ashur, A, Radix Iterative Parallel Modular Multiplier, Accepted for publication in the Journal of Circuits and Systems, and Computers.
9. Mekhallalati, M, Ibrahim, M K, Ashur, A, Novel Radix Finite Field Multiplier for GF(2^N), Accepted for publication in the Journal of VLSI Signal Processing.
10. Gschwind, M and Salapura, V., A VHDL design methodology for FPGAs, Field Programmable Logic and Applications, 5th International workshop, FPL95, Oxford, UK, pp209-226.

An FPGA Implementation of a Matched Filter Detector for Spread Spectrum Communications Systems

T. Mathews, S.G. Gibb, L.E. Turner, P.J.W. Graumann and M. Fattouche

Department of Electrical and Computer Engineering, University of Calgary, 2500 University Dr. NW, Calgary, AB, CANADA, T2N 1N4

Abstract. The implementation of a matched filter detector (MFD) for use in direct sequence spread spectrum communications is described. First, a brief overview of spread spectrum communications is given, leading to a look at why a matched filter synchronizer is superior to the more commonly used sliding correlator synchronizer. The design requirements for a specific MFD is then given, and the reasons for using an FPGA in this application are examined. Both bit–serial and bit–parallel versions of the circuit were designed. The design process is briefly described and details of individual modules within the design are given. The matched filter detector was implemented in an early Xilinx 4005 FPGA sample. On this older, slower device, the bit–parallel circuit correctly operates with 4-bit input data streams at a rate of 17 MHz; the bit–serial circuit operates at an input rate of 3.4 MHz.

1 Introduction

Spread spectrum communications is the basis for transmitting data in a number of familiar applications, including the Global Positioning System (GPS) and wireless communication products. In essence, spread spectrum techniques transmit signals in a bandwidth much wider than actually necessary to transmit the information. This seemingly inefficient use of bandwidth results in a number of benefits. These include a relatively low signal power spectrum, allowing the spread spectrum signal to peacefully coexist with narrow-band communications; resistance to interference, unintentional and otherwise; and a relatively high immunity to multipath-related problems. Furthermore, spread spectrum communications facilitates selective addressing (targeting communications to specific receivers within a group) and multiple access (the same bandwidth being used to communicate more than one spread spectrum signal).

All spread spectrum techniques use a repeating pseudo-random code sequence to spread the spectrum of the data signal to be transmitted. One of the most popular spread spectrum methods is called 'direct sequence'. In this technique, each baud $b(t)$ of the data signal is multiplied by the pseudo-random code sequence $p(t)$, as shown in Fig. 1. $p(t)$ is a periodic sequence which is divided into n chips (or bits), and the chip time c of the sequence is such that $n \times c$ is equal to

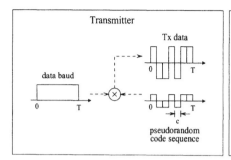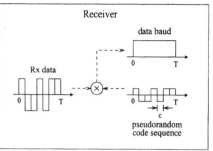

Fig. 1. Generation and Recovery of a Direct Sequence Spread Spectrum Signal

the baud time T of the data signal. This procedure, in effect, changes the baud pulse shape to that of the pseudo-random code sequence, thereby widening the spectrum of the data signal. At the receiver, the original data is recovered by multiplying each received baud by the pseudo-random code sequence.

However, for the data to be recovered, the receiver's pseudo-random code sequence and the received data must be correctly synchronized. That is, the first chip of the code sequence must be time aligned with the first chip of the received data pulse. Establishing synchronization is by no means a trivial task; in fact, more energy and resources have been invested in developing and improving synchronization techniques than to any other aspect of spread spectrum communication [1].

The simplest and most popular method of establishing synchronization is with the use of a sliding correlator. In this circuit, a received baud is correlated with a local version of the pseudo-random code sequence (i.e. $k = \int_0^T r(t) \cdot p(t) \cdot dt$). If the resulting value exceeds a specified threshold, the code is assumed to be synchronized with the data. Otherwise, the code is shifted in time by half a chip, and is correlated with the next baud. This procedure is repeated until the correlation value equals or exceeds the specified threshold. Once synchronization has been established, the transmitted data can be recovered directly from the correlator output.

In the worst case, a sliding correlator will require correlation with $2n$ code sequences to establish synchronization. During this time, no useful data can be transmitted; instead, $2n$ bauds of preamble must be sent to enable synchronization. Furthermore, strategies to maintain this lock (tracking) must also be implemented. In case synchronization is lost, the synchronization procedure must be repeated periodically. This requires another $2n$ bauds of preamble to be sent, which may be a significant overhead for the actual data transmission. Since establishing lock is a lengthy procedure, the receiver must lock with the first correlation that exceeds the threshold; it can not afford to try other code phase positions to see if an even better synchronization can be achieved (due to multipath conditions, for example).

A much better, though more complex, approach is to use a matched filter to process the received data. A matched filter's impulse response is the complex conjugate of the pulse it is matched to. Thus, the impulse response of a matched filter matched to a specific pseudo-random code sequence is simply the sequence reversed in time. For use with a matched filter, the received data is sampled at twice the chip rate (thus, there are two data samples per chip) and this sampled data stream is fed into the matched filter. The output of the filter will peak when the data is matched to the filter; this indicates that the data and the code sequence are synchronized. Synchronization can be recognized by comparing the filter output against a specified threshold, as is done with the sliding correlator.

In the worst case, a matched filter must process $2n$ *chips* before finding a match, as opposed to $2n$ code sequences for the sliding correlator. If for some reason the synchronization slips by a few chips after it is first established, the peak in the filter's output will simply appear a few chips later; no tracking strategy is necessary. As it processes the data stream, the matched filter output will peak for every match it finds. Hence, it is not costly to find the best match under multipath conditions. Furthermore, since all of the peaks arising from multipath conditions are readily available in the matched filter output, a rake receiver [2] can be implemented. A rake receiver coherently combines multiple signal paths to yield a very robust communication link.

Hence, a matched filter is much superior to a sliding correlator for establishing synchronization in a direct sequence spread spectrum communication system. Its principle drawback is that it is a much more complex circuit to implement. In this paper, we describe the implementation of a matched filter detector for use in a quadrature differential phase shift keying (QDPSK) direct sequence spread spectrum communication system. This work was undertaken as part of a feasibility study for a real application. First, the specifications of the circuit to be implemented are given. These were determined by the specific application the matched filter detector was intended for. Next, the details of implementation are discussed. This includes a overview of the CAD software used, as well as the specifics of circuit implementation. Finally, the results of preliminary testing are presented.

2 Design Requirements

A block diagram of the matched filter detector (MFD) to be designed is given in Fig. 2. The MFD is supplied with two 4-bit wide streams of two's complement data, representing the I and Q channel information from a QDPSK signal. The chip rate of the transmitted signal is 10 MHz. The received signal is sampled such that there are two samples per chip; consequently, the input data rate to the MFD is 20 MHz on both the I and Q channels.

The MFD generates decoded signals for both the I and Q channels. In addition, a magnitude signal is to be generated based on the I and Q channel filter outputs. This magnitude is a simple approximation, given by

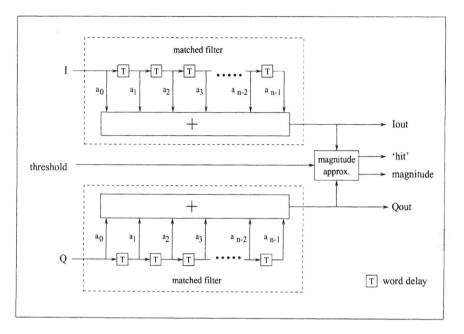

Fig. 2. Block Diagram of the Matched Filter Detector

$M = max\{|I|, |Q|\} + \frac{1}{2}min\{|I|, |Q|\}$. A 'hit' signal must also be asserted if the magnitude is greater than a user–specified threshold.

The pseudo-random code to be used has 11 or 15 chips, depending on the mode of operation. Thus, the matched filters used must have 15 taps, reducible to 11 taps on demand. Finally, the tap weights of the filters (± 1) must be individually setable on demand, so that the filter can be made to match any 11 or 15 chip pseudo-random code pattern.

3 Implementation

3.1 Why Use an FPGA?

In many filtering applications, a specialized digital signal processing micropro-cessor (DSP) is used as the basis of implementation. Most general purpose DSPs require a sequential algorithm implementation, whereas the MFD is tailor–made for concurrent processing. DSPs are now available which permit parallel pro-cessing, but the concurrency is limited and comes with a relatively high price tag. Furthermore, all DSPs have the resource and processing overhead of the inherent instruction fetch – data fetch – data store sequence. In addition, DSPs are generally optimized for performing fractional multiplications and divisions. The matched filter discussed here, however, has tap weights of ± 1 and thus requires additions and subtractions only. Thus, using a DSP to implement the MFD would be an inefficient use of resources.

An FPGA, on the other hand, provides an ideal platform for implementing the MFD, especially at the prototype stage of development. A design optimized for performing additions and subtractions can be implemented without the initial cost of doing a full custom VLSI design. Also, a concurrent algorithm implementation can be readily achieved; concurrency is only limited by the resources available in the particular device selected. Furthermore, the circuit can also be easily modified to make system design changes or correct for errors encountered during testing. Finally, the relatively inexpensive FPGAs currently available can easily handle the required data throughput if a heavily pipelined design is implemented.

A Xilinx 4005 FPGA was selected as the platform for implementing the MFD circuit. This device has sufficient resources to accommodate the many pipelining stages required. In addition, Xilinx 4000 series devices includes fast-carry logic, which further facilitates high speed design.

3.2 The Design Process

A bit–parallel version of the MFD was first designed and verified in software using Logsim[3], an easy to use text–based digital logic netlist specification and event–driven simulation package. For comparison, a bit–serial version of the circuit was also generated. This version was designed and implemented using DFIRST, a register–transfer level language based on Denyer and Renshaw's FIRST[4]. The bit–serial design was verified using DSIM, an event driven simulator for digit–serial designs.

Once the designs were verified in software, each circuit description was translated into macromodules specified in Xilinx's XNF format. This was accomplished using Trans [5–7], a hardware compiler that converts a variety of high level circuit description formats into a number of different gate level digital logic netlists. Trans is also capable of performing optimizations such as converting shift registers into serial RAM units; combining multiple instances of parts with identical input nets but different output nets; and making use of flip-flops in unused I/O blocks.

After translating and optimizing with Trans, macromodules which could take advantage of the Xilinx 4000 series fast–carry logic were re-implemented using Viewlogic Powerview/Workview schematic capture software. This was done for the bit–parallel design only, as bit–serial designs are not suited to using the Xilinx fast–carry logic.

Finally, each design was partitioned, placed and routed for the Xilinx 4005 FPGA using Xilinx's software tools.

3.3 Circuit Details

Bit–Parallel versus Bit–serial Design Apart from the obvious difference in how the data is routed through the design, there are two significant differences between the bit–parallel and bit–serial versions of the MFD. First of all, the serial

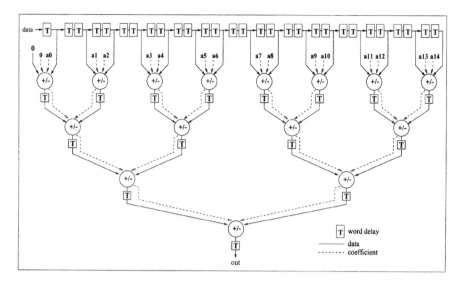

Fig. 3. Block Diagram of a Matched Filter

design requires parallel–to–serial and serial–to–parallel converters to convert the input and output data to the appropriate form.

The second difference relates to the data word size. In the bit–parallel circuit, if a module requires an additional output bit to preserve data accuracy, it is simply added. In synchronous bit–serial designs, however, it is very difficult to 'grow' the wordlength in this manner. Consequently, the maximum wordlength required in the circuit must be determined, and all of the bit–serial modules must use this wordlength. This means that the input data must be padded out to this maximum wordlength as well.

Matched Filters The MFD uses two identical matched filters to process the I and Q input data streams. A block diagram of a matched filter is shown in Fig. 3.

The input data stream has been sampled to yield two samples per received chip, which for convenience we will call sample A and sample B. Thus, the 20 MHz input data stream could be thought of as two separate interlaced 10 MHz streams — one corresponding to the A chip samples, and the other corresponding to the B chip samples. In calculating its output, the matched filter must not mix these two streams. This separation of the two streams is accomplished using two delays between each tap in the delay chain of the filter.

The heart of the matched filter circuit is the tree of adder/subtractor modules. In these modules data is represented using a two's complement word *as well as an additional weighting factor* of 1 or −1. To get the actual value of the data, the two's complement word must be multiplied by the weighting factor. Thus

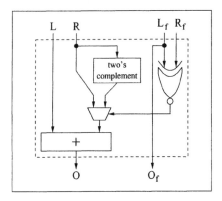

Fig. 4. Block Diagram of a Adder/Subtractor Module

the value +3 could be represented using a word value of +3 and a coefficient of +1, or as a word value of −3 and a coefficient of −1. Similarly, the value −2 is represented as a word value of −2 and a weighting factor of +1, or a word value of +2 and a coefficient of −1. The use of this additional weighting factor simplifies circuit implementation.

The block diagram of a single adder/subtractor module is shown in Fig. 4. In the bit–parallel implementation, each adder/subtractor module takes two m-bit two's complement words (L and R) with their associated weighting factors (L_f and R_f) and generates an $m + 1$-bit two's complement word and its weighting factor. The input wordlength m grows from four in the uppermost level of the tree structure to seven in the final level, in order to preserve accuracy as the filter output value is accumulated. The bit–serial version of the adder/subtractor module is identical to the bit–parallel version except that the input and output data words are eight bits in length.

The outputs of the adder/subtractor module are given by

$$O = L_f \cdot L + R_f \cdot R = L_f(L + \frac{R_f}{L_f}R) \tag{1}$$

$$O_f = L_f \tag{2}$$

In hardware, the adder/subtractor module is implemented as an adder that optionally two's complements its second operand before performing the addition. This two's complement operation is performed based on the value of R_f/L_f, which is equivalent to $\overline{R_f \oplus L_f}$. This scheme of using additional weighting factors avoids the situation of having to perform a two's complement operation on *both* operands if their weighting factors are negative.

At the highest level of the adder/subtractor tree, the weighting factors associated with the input data to each adder/subtractor module are simply the weights of the associated filter taps. These weighting factors are binary input signals, providing a simple way of specifying the tap weights and changing them

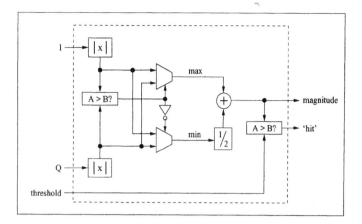

Fig. 5. Block Diagram of Magnitude Approximation Module

on demand. Subsequent adder/subtractor stages use the output coefficients of the previous stages as their input coefficients.

The reduction of filter tap length on demand is accomplished by tying the reduction indicating signal to the 'reset' pins of the flip-flops pertaining to the affected taps and any relevant adder/subtractor stages. When the smaller tap length is required, the reset pins are held asserted, forcing their outputs to zero and thereby preventing data from passing through the flip-flops. When the signal is removed, the flip-flops again allow data to flow through.

In the bit–parallel implementation, a pipeline stage was added after each adder/subtractor module to reduce the critical path through the circuit. This is necessary for the circuit to operate at the high data rate required. The bit–serial implementation is inherently pipelined.

Magnitude Approximation The magnitude approximation module is comprised of absolute value, comparator, and magnitude approximation circuits. A block diagram of the power detector module is given in Fig. 5.

The absolute value circuit conditionally takes the two's complement of its input value based on the input's sign.

The comparator circuit is only required to indicate which of two input values is greater. In the bit–parallel version of the circuit, an effective method of implementing this is by using a subtractor, with the borrow signal coming out of the most significant bit of the subtractor indicating which input word is greater. This circuit is particularly suited to implementation in the Xilinx 4000 series FPGA with its fast–carry logic.

Having performed the absolute value operation and then determined which of the two input values is the greater, the magnitude approximation is the sum of the lesser value right-shifted by one bit, and the greater value. The result is an unsigned eight bit value. This value is then compared with a user–supplied threshold using another comparator circuit to determine the 'hit' signal.

In the bit–parallel version, pipeline stages are needed after the absolute value circuits, the first comparator circuit, and the summation circuit to permit the desired data throughput rate. These pipeline stages are inherent in the bit–serial design. For both designs, additional pipeline stages have to be added after the matched filter modules to insure that the I and Q data outputs are time–aligned with their magnitude approximation.

4 Testing and Results

To test our hardware implementation, a data generator that produced data sequences in the required form and data rate was implemented in a second Xilinx 4005 FPGA. This was connected to the FPGA holding the MFD circuit and the output data generated by the MFD was carefully scrutinized using a logic analyzer. The Xilinx 4005 FPGAs used are not speed graded. We estimate the logic block propagation delay is approximately 6 ns (i.e., the equivalent of a Xilinx 4000 series –6 part).

Proper operation of the bit–parallel circuit was observed for data speeds as high as 17 MHz. This speed appears to be limited by our testbed and not the circuit. With the faster XC4000 series FPGAs now available, we expect that the circuit will easily operate at data rates exceeding 30 MHz. The bit–serial version's throughput was 3.4 MHz, approximately one fifth that of the bit–parallel version. This is better than the one eighth of the bit–parallel throughput we were expecting; the enhanced performance is likely due to the more intensive pipelining inherent in bit–serial designs.

However, the bit–serial implementation was much larger than what we had expected. The bit–serial design uses approximately two thirds the resources that the bit–parallel design uses; we had expected closer to one eighth the resource requirements, since the data path width was one eight the maximum path in the bit–parallel design.

On further consideration, we now realize that this particular circuit is a bit–serial designer's nightmare (a bad dream, at least...). First of all, there is a considerable resource overhead in the parallel–serial format conversions required in the serial circuit. Secondly, the structure of the Xilinx 4000 series configurable logic blocks (CLBs) does not lend itself to efficient bit–serial implementation of the adder–subtractors that are the heart of the matched filter. Whereas an n-bit bit–serial adder uses $1/n$ times as much hardware as its bit–parallel counterpart, the bit–serial adder–subtractor uses approximately $3/n$ times the resources of the bit–parallel version. Finally, in the bit–serial version, the data wordlength must be as large throughout as may be required in even the smallest portion of the circuit. This means that resources are wasted in portions where this large data wordlength is not required. In contrast, the bit–parallel implementation 'grows' its wordlength as necessary, using a minimum of resources to preserve data accuracy.

5 Conclusions

In conclusion, the FPGA implementation of the matched filter detector has been shown to be both feasible and tractable. In general, the bit–parallel version of the circuit performed better than its bit–serial counterpart. This particular circuit is not well suited to a bit–serial design, given the poor mapping of the bit–serial modules to the device structure and the growing wordlength required in the design.

As future work, the MFD could be implemented using a dynamically reconfigurable FPGA. In such an implementation, every time the filter tap length or tap weights are changed the MFD would be dynamically reconfigured accordingly. Every instance of the matched filter could be treated as a fixed tap weight filter. Therefore, adders and subtractors could be used instead of the more complex and cumbersome adder/subtractor modules that are currently necessary. This should result in smaller circuits for both bit–parallel and bit–serial design methods.

6 Acknowledgements

We would like to thank E. Patton for his assistance during this work. Funding for this project was provided by Wi-LAN Inc., Micronet, and NSERC of Canada.

References

1. Robert C. Dixon. *Spread Spectrum Systems with Commercial Applications*, page 220. John Wiley and Sons, New York, NY, 1994.
2. Edward Patton. Hardware implementation of the digital rake transceiver. M. Sc thesis, University of Calgary, Department of Electrical and Computer Engineering, August 1993.
3. B. Kish, J.M. Bauer L.E. Turner, and R. Wheatley. Logsim user's guide. Technical report, University of Calgary, Department of Electrical and Computer Engineering, 1989.
4. Peter Denyer and David Renshaw. *VLSI Signal Processing: A Bit Serial Approach*. Addison-Wesley, 1985.
5. L.E Turner and P.J.W. Graumann. Rapid hardware prototyping of digital signal processing systems using field programmable gate arrays. *International Workshop on Field Programmable Logic*, August 1995.
6. P.J. Graumann and L.E. Turner. Specifying and harware prototyping of dsp systems using a register transfer level languae, pipelined bit-serial arithmetic and fpgas. *Second Canadian Workshop on Field Programmable Devices*, June 1994.
7. I.W. Barker, P. Graumann, and L. Turner. Trans user's guide. Technical report, University of Calgary, Department of Electrical and Computer Engineering, 1993.

An NTSC and PAL Closed Caption Processor

Sayan Teerapanyawatt and Krit Athikulwongse
Electrical Engineering Department,
Chulalongkorn University, Bangkok, Thailand.

Abstract. This paper present an application of FPGA on Closed Caption data processing. With top-down design methodology, start from a Closed Caption decoder abstraction, drill down to a gate-level circuit then implementation does under bottom-up scheme. The Xilinx and Viewlogic software tools are use to implement Closed Caption processor. This processor work on both NTSC and PAL TV signal. A prototype set-top decoder can be build with only a few additional component such as memory (RAM and ROM), comparator and video overlay synchronizer.

1. Introduction

Closed caption TV system is provide a short message to support the information carried by the main audio channel. The service usually accommodates the hearing impaired persons or viewers who can not understand the language used on the primary audio channel. The message are carried in digital forms on the analog video program signal during its vertical blanking interval (VBI) which does not carry any picture or program information. This technique does not interfere with a normal viewing of the video program. The TV set must have either a built-in or set-top caption decoder in order to view these data as they are overlaid on the screen. Caption data are meant to be used as a visual depiction of the soundtrack of the video program and therefore is highly beneficial to the hard hearing or deaf persons. They must be carefully placed to identify speakers, on and off screen sound effects, music and laughter. When not in use, they can be closed or turned off so as not to unnecessarily obscure the video image. Closed captioning was first introduced in the United States of America more than 40 years ago. Since then, it has evolved to become a defacto standard for closed captioning in line 21 of the NTSC video signal. On the other hand, closed captioning in countries that use PAL video is usually done as one of the teletext services, even though teletext data are difficult or impossible to record in ordinary VHS format. Hence an alternative method for closed captioning in PAL video based on the NTSC line 21 caption encoding format will be investigated in this paper. In particular, a trial system using PAL line 22 incorporating the same encoding format as NTSC line 21 is implemented. The system can support English and other language e.g. Thai on different channels, each of which uses only 7 bits per character. An English-Thai closed caption decoder is also developed to be used as a set-top device for ordinary TV sets that no build-in decoders. The decoder main chip is synthesized by using Xilinx FPGA that combines the functions of a processing unit that performs signal conversion, decode incoming data and language information management and a display unit that generates characters with appropriate signal to overlaid on TV screen, the latter of which is called an Thai-English on screen display (OSD) generator.

2.The Closed caption format

The NTSC Closed Caption are transmitted during the vertical blanking interval on line 21 of both field [1] of composite video signal. Using PAM (Pulse Amplitude Modulation) reduced amplitude NRZ (Non-Return to Zero) format. Fig.1 shows the Closed Caption signal waveform. This information are coded into digital data signal, a logic one is set to 50 IRE and logic zero is set to black level (0 IRE). The baud rate is 0.5034 MHz. The synchronization bit pattern called clock run-in, starts at 10.074 microsecond after the start of the line. The clock run-in is a symmetrical sine wave with its maximum and minimum amplitude being equal to the logic one and zero level respectively of the encoded data. The clock run-in is also in phase with all of logic level transmissions of the start and data bits. The three start bits follow the same specifications as the data bits, but are always defined to be "001". The two ASCII characters are transmitted every line. All characters use an eighth bit for parity. The first character is preceded by start bit. The last data bit must be end at least 2.22 microsecond before the end of line.

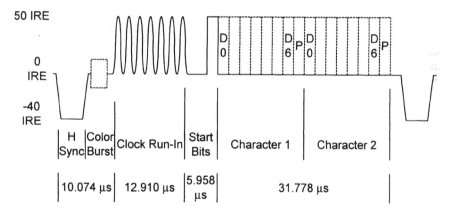

Fig.1 The Closed Caption waveform.

These data coded between 10H to 1FH are called "Non-printing character" when the others which coded between 20H to 7FH are called "printing character ". The all captioning text related to video program characters are using printing characters. While control code pair are using non-printing character on the first character, immediately following by printing character on the second byte. The control code pair must be transmitted in a single scan line and usually transmitted twice in succession to help insure correct reception of the control instruction. Only the valid data will be process. The invalid data will be ignore. Invalid data is any data that fails to pass a check for odd parity, or which, having passed the parity check is assigned on function. There are three type of control codes pair used to identify the format, location, attributes, and display of characters: Preamble Address codes, Mid-Row codes, and Miscellaneous control codes.

3. Architecture

The purpose of this processor is to be integrated Closed Caption data decoding and video signal processing function, which the existing Closed Caption integrated circuit cannot support. The mainly difference functions from the others are PAL video signal processing. On the advantage of using VHDL and FPGA for fast prototyping and partitioning, this processor was created under top-down design concept and bottom-up implementation. The fig.2 shows the set-top Closed Caption decoder block diagram. The Closed Caption Processor cooperate with a few external circuits such as Data Slicer , Sync. Separator, memory and video overlay synchronizer. Fig 3. Shows the Closed Caption processor's block diagram. This figure can be categorized into three major part: input, process, and output part. The input part consist of Sync. Processor and caption data register. The output part is on screen display generator. All information are done by the central processing unit.

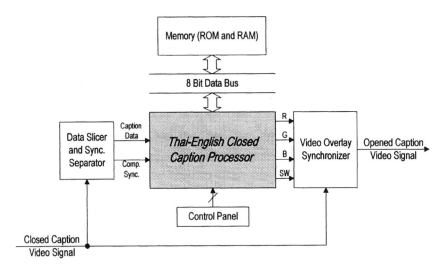

Fig. 2 The Prototype Closed Caption Decoder

The Sync. Processor receive the composite synchronization signal and extracts the component of video timing signal supply to others block. The data register samples sliced data into shift register. It's parallel output fed to a processing unit whose structure resembles that of a microprocessor. This processing unit is responsible for decoding control codes and user inputs by a program of instructions stored in external ROM memory. Caption data are also processed and passed to appropriate external RAM memory. The on screen display generator function block converts data on memory to RGB signal represent the character fonts which overlaid the program picture.

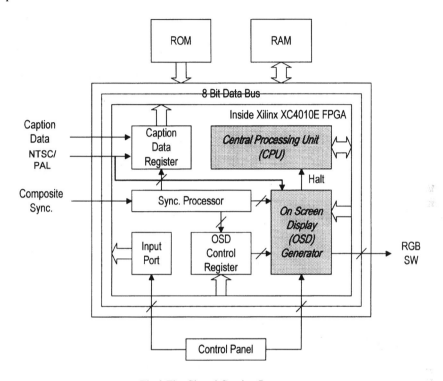

Fig.3 The Closed Caption Processor

4. Implementation

These Closed Caption processor are implemented on a single Xilinx 4010E with 12 MHz system clock, using 'Viewsynthesis' the synthesis tool of Viewlogic and XACT 6.0 Xilinx development system software. As mention above, the implementation done by bottom-up scheme. Each function blocks are individual written, synthesis, simulation and test. Here is brief of implementation:

The Sync. Processor (Fig.4)

Vertical sync The regeneration of signal represent to the vertical synchronization pulse is done by 8 bits timer. This timer measure time during logic zero of the Composite Sync. The period which longer than 21.33 nS. mean for logic zero of signal represents to the vertical synchronization pulse.

Horizontal sync These pulse which in phase with falling edge of Composite Sync generate by the falling edge detector circuit.

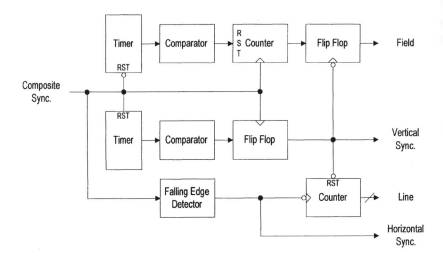

Fig 4. Sync Processor block diagram

The Central Processing Unit (Fig.5)

The concept of design is a simple general purpose CPU which construct of instruction register, control & decode unit, register file, arithmetic logic & shift unit, address register, and data out register. It's instruction support the Closed Caption control code operation.

Instruction Register The 16 bits register ,store the instruction from the 8 bits wide data bus. The instruction feed to the control and decode unit ,simultaneously feed the instruction constant to the Y bus to be an operand.

Control and Decode Unit This unit provide the control signal to manipulate the other part of the CPU. It also give the Read/Write (RD/WD) to control direction of data flow of the external memory.

Arithmetic, Logic and Shift Unit(ALU) This ALU implemented by Xilinx's supply library. The fast carry logic feature on the library allow fast arithmetic operation than the VHDL synthesis's circuit which not support this feature.

Register There are three type of register: Register file , Address Register and Data out Register. The Register file consist of 16 of 8 bits register. Registers name "re" and "rf" are joined in order to be the 16 bits instruction pointer. The Register File is implemented by the Xilinx library RAM module "16x8". The number of CLB in case of use the RAM module smaller than the others use of D flip-flop or can The Address Register (16 bits) and Data out Register (8 bits) are hold address and data for the external memory.

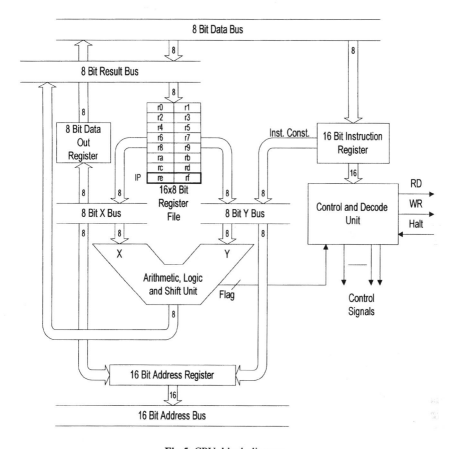

Fig 5. CPU block diagram

The On Screen Display generator (Fig.6)

These block consist of OSD time, address generator, display code buffer, display font buffer, background & foreground control and dot shifter. The two major considered point of this part are displaying of dot jitters problem and character windows size on difference video signal. The dot jitters problem can be solve by re-synchronize display dot clock with horizontal sync. pulse every scan line. The difference video signal cause of difference windows size. All counter in the OSD timer were use set/reset counter. To support FCC standard on NTSC video signal each character display using 12 by 12 matrix on 16 by 13 windows and available from line 43 to 238. For PAL video signal the character display on 12 by 15 matrix on 16 by 16 windows which start form line 40 to 296.

Memory

The single port RAM (62256) contain the three pages of 2 Kbytes. Page one and two are display memory and non-display memory of caption data. The last page keeps data service information. Each page is arranged as 34 columns by 15 rows of two bytes space. The second byte is specifically used for Thai vowels. The ROM memory contain the FPGA configuration program, user program and character fonts. The fonts take 16 Kbytes space. This is enough to contain two set of 128 fonts each of which has a size of 16*32 dots. The 32 dots height is used for both fields to allow high resolution character display. The first set of 128 fonts is the ASCII character set. The second set of 128 font is the Thai Standard Character set with mixed pattern of Thai vowels.

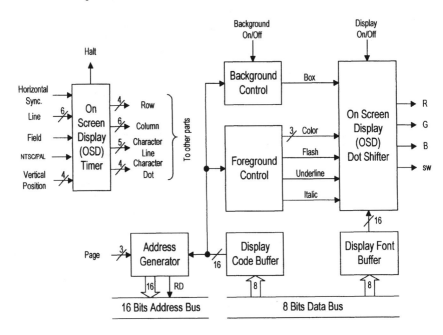

Fig.6 The OSD block diagram

5. Conclusion

This paper present one of FPGA application on television signal processing. The Top-down design methodology and bottom-up scheme are use. The Closed Caption application bring the new channel of communication to the hearing impaired persons on PAL TV system country, especially the Thai deaf who's learning English from video. The limitation of this prototype is expected a clear captioning video signal. It need to be add the signal conditioning circuit as low pass filter for noisy video signal.

6. Acknowledgment

This work is supported by The National Research Council of Thailand (NRCT) and National Science and Technology Development Agency, Ministry of Science Technology and Environment.

References

[1] Engineering Dep., Electronic Industries Association, EIA-608 Recommended Practice for Line 21 Data service. Sep. 1994

[2] Digital Data on the Spacelab's Analog Video Channel, IEEE Aerospace and Electronic System Magazine., Vol. 8 Iss.5, 42-50, May 1993.

[3] Spicer Smith D.L., Hanby, W.D.C., Barlow P., Techniques for Transmitting Use of The Vertical Blanking Interval on BBC Distribution. IEEE Colloquium on Time between Picture - The Vertical Blanking Interval., p. 2/1-4,1994

[4] Benson K.B., Whitaker J.C., Television Engineering Handbook McGraw-Hill, NY.,1992

[5] Xilinx Inc., San Jose, California. The Programmable Logic Data Book,1996

[6] Hurley,N.F., A Single Chip Line 21 Captioning Decoder, IEEE Transactions on Consumer Electronic., Vol. 38 No.3, Aug. 1992.

[7] Moller U., Berthin W., Schwendt R., A Single Chip Solution for Closed Caption Decoder, IEEE Transactions on Consumer Electronic., Vol. 38 No.3, Aug. 1992.

A 800Mpixel/sec Reconfigurable Image Correlator on XC6216

Tom Kean and Ann Duncan

Xilinx Development Corporation, 52 Mortonhall Gate, Edinburgh, EH16 6TJ, Scotland

Abstract. A high performance image correlator design for the XC6200 is discussed. The design correlates a 16x32 pixel template against a larger image, loading 16 new pixel values in parallel, at a clock rate of up to 50MHz. The dynamic reconfiguration capabilities of the device are exploited to rapidly reconfigure between hardwired match image templates at the logic level.

1. Introduction

The Xilinx XC6200 is the first commercial FPGA specifically designed to work in close co-operation with a microprocessor [1]. To the microprocessor, the XC6200 appears as a block of RAM; configuration data and registers within the user design appear within this memory space. XC6200 application development and debug is supported by XACT6000 CAD software package and by a PCI bus resident development system [2]. This board is available from Xilinx's partner companies. The XC6200 architecture is derived from the Algotronix CAL technology which Xilinx acquired in 1992 [3][4].

Image processing functions lend themselves very well to hardware acceleration, due to the large volumes of data involved. Correlation is an example of one such computationally intensive function [5]. Applications are found in industrial inspection and target recognition.

Image correlation is performed by passing a template over a binary image and determining, at each pixel position, if a match has been found. The design described uses a 32x16 one bit pixel match image and a 32x16 pixel mask image to construct the template. The match image is a small image to be located. The mask image allows masking out of background regions in the match image. Only unmasked pixels will be correlated. If, at any point in the image, the number of matching pixels exceeds a threshold, a Hit is detected.

The correlator is implemented as a pipelined design. Image data is fed in 16 rows at a time into a 32 stage pipeline. At each clock step the pixel data in each of the 512 registers is compared to the template.

The match and mask images are hardwired into the logic. See Figure 1. The ability to dynamically reconfigure gates means that the image template can be rapidly reconfigured an unlimited number of times. Partial reconfiguration and fast reconfiguration times make XC6200 an ideal platform for this application.

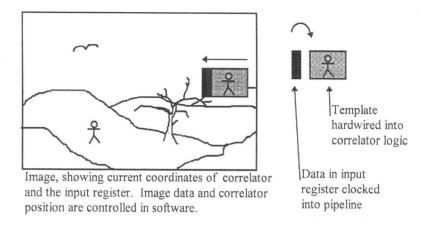

Image, showing current coordinates of correlator and the input register. Image data and correlator position are controlled in software.

Template hardwired into correlator logic

Data in input register clocked into pipeline

Fig. 1. Image pushed into correlator pipeline in rows of 16

2. Physical Implementation

Figure 2 shows the hardware components that constitute the 32x16 correlator. The design comprises the following:

- 16 bit input register
- 32x16 bit correlator block programmed with mask and match
- Threshold block
- 1 bit hit register
- CBUF clock: generates a clock pulse on each write to the input register

The image data is written in columns of 16 pixels to the input register via the control store interface [1]. On each write to this register, the CBUF generates a clock pulse clocking the design and shifting the columns of data into the pipeline. The matching pixels in the pipeline are summed in a tree of custom adders [6].

Figure 3 shows a 3 pixel correlator. The gates inside each dotted box perform selection on the pixels depending on the match and mask data. All of these gates can be eliminated from the design if it is possible to reconfigure the gates within the three bit adder, minimising the adder logic according to the match and mask data for each pixel [7][8].

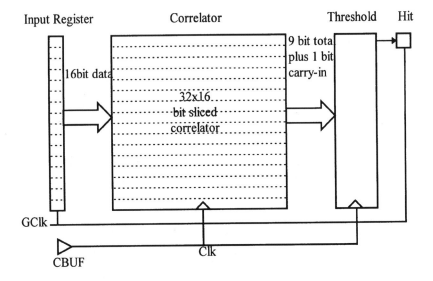

Fig. 2. Architecure of correlator design

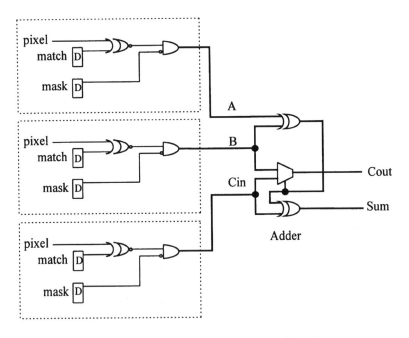

Fig. 3. A 3 pixel correlator without reconfiguration

Figure 4 is the base configuration for the lowest level block of the design showing a reconfigurable 2 stage adder. The adder inputs come from a section of the pipeline. Operation is as follows. Pixel data is clocked through the shift register. Summing of the number of matched pixels takes place over two clock cycles. The output from ABXOR is delayed for one clock cycle and then operates as in a standard adder design when the three pixels being correlated lie in the registers. The adder sums the number of matched pixels, the sum has a one clock cycle delay, the carry a delay of two. When placed and routed using XACT6000, this block occupies 6 function units, the design in Figure 3 would require 12. The circuit in Figure 3 would also operate at a lower speed due to there being more gates in the critical path.

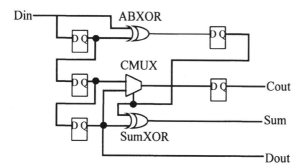

Fig. 4. 3 pixel correlator showing shift register and 3 reconfigurable gates.

3. Gate Reconfiguration

Figure 4 shows the three pixel reconfigurable adder. As discussed, correlation, or matching against different pixel patterns, is achieved by making each gate of this adder reconfigurable, basing the new gate on the mask and match data. For example, if the pixel on the input, Din, were masked and the pixel in the top flip-flop corresponded to a match of 1, the gate ABXOR would be configured to be a buffer allowing a value of 1 to pass through to the next stage of the adder one clock cycle later. If the pixel in the first flip-flop matched against 0, ABXOR would be configured to an inverter. Similarly, gates and their net connections are evaluated for CMUX and SUMXOR. Say for example the template for these 3 pixels is as follows [*, 1, 0] where the * signifies that the first pixel is masked out. The pixel on Din is masked, and the pixel in the top flip-flop is matched with 1. One clock cycle later, all pixels will have shifted one step along the pipeline. Therefore, at T+1, the pixel in the middle flip-flop matches with 1 and the pixel in the bottom flip-flop matches with 0. During reconfiguration. the CMUX is replaced by a MUX with an inverter on the second input and the SumXOR with an XNOR gate. See Appendix A for all possible reconfigurations of mask and match combinations.

The correlator is constructed hierarchically from these low level blocks. Two blocks are chained together to make a 6 bit shift register. See Figure 5. The sum and carry bits from these two blocks are then summed to give a matched pixel total for the 6 bits. As a space saving trick here, an extra pixel is taken as a carry in, making a seven pixel correlator block [6]. The carry in pixel must be delayed by one clock cycle to synchronise timing with the sum output from the 3 pixel correlator to which it is being added. The gate which takes the carry in is reconfigurable, another ABXOR but with only one of the inputs being a pixel value the other is assumed always to have a mask value of 0 and a match of 1.

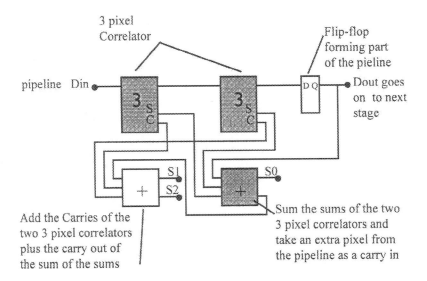

Fig. 5. A 7 pixel correlator built by connecting two 3 pixel correlators in series. Shaded blocks contain reconfigurable gates.

Two of these 7 pixel correlators are combined in parallel, their totals summed, plus an extra pixel on the carry in, delayed by two, to make a 15 pixel correlator, see Figure 6. Two 15 pixel correlators are combined in series with an extra carry in delayed by three clock cycles, to make a 31 bit correlator; and so on, until a 511 pixel correlator is constructed, leaving one pixel untested.

The sum from the 511 pixel correlator plus the 512th pixel are fed into the Threshold block.

In total, if comparing the reconfigurable design against the alternative shown in Figure 3, a total of 2048 function units are saved. The final design uses 2343 function units in all.

4. Threshold Block Logic

The threshold value determines the quality of the match of the image with the template. Logically, the threshold block compares the number of matches with a preset threshold t, and if it is greater, sets the *Hit* flag. For a perfect match the threshold value would be 32x16=512. In practice, the threshold block operates by adding the total number of matched pixels plus the extra carry in pixel to a constant value and generating an overflow when the threshold is exceeded. This means programming the threshold block with a constant value, 512-t, where t is the threshold value. This is done by performing a single threshold register write before correlation of the image begins. The overflow or carry appears in the single bit *Hit* register.

Of course, the threshold register could be removed and the gates within the comparator changed to be reconfigurable, resulting in a saving of a further nine function units.

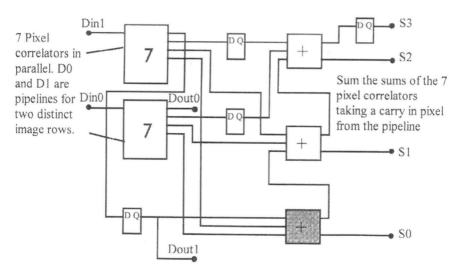

Fig. 6. A 15 pixel correlator constructed from two 7 pixel correlators in parallel.

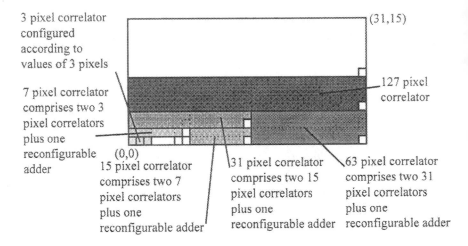

3 pixel correlator configured according to values of 3 pixels

7 pixel correlator comprises two 3 pixel correlators plus one reconfigurable adder

(31,15)

127 pixel correlator

(0,0)

15 pixel correlator comprises two 7 pixel correlators plus one reconfigurable adder

31 pixel correlator comprises two 15 pixel correlators plus one reconfigurable adder

63 pixel correlator comprises two 31 pixel correlators plus one reconfigurable adder

Fig. 7. The correspondence between template pixel values and reconfigurable gates

Figure 7 shows how the reconfigurable gates map to the pixels in the mask and match template images.

5. Performance Measures

There are three possible modes of operation.

1. XC6216 as part of a purpose built hardware system for image processing.
XC6216 would reside on a board with frame grabber and control logic. Data input would be via the array pins. According to timing analysis of the design using XACT6000, the maximum clock speed to the design is 50MHz if data is fed direct to IOBs. This would enable correlation of a 512x512 image in 5.2ms; a maximum frame rate of 190Hz.

2. A system where data is fed to the XC6216 design via the microprocessor interface, from local memory.
Speeds of 20-25MHz would be attainable giving a worst case correlation time for a 512x512 image of 13ms. This equates to a frame rate of 76Hz.

3. As part of a development system where image data is stored on the PC.
Timing is dependent on PCI bus performance and the control software.

Actual values measured from the software control program give correlation times using the XC6200, of 0.69s. The software control and PCI interface increase the delay from the calculated value. This compares with a value of 38s for a software correlator running on a 133MHz pentium. To improve speed, *Hit* could be wired to an interrupt signal and image data could be stored in the on-board RAM.

5.1 Speed of Initialisation

In performing reconfiguration, time taken is dependent on the method of reconfiguration. In the most simple method, also the most time consuming, each address/data pair is written individually from the control program. It is also possible to perform batch reconfiguration.

Total time taken to change the mask and match image for this design is given by the following equation:

$$T = G \times 2Nclk \times \frac{1}{Fclk}$$

where T is the total reconfiguration time, G is the number of gates, $Nclk$ is the number of Clock cycles taken for a write of one address/data pair and $Fclk$ is the frequency of the clock. In this design, there are 512 reconfigurable gates, each of which requires two address/data pairs for gate reconfiguration. On a development system board, five clock cycles are needed to write one address/data pair.

$$T = 512 \times 2 \times 5 \times \frac{1}{33 \times 10^6} = 0.155ms$$

5.2 Performance with Multiple Templates

Given that correlator reconfiguration with a 32x16 mask and match template requires less than 0.2ms and correlation over a 512x512 image takes approximately 0.69s, reconfiguration time between templates can be regarded as insignificant.

6. Summary

For the problems of image correlation: large amounts of data, intensive computation and the desire, in most applications, to frequently change mask and match images, advantages of XC6200 are clear. It combines the flexibility of rapid partial reconfiguration for changing between image templates (times of less than 0.2ms calculated), with high density on this kind of stuctured, heavily pipelined application.

Performance was evaluated for standard images of dimensions 512x512. A correlation time of 5.2ms was calculated. Time measured for a development system board with image data transfer from the PC to XC6216 via the PCI bus, was 0.69s. This compares with a correlation time for software running on a 133MHz pentium, of 38s.

References

[1]Xilinx Inc, "XC6200 FPGA Family Advanced Product Description", available from Xilinx Inc. 2100 Logic Drive, San Jose, CA

[2]Wayne Luk and Nabeel Shirazi, "Modelling and Optimising Run Time Reconfigurable Systems", Proc IEEE Symposium on FPGAs for Custom Computing Machines, Napa, CA 1996.

[3]Tom Kean, "Configurable Logic: A Dynamically Programmable Cellular Architecture and its VLSI Implementation", PhD Thesis CST62-89, University of Edinburgh, Dept of Computer Science.

[4]Tom Kean and John Gray "Configurable Hardware: A new paradigm for computation", Advanced Research in VLSI, Proc. Decennial Caltech Conference, MIT Press 1989.

[5]Brian Von Herzen, "250MHz correlation using high-performance reconfigurable computing engines", Proc SPIE - The International Society for Optical Engineering, vol. 2914, pp34-43.

[6]Parallel Counters, Earl E. Swartzlander Jnr. IEEE Trans on Computing C-22, pp1021-1024 (1973)

[7]Tom Kean and John Gray, "Configurable Hardware: Two case studies of micro-grain computation", Systolic Array Processors, Prentice Hall, 1989. Edited by McCanny, McWhirter and Swartzlander, pp310-319.

[8]P.W. Foulk and L.D. Hodgson, "Data folding in SRAM configurable FPGAs", Proc IEEE Symposium on FPGAs for Custom Computing Machines, Napa, CA 1993

Appendix A
Gate reconfigurations depending on match and mask values

Gate	Condition	Reconfiguration	Net connections
ABXOR	Match A = Match B	XOR2	A, B
	Match A != Match B	XNOR2	A, B
	Match A = 1, B Masked	BUF	A
	Match A = 0, B Masked	INV	A
	Match B = 1, A Masked	BUF	B
	Match B = 0, A Masked	INV	B
	A, B Masked	GND	
RCMUX	Match C, Match D = 1	M2_1	C, D, S
	Match C = 0, Match D = 1	M2_1B1	Cinv, D, S
	Match C = 1, Match D = 0	M2_1B1B	C, Dinv, S
	Match C, Match D = 0	M2_1B2	Cinv, Dinv, S
	Match C = 1, D Masked	AND2B1	C, S
	Match D = 1, C Masked	AND	S0, D1
	Match C = 0, D Masked	AND2B2	Cinv, Sinv
	Match D = 0, C Masked	AND2B1	S, Dinv
	C, D Masked	GND	
SUMXOR	Match D = 1	XOR2	D, S
	Match D = 0	XNOR2	D, S
	D Masked	BUF	S

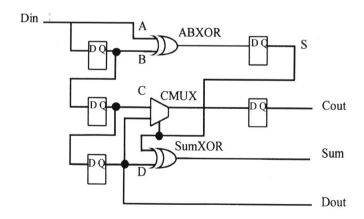

A Reconfigurable Coprocessor for a PCI-based Real Time Computer Vision System

Ferran Lisa, Faustino Cuadrado, Dolores Rexachs, Jordi Carrabina

Universitat Aurònoma de Barcelona, Departament d'Informàtica
Campus U.A.B., edifici C
08193 Bellaterra (Barcelona), SPAIN
e_mail: ferran@cnm.es

Abstract. This paper describes a reconfigurable coprocessor based on an Altera CPLD, specifically designed for a real-time computer vision system. An overview of the system is given and the architecture of the coprocessor is described, discussing the utility of its distributed memory organization for image processing applications.

1. Introduction

Real-time Image Processing (IP) and Computer Vision (CV) have always been very interesting application areas for VLSI and parallel processing because of its high computing requirements.

Consider a real-time CV system dealing with color images with spatial resolution of 512 x 512 pixels and color resolution of 24 bits per pixel, and suppose images are being acquired at a standard frame rate of 30 images per second. This system requires operating at a bandwidth of 24Mbytes per second, and processing data in different ways. That leads to the need of *parallelism* in hardware implementations of CV systems, as well as high performance buses for image transfer.

The hardware platforms used for this kind of applications have evolved through the history of parallel computing, using different architectures (meshes of processors, hypercubes, pyramids, ...). Each of them has some advantages related to different kind of computer vision tasks. But computer vision problems span a broad spectrum of algorithms that combine computations with specific hardware requirements, which demand architectures that support such heterogeneity.

The devices used as processing elements in these systems have evolved through the history of CPUs (CISC, RISC) leading to more specialized processors like DSPs, and is now using FPDs (Field Programmable Devices) as customizable processors.

The advances in VLSI technology made possible the existence of *FPDs* with enough amount of computational resources, that are either reconfigurable processing elements (FPGAs and CPLDs), or reconfigurable interconnection elements (FPICs). The combination of these two kinds of devices makes possible to develop fully reconfigurable systems that can be adapted to the architecture that best matches a given CV algorithm, achieving a computing power close to supercomputers at a fraction of its cost.

The aim of our research is the use of FPDs combined with the high-performance PCI bus to develop the hardware of a complete real time computer vision system, suitable both for scientific purposes and commercial applications. To do so, the goal we try to keep in mind is to maximize both flexibility and efficiency of the system in order to achieve the high throughput required in different kinds of CV applications.

In this paper we give an overview of our system and introduce our reconfigurable processing board based on an Altera CPLD. The rest of the paper is organized as follows: section 2 discusses the main issues in the design of our computer vision system, section 3 gives a description of our reconfigurable coprocessor and section 4 gives an overview of the library of elements we are currently developing for customizing the system, finishing with the conclusions in section 5.

2. Our Real Time Computer Vision System

In order to built the whole system we keep in mind to be able to use, at the same time, low cost but fast enough common resources (PCI-based PC, standard cameras) together with specialized communications around central processing devices, the reconfigurable ones. This scheme will increase throughput and maximize the number of IP operations allowed by real time conditions.

2.1 PCI Bus

PCI bus was chosen because of its properties and its growing popularity in commercial applications. Among the most interesting features of this local bus it is interesting to mention its high bandwidth up to 132 Mbytes/sec on 32bit/33Mhz systems (528 Mbytes/sec. will be soon available for 64bit/66Mhz systems), its independence of the CPU, its configurability and expandability, and its master-slave behavior [1]. Figure 1 shows an example IP system using PCI bus.

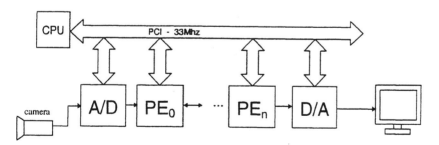

Fig. 1. Sample PCI-based image processing system

This system is composed of a set of PCI boards interconnected through a local bus in a pipeline-like scheme. The PCI bus is devoted to configuration and control purposes while the local bus is used for fast transfer of images between processing stages. The processing boards will be composed of different kinds of devices, including FPDs and DSPs and will be capable of configure to implement the different processes in a CV algorithm.

2.2 A PCI-based Acquisition Board

The first achievement of the project was an acquisition board developed last year [2]. This board is capable of acquiring images in real time from different sources in standard formats (PAL, NTSC, SECAM), transfer them through the internal bus to other boards specifically designed for image processing, and through the PCI bus to the system memory. The board is capable of producing several output formats in synchronous and asynchronous modes, among them RGB 5+5+5, RGB 8+8+8, YUV 4:2:2 and eight-bit monochrome.

An important feature of this board is its specialization. It was designed to maximize the efficiency of the whole system, so the features specific to some kind of applications were avoided. For example, it hasn't on-board memory, because in a lot of real-time CV applications, it has no sense to store the image before a complex processing, and can be done on other boards when required.

It has only a small set of preprocessing operations (point operators), like contrast and bright control, and it is capable of scaling images to different resolutions and sizes, allowing the system to dynamically define an acquisition window. This feature is very interesting for some CV applications like tracking, where after locating the object of interest only a small fraction of the image is needed.

2.3 Processing Elements

As stated in previous sections, we try to maximize the flexibility and efficiency of our system, so we need to design processing elements capable of configuring to implement the tasks specific of different applications and the data to be processed in different stages.

As an example, the resolution of the pixels or the size of the images can change from one application to another, and so can do the weights in a convolution mask.

This makes very interesting the use of FPDs but, on the other hand, it has been shown [3][4] that FPDs have limitations when dealing with complex tasks, in particular when implementing operations with a high number of products.

This kind of tasks is present in some CV algorithms like the ones that make frequency-domain analysis of the images. In this case the system requires implementing floating point multipliers for the FFT transforms. For these applications our system will have additional processing boards containing "of the shelf" devices like floating point multipliers or DSPs.

3. A Reconfigurable Image Processing Board

Image processing has been always a good focus of attraction for programmable logic developers. A lot of specific purpose IP and CV systems, like the one in [13], have been designed using FPDs, showing the utility of these devices for these kind of applications. These systems use FPDs as processors customizable for different tasks, but its flexibility in the processing elements is dramatically limited by rigid communication schemes between them and other elements, like memory modules or the host CPU. This make these systems only suited for a limited class of algorithms with similar I/O requirements.

In the last few years, a big effort has been done in the development of general purpose FPD-based systems. These so-called FCCMs (FPGA based Custom Computing Machines) have also been used for image processing purposes, like in the case of the VTSplash system [5].

Only a few systems [6][7] have been specifically designed to act as general purpose IP systems. Though, these systems still lack of the flexibility required to give support to a wide number of IP and CV applications.

In this section we give a description of a new board to be used as part of our reconfigurable real-time computer vision.

3.1 Processing Elements

The architectures of FPGAs and CPLDs have evolved in the last decade increasing performance and density. Each kind of device followed a different approach in the architecture of a FPD, but their distinction is becoming less clear every day [8].

FPGAs traditionally had a higher density but a lower predictability than CPLDs because of their segmented routing, but FPGA vendors are trying to improve their devices by experimenting with FPGAs partitioned into blocks with dedicated routing resources.

CPLDs, on the other hand, have evolved from the integration of several copies of the first 22v10 SPLDs, to more complex architectures, similar in some cases to the LUT-based FPGAs.

We decided to use an Altera Flex10k100 CPLD, mainly because its high pinout count (403 user IO pins), which make it suitable for the kind of architectures we have in mind. Using CPLDs instead of a FPGAs is attractive in order to skip the problems of routing delays reported by different authors [7]. Other reasons of using this device are its embedded memory (24Kbit in its EAB blocks), its high capacity (100.000 typical gates) and the good CAD tools available at low cost for universities in the Europractice program.

The high gate and pin count make possible the design of a board with a single device implementing different processing elements in an Image Processing chain. This is also very interesting in order to avoid the architectural and functional limitations present in other systems containing processing elements implemented into different FPDs (i.e. fragmentation problems).

3.2 Distributed Memory Architecture

According to the available memory access modes, custom computing machines can be classified into different classes [9]: some CCMs can only have access to a local memory, while in others only the access to the system memory is possible. The architecture of the PCI bus allows the development of systems that use a local memory and system memory concurrently with the host CPU. To do so, our board has a master/slave PCI interface by AMCC, capable of transferring data in different modes between the system memory and the on-board memory at the maximum bandwidth available in PCI (132 Mbytes/sec).

The structure of the local memory is normally very simple. In most of the cases there exists one or two RAM modules per each FPD. A scheme composed of two interleaved memory modules is very useful for real time IP systems, because an incoming frame can be stored in a module while processing the previous one, stored in the other.

In these systems, images can only be accessed as a long vector of pixels, normally organized in row-scan format. Though this, some developments from parallel computation for image processing [10] showed the that it is very useful the organization of images into different memory blocks.

In order to give flexibility to the organization of images in memory, our board has parallel access to eight 128Kx8 chips of high speed (20ns of access time) static RAM.

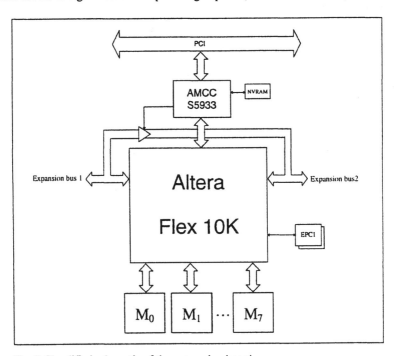

Fig. 2. Simplified schematic of the processing board

The two expansion buses are 32 bit-wide and are used to transfer images from the acquisition board and between different processing boards (see figure 1). By using these expansion buses the system can configure different interconnection schemes without using crossbars or other switching devices with limited flexibility and bandwidth.

Expansion buses are interconnected through a driver controlled by the PCI interface, in order to allow the system to transfer images through boards when they are in the configuration phase. Using this feature, a system composed of two boards can reconfigure one board without breaking the processing chain.

In order to give flexibility to the configuration of the CPLD, three different modes are available. It can be configured either from Altera EPROMs (two EPC1 devices), from a serial external connector (Bit-blaster) or directly from the PCI bus through the AMCC controller.

4. A Library of Parametrised Modules

In order to speed up the design of implementations of CV applications, a library of parametrised modules is incrementally being developed in VHDL. This library is divided into two sets: system control modules and image processors.

System control modules are devoted to control the dataflow between processing boards or other devices in the board, like memory chips or the PCI controller, and finally to synchronize and control the different processing modules.

As an initial set of image processing modules, we are working in the implementation of point operators, convolutions, histograms, mathematical morphology operators, and image labeling.

4.1 Memory controllers

Memory organization is a key element in our system. In order to give support to different classes of image processing operations while maximizing its efficiency, we are developing parametrised memory controllers with parameters like size or resolution of images and its organization in memory.

One of the main drawbacks of parallel computers for computer vision is its high inefficiency. Every CV application has its own requirements in the size of the images and the resolution of the pixels, and an efficient CV architecture should adapt to them. Systems based on FPDs can overcome this problem by configuring the sizes of the buses and the architecture of the processors. In our case, images come from a versatile acquisition board capable of digitizing images in different formats, color or monochrome, and these images are processed by our system producing several intermediate images, for instance containing different features of the original one.

Memory controllers are responsible of the conversion between pixels used by the processing side, and the bytes stored in the memory side (figure 3). This conversion involves masking bits in a word or accessing different words, when pixel resolution is other than eight bits.

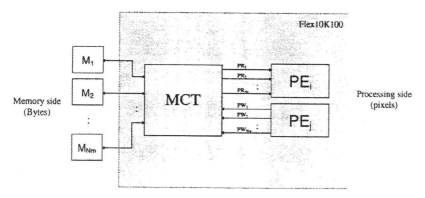

Fig. 3. Simplified Scheme of a Memory Controller

Another parameter completely dependent on the application is the organization of images into memory blocks. The library we are developing takes into account four different organizations: linear, rows, columns and partitioned.

- *Linear:* an image is stored in a single memory module in row-scan format.
- *Rows/columns:* an image is stored in k modules accessed as a word of k·8 bits containing a set of pixels in the same row or column.
- *Partitioned:* images are partitioned and distributed into blocks, in order to be processed in parallel by different processors in a "divide and conquer" style.

Combining different memory controllers, the system can be configured to implement an heterogeneous pipeline with different memory organizations at each stage. For example, image organization by rows or columns is very useful when implementing low-level neighborhood operations like convolution filters. Figure 4 shows a very common processing scheme for the convolution of an image with a 3x3 mask [12]. Two rows of the image are stored into two FIFO memories, operating as delay lines in order to simultaneously access 3 pixels of the same column.

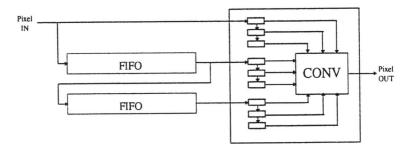

Fig. 4. FIFOs in an implementation of a 3x3 convolution.

FIFOs in previous figure, and its associated latency, can be avoided distributing images in a memory composed of different blocks, thus allowing parallel access to the tree rows involved in the process. Using our memory controllers, we can efficiently implement the convolution for a mask of up to 7x7 pixels.

Conclusions

The paper presents a reconfigurable coprocessor to be used as part of a PCI-based real time computer vision system. This coprocessor has been provided of a distributed memory scheme in order to allow it to be configured into different organizations for the storage and processing of images. This capability will allow us to experiment with different parallel implementations of image processing algorithms.

To speed-up the development of these implementations a library of parametrised modules for system control and image processing is currently being developed.

Acknowledgments

This research is supported by "Comision Interministerial De Ciencia y Tecnología", project TIC95-0230-C02-02.

References

1. PCI Local Bus Specification. Revision 2.1., 1995.
2. F. Lisa, J. Noguera, F. Cuadrado, D. Rexachs, J. Carrabina; "A PCI-based real time Image Acquisition and Processing System"; VII Simposium Nacional de Reconocimiento de Formas y Analisis de Imágenes; Barcelona, Spain; April 1997
3. O.T.Albaharna, P.Y.K.Cheung, T.J. Clarke; "On the Viability of FPGA-Based Integrated Coprocessors"; Proc. Of the 1996 IEEE Symposium on FPGAs for Custom Computing Machines.
4. R.J.Petersen, B.L.Hutchings; "An Assessment of the Suitability of FPGA-Based Systems for Use in Digital Signal Processig"; Proc. Of the 1995 Intl. Workshop on Field Programmable Logic and Applications
5. P.M. Athanas, A.L. Abbott; "Real-Time Image Processing on a Custom Computing Platform"; IEEE Computer, Feb. 1995, pp 16-24
6. S.C.Chan, H.O.Ngai, K.L. Ho; "A programmable image Processing system using FPGAs"; Int. J. Of Electronics, 1993, vol. 75, No. 4, pp 725-730
7. P. Dunn; "A Configurable Logic Processor for Machine Vision"; Proc. Of 1995 Int. Workshop on Field Programmable Logic and Applications.
8. S.Brown, J. Rose; "FPGAs and CPLD Architectures: a Tutorial"; IEEE Design and Test of Computers, Summer Issue, 1996.
9. R. W. Hartenstein, J. Becker, R. Kress; "Custom Computing Machines vs. Hardware-Software Co-Design: From a Globalized Pint of View"; Proc. Of the 1996 Intl. Workshop on Field Programmable Logic and Applications.
10. H. Hwang, D. Kumar-Panda; "The USC Orthogonal Multiprocessor for Image Understanding"; in Parallel Architectures and Algorithms for Image Understanding; V.K. Prasanna Kumar ed.; Academic Press, 1991
11. D. Benítez-Díaz, J. Carrabina; "Modular Architectures for Custom-Built Systems Oriented to Real Time Artificial Vision Tasks"; Proceedings of EUROMICRO - Design of Hardware / Software Systems; 1995
12. D. Ridgeway; "Designing Complex 2-Dimensional Convolution Filters"; The Programmable Logic Data Book; Xilinx, 1994
13. G. Knittel "A PCI-compatible FPGA-Coprocessor for 2D/3D Image Processing" 1996 IEEE Symposium on FPGAs for Custom Computing Machines

Real-Time Stereopsis Using FPGAs

Paul Dunn and Peter Corke

CSIRO Division of Manufacturing Technology, Locked Bag No. 9, Preston Vic., 3072
Australia

Abstract. This paper describes the use of configurable logic boards to
carry out real-time stereo image matching. The main factors in achieving this high performance are the elimination of instruction and data
fetching overheads associated with conventional processors and the use
of a non-parametric matching measure. Many reported stereo-matching
systems are based on numerically intensive measures such as normalized
cross-correlation which is expensive in terms of logic. Non-parametric
measures require only simple, parallelizable, functions such as comparators, counters and exclusive-or, and are thus better suited to FPGA
implementation. Further advantage is obtained by being able to overlap
processing with image input and output, and parallel computation of
disparities.

1 Introduction

Recently we have been working in the area of mine automation where sensing of
complex and time-varying three dimensional environments, is highly important.
Direct 3D sensing of a scene at a rate of tens of frames per second provides
information that is difficult or impossible to derive from 2D vision processing.
Stereopsis is one of many techniques for 3D imaging.

2 The Census Algorithm for Stereo Matching

Area-based stereo matching involves computing the *similarity* between a region
in the left image and a region in the right image[1]. For a particular region in the
right image, the left image region is moved horizontally and the best matching
region in the left image is found. The displacement between the right image
region and the best match is the *disparity* and is inversely related to range. This
procedure is repeated for every pixel in the right image. For a pair of $n \times m$
images, with a matching window of $w \times w$ pixels, and d disparity evaluations
the total time is $O(nmdw^2)$. Optimisations[2] can reduce this to $O(nmd)$ at the
expense of greater intermediate storage. To achieve a performance increase it
is necessary to reduce the time required to compute the *similarity* between two
image regions, compute similarity in parallel, or both.

---

[1] We assume that the images are in correspondence, that is, the epipolar line is parallel
to the scan line.

The simplest, and most intuitive, similarity or matching measures are the sum of absolute differences

$$\text{SAD}(x,y) = \sum_{(i,j)\in W} |T(i,j) - I(x+i,y+j)| \tag{1}$$

and the sum of squared differences

$$\text{SSD}(x,y) = \sum_{(i,j)\in W} \{T(i,j) - I(x+i,y+j)\}^2 \tag{2}$$

where W is a set of coordinates defining the matching region, T is the template, and I the image in which the patch is moved. Both these measures decrease with degree of similarity and will be zero for an exact match. More complex measures, which are invariant to mean intensity in the regions, include normalized cross-correlation

$$CC(x,y) = \frac{\sum_{(i,j)\in W} (T(i,j) - \mu_T)(I(x+i,y+j) - \mu_I)}{\sqrt{\sum_{(i,j)\in W} (T(i,j) - \mu_T)^2}\sqrt{\sum_{(i,j)\in W} (I(x+i,y+j) - \mu_I)^2}} \tag{3}$$

Such measures are commonly used in stereo-matching but are reliant on high-speed arithmetic capability for implementation. For FPGA implementation these methods would be expensive in terms of gate and look-up table RAM usage. An alternative is to use a non-parametric similarity measure[8] such as *census* or *rank*. Consider the small image patch

$$I = \begin{bmatrix} 79 & 83 & 87 \\ 81 & 85 & 86 \\ 83 & 86 & 84 \end{bmatrix} \tag{4}$$

we can represent the local intensity pattern by a corresponding window where each bit encodes the intensity of the corresponding neighbourhood pixel relative to the center pixel.

$$I = \begin{bmatrix} 0 & 0 & 1 \\ 0 & x & 1 \\ 0 & 1 & 0 \end{bmatrix} \tag{5}$$

This bit pattern can then be represented as a bitstring, in this case the binary number 00101010, known as the *census value*. The census value thus encodes the local intensity pattern and is immune to intensity offsets or gradual gradients. Another non-parametric measure is *rank* which is simply the number of 1 bits in the census value, 3 in this example.

The matching metric is the number of clear bits in the exclusive-or of the two bit masks. To improve the statistical base for matching it is common to use a window about the pixel of interest. In this implementation we compute the summation of the number of zero bits from the per pixel exclusive-or operation over a square window around the centre pixel — this is analogous to the use of a window in more conventional stereo matching methods. Simulations using variations on the size of the correlation window show that match quality decreases rapidly with window size, as is also the case for more conventional correlation matching methods. We have found empirically that an 11 × 11 window works well for the class of images we are processing.

A further advantage of the census method is that because all intensities are relative to the center pixel it is invariant to overall changes in intensity or gradual intensity gradients. For correlation based matching this robustness is achieved by normalisation and subtraction of the mean intensity.

3   The Configurable Logic Processor

The CLP[2][1] is a VMEbus circuit board containing several FPGAs which can be configured via the VMEbus to carry out image processing operations or associated control tasks. The board includes digital input and output ports and associated timing compatible with the Datacube[3] 10 MHz MAXBUS digital video format. A simplified block diagram of the CLP is shown in figure 1.

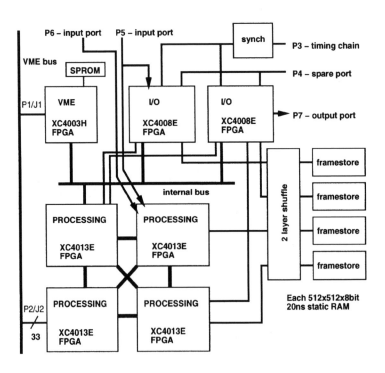

Fig. 1. CLP block diagram

The board contains four 512 × 512 8-bit frame stores which can be switched

---

[2] Configurable Logic Processor, named before the term became generic

[3] Datacube Inc., Danvers, MA.

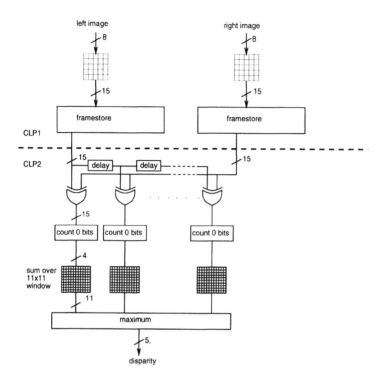

Fig. 2. Data flow through the FPGA system.

between various FPGAs. These use 20 ns static RAMs and are connected via a shuffle network to four of the FPGAs. All address, data and control signals are separate so as to allow independent access to each frame store.

4 A Configurable Logic Implementation

Our implementation has three major stages:

1. Compute census values over all 5 × 5 windows in each image.
2. Compare two census values using the exclusive OR operation and count the number of zero bits resulting. The sum of the number of matching bits over a larger, 11 × 11, window provides the similarity measure.
3. Find the maximum of the array of computed similarity measures, the index being the disparity.

Our stereo-matcher is implemented for an image size of 256 × 256 and two CLP boards are required. Figure 2 shows a general block diagram and the data flow. The first board writes the incoming left and right image pixel streams into frame stores, then reads the frame stores to form 15-bit census values for a 5 × 5 window around each pixel. These values are stored in two more frame stores.

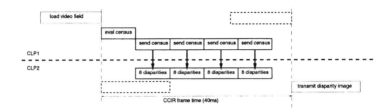

Fig. 3. Timing diagram for the CLP implementation.

The census values are then read in raster order 11 lines apart for both left and right images and the four resulting data streams are passed to the second board for disparity computation.

The second processing board compares the data streams from the left and right images at each of eight horizontal offsets (disparities) and stores the maximum match value and the offset at which it occurred. Four such passes are used to build up a total disparity range of 32 and the resulting disparity image is then written to an output port.

Image input and output can be overlapped with processing so the entire operation takes 5 passes where each pass occupies 6.8 ms. Total time is 34 ms which is less than the CCIR video frame time, see Figure 3.

4.1 Programming the FPGAs

Two compilers have been developed for FPGA programming. A circuit generation language, CCGL[1], was developed first and is based on C language syntax. It was developed before VHDL was available and has continued to be used. It has proved to be an enormously flexible and concise way of expressing designs while still allowing optional control of placement and timing. It does however lack the formalism of VHDL. Unfortunately it still requires the user to be accomplished in circuit design.

A fairly complete OCCAM2 compiler has been developed along the lines suggested by Ian Page and Wayne Luk [5]. With this a parallel algorithm can be simply encoded and the compiler will generate CCGL code.

The OCCAM2 compiler has proved worthwhile in coding complex control sequences and conditional logic, logic which would be quite laborious and error prone in CCGL. It has also proved useful for rapid prototyping of simple algorithms which can be compiled and working correctly in minutes rather than weeks.

In implementing the stereopsis algorithm it was not possible to use the OCCAM2 compiler. Detailed circuit design and minimisation and careful timing control was needed to fit the complex and difficult pipeline code into an architecture which did not suit the problem well.

4.2 Computing the Image Census

The left and right images are scanned by a 5×5 window, and a selection of pixels in the window are compared with the central pixel to form a census value representing that image patch. The 5×5 window has 25 pixels and would require a 24-bit value to represent a full census. To reduce logic requirements downstream we have used a partial census of 15-bits.

The processor boards have two clocks, c1 at 10 MHz and c4 at 40 MHz, and the video streams are at 10 MHz. To enable selected edges of c4 to be used within the c1 period, signals c4_p0 to c4_p3 are generated, each selecting one of four phases of c4.

The formation of the 5×5 window is shown in figure 4. Five short shift registers hold each row of the window, and are shifted left as the window moves to the right. At each shift it is necessary to load the five shift registers with window right-edge pixel values from the framestore. At the scanning rate of 10 MHz we can extract only 4 pixels from the framestore per cycle. The fifth line is obtained from a one-line-time delay line built from a 256×8 on-chip RAM.

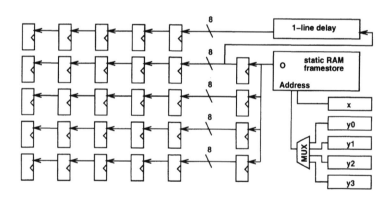

Fig. 4. Frame store read window

The requirement for four accesses to the external static RAM per pixel cycle, that is a 25 ns cycle, imposes some restrictions on the implementation. The FPGAs are 3 ns components and the static RAMs have a 20 ns read cycle. To minimise skew the addresses are clocked into output flip-flops at the FPGA output buffers. The four addresses cannot be simply multiplexed into the output flip-flops as the time constraints for the logic cannot be met. Instead the four addresses are clocked in parallel every 100 ns into a four-stage 8-bit shift register which is then shifted into the output register every 25 ns. The smaller two-input multiplexers required do meet the time constraints. This is shown schematically in figure 5 and is coded simply as

```
for (i=0 ; i<=2 ; i++)
    sr[i] = reg(8){D=mux2_n(8){c4_p3, sr[i+1], y[i]}, C=c4};
sr[3] = reg(8){D=y[3], C=c4};
mem_address[7..0] = x;
mem_address[15..8] = sr0;
outff_con("mem_address", "c4", 0, 15);
```

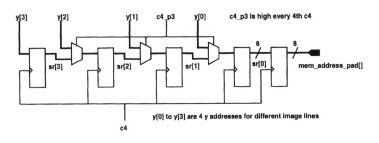

Fig. 5. Address multiplexing for frame store static RAM

The one-line-time delay to form the 5th line is simply an internal SRAM module whose output is clocked into a register.

Fifteen comparators, operating in parallel, compute the partial census value for each of the input images at 10 MHz, the resulting 15-bit values being stored in a frame store in two successive writes at 20 MHz as they are generated.

In the following four passes the same circuitry, with different address mapping, simultaneously reads two 15-bit census values from both left and right census value frame stores at addresses 11 lines apart. The resulting data is streamed at 20 MHz across a 30-bit backplane connection[4] to the second processing board.

4.3 Averaging and Matching

Figure 6 shows the use of two streams of census data from each of the left and right images. The two streams are associated with the top and bottom of the averaging window. The window width is formed by 11-pixel delays. Census values from the left and right images are compared using an exclusive-or, the resulting zero bits being counted. The sum of the bit counts over an 11×11 window is formed by running sums in x and y[2]. The right image data is also delayed by one pixel so that data at another disparity can be processed simultaneously. Three other FPGAs also receive these data streams, but with the right image offset a further two pixels for each one, giving a total of eight different disparities computed in parallel.

A function to maintain a running sum along a row

---

[4] via the VMEbus P2 connector

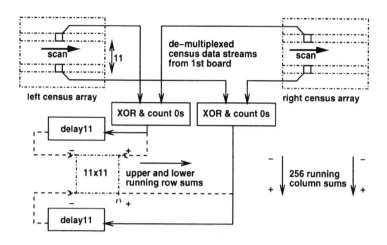

Fig. 6. Averaging window data streams

```
sig
row_run_sum {O[8] <- L[15], R[15]} {
      sig      sum[8], diff[8], match_new[4], match_old[4];

      match_new = count{L, R};
      match_old = delay (11) {match_new};
      sum = add{match_new, O};
      diff = sub{sum, match_old};
      O = reg{D=diff, C=c1};
}
```
is called four times, at the top and bottom of the averaging window for each of
two image offsets. The input signal arrays are the census input streams. Function
count{} performs an exclusive-or of its two input signal arguments and returns
a 4-bit count of the number of zeros (bit matches). The running sum is kept in
register **O** by adding the new component and subtracting the delayed component.

A similar function
```
sig
col_run_sum {O[11] <- B[8], T[8], RAM_INDEX[8]} {
      sig      sum[11], diff[9];

      sum = add{B, O};
      diff = sub{sum, T};
      O = sram(11, RAMWIDTH){D=diff, A=RAM_INDEX, WE=1, WCLK=c1};
}
```
maintains the column running sum. Here the running sum must be kept for each
x coordinate, so an internal synchronous RAM module is used. B[] and T[] are
the bottom and top running row sums for the averaging window. This function
is called twice, once for each image disparity.

4.4 Disparity Determination

At each major clock cycle each processing FPGA produces two running match sums. These are pipelined through a comparison tree which determines the maximum match value and the image offset at which it occurred. The results are stored in a frame store for comparison with corresponding values on further passes, each pass offsetting the left image by a further 8 pixels. Where a new maximum match value exceeds the stored value, it is replaced and the new disparity stored with it.

5 Performance

The disparity image has very slightly more noise visible than that produced by cross-correlation.

The reported performance of a number of stereo vision systems are summarized in Table 1. A useful comparison metric is the figure of merit (FOM), the fundamental time to perform one matching operation and is evaluated by

$$FOM = \frac{T}{nmd}$$

where the image size is $n \times m$ pixels, d disparities evaluated in a total time of T seconds. The L/R column indicates whether back matching has been included.

Group	$n \times m$	d	T (s)	FOM (ns)	L/R	comments
SRI SVM[3]	160 × 120	16	74 ms	240	yes	small DSP based
CMU iWarp[6]	240 × 256	16	64 ms	65		64 processor iWarp computer
CMU STORM	256 × 256	32	2.82	1370		Sparc 10
INRIA software[2]	256 × 256	32	59	28000	yes	Sparc 2, forward-back match
INRIA stecor	512 × 512	95	61.5	2470	yes	Sparc 5†
INRIA DSP[2]	256 × 256	32	9.6	4600	yes	4 M96002 DSP
INRIA hardware[2]	256 × 256	32	0.28	134	yes	PeRLe-1 board
JPL[4]	64 × 60	32	0.6	1630	yes	Datacube + 68040
match	512 × 512	95	189	7600	no	Pentium Pro (150 MHz), naive matching†
match	512 × 512	95	26.8	1080	no	Pentium Pro (150 MHz), optimized†
DMT CLP	256 × 256	32	34 ms	16	no	custom FPGA system†
PARTS engine	320 × 240	24	23.8 ms	13	no	custom FPGA system

Table 1. Summary of reported stereo vision system performance. The † indicates measurements conducted by the authors.

Two other FPGA-based systems are included in this table. The first was developed at INRIA[2] and based on the PeRLe-1 board developed at DEC-PRL in France. That board contains 23 Xilinx XC3090 FPGAs. In the stereo vision application the PeRLe-1 board performs the matching operation using the conventional cross-correlation method and table lookup logarithms are used to evaluate the divisions and square-root operations required. The second FPGA-based system, described in a very recent paper [7], uses 16 Xilinx XC4025 FPGAs and 16 one-megabyte SRAMs on a PCI card. The stereo implementation uses the census algorithm.

6 Conclusions

This paper has described the implementation of a real-time stereo matching system based on configurable logic and a non-parametric similarity measure. The frame rate performance achieved at this image size is impressive and is born out by comparison of the figure of merit with other reported systems.

Windowing is a fundamental task in most image-processing algorithms and this requires either large amounts of on-chip storage or very high-speed access to off-chip storage. Our next configurable logic processor has an architecture which facilitates parallel access to frame store windows.

Implementation difficulties encountered in this application highlighted the need to maximise inter-FPGA connections and external RAM modules.

Despite the availability of a high level parallel programming language, this algorithm was sufficiently demanding that the implementation was only possible at a circuit design level.

References

1. P. Dunn. A Configurable Logic Processor for Machine Vision. *Proc. 5th International Workshop on Field-Programmable Logic and Applications*, W. Moore and W. Luk, Eds., LCS 975, Springer, pp68-77, 1995.
2. O. Faugeras, B. Hotz, H. Mathieu, et al. Real time correlation-based stereo: algorithm, implementations and applications. Technical Report 2013, INRIA, August 1993.
3. Kurt Konolige. Small Vision Module, SRI International, http://www.ai.sri.com/ konolige/svm/index.html.
4. L. Matthies, A. Kelly, T. Litwin, and G. Tharp. Obstacle detection for unmanned ground vehicles: a progress report. In *Robotics Research: the Seventh International Symposium*, G. Giralt and G. Hirzinger, Eds. Springer, 1996.
5. I. Page and W. Luk. Compiling Occam into FPGAs. *FPGAs*, W. Moore and W. Luk, Eds., Abingdon EE&CS Books, pp271-283, 1991.
6. Jon A. Webb. Implementation and performance of fast parallel multi-baseline stereo vision. In *Computer Architectures for Machine Perception*, New Orleans, December 1993.
7. J. Woodfill and B. Von Herzen. Real-Time Stereo Vision on the PARTS Reconfigurable Computer. *IEEE Workshop on FPGAs for Custom Computing Machines*, pp242-250, April 1997.
8. R. Zabih and J. Woodfill. Non-parametric local transforms for computing visual correspondence. *Proc. 3rd European Conf. Computer Vision*, Stockholm, May 1994.

FPGAs Implementation of a Digital IQ Demodulator Using VHDL

C. C. Jong, Y. Y. H. Lam and L. S. Ng

School of Electrical and Electronic Engineering
Nanyang Technological University
Nanyang Avenue, Singapore 639798

Abstract. This paper presents the implementation of a digital IQ demodulator in FPGA devices. A DSP-based algorithm representing the demodulator was described in behavioural VHDL, which was used for simulation at the functional level. During the simulation, not only the functions were verified, the finite word-length effect and the coefficient quantization error were investigated as well. The VHDL description was then synthesised by using behavioural synthesis software. Different architectures were explored and manual partition was done on the behavioural description so that the selected architecture could fit into five reprogrammable FPGA devices. Post-routing simulation and hardware testing were carried out to verify the final implementation.

1 Introduction

Behavioural synthesis of digital systems [1,2] allows different design architectures to be explored in a short period of time. Using the behavioural synthesis technique together with reprogrammable FPGA devices, rapid prototyping of integrated digital systems can be achieved. This paper describes the implementation of a digital IQ demodulator using such approach. A DSP-based algorithm representing the behaviour of the demodulator was first coded in VHDL. The VHDL description was simulated to verify the correctness of the functionality. At the same time, the finite word-length effect and the coefficient quantization error were investigated before synthesis. Various design architectures were explored and partition of the system was done during the synthesis stage because the whole design was found to be too large to fit into one single FPGA device. The partition was done manually at the functional level and the final implementation of the design consists of five FPGA devices each of which has ten thousand equivalent gates. The design was tested and the results obtained were satisfactory.

The paper details the implementation of the IQ demodulator and it is organised as follows: Section 2 outlines the digital IQ demodulator implemented. Section 3 presents the VHDL description and the simulation. The coefficient quantization error is also investigated. Section 4 describes the synthesis results and the manual partition. Section 5 discusses the set-up for testing the design and Section 6 concludes the paper.

2 Digital IQ Demodulator

The demodulator takes as input a stream of 16-bit data in 2's complement integer format and generates as outputs two demodulated signals (In-phase and Quadrature-phase). It consists of two mixers (also called product modulator) and two low pass FIR filters. The input data and a signal from a local oscillator are fed into the mixers, which are basically signal multipliers, to produce two components which are the sum and the difference in frequency of the two signals. Each of the signals is then fed into one filter where the upper component of the signal is removed. Fig. 1 shows the functional block diagram of the IQ demodulator.

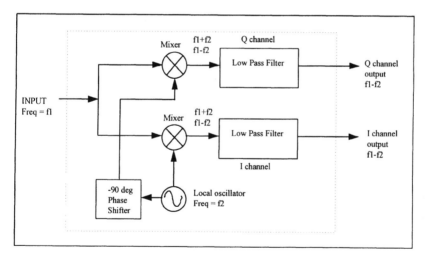

Fig. 1. Functional block diagram of the IQ demodulator

The digital output of the local oscillator is a stream of data consisting of (0, +1, 0, -1, 0, +1, 0, -1, 0, +1, 0, -1, ...), which is a 4-point representation of a sinusoidal wave. The mixing is done by multiplying the input data with the corresponding data from the oscillator. If the input data are (D_0, D_1, D_2, D_3, D_4, ...), the I-channel mixer output stream is (0, -D_1, 0, +D_3, 0, -D_5, ...) and the Q-channel mixer output stream is (+D0, 0, -D2, 0, +D4, 0, -D6, ...). The filter used is a 21-tap FIR filter[3] which requires 21 coefficients. The output is obtained according to the following formula:

$$Y_i = \sum_{k=0}^{20} (C_k \times D_{i-k})$$

where Y_i is the ith output sample,
         D_i is the ith input sample,
         C_k is the FIR filter coefficients.

3 Behavioural Description and Simulation

The function of the IQ demodulator was modelled first in C and then coded in VHDL at behavioural level. Fig. 2 shows the VHDL codes of the mixer, where *in16* is the current input data. Both the C codes and the VHDL codes were simulated to verify the correctness of the functions. The C model was used as a reference when the finite word-length effect and the coefficient quantization error were investigated.

```
variable s : integer range 0 to 3;
case s is
when 0 =>
        I(20) := (I(19) + (-1*in16))/2;
        Q(20) := in16_temp;
        s:=s+1;
when 1 =>
        I(20) := (-1*in16_temp);
        Q(20) := (Q(19) + (-1*in16))/2;
        s:=s+1;
when 2 =>
        I(20) := (I(19) + in16)/2;
        Q(20) := (-1*in16_temp);
        s:=s+1;
when 3 =>
        I(20) := in16_temp;
        Q(20) := (Q(19) + in16)/2;
        s := 0;
end case;
in16_temp := in16;
```

Fig. 2. VHDL code of the mixer

3.1 Finite word-length effect and quantization error

In the 21-tap FIR filter, there are 21 coefficients ranging from 0.0143308141 to 0.2436238396. Therefore, it is needed to store a total of 42 such coefficients. They are represented by integers in the VHDL description because only integer multipliers are available. The integer representation was achieved by multiplying the coefficients with 10^n and discarding the decimals. Hence, n represents the number of decimal precision. Simulation was performed on the C model with full precision for n = 10. In VHDL, different word lengths were used to represent the integers and simulation was performed to study the effect. ArcTan(Q/I) was plotted against time and linear regression technique was applied to measure how well the phase plot could fit a straight line. Fig. 3 shows the phase plot of the simulation result with n = 5 and Table 1 compares the simulated results.

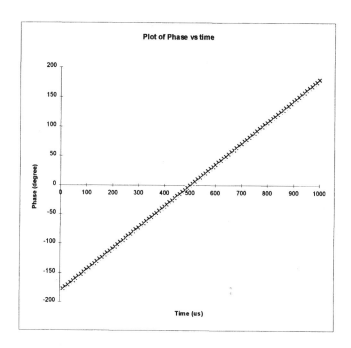

Fig. 3. Phase plot of simulated results

Table 1. Comparison of simulation results

Source Code	C	VHDL	VHDL
Number of decimal precision	10	5	3
RSQ	1.0	1.0	1.0
y-intercept (Degree)	-180.821°	-180.827°	-180.853°
Error for y-intercept (Degree)	Reference	0.006°	0.032°
Gradient (Degree/µsec)	0.360001	0.36	0.360003

Where RSQ = Square of PEARSON product moment correlation coefficient.

3.2 Choice of Word Length

The simulation results in Table 1 show that the design with 5 decimal point precision has a slightly better accuracy. However, the one with 3 decimal point precision is a better choice due to two reasons. First the result is still good and second its size is much smaller. With this choice, 9 bits are required to store one coefficient instead of 16 bits as in the design with 5 decimal point precision.

4 Synthesis and Partition

The design was synthesised into FPGA devices. To explore various possible architectures, Behavioral Compiler[4] and FPGA Compiler[5] were used. The FPGA device available was the Xilinx XC4010E[6] which contains 400 Configurable Logic Blocks (CLBs). As the functional description of the design contains 42 multiplications and 40 additions, there are many choices on the architectures varying from a single multiplier and a single adder to 42 multipliers and 40 adders. However, it was found that a simple design consisting only one multiplier would require almost 200 CLBs. Therefore, the architecture exploration was limited to those with either one or two multipliers. Table 2 shows some of the synthesis results.

Table 2. Comparison of synthesis results

Design	Decimal Precision	No. of multipliers	Circuit Size (CLBs)
1	5	1	2009
2	5	2	2365
3	3	1	1873
4	3	2	2069

It can be seen from Table 2 that even the smallest design needs almost 2000 CLBs and cannot be fitted into a single XC4010E. Therefore, the design has to be partitioned. By investigating the size of different modules, the design was partitioned manually at the behavioural level into five modules. The five modules are: the mixer module, the I Channel multiplier and adder module, the Q Channel multiplier and adder module, the I Channel coefficient and register module and the Q Channel coefficient and register module. Fig. 4 shows the block diagram of the final partition where each module is implemented with one XC4010E device. The exchange of data among the modules is controlled by an extra control signal. The control signal IQ_RDY signals the data from the mixer is ready and it is used by the register modules to read in the data. The START signal triggers the multiplier module to read in the data and start the multiplication. As a result, the number of control signals is kept to be the minimum and at the same time the synchronisation of data is maintained.

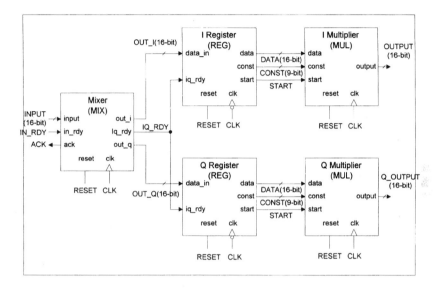

Fig. 4. Final implementation after partition

5 Hardware Test

The final implementation was simulated again after synthesis. The results obtained from the post synthesis simulation were the same as those obtained from the functional level simulation. The hardware implementation was then built and tested. Fig. 5 shows the major components in the test set-up.

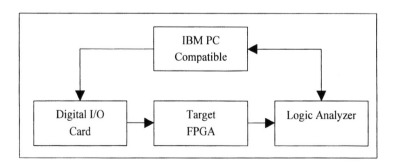

Fig. 5. Hardware test set-up

The testing process is controlled and monitored by an IBC compatible PC. The input data stream is generated by a digital I/O card plugged into the PC. The output of the demodulator is monitored by a logic analyzer and the data are collected and stored in the PC. A software program was developed using LabWindow/CVI to control the testing as well as to capture and display the test results. Fig. 6 shows an example of a test result displayed in the Lab/Window environment, where the input was a pure sinusoidal wave. The test results were compared byte by byte with those obtained from the behavioural level simulation. It was found that they were the same except that there were a few bytes with a different least significant bit.

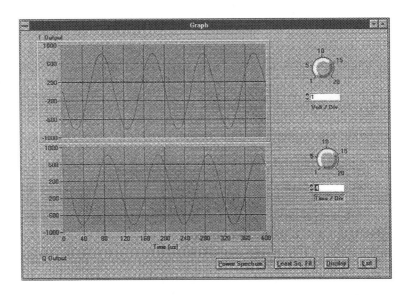

Fig. 6. Test results displayed in LabWindow environment

6 Conclusions

The implementation of a digital IQ demodulator in FPGA devices was achieved and presented. The DSP algorithm representing the demodulator was coded in behavioural VHDL and synthesised into FPGA devices by using behavioural synthesis software. Several design architectures were explored and problems of data accuracy were investigated before the final architecture was chosen. Design partitioning was carried out at the behavioural level and different design partitions were studied. The final implementation was built and tested. The test results were the same as those obtained from the simulation. In this project, a technique for rapid prototyping of DSP-based ASICs in reprogrammable FPGA devices using behavioural synthesis approach was experienced and the problems encountered were investigated and solved.

References

1. Sjoholm S., Lindh L.: VHDL for Designers. Prentice Hall Europe (1997)
2. De Micheli G.: Synthesis and Optimization of Digital Circuits. McGraw-Hill Inc. (1994)
3. Ifeachor E. C., Jervis B. W.: Digital Signal Processing: A Practical Approach. Addison Wesley (1993).
4. Synopsys: Behavioral Compiler User Guide. Synopsys Inc. (1996)
5. Synopsys: FPGA Compiler User Guide. Synopsys Inc. (1996)
6. Xilinx: The Programmable Logic Data Book. Xilinx Inc. (1996)

Hardware Compilation, Configurable Platforms and ASICs for Self-Validating Sensors

Ian Page

Oxford University Computing Laboratory, Parks Road, Oxford OX1 3QD, U.K.

Abstract. We describe the use of hardware compilation techniques to provide a flexible interfacing and computing platform for building self-validating (SEVA) sensors. SEVA technology has been developed at Oxford as a generic method of supplying a real-time guarantee on the quality of a physical measurement as well as the measurement itself. We describe the reconfigurable hardware platforms that have been constructed to support this work and the development tools and techniques used to generate new sensor implementations quickly and reliably. We further describe the compilation of SEVA systems into RISC-based ASIC technology for mass market applications.

1 Introduction

Self-validating (SEVA) sensor technology has been developed at Oxford over the last nine years as a generic method of supplying a real-time guarantee on the quality of a physical measurement as well as the measurement itself. A major benefit of the approach is that it gives a system-level framework for building fault-tolerant control systems. Such systems require an intimate knowledge of the state of all sensors and effectors both during normal operation and especially in abnormal circumstances so that the appropriate system-level response can be made. The sort of application areas which are intended include chemical process plants, oil rigs and aircraft as well as much smaller scale systems.

These application areas are already being transformed by the introduction of new technology, and 'smart sensors' in particular. For instance, a traditional temperature sensor might employ a thermocouple element and analogue circuitry to turn the varying resistance into a 4-20mA current (representing an arbitrary device-independent scale of 0-100) for onward transmission to the controller at some remote location over dedicated analogue wires. A corresponding 'smart sensor' implementation might employ a local microcontroller to perform self-calibration, maintain an operational history, digitize the thermocouple value locally, process it to obtain a true temperature reading and pass this reading on to the controlling computer over a digital network.

A SEVA sensor will in addition perform detailed checks on the physical device, for example by periodically driving current into a thermocouple and using it as an output device. By examining the response, it can be determined if the thermocouple is open or short circuit, or has lost thermal contact with the substrate being monitored. A SEVA system contains a comprehensive model of the sensor and its environment. For example it might be capable in real-time of detecting a fault in one physical sensor from an ensemble, inverting the model equations with respect to that sensor and continuing to give a 'best efforts' reading with a new, but somewhat poorer, guarantee of quality.

Over a similar timescale the Hardware Compilation Research Group working at Oxford University Computing Laboratory had been developing techniques for compiling programs into hardware implementations. The implementations are realised using FPGA (Field Programmable Gate Array) chips on various hardware platforms that we have constructed. Hardware compilation is the term we use to describe the fully automatic implementation of hardware from a program. It differs from high-level hardware description languages such as VHDL in that the intended users of the language are programmers, not hardware designers. The hardware compilation techniques are described in [8, 9] and other publications.

In 1994 the two groups began to collaborate to see if hardware compilation techniques could be used to produce prototypes of industrial sensor and control systems both quickly and reliably. The results of the collaboration have been wholly positive for both groups and have led to very much faster prototyping and increased re-use of design effort between applications.

2 Hardware Compilation and the Handel-C Language

The Handel-C language has been developed at Oxford for describing programs which are (usually) compiled into hardware. Handel-C is a small subset of C, extended with the parallelism and communication constructs of Hoare's CSP and expressions of arbitrary bitwidth. The language is necessarily different from C itself because of the different nature of hardware and software implementations.

C is not adequate for our task because it is a sequential language. This means that algorithmic parallelism can't be expressed in C programs. However it is in the very nature of hardware implementations that they achieve their speed through exploiting parallelism. It is beyond the state of the art automatically to introduce the appropriate parallelism into a sequential program. For this reason we expect the programmer explicitly to denote the parallelism that is appropriate for the desired hardware implementation.

Programs, written in Handel-C, are mapped into hardware at the level of netlists by a series of transformations [9]. The implementations fully exploit the explicit parallelism of the programs themselves and serial execution is only introduced where explicitly required. We have successfully used the system to compile and execute applications in real-time video, machine vision, data compression and other applications, including a real-time video object detection and tracking system [10].

3 Reconfigurable Hardware Platforms

The majority of industrial sensors are powered over the communication medium itself. These devices tend to be constrained by cost and power consumption. The most sophisticated sensors enjoy separate power supplies; in this smaller market segment, functionality is more important than power consumption or cost.

Partly to address this range of issues, two separate architectures were developed. The first architecture is a low-cost ISA bus FPGA-only card aimed at simple validation applications. Each card can interface to a daughter-card holding application-specific circuitry and components. The second architecture is a transputer-based card for the more sophisticated applications.

It is clearly desirable to minimise the specialist knowledge required to develop applications. We have been tackling this issue on a number of fronts. First and most obvious is the use of a programming language rather than a hardware description language to design the hardware processing units. Secondly we have developed a library of interfaces for commonly used input/output chips. Thirdly we have built higher level Handel library routines (including RMS-DC conversion, high accuracy frequency measurement and noise filtering). We also use National Instrument's Labview for front-end prototyping and our long term goal is to provide all host functionality within Labview, so that any user will be provided with a well-documented and supported front-end tool.

To date the two groups have collaborated on the development of three successful SEVA applications using this technology: a self-validating thermocouple [1], a dissolved oxygen [3], and a Coriolis mass flow [6] metering system. Although the mass flow meter is the most complex and impressive system, it has been particularly gratifying to be able to demonstrate that an entire sensor validation application (the thermocouple) can be coded entirely within Handel and implemented within FPGAs.

In the following sections we give details of the various hardware platforms that we have developed to support this work. Some of these platforms are now available from commercial suppliers which is a good indication that the research work we have been doing has been well focussed and is industrially relevant. We also detail ongoing work which we hope will give us a route into single chip implementations of even quite complex self-validating sensors.

3.1 The 'Valcard' Reconfigurable Platform

The Valcard system was our first-generation reconfigurable platform. It is an ISA bus card which hosts one or two 3000 series Xilinx FPGAs, usually XC3195A parts. It has an internal 16 bit bus which is used for communication between the FPGAs and with the daughter-card, which holds application-specific circuitry. It provides hardware support for interfacing with the PC bus, and Handel routines are available for creating software ports in the PC I/O address space, allowing the transfer of data between PC host software and Handel applications.

The board carries a pair of daughterboard connectors. Typically, a small and very simple daughterboard would be constructed for each new sensor or control interface. This board would carry the application-specific components which could not be part of the general-purpose motherboard, such as AD converters, perhaps a clock, level converters, and external connectors. The FPGA chips on the motherboard usually hold the kernel algorithms for handling the sensors, processing the resulting data, and communicating results back to the host PC. The Valcard system is designed to be cheap, easy to manufacture, and reusable. Further details on this card and all our other hardware platforms can be seen at the world-wide web site for the Oxford Hardware Compilation Group.

Valcard was used in two SEVA applications. The thermocouple project [1] demonstrated that a complete, if simplified, SEVA sensor (including measurement and uncertainty calculation, and fault detection and compensation) can be defined in Handel code and implemented within FPGAs. This implementation proved to be a powerful demonstration to industrial audiences that SEVA is implementable within commercial technology.

The other application is a dissolved oxygen sensor [3]. Here, Valcard provides a simple interface via the PC ports to data from a dissolved oxygen probe and the larger validation program was implemented primarily in software. The interface requirements included controlling a sigma-delta A-D converter, providing a drive current to stimulate the transducer, and operating switching circuitry to control an active diagnostic test.

The availability of the Valcard FPGA board has also stimulated other groups to use it, for instance in a high-speed decoding project in the Engineering Science Department. To show its versatility, we have also demonstrated a number of Handel-C applications which directly generate video signals from the FPGA.

3.2 The RC1000-II Reconfigurable Platform

A second-generation version of Valcard has recently been designed and is being offered for sale commercially by Embedded Solutions Ltd. as the RC1000-II. It differs from the prototype card by being based on the Xilinx 4K part and by carrying a daughterboard interface to the international Industry Pack standard. The intended application areas of the board are high performance signal and control processing, integer/logical co-processing for the PC, legacy hard logic system replacement, and algorithm modelling and architecture design.

The RC1000-II is a standard half size ISA bus card equipped with a Xilinx XC4000E series FPGA in PLCC-84 format. This is socketed, and may be replaced by any pin-compatible FPGA from the Xilinx XC4003E to the XC4010E device range. It is programmable with the Handel and Xilinx XACTstep tools, and a PC provides server microprocessor support. An on-board serial PROM can be used to configure the FPGA for dedicated run time applications. Otherwise the FPGA can be configured over the ISA bus using the PC-based interface programs provided.

The board is equipped with an Industry Pack mezzanine daughter-board slot carrier directly interfaced to the FPGA. The Industry Pack interface is essentially a 16 bit, memory-mapped interface working at either 8 or 32 MHz. A choice of over 400 I/0 modules are commercially available for Industry Pack, covering the needs of most prototyping and low volume bespoke application needs. This mezzanine format is ideal for prototyping custom I/O or additional hardware assist circuitry for the FPGA, such as SRAM or graphics accelerators. In addition general-purpose input/output capability is provided by a 50-pin SCSI-2 style panel connector connected to the FPGA.

The board has a programmable clock which operates over the range 400KHz to 100MHz and can be configured from the host PC. In addition the 33MHz ISA clock, and the Industry Pack 8MHz and 32MHz clocks are available for clocking the FPGA.

In a typical sensor application, this board would be equipped with a suitable Industry Pack module or the general-purpose digital input/output connector would be used to connect to the physical sensors. A control program would be written in Handel-C and compiled to a netlist by the Handel-C compiler. The Xilinx tools would then place and route this netlist and produce a bitmap file for loading into the FPGA. For in-PC operation, the user would probably want to write a PC program which loaded the bitmap into the FPGA over the ISA bus and then went into the main application code. This code would interface with the Handel-C program running in the FPGA to obtain the sensor data after processing. Data exchange between the PC and the FPGA is mediated by a set of software routines and can be run as either an interrupt or a polling interface.

3.3 The 'ValTram' Reconfigurable Module

The 'ValTram' system [7] is based on the Tram module standard originated by Inmos Ltd. It was designed to support the rapid implementation of smart, self-validating sensor systems and conventional control systems through the use of FPGAs and hardware compilation. The ValTram system consists of two daughter boards each of which simply plug into any industry-standard Tram motherboard, or into each other, or into a custom-designed motherboard. These modules are also now available in commercial form from ESL Ltd.

The smaller ComTram (Communication Tram) module is a size 1 Tram which converts a standard transputer 20Mbit/sec serial link into two parallel buses. These buses are linked to as many FPGA modules as necessary by ribbon cable. One bus is used to configure the FPGAs and the other is used to communicate between the host transputer(s) and the FPGA modules once configured. The configuration bus can also be used as an inter-FPGA communication bus.

The larger (size 2) FPGA-Tram module has sockets for up to two Xilinx X3000-series FPGAs. As well as linking to a ComTram module via the ribbon cable buses, it also has a daughterboard connector on which the sensor or control interface can be mounted. Examples of daughterboards which have been constructed range from simple sigma-delta analogue-digital converters up to moderately large interfaces which have their own reprogrammable FPGAs on board.

The Coriolis mass flow meter is an example of one of our more complex sensor systems. The base is a standard, commercial, 10-slot, PC Tram motherboard. The rest of the system is stacked on this mother board and consists of i) three standard, size-1, transputer-based, 4Mbyte compute Trams, ii) one ComTram module, iii) three FPGA-Tram modules driven from the one ComTram, and iv) three application-specific daughterboards on the FPGA-Trams, two of which happen to have their own FPGAs on board. The application is quite a complex one as can be judged from the fact that the implementation has three separate clocks simultaneously active, two active buses, and a total of 5,000 lines of Handel code mapped into hardware implemented in the eight FPGAs.

When running an application, the transputer link is used initially to download FPGA configuration data, and then subsequently to provide data transfer between the Valcard modules and the transputer. Data transfer is controlled by the transputer, using a simple protocol. Bus control is itself almost entirely implemented via Handel, and so can be redefined to suit the requirements of each application. The implementation has been very successful and a commercial partner is currently funding extended industrial trials.

3.4 Aspire Project and ASIC implementations

Our reconfigurable hardware platforms have proved suitable for rapid prototyping of complex sensor systems. However they would usually be too expensive for deployment for industrial control purposes, except perhaps for high end, self-powered systems such as the mass flow meter. For low end, high volume, cost sensitive applications it is necessary to find alternative implementation techniques.

One easily available and attractive route is simply to take the configuration constructed for the FPGA implementation and to transfer it to a commercial gate array technology. In our case the move to Xilinx HardWire technology is particularly simple as functional equivalence is guaranteed by the vendors. The natural implementation of the core of a SEVA sensor would then be a microprocessor system connected to suitable memory components and to the gate array chip(s). Depending on system complexity the processor system might be anything from a single microcontroller to an array of DSP processors. However, in the low cost, simple implementations which are likely to make the earliest impact on the mass market, it is likely that a high-performance microcontroller would be an adequate basis.

It is obviously attractive to consider single chip implementations for high volume systems. This necessitates having a microcontroller core, memory resources and the application-specific circuitry on the same silicon. We have already begun a collaboration with ARM Ltd. and Atmel-ES2 under the European Union OMI (Open Microprocessor Initiative) to develop a route from our language directly into ARM-based ASICs. This 'Aspire' Project has the ultimate target of the compilation process being an ASIC which has an ARM processor core, memory resources, network communications, and the designer's application-specific hardware on the same chip. Given the high MIPS/area and MIPS/watts ratings of the ARM devices, the resulting ASICs should have exactly the right characteristics for the control industry.

The goal of the Aspire project is rather wider than the compilation of RISC-based ASICs however. It is to build a 'shrink-wrap' product for hardware-software co-designers which will allow them easily to design ASICs for embedded computing systems based on ARM processor technology. The Aspire package will consist of a reconfigurable processor system on a PCI-based board, support software including a hardware compiler, and user documentation to enable a programming-literate designer to design such embedded ASICs.

A block diagram of the reconfigurable processor system is shown in Fig. 1. It will be capable of being installed in any standard PCI-based host system. It will be based around the AT91M40200 chip [2] from Atmel-ES2 which is their new microcontroller version of the ARM700TDMI. The box on the left represents the chip boundary of this part. The 'Peripherals' box contains standard timer and interrupt controller modules. The processor, peripherals and small on-chip memory communicate using ARM's on-chip AMBA bus.

Memory resources are expanded on the board-level bus with a bank of DRAM and flash ram. The flash ram will be particularly important for stand-alone uses of the board and it is being designed so that it can operate quite sensibly independently of any PCI host system.

The reconfigurable hardware will consist of three or four large, fast Xilinx 4k series parts one of which supports the PCI interface chip. The FPGAs are bussed together as shown and expansion is provided by daughter-board and back-panel connectors. It is envisaged that an early daughterboard will be one with further FPGA chips.

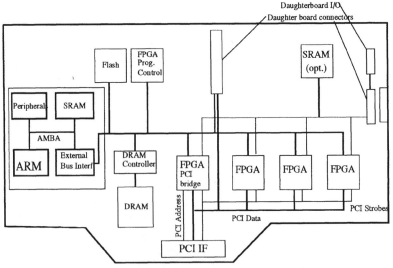

Fig. 1. The Aspire ASIC Prototyping Board

Our support software will compile Handel-C programs into FPGA configurations and will be able to merge these configurations with C or C++ programs for the microprocessor. We have built a co-simulation system which runs C code on the Armulator against a software simulation of the hardware generated by our compiler. This enables us to prototype complex ARM-based systems on a standard workstation with no special hardware. This is clearly important to support SMEs who want to build, or evaluate the building of, ARM-based systems.

After software-based co-simulation, the next step might usually be to implement the same design on the PCI board. We have already taken a number of small hardware/software codesigns through our Handel-C compiler and have them working on the ARM PID7T and PIE boards which we are using as prototypes until the PCI board is completed (in 4Q97). The PCI board is now being designed by ARM to an architectural plan produced jointly by all Aspire partners.

When the user has a co-design operating satisfactorily on the PCI card or similar reconfigurable platform, we produce design files for shipment to Atmel-ES2 from which they can manufacture a single custom ASIC which includes the ARM core together with the application-specific hardware generated from the C-like program. The overall architecture of the chip produced by this process is shown in Fig. 2.

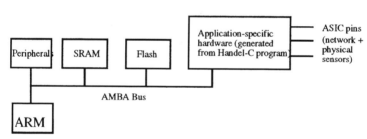

Fig. 2. The Aspire ASIC : General Architecture

We have recently demonstrated a complete flow from our hardware compilation system into ES2's ASIC flow. The result is a hardware design which incorporates our application-specific hardware equivalent to the original Handel-C program, an ARM700TDMI processor core and the interface hardware between the two. This has resulted in satisfactory hardware simulations of the entire system within the ES2-Atmel EDA environment. The only stages we have not yet tested out are final placement and routing and actual manufacture and test. We expect these to be straightforward and so the next major step will be the manufacture and demonstration of an ARM-based ASIC from our Handel-C flow. We are currently planning this design but we have already settled on making it a video-based application.

4 A Large Scale Application

Here we present an example of our Handel-C software and SEVA platforms in use independently of the Oxford groups. Dr. Timo Korhonen at KEK (High Energy Physics Accelerator Research Organization, Tsukuba, Japan) is building a large scale position control system for their Accelerator Test Facility.

This is a positron-electron collider being built to evaluate some of the design considerations for a proposed large linear collider. It consists of 1.54 GeV S-band Injector feeding a damping ring which has a racetrack shape with a circumference of 138.6 m.

The damping ring uses up to 40 ValTram systems to implement dynamic stabilisation of the beam path via the movable alignment tables which carry the beam steering magnets. The positioning of the tables is done with an accuracy better than 20 microns and it is important to remove the effects of traffic-induced vibration and other earth tremors on the apparatus.

It was very pleasing that within four weeks of receiving the system, Timo had interfaced our reconfigurable hardware to the controllers for moving the large magnet-carrying tables, had written Handel-C programs to implement the control algorithm and had the whole system working. His comment that "after two or three days, I completely forgot that it was hardware that I was developing" was especially good to hear.

Even using rather slow Xilinx place and route software, he was still able to test a number of new hardware configurations within a working day. This flexibility and rapid implementation is especially important in an experimental set-up such as this, when the best control algorithm cannot be known in advance of building the system. He is now planning to use this flexibility to enhance the sensing and control systems beyond original brief because it is easy to do so.

5 Conclusions and Future Work

Over the next two years we will be designing and building new hardware platforms to support the next generation of self-validating sensor work. For this next phase of work we have identified Fieldbus [5, 4] and PCI bus as important interfaces that need to be supported by our new platforms. We are already confident that these research developments will also result in commercial exploitation.

We are extending the Handel language to raise its level of abstraction while retaining its properties of accurate specification of the temporal properties of programs. We are recoding the compiler to take advantage of the knowledge accumulated by the group through the use of the previous compilation system over the past five years.

Our experience with developing sensor applications has suggested that additional hardware components could be also described within the Handel framework. These include the new generation of 'analogue' FPGAs (e.g. the EPAC chips from IMP Inc.), which are programmed to provide analogue signal conditioning, and programmable inter-connect chips (e.g. Aptix). We hope to extend our compilation tools to target these and other types of programmable hardware components.

We hope to develop some much higher bandwidth smart sensors (not necessarily self-validating sensors) which use video from one or more cameras and execute quite complex algorithms in hardware and software to produce the sensor outputs. One reason for choosing applications such as video-based detection, tracking, and identification is to improve our technology through meeting these challenges. Another is the observed power of video-based demonstrations to effectively carry the message of what our technology is and what it can do.

Acknowledgements

I would like to thank all the members of the SEVA group in the Department of Engineering Science at Oxford. I would also like to thank Dr. Timo Korhonen of KEK for his permission to quote details of his work in this paper. Part of the SEVA work has been undertaken as part of a previous grant (GR/34436) and part is being undertaken under a current grant (GR/L41547) for which I wish to acknowledge the support of the UK EPSRC. The Aspire project is supported by the EU OMI office and by ARM Ltd. and Atmel-ES2 which I also wish to acknowledge.

References

1. M. Atia, J. Bowles, D.W. Clarke, M.P. Henry, I. Page, J. Randall and J.C. Yang. A self-validating temperature sensor implemented in FPGAs. In W. Luk and W. Moore, editors, *Field Programmable Logic and Applications*, Lecture Notes in Computer Science. Springer Verlag, 1995.
2. Atmel ES2, Zone Industrielle, 13106 Rousset Cedex, France. *AT91 Series Microcontrollers - Product Overview.*
3. D.W. Clarke and P.M Fraher. Model-based validation of a DOx sensor. *Control Engineering Practice*, 4(9):1313–1320, 1995.
4. M. Henry. Fieldbus and sensor validation. *IEE Computing and Control Journal*, 6(6):263–272, 1995.
5. M. Henry. Smart field devices and sensor validation. *IEE Computing and Control Journal*, 6(6):263–272, 1995.
6. M. Henry and D.W. Clarke. The self-validating sensor: rationale, definitions and examples. *Control Eng. Practice*, 1(4):585–610, 1993.
7. M.P. Henry, N. Archer, M. Atia, J. Bowles, D.W. Clarke, P. Fraher, I. Page, J. Randall and J.C. Yang. Programmable hardware architectures for sensor validation. *Control Eng. Practice*, 4(10):1339–1354, 1996.
8. C.A.R. Hoare and Ian Page. Hardware and software: the closing gap. *Transputer Communications*, 2(2):69–90, June 1994.
9. I. Page. Constructing hardware-software systems from a single description. *Journal of VLSI Signal Processing*, 12(1):87–107, 1996.
10. Ian Page. Video motion tracking using Harp and Handel: A case study. In *Proc. 7th Annual Advanced PLD and FPGA Conference*. Miller Freeman PLC, London, SE18 6QH, May 1997.

PostScript™ Rendering with Virtual Hardware

Satnam Singh, John Patterson, Jim Burns and Michael Dales

The Department of Computing Science
The University of Glasgow
Glasgow G12 8RZ, United Kingdom

Abstract. This paper presents the techniques being developed to design a high speed PostScript rendering engine using virtual hardware. The problem of rendering PostScript is decomposed into a series of tasks which can be performed in sequence on one FPGA using virtual hardware.

1   Introduction

A proto-type system called FPGA/PS for accelerating the rendering of PostScript™ [1] printer jobs using FPGA technology is presented. We argue that PostScript rasterisation is a task highly suited to implementation on an FPGA subsystem connected to a host workstation. We describe a sequence of compute intensive operations that are realised by just one FPGA on an off the shelf system which is dynamically reconfigured for each stage in the processing pipeline.

A novel aspect of this system is that it exploits dynamic reconfiguration of Xilinx XC6200 FPGAs [17] to swap in various circuits for coordinate transformation, rasterisation, image processing, colour correction and compression. Together these circuits can not fit onto one XC6216 FPGA (64 by 64 cells, 16,000 equivalent gates). By swapping in only the circuit needed at a particular moment, we can realise a large virtual circuit on a limited physical resource. We call this approach *virtual hardware*.

PostScript is a page description language that specifies what is to be drawn on a printed page. Many printers accept PostScript files and interpret them to generate a rasterised image. Laser printers typically rasterise to a resolution of about 600 dots per inch (dpi) but production quality output for publishing requires a far greater resolution. Rasterising colour images for publishing at 5080 dpi can takes hours of processing on dedicated Raster Image Processors (RIP). This paper describes a project with the objective of realising the compute intensive aspects of PostScript rendering on FPGAs. Once completed the proposed system should deliver high speed PostScript rendering at a fraction of the cost of today's dedicated RIP systems.

2   System Overview

The work flow through the FPGA/PS system is shown in Fig. 1. The system's input is a Postscript job. Rather than interpret the page description commands in the Post-Script job we convert the PostScript to a Portable Document Format (PDF) file using an off the shelf package, Adobe Distiller™. The PDF form of the PostScript commands are then rasterised using the virtual hardware subsystem. Finally the level 2 PostScript bitmap is sent to the printer via a fast network connection.

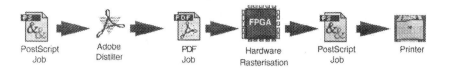

Fig. 1. The FPGA/PS System

PostScript Portable Document Format contains a static description of what is to be rendered on each page. There are no variables or programming language constructs like function definitions or loops. This makes PDF a far more suitable input for the core graphics rendering system. Our system uses Adobe's Acrobat Distiller package to translate PostScript jobs into PDF. We then pick out the graphics commands for each page and render the compute intensive tasks on hardware and the remaining graphics operations in software.

3   The PostScript Language

PostScript was developed by Adobe Systems as a page description language and it is commonly used to communicate with printers. It is essentially a programming language that uses variables and functions and a stack based mechanism to *calculate* the commands needed to specify the contents of a page. Variants of PostScript (called Display PostScript) have been used as the primary graphics programming language for some computer displays, notably the News system from Sun and on the early NeXT computers. Computer displays are typically around 80 dpi which makes software PostScript rasterisers viable. We concentrate on the acceleration of PostScript rendering for high quality printed output which is far more compute intensive.

The first stage in processing a PostScript job is to interpret the PostScript to yield a sequence of graphics objects which have to be realised in a grid of squares (each square corresponds to a pixel on a display or a dot produced by a printer). A simple PostScript file is shown below and the picture shown in Figure 2 is drawn by these commands.

```
0 0 moveto
0 90 lineto
stroke
50 50 40 0 360 arc fill
100 10 150 {dup 0 moveto 90 lineto stroke} for
newpath 160 50 moveto 200 90 230 10 260 50 curveto stroke
/Helvetica findfont
10 scalefont setfont
160 10 moveto
(A PostScript Example) show
showpage
```

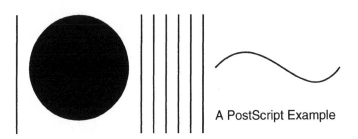

Fig. 2. Sample PostScript Output

The commands are used to build a *path* which describes the motion of a notional pen laying paint on the output page. The first moveto command states that the drawing path should start at (0,0). Literals are pushed onto a stack and PostScript™ commands work by removing their arguments from the stack. The lineto command adds the point (0, 90) on the path. The stroke command connects the points on the drawing path with a straight line i.e. a line from (0,0) to (0,90) which is the left most vertical line in Figure 2. The exact shape of the line is determined by some global graphics context information. This specifies properties like the width of a line, its cap style and colour.

The fourth line specifies an arc centred at (50, 50) of radius 40 that sweeps from 0° to 360°, i.e. a complete circle. The fill command is used to render the arc by filling in the circle. The fifth line gives an example of a drawing specified algorithmically in PostScript. The stack is assumed to contain the initial value of the loop counter (100), the increment value (10) and the final value (150). The body of a for loop is shown in braces and is executed for each value in the specified range i.e. 100, 110, 120, 130, 150. The body of the loop duplicates the loop control value x, uses that value to render a line from $(x,0)$ to $(x,90)$. The remaining commands specify a cubic-Bézier curve, select a font, and render text. The contents of a page are not actually rendered until a showpage command is encountered.

Rather than writing our own PostScript interpreter or modify an existing one for hardware acceleration (a major undertaking) we have decided to make use of another new technology from Adobe called Acrobat which provides support for Adobe's Portable Document Format (PDF) [2].

4   The Portable Document Format (PDF)

PDF is essentially PostScript with extensions to facilitate better interactive file viewing. PostScript is page dependant and so the whole file may require interpretation to resolve variables towards the end of the file. This can often be very slow. In contrast PDF files contain a reference table which points to the graphics *streams* which contain drawing or *path operators*. The index table and the lack of functions and variables ensure the greater performance of PDF viewing. PDF and PostScript use the same imaging model i.e. the page is a grid of dots, which means that any PostScript program can be described using the less complex PDF format.

Our system employs the Adobe Distiller system, called using a programming control interface, to convert the PostScript file into a PDF file. This is a relatively swift process because no rasterisation takes place. After some colour correction information, the PDF file specifies the graphics stream to be rendered. Each of the coordinates for graphics commands is in user space and must be further manipulated by applying the *current transformation matrix* (CTM) to calculate the actual coordinates to be used for rendering on the printed page. Coordinates are multiplied by the CTM to effect translation, scaling, skew and rotation. PDF also supports text space, character space, image space, form space and pattern space.

PostScript and PDF contain a rich collection of commands for specifying colours, for applying transformations to coordinates and for processing bitmaps. PostScript also provides support for colour calibration and compression.

5   Rasterising PostScript

The FPGA/PS system is design to accelerate the following computations that are required for PostScript rendering:

- **Matrix-based coordinate transformations.** Every coordinate in the PDF file must be transformed with the CTM matrix.
- **Rasterisation of lines.** We have hand designed a line rendering circuit, and we are refining it further to improve its performance. In the proto-type system, this is the only circuit that actually performs rasterisation. Factors to take into account include line thickness, cap-styles, line colour and anti-aliasing.
- **Rasterisation of Bézier curves.** These curves are rendered by calculating a series of lines that approximate the curve (depending on the slope of the curve) and then using the line rasterisation circuit to actually render the curve.
- **Rasterisation of fonts.** We use descriptions of fonts based on.
- **Anti-Aliasing.** Jaggies from lines, curves and fonts are smoothed out.
- **Colour correction.** Rendering algorithms typically specify colours using the RGB model. To allow for faithful reproduction on printed output, we are working on colour correction support which will read device profiles from the Post-Script/PDF file and perform compute intensive colour correction calculations on the FPGA subsystem.
- **Bitmap Compression.** The output of our system is a Level 2 Postscript bitmap. The reduce the size of the bitmap (thus reducing the time it takes to communicate the bitmap to the printer) we will apply LZW compression, implemented on the FPGA.
- **RAM Bitmap to SCSI Output.** The generated PostScript bitmap file will be copied from RAM directly to the SCSI port and then sent to a production quality printer. This step avoids the need to save a large bitmap file to secondary storage.

Other aspects of PostScript are realised in software, either because some tasks are outside the original specification of the proto-type system or because some tasks are not suitable for hardware implementation. Recent advances in microprocessor speed and functionality have made certain kinds of hardware acceleration no longer viable. We have been careful to identify tasks which still benefit from hardware acceleration and to use microprocessors whenever appropriate.

5.1   Pre-Processing PDF

The original PostScript path description commands are read from the PDF file produced by Acrobat Distiller by a software parsing program. The PDF file is read sequentially and each path operator together with its operands are stored in a list. The elements of the list cannot be reordered or have duplicates removed because the PostScript imaging model specifies that *paint* is applied to the page in layers, so the last operation may cover all lower layers. Some speed up can be achieved by looking ahead and calculating which hardware circuit is required and for how many operations.

5.2   Coordinate Transformations

Multiplication of every coordinate by a six element transformation matrix by the microprocessor would take much longer than the rasterisation of some graphics commands like thin line rendering. However, for more complex rendering operations, if these multiplications were carried out using the arithmetic units of a Pentium processor, then the time taken in software to perform the matrix transformation might be rather small. Furthermore, it would yield more space on the FPGA to optimise rendering circuits (space which would have been taken up with a matrix transformation circuit used only briefly at the beginning of each rendering command). Alternatively one could 'swap out' the matrix circuit after each use, but our intuition is that the swapping overhead will be too large to merit this approach. Currently we are investigating two approaches: (i) perform the matrix multiplications only in software; (ii) have always-resident matrix multiplication hardware on the FPGA. After profiling we shall adopted the technique which delivers the best performance.

We have adopted a fixed page size (A4) and an upper limit of 1693 dpi so we have known ranges for x and y coordinates (13544 for x representable in 14 bits, 20316 for y representable in 15 bits). For simplicity and ease of conversion, we have adopted 16-bit signed arithmetic possible for both x and y coordinates. The scaling of the floating point values in the PDF file to 16-bit signed values is performed in software.

5.3   Rasterising Lines

Hardware implementations of circuits for scan converting lines are common, and we base the design of the line rasteriser used in FPGA/PS on the standard midpoint line scan-conversion algorithm. This is a variation of the classic line drawing algorithm developed by Bresenham [3][4] which uses only integer arithmetic. This algorithm is suitable for rasterising "thin" lines which have no special cap style. The algorithm can be used to help rasterise lines of a given thickness in conjunction with a specialised circuit (based on a counter) which draws horizontal or vertical lines.

The circuit that corresponds to this algorithm can be easily designed using shifts and adds to set up the initial register values and then using text-book designs for the comparator and adder circuits. Some additional circuitry is required to normalise and reflect the line into the working range of the midpoint line circuit.

The line rasterisers output is a pair of coordinate values that specify a dot on the printed page and a colour value for that dot. These outputs are connected to a circuit which translates the x and y coordinates into an address into the on-board 512K fast SRAM on the XC6200/PCI board. The 16-bit colour value is written at the given address. The address management circuit is always resident and it has an interface laid to facilitate direct connection to a variety of rendering circuits that can be swapped into the core of the FPGA.

5.4 Rasterising Circles

Arc commands in PostScript (of which circles are a special case) are translated into Bézier curves in PDF. This means that certain kinds of curves are represented approximately with Bézier curves and circles are known to be difficult to represent adequately with Bézier curves. We aim to spot circles that have been translated into four Bézier curves and then use a specific circle rendering circuit.

Circles are drawn with an implementation of the midpoint circle scan-conversion algorithm with generates 1/8th of a circle outline. This circuit can be used in conjunction with a register containing the colour of the circle and an address generation circuit to render a circle into the 512K on board SRAM.

A structural VHDL description was also produced and this was used to specify the layout of the circuit on an XC6216 FPGA, shown in Fig. 3. The circle circuit itself occupies less than 10% of the chip area. A reflection circuit is also implemented which can reflect the coordinate computed in one octant of the circle to obtain 7 other points on the circle.This version of the circle circuit renders the circle into a small VRAM which is implemented on the FPGA itself.

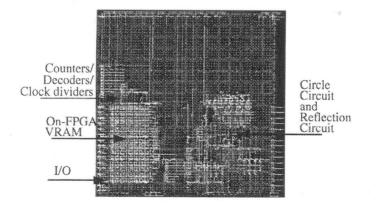

Fig. 3. A circle circuit on an XC6216 FPGA with on-chip VRAM, clock dividers, decoders, reflection and I/O

The structural VHDL was translated into EDIF using our own VHDL to EDIF compiler which understands VHDL attributes specifying pragmas like layout and wiring criticality (using the standard Xilinx RLOC-style pragmas).

5.5   Rasterising Cubic Bézier Curves

Rendering curves has been identified as the most important graphics operation for hardware acceleration. Industrial scale technical drawings can contain hundreds of thousands of curve segments that describe the form of surfaces e.g. car bodies. Furthermore, PostScript font outlines are specified with Bézier curves.

Rasterising curves is computationally intensive compared to rendering lines or pre-calculated fonts. PostScript supports several commands that specify curves (arcs and Bézier curves) but PDF only supports Bézier curves (arcs can be represented with Bézier curves).

Bézier curves are an instance of parametric cubic curves which specify a curve segment by its end points and two control points which are used to guide the shape of the curve. This makes Bézier curves suitable for *ab initio* design of free-form curves because the control points provide the intuitive feel for the flow of the curve. The bounding bow of a Bézier curve is always inside the convex hull of the control points making clipping operations trivial. The curve can be efficiently drawn using differences. We have already used a differencing technique to design and implement a circuit for rendering spheres on FPGA hardware [12].

Another issue to how to determine how many line segments to use to render a given curve. The steeper the curve, the more line segments are required for a reasonable approximation. A standard collection of adaptive techniques for solving this problem in software can be readily applied to FPGA hardware. Furthermore, since many adaptive calculations involve repeated multiplications with known coefficients, we can exploit a dynamically reconfigurable multiplier design [8] for the XC6200 FPGA. This is a very compact table-based multiplier that can be rapidly reconfigured.

5.6   Rasterising Fonts

Rendering fonts typically requires rasterising a font at a particular size to generate a bitmap (held in a font cache) and then copying this bitmap to a portion of the image buffer. The shape of a PostScript font is described with a series of Bézier curves (see Section 5.5). We propose to attempt font rendering in hardware, but we suspect that because of the large amounts of data that has to be moved from the host memory to the FPGA subsystem memory, it may be more economical to perform font rendering in software and use hardware support to accelerate the copying of font bitmaps (e.g. a DMA device). This is another case where software preprocessing might be more appropriate. Since Bézier curves are part of the core system, however, we plan to experiment with hardware font rendering with a view to profiling it and assessing its suitability for FPGA implementation.

5.7   Anti-Aliasing

Rasterisation of continuous smooth objects like lines, circles and fonts results in jaggies or staircasing. This effect is called *aliasing*. This problem occurs because the colour of the target object is simply copied to the target pixels, leaving rough edges. The are a variety of anti-aliasing techniques which smooth out jaggies, and we propose to use the Gupta-Sproull method because it provides an efficient incremental algorithm, even though some fractional arithmetic is required. In the first proto-type of the system no anti-aliasing will be implemented.

5.8 Colour Correction

A colour picture displayed on a computer monitor when printed on a colour printed might be drawn with very different colours. Different devices have different printing characteristics and use different colour models. Computer displays use the RGB (red, green, blue) model which is an additive method describing how to mix light. Colour printing usually uses the CMYK (cyan, magenta, yellow and black) model which describes the amount of colour ink applied to each dot.

The International Colour Consortium (ICC) has defined a device independent colour space known as LAB colours which is influenced by the colour perception of the human eye and can capture all visible colours. The L-value describes the brightness (between 0 and 100) and the A and B coordinates describe the distance of a colour from a reference white point.

We could perform colour correction dynamically, manipulating each colour value on the fly before it is written to the 512K image data SRAM. However, we expect it to be more economical to apply colour correction to the final RGB image data for a page. Instead of generating a colour corrected RGB version of the image data, we propose to generate a CMYK colour corrected compressed PostScript binary image which will be output directly to the printer.

5.9 LZW Compression

The output of the rasterisation process is a colour corrected CMYK 1693 dpi image in the host's memory space. In order to speed to transfer of the image data to the printer, a circuit for performing LZW compression will be swapped into the FPGA (binary level 2 PostScript supports LZW compression). This will compress the image data and send it directly to the SCSI interface. Once again, the FPGA resource is being reused with a small reprogramming cost (dynamic reconfiguration with a fixed circuit) for yet another task which accelerates the printing process. An FPGA solution allows us to implement a parallel, pipelined LZW engine which should be competitive against an MMX style microprocessor running LZW in software.

6 Design Flow

All the circuits for the FPGA/PS system have been designed in structural VHDL and translated into EDIF netlists by our own VHDL to EDIF compiler. This compiler understands attributes for the specification of layout as well as specifying other useful information e.g. wiring criticality.

The EDIF netlist is consumed by the XACT Step6000 software which reads in the netlist and performs place and route, generates timing information and produces the programming data for the XC6216 device used.

7 Rasterising PostScript in Software

The rapid increase in processor performance is making many kinds of graphics accelerators obsolete as the main CPU takes on tasks that used to be performed in hardware. Are processors becoming so powerful that it no longer makes sense to invest in hardware solutions for PostScript rasterisation?

The Intel Pentium architecture has been enhanced with new MMX [11] instructions designed to improve the performance of multi-media applications. These include SIMD type instructions that allow eight 8-bit values to be independently manipulated in 64-bit registers and saturation arithmetic which is useful for combining colour values. Other useful operations include saturated arithmetic which is useful for combining colour values. Rather than compute with microprocessors, we aim to exploit their strengths whenever and only resort to FPGA hardware implementation when there is a clear indication of improved performance from a particular architecture.

8 Limitations

The system is described is a proto-type which is still under development. Important aspects of rasterisation have been left to software implementation e.g. clipping. We have fixed the page size to A4 and the dpi to 1693. Although jobs containing PostScript operations that we support in hardware should be processed faster, other types of jobs may be rasterised slower than conventional RIP systems.

If we can establish the viability of the basic concept of using virtual hardware to implement a series of compute intensive image rendering and manipulation tasks, then we shall seek further support to develop a fully-featured commercial PostScript rendering system. This would include supporting more graphics operations in hardware e.g. clipping as well as designing more optimised circuits rather than using the straight forward implementations of arithmetic units as we do now.

9 Summary

PostScript rasterisation for high quality colour output is often performed by a dedicated workstation and can take many minutes or hours to perform. We have shown a decomposition of the problem and mapped it to an FPGA sub-system architecture which works in concert with a microprocessor over a PCI bus. The proto-type is still under development, but we are confident, based on initial designs implemented, that the system will perform competitively with expensive RIPs. The FPGA sub-system is an off the shelf card which can be bought for around $1000US, whereas similar RIP systems cost several thousands of pounds.

The concept of *virtual hardware* is used to dynamically reprogram the FPGA to realise different tasks. On-going research is investigating how to effect the swapping in of circuits at high speed and in a controlled manner. For example, the on-board memory could be used to cache designs, and design information could be compressed before being sent over the bus.

Acknowledgements

The authors would like to thank Graham Borland at Alternate Design & Publishing for information about industrial RIP systems. We wish to acknowledge the support of Xilinx Corp. for software, hardware and technical support. This project is funded by the UK Engineering and Physical Sciences Research Council grant GR/K82055.

References

1. Adobe Systems Incorporated. PostScript Language Reference Manual. Addison-Wesley, 1985.
2. Adobe Systems Incorporated. Portable Document Format Reference Manual. Version 1.2. 1996.
3. J. E. Bresenham. *Algorithm for Computer Control of a Digital Plotter*. IBM Systems Journal, 4(1), 1965.
4. J. E. Bresenham. *A Linear Algorithm for Incremental Display of Circular Arcs*. Communications of the ACM, 20(2). February, 1997.
5. H. Eggers, P. Lysaght. H. Dick and G. McGregor, *Fast Reconfigurable Crossbar Switching in FPGAs*. In, R. W. Hartenstein, M. Glesner (Eds.) Field-Programmable Logic—Smart Applications, New Paradigms and Compilers, Springer Verlag, Germany, 1996, pp. 297-306.
6. James G. Eldridge, Brad L. Hutchings. *RRANN: The Run-Time Reconfiguration Artificial Neural Network*. IEEE Custom Integrated Circuits Conference. 1994.
7. J. D. Hadley, B. L. Hutchings. *Design Methodologies for Partially Reconfigured Systems*. FCCM'95. IEEE Computer Society, 1995.
8. T. Kean, B. New, B. Slous. *A Multiplier for the XC6200*. Sixth International Workshop on Field Programmable Logic and Applications. Darmstadt, 1996.
9. H. T. Kung. *Why Systolic Architectures*. IEEE Computer. January 1982.
10. Charles E. Leiserson. *Systolic and Semisystolic Design*. IEEE Conference on Computer Design/VLSI In Computers (ICCD'83). 1983.
11. Alex Peleg, Sam Wilkie and Uri Weiser. *Intel MMX for Multimedia PCs*. Communications of the ACM. Vol. 40, No. 1, January 1997.
12. Satnam Singh and Pierre Bellec. *Virtual Hardware for Graphics Applications using FPGAs*. FCCM'94. IEEE Computer Society, 1994.
13. Satnam Singh. Architectural Descriptions for FPGA Circuits. FCCM'95. IEEE Computer Society. 1995.
14. M. Sheeran, G. Jones. *Circuit Design in Ruby*. Formal Methods for VLSI Design, J. Stanstrup, North Holland, 1992.
15. Michael J. Wirthlin and Brad L. Hutchings. *A dynamic instruction set computer*. FCCM'95. IEEE Computer Society. 1995.
16. Michael J. Wirthlin, Brad L. Hutchings. *Improving functional density through run-time constant propagation*. To be published in FPGA'97.
17. Xilinx. *XC6200 FPGA Family Data Sheet*. Xilinx Inc. 1995.

P4: A Platform for FPGA Implementation of Protocol Boosters

Ilija Hadžić and Jonathan M. Smith
ihadzic@ee.upenn.edu, jms@cis.upenn.edu

Distributed Systems Laboratory, University of Pennsylvania*

Abstract. Protocol Boosters are functional elements, inserted and deleted from network protocol stacks on an as-needed basis. The Protocol Booster design methodology attempts to improve end-to-end networking performance by adapting protocols to network dynamics.

We describe a new dynamically reconfigurable FPGA based architecture, called the Programmable Protocol Processing Pipeline (P4), which provides a platform for highly-flexible hardware implementations of Protocol Boosters. The prototype P4 is designed to interface to an OC3 (155 Mb/s) ATM link and perform selected boosting functions at this line rate.

The FPGA devices process the data stream as a pipeline of processing elements. Processing elements are downloaded and activated dynamically, based on policies used by the controller to choose configurations. As modules become unnecessary they are removed from the pipeline chain.

1 Introduction

Network protocols are designed to meet application requirements for data communications, including security, reliability and performance. The dominant design and implementation process for protocols begins by enumerating the requirements for the protocol, and then designing a protocol that provides the necessary features end-to-end[SRC81]. Optimization consists of identifying common cases and implementing fast paths for these cases; TCP/IP is a case study[CJRS89]. The resulting protocol is robust end-to-end and typically provides good performance. However, extremely poor performance can result when the assumptions permitting fast path execution are not met.

1.1 Protocol Boosters

"Protocol Boosters"[FMS96] are protocol elements intended to be transparently inserted into and deleted from protocol stacks on an as-needed basis, e.g., in response to unanticipated or changing network characteristics.

---

* This research was supported by DARPA under Contracts #NCR95-20963 and #DABT63-95-C-0073, the AT&T Foundation, the Hewlett-Packard Corporation, the Intel Corporation and the Altera University Grants Program.

A *policy* associated with the booster is used to selectively insert, delete and invoke the protocol functions. For example, a Forward Error Correction code (FEC) might be used over a wireless data link to bring its error behavior into an acceptable operating range, without using the FEC end-to-end [McA95]. The error performance of the subnet is thus "boosted" to an acceptable level to improve end-to-end performance. Figure 1 shows a booster used in a network, in this case boosting a subnet between an end-host and a router.

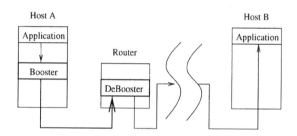

Fig. 1. Boosting a link or subnet

A *transparent* booster does not modify the packet it boosts. For example, an FEC booster may send FEC packets *in addition to* the data packets it encodes. Non-transparent boosters, on the other hand, modify data packets. For example, a compression booster might be used to reduce the bandwidth requirements.

The recent discussion of boosters in Feldmeier, *et al.*[FMS96] restricts boosting to transparent boosters to cope with dynamic routing. Non transparent boosters generally need a static route (e.g., an ATM VC) for successful implementation. Our platform is capable of implementing either type of booster.

1.2 Previous Work

The Protocol Booster idea has been investigated in a software implementation [MCS97], where a FreeBSD TCP/IP protocol stack was boosted with various functions. The major finding was that the operating system overhead to support boosters was minimal; the major cost of boosting was the processing cost of the boosting algorithm(s) applied to the data stream. This suggested that any boosting scheme requiring high performance would need a way to first provide the flexibility required by design methodology, and second, provide this flexibility with high processing performance.

Field-programmable gate array (FPGA) devices offer the degrees of flexibility and performance required for protocol boosting. Our FPGA-based architecture, the Programmable Protocol Processing Pipeline (P4) provides a new platform for Protocol Boosters hardware implementation.

The P4 operates on streams of Asynchronous Transfer Mode (ATM) cells, and serves as a generic platform for running Protocol Boosters over the ATM infrastructure. This paper's primary focus is the P4's logical structure, and describes the architectural solutions used in realization of a P4 prototype. Although the P4 is designed to operate in the ATM environment, the concept is applicable in any other protocol.

The next section, Sect. 2, presents the architecture of the P4, and briefly describes its operation. Section 3 details the functional modules and their operation. Section 4 points out questions associated with the implementation and usage of the P4 architecture and outlines the the longer-term project goals. Section 5 concludes the paper with a summary of what the architecture proves, and next steps in our research.

2 Basic Architecture

Figure 2 shows a block diagram for the P4 architecture. The core of the architecture is a pool of Altera Flex8000 family[Alt96] FPGA devices, acting as processing elements (PEs), and a switching array that provides interconnections among processing elements.

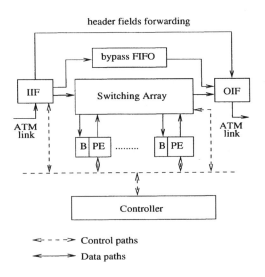

Fig. 2. Block diagram of the P4 Architecture

PEs typically form a pipeline chain via the switching array. Each PE implements one function in the pipeline chain. For example, if the FEC was to be added into the protocol stack, the available FPGA device would be configured to implement the encoder on the transmitter side; on the receiver side, one of the PEs would be configured as the decoder.

Associated with each PE is a FIFO buffer (B). A PE reads the data from its FIFO buffer, performs its processing, and writes into the FIFO buffer associated with the next device in the chain. Connection to the next device is achieved via the switching array.

Data from the ATM link are received in parallel (octet by octet), at 19.44MHz, making the board capable of processing the cell stream in real time at 155Mb/s. In our current experimental setup, we use the HP75000 Broadband Test-system equipped with the Network Impairment Emulation Module[DGH+97]. The P4 board connects to the parallel line interface of the HP75000[Hew94]. The system can generate test traffic patterns and introduce impairments (cell losses, bit errors etc.) and thus emulate the conditions encountered in real networks.

While the current version of the P4 board does not have a Sonet/SDH interface, timings of the parallel interface are compatible with the timings of commercially available framing devices[Tex95], making future addition of such a Sonet/SDH interface straightforward. At this stage, we are primarily interested in the implementation of the P4 functionality and the role it can play in protocol processing.

3 Functional Elements

3.1 Input Interface (IIF)

The input interface (IIF) of Fig.2 provides initial processing of the cells. It is realized by a specifically configured FPGA device downloaded on powerup. The IIF is the first device in the chain and it does not have a FIFO buffer preceding it. It connects to the parallel cell stream which may come either from the test equipment or a Sonet/SDH framing device.

The IIF scans the input stream for cells belonging to the processed virtual circuit and stores them in the FIFO buffer associated with the first PE in the pipeline chain. All other cells are stored in the bypass buffer together with the header fields. Thus the IIF selects the virtual circuit to boost assuming that the VPI/VCI corresponds to a single application (a realistic assumption for an ATM network).

Headers of cells on the boosted channel are stripped off and payload data are stored into the FIFO buffer. Header information is passed directly to the output interface (OIF) via a special data path, bypassing the PEs.

There are two reasons for stripping off the header fields:

1. Having processing elements deal with only the payload reduces their complexity, saving space in the FPGA device and enabling us to put more *processing* functionality in a PE.
2. Some processing algorithms may change the size of the actual data (e.g. data compression). In such cases it is clearly simpler to remove the header before the processing and reconstruct cells at the output.

3.2 Output Interface (OIF)

The output interface (OIF) of Fig.2 complements the IIF. The device reads from the FIFO buffer associated with it, creating a boosted cell stream. If the buffer is empty, the OIF will try to read the cell from the bypass FIFO. If both buffers are empty, the OIF will not generate a cell. The IIF and OIF operation is illustrated in Fig.3.

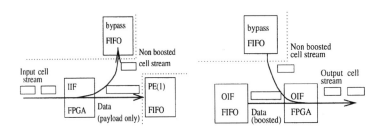

Fig. 3. The IIF/OIF operation

The OIF can be delayed by any processing element in the chain. For example, if one of the processing elements performs data compression, it would typically request a delay in cell generation if it produced output shorter than 48 bytes before new data appeared. Delaying the cell for a limited amount of time might avoid padding the payload with idle data and improve the bandwidth utilization. However, a cell must not be arbitrarily delayed; we will study tradeoffs across a selection of traffic types.

3.3 Switching Array

The *switching array* improves flexibility of the P4 architecture as functions in the PEs are not necessarely commutative. If a new processing element must be activated, it can be downloaded into any available FPGA device. The switching array will allow placing of the processing element at the appropriate point in the pipeline chain regardless of the actual FPGA device used; it serves to virtualize pipeline ordering.

3.4 Processing Element (PE)

The Processing Element (PE) is the generic part of the architecture. It is a combination of a FIFO buffer and an FPGA device. The output of the FPGA device is the set of control and data signals compatible with the FIFO buffer input. Processing consists of reading from the FIFO buffer, processing the data and writing it into the FIFO buffer associated with the next PE in the chain.

In the minimal case (no PE active), the IIF reads the data from the link and writes into the FIFO buffer of the OIF. The boundary between PEs is the FIFO data and control bus interface. Since this is well defined for the given FIFO[Cyp95] it is simple to insert and remove PEs in the P4's chain.

3.5 Controller

From the controller's perspective, PEs are I/O devices. The controller has a pool of FPGA configurations with which it can implement a processing algorithm. Depending on the policy and network conditions, it chooses the appropriate set of configurations and downloads the FPGA devices on demand. Downloading is managed by the controller, and the FPGA is in the passive configuration mode[Alt94]. Using the switching array, the controller creates the chain with processing elements ordered appropriately.

3.6 P4 Role and Placement in a Network

The P4 board is intended to work as an "edge" device for the boosted portion of a network cloud as shown in Fig.4. Although the prototype is designed to operate on the stream of ATM cells, the same concept can be applied to any type of packet stream.

The P4 can work as a standalone unit with its own controller or as part of a switch or other network element. In the latter case the P4 can be an integral part of the switch and thus managed by the network element's controller. A single controller may be responsible for multiple P4 boards.

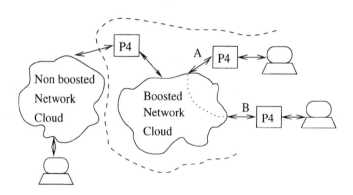

Fig. 4. P4 as the part of the network

In Fig.4, the boosted portion of the network has a P4 board associated with all access points. If a user connected to the P4 enhanced access point communicates with another user connected to the P4 enhanced access point, their communication can be point-to-point boosted provided that there is a route entirely through boosted portion of the network. An important property of the P4 as used in this mode is that if the route has to pass through the non-boosted portion of the network, the P4 board at the boundary will act as a conversion unit by deboosting the outgoing traffic and boosting the incoming traffic. This enhances the interoperability of the scheme.

4 Research Questions

The P4 board will aid exploration of the design space for hardware implementation of Protocol Boosters. It should process the 155Mb/s ATM cell stream in real time, and allow dynamic reconfiguration with minimal delay. This section describes the performance metrics for the architecture.

4.1 Protocol Processing Performance

One goal of Protocol Boosting is end-to-end performance improvement via adding a parasitic module to the original protocol[FMS96]. As shown with the software prototype of Protocol Boosters[MCS97], some functions can be efficiently implemented in software resulting in better end-to-end performance. However, functions such as data encryption significantly degrade throughput while adding their functionality (e.g., privacy of communication).

Preliminary measurements with the P4 board using the convolutional FEC booster have shown that the processing delay is minimal compared to the cell transmission delay. Due to the pipelined nature of the architecture, link throughput is preserved[2] and the processing overhead is reflected in latency through the P4. The limit on the pipeline depth depends on the existing propagation delay and the total delay acceptable by the application.

4.2 Resource Management Issues

A second important issue is the functional "agility" of the P4. By agility we mean latency for reaction to the network anomalies. For example, consider the case where we want to address load-induced congestion by adding data compression to the link protocol (provided we can successfully compress!). At a link speed of 155Mb/s the time to download the FLEX8K device[Alt96] can be too long to successfully address the problem of congestion.

For some applications, agility is not a critical issue. For example, a user is not likely to notice the slight delay in adding FEC to audio traffic to improve sound quality. Thus, there are applications which require very fast reaction ("agility") from the protocol booster and there are also applications where the low agility can be tolerated.

Use of the P4 architecture for non-trivial Protocol Boosters implementations raises the question of management of limited resources (namely FPGA devices which implement the processing elements) to achieve statistically (i.e., almost always) high agility. Resource management in the P4 architecture is analogous to cache management in a general purpose computer system. A cache controller takes advantage of the correlation between accesses to nearby memory locations and statistically reduces the memory access time. In a similar fashion, we can

---

[2] It should be understood that if the booster generates additional cells, the input data can not arrive at the maximum bit rate. However this limitation is not due to the P4 processing overhead but the link capacity.

achieve statistically improved agility of the Protocol Booster running on the P4 architecture.

A well-designed management algorithm will try to keep the agility sensitive configurations pre-loaded in the FPGA devices and clear them only if deemed necessary (i.e., if an FPGA device is needed for some other algorithm and there is no free device available). Resource management algorithms and statistical properties of the events in the network that can be used to trigger the activation of the specific Protocol Booster will be major (and we believe generalizable) results of this research.

4.3 Policy Modules

An important part of the Protocol Booster is the policy module. The policy module decides when a function should be activated based on its observations of the data stream.

In this paper, we have focused on the issues of implementing processing functions (i.e., mechanism modules) in the P4 architecture. As the controller decides which function to download, enforcing the policy is the task of the controller.

The controller can delegate some of this role to a PE by configuring it to monitor the link instead of performing the actual processing. The controller has full access to the PE and can communicate with it via its data bus, and thus some PEs can be programmed to monitor the events on the link and provide the controller with information crucial to policy decisions.

Thus, the P4 controller *enforces* the policy but PEs can perform functions which are part of both policy and mechanism module.

4.4 Need for Synchronization

Boosting a portion of the network means adding some processing on one side and restoring the original data on the other side. For convenience, assume a non-transparent booster and call these functions boosting and deboosting respectively. The booster needs to be synchronized with the debooster[3], that is, the debooster must know when to engage. This requires a signalling protocol, either explicit or implicit.

Consider the infrastructure shown in Fig.4, and suppose that side A decides to load a booster and that its matching debooster must be loaded on side B. Side A needs a method of in-band signalling to inform side B when the booster is to be activated. As a simple scheme for testing the P4 architecture prototype, we will use in-band signalling with control cells in the same virtual channel as the data, and use the payload-type field[ATM94] to distinguish signalling and user data.

---

[3] This is not strictly required for transparent boosters[FMS96], but of course they are more effective when synchronized!

4.5 Extension to Multiple Virtual Circuits

The architecture has been described assuming single virtual channel (VC) processing. The current prototype is limited to one VC, as the focus is on the basic P4 architecture. In later versions we plan to implement multiple VC processing on one board.

Besides the pipeline topology, the switching array can connect processing elements in the parallel pipelines and provide different processing on different VCs. It is also straightforward to modify the IIF to use VPI only instead of the VPI/VCI combination to identify the boosted cell stream.

An interesting approach (suggested by Bill Marcus of Bellcore) is to use "context switch"-like processing in the PE. The state of the PE can be stored away in the content addressable memory (CAM) and VPI/VCI can be used as an index to the state information in the CAM. A PE can thus switch state accordingly and apply the same algorithm to multiple concurrent virtual channels.

5 Conclusion

We have presented an architecture which has the software-like flexibility required for the Protocol Boosters design methodology, while providing the performance expected from a hardware implementation. However, the P4 is more generally useful; for example, as an array of downloadable devices which perform protocol processing, it might function as a high-performance element in an *Active Network*[SDG+96].

The P4 prototype raises several issues in networking subsystem design. On one hand, it is a generic architecture with a limited number of resources (i.e., the processing elements) which need a good management policy. On the other hand it is a platform which allows the implementation of certain functions in hardware and gives us a new tool for addressing the balance between hardware and software in protocol implementation. It is particularly important for evaluating these engineering tradeoffs that the generic structure of the P4 allows fine-grained control of the boundary between functions implemented in hardware and software.

The prototype is operating as a wirewrap card. Next is PCB implementation, and evaluation of various processing functions that will run on the P4 board. The construction of the testbed involving the ATM subnet enhanced with the P4 boards at the entry points will allow us to outline the domain of applicability of P4-like architectures for protocol processing, and evaluate algorithms for dynamically allocating the hardware resources based on application demands and network dynamics.

6 Acknowledgements

Bill Marcus of Bellcore provided a number of helpful comments in addition to the multiple VC scheme (Sect. 4.5). Tony McAuley, also of Bellcore, helped clarify our thinking on the roles of transparency, routing and booster state. Dave Feldmeier of MUSIC Semiconductors made some useful suggestions. Tyler Arnold has been investigating the implementation of a convolutional FEC booster for use in the P4.

References

[Alt94] Altera, Corporation, 2610 Orchard Pkwy., San Jose, CA 95134. *Configuring the FLEX8000 devices - Application Note 33*, 3 edition, May 1994.

[Alt96] Altera, Corporation, 2610 Orchard Pkwy., San Jose, CA 95134. *FLEX8000 Programmable Logic Device Family - Data Sheet*, 8 edition, June 1996.

[ATM94] ATM Forum. *ATM User Network Interface Specification, version 3.1*, September 1994.

[CJRS89] David D. Clark, Van Jacobson, John Romkey, and Howard Salwen. An analysis of tcp processing overhead. *IEEE Communications Magazine*, 27(6):23–29, June 1989.

[Cyp95] Cypress Semiconductor Corporation, 3901 N. 1st St, San Jose, CA 95134. *CY7C421 Data Sheet*, January 1995.

[DGH+97] R.W. Dmitroca, S.G. Gibson, T. R. Hill, L. M. Morales, and C. T. Ong. Emulating atm network impairments in the laboratory. *Hewlett-Packard Journal*, 48(2):45–50, April 1997.

[FMS96] D. C. Feldmeier, A. J. McAuley, and J. M. Smith. Protocol boosters. *submitted to IEEE JSAC Special Issue on Protocol Architectures for the 21st Century*, 1996. U. Penn CIS TR MS-CIS-96-34.

[Hew94] Hewlett - Packard, IDACOM Telecommunications Operation, 4211 95 Street, Edmonton, Alberta, Canada. *Optical Line Interface User's Guide*, 3 edition, June 1994.

[McA95] A. J. McAuley. Error control for messaging applications in a wireless environment. In *INFOCOM 95*, April 1995.

[MCS97] A. Mallet, J. D. Chung, and J. M. Smith. Operating systems support for protocol boosters. In *Proceedings, HIPPARCH Workshop*, June 1997.

[SDG+96] J. M. Smith, D.J.Farber, C. A. Gunter, S. M. Nettles, D. C. Feldmeier, and W.D. Sincoskie. Switchware: Accelerating network evolution (white paper). Technical report, University of Pennsylvania, URL: http//www.cis.upenn.edu/ jms/white-paper.ps, June 1996.

[SRC81] J. H. Saltzer, D. P. Reed, and D. D. Clark. End-to-end arguments in system design. In *Proceedings of the 2'nd IEEE International Conference on Distributed Computing Systems*, pages 509–512, April 1981.

[Tex95] Texas Instruments, P.O. Box 655303, Dallas, TX 75265. *TNETA 1500, 155.52Mb/s Sonet/SDH ATM receiver/transmitter*, July 1995.

Satisfiability on Reconfigurable Hardware

Miron Abramovici

Bell Labs - Lucent Technologies

Murray Hill, NJ 07974

Daniel Saab

Case Western Reserve University

Cleveland, Ohio 44106

Abstract. In this paper we present a new approach to implement satisfiability (SAT) on reconfigurable hardware. Given a combinational circuit C, we automatically design a SAT circuit whose architecture implements a branch-and-bound SAT algorithm specialized for C. A major novel feature is that both the next variable to assign and its value are dynamically determined by a backward model traversal done in hardware. Our approach relies on fine-grain massive parallelism.

1. Introduction

The combinational satisfiability (SAT) problem - given a combinational circuit, find a vector that sets its output to 1, or prove that no such vector exists - has many CAD applications. For example, Figure 1a illustrates a formal verification method where the goal is to check whether two circuits are equivalent. If SAT succeeds in setting $Diff=1$, then the obtained vector t sets at least one pair of corresponding primary outputs (POs) to different values; otherwise, the circuits are equivalent. In Figure 1b, the problem is to solve a system of boolean equations of the form $f_i(x_1, x_2,...,x_n) = a_i$, $(i=1,2,...m)$, where x_j are boolean variables, f_i are boolean functions, and a_i are boolean constants. If SAT succeeds in setting $Solve=1$, then the obtained vector x sets $f_i = a_i$ for every i, and hence is a solution of the given system; otherwise no solution exists. CAD problems that require solving large systems of boolean equations include timing verification[9][14], routing and routability analysis[19], and fault diagnosis[20]. SAT is also related to automatic test-pattern generation (ATPG) algorithms, as it can be viewed as the problem of generating a test for a stuck-at-0 fault on a PO. An important component in ATPG is the *line justification* problem, which deals with setting an internal signal to a given value. This corresponds to SAT on a subcircuit. Also, the entire ATPG problem may be formulated as a SAT problem[13][16]. Logic synthesis[4] is another CAD area with many SAT applications.

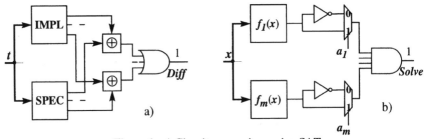

Figure 1. a) Circuit comparison using SAT
b) Solving a system of boolean equation using SAT

The above formulation of the SAT problem in terms of an arbitrary circuit is more general than the conventional one dealing with a formula in conjunctive normal form, which corresponds to a restricted class of circuits, namely two-level OR-AND circuits with inversions allowed only at the primary inputs (PIs). It is well-known that SAT is an NP-complete problem[8]. Applied to complex VLSI circuits, SAT algorithms have long run-times. Thus speeding up SAT will result in improving the efficiency of many CAD algorithms relying on SAT.

Because of the computational complexity of SAT and SAT-related problems, several approaches have tried to use implementations based on reconfigurable hardware. Figure 2 illustrates the general data flow of such an approach, whose goal is to speed up an algorithm *ALG* working on a given circuit *C*. A mapping program processes *C* to generate the model of a new circuit *ALG(C)*, which executes *ALG* for *C*. Since *ALG(C)* will be much larger than *C*, and because it will be used only once, it would not be economically feasible to implement it in a conventional way. The solution is to *use reconfigurable hardware to "virtually" create the ALG(C) circuit*; then the algorithm is executed by emulating this circuit. In contrast to a hardware accelerator for *ALG* (for example, a simulation accelerator), where the same special-purpose hardware processes different circuits, in this approach *the ALG hardware is designed specifically for a single circuit*. The advantage is that *ALG will run at emulation speed, without incurring the cost of building special-purpose hardware*. (Of course, mapping to reconfigurable hardware has been applied to many other computationally difficult problems; for example, graph problems such as finding independent sets[7] and transitive closure[3].)

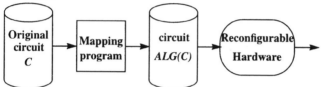

Figure 2. General data flow to speed up algorithm *ALG* for circuit *C*

One such approach was developed by Hirose *et al.* for combinational ATPG[12]. Tests are generated by simulating an *ATPG(C)* circuit on a simulation hardware accelerator. (While simulation is less efficient than emulation, since it involves an additional layer of interpretation, reconfigurable hardware was not available at that time.) While it obtained significant speed-ups, fault coverage results were worse than those achieved by software ATPG, because its vector generation mechanism is constrained by a fixed-order exploration of the search space. That is, if *A* is the first input, *A*=1 will always be tried first, even if *A* cannot influence the detection of the target fault; this results in a lot of unnecessary backtracking, which in turn leads to aborting the search when is becoming too expensive.

The SAT method of Suyama *et al.* [17] was the first to create a *SAT(C)* circuit, which is downloaded into an emulator for execution. Although this method has been reported to be much more efficient than a general SAT procedure implemented on a general-purpose computer, its applicability to CAD is limited because the generated vectors are always fully specified. For example, in a circuit with 100 PIs, all the PIs will be assigned binary values, even it may be sufficient to assign only 10 to set the PO to 1. This overspecification is detrimental in most CAD applications, where minimizing the number of specified PIs is advantageous. For example, in ATPG, full specification would preclude test set compaction[1], and may preclude the generation of the obtained vector by a different circuit. In circuit verification, after SAT generates a vector that creates a mismatch between the compared circuits, the source of the error must be located. This process would be more difficult if most of the specified PI assignments are not necessary to detect the mismatch. Like [12], the efficiency of this method is also limited by the use of a fixed-order exploration of the search space, although an improvement is obtained by skipping variables for which a value change is not needed in the current state.

In this paper, we present a new SAT method using reconfigurable hardware. In contrast with previous work, *in our approach PIs may be assigned in any desired order.* The backtracking mechanism supporting this feature relies on a hardware stack. *Both the next PI to assign and its value are dynamically determined* based on the current values in the circuit. This process is implemented using a separate model that allows a backward traversal of the circuit. Our SAT circuit relies on *fine-grain massive parallelism* to such an extent that all the operations performed between successive decision steps in the algorithm are completed in a single clock cycle.

The remainder of this paper is organized as follows. Section 2 discusses the architecture of the SAT circuit. Section 3 reports our experience with the prototype implementation, and Section 4 presents conclusions and future work.

2. The SAT Circuit

2.1. The Search Strategy

The main criteria in selecting the search strategy to be used in the SAT circuit are efficiency and suitability for a simple hardware implementation. Since the SAT problem is equivalent to the ATPG line justification problem, we have adopted a justification technique similar to that used in the PODEM[11]

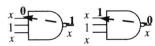

Figure 3. Backtracing through a NAND gate

ATPG algorithm. Central to this algorithm is the concept of *objective*, which is a desired assignment $l=v$ of value v to line l, which currently has an unknown value x. Initially all lines are set to x. An objective may be achieved only by PI assignments. A *backtrace* procedure propagates an objective $l=v$ along a single path from l to a PI i, where all the lines along the path have value x, and determines a PI assignment $i=v_i$ that is likely to contribute to achieving $l=v$. Figure 3 illustrates backtracing through a NAND gate (objectives are shown in bold). The justification procedure simulates the PI assignment obtained by backtracing the desired objective. Note that all values are obtained only by simulating PI assignments, hence there are no unjustified values. This allows a simple backtracking mechanism based on simulating the complementary assignment $i=v_i$.

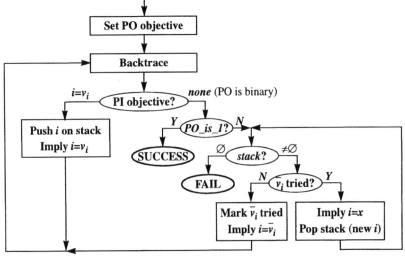

Figure 4. Flowchart of the algorithm

The SAT algorithm whose flowchart is shown in Figure 4 starts by setting the initial objective to set the PO to 1. The backtracing process maps the PO objective into a PI objective $i=v_i$. Then i is pushed on the decision stack and the value v_i is assigned to i and simulated. This process continues until no more objectives exist, which happens only when the PO value has become binary. Then if the PO value is not 1, a dead-end has been reached which requires backtracking. If the stack is not empty, the last PI assigned i is at the top of the stack. If the value $\bar{v}_i$ has not yet been tried, the complementary assignment $i=\bar{v}_i$ is simulated. If both v_i and $\bar{v}_i$ have been unsuccessfully tried, the algorithm implies $i=x$ and pops the stack to look for a previous assignment not yet reversed. At this point an empty stack indicates that no untried alternative decision still exists, therefore the algorithm has failed because the PO may never be set to 1.

2.2. The Basic Idea

The main problem in implementing such an algorithm in hardware is processing of objectives, which propagate only backward through the circuit, while a hardware model (of the type used in emulation) can only propagate values forward. The key idea that has solved this dichotomy is to *have two distinct models of the circuit: a forward model for*

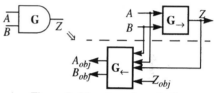

Figure 5. Mapping of a gate G

propagating values and a backward model for propagating objectives. (A similar dual model has been used in the design of a fault simulation algorithm for reconfigurable hardware[2].) As illustrated in Figure 5, every gate **G** in the original circuit is mapped into an element $\mathbf{G}_\rightarrow$ in the forward network and an element $\mathbf{G}_\leftarrow$ in the backward network. Here A, B, and C propagate the values (0, 1, or x) of their corresponding signals, while A_{obj}, B_{obj}, and C_{obj} carry their objectives. Note that the propagation of objectives requires the knowledge of values; for example, an objective may propagate only on a line with value x.

2.3. Architecture

Figure 6 shows the block diagram of the SAT circuit. The backward network has one PO for every PI of the forward network and receives the initial objective for the PO of the original circuit. The PO objective is asserted as long as the PO value is x. The computation of objectives in the backward network relies on the signal values computed in the forward network. The PI Decision Block is the "brain" of the system: it maps the PI objectives produced by the backward network

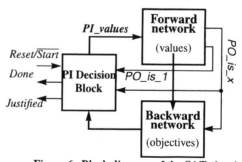

Figure 6. Block diagram of the SAT circuit

into PI values that propagate in the forward network, and implements the entire decision and backtracking process of Figure 4. *Done* is the completion signal and *Justified* indicates whether the SAT problem has been successfully solved. If Justified is 1, *PI_values* is the vector that sets the PO to 1. The following subsections present the various blocks of Figure 6 in more details.

2.4. The Forward Network

Table 1 shows the coding used in the forward network for the 3-valued logic. To represent the value of a signal A we use two bits, A_0 and A_1, where A_0 (A_1) means "A is 0 (1) or x"; then $A_0=A_1=1$ means "A has value x".

Table 1: Value coding

Value	Code	
A	A_0	A_1
0	1	0
1	0	1
x	1	1

Table 2 illustrates the mapping process for the forward network for AND and OR gates. The equations for Z_0 and Z_1 are easy to derive: for example, Z is 0 or x when A is 0 or x, OR B is 0 or x ($Z_0=A_0+B_0$), and has value 1 or x when both A AND B are 1 or x ($Z_1=A_1 \cdot B_1$). The mapping for an inverter with input A and output Z is given by $Z_0=A_1$ and $Z_1=A_0$, which shows that negation is realized without logic, just by swapping A_0 and A_1.

Table 2: Mapping for forward network

2.5. The Backward Network

To propagate objectives in the backward network, we need to distinguish among three situations: a line with a 0-objective, a line with a 1-objective, and a line without any objective. We use two bits to code these three "values", as shown in Table 3. One of the bits is the *objective flag*, which denotes whether there is an

Table 3: Objective coding

Objective	Flag	Value
$\varnothing$ (none)	0	-
0	1	0
1	1	1

objective for a line; when the flag is 1, the other bit gives the *objective value* (0 or 1).

Table 4 summarizes the objective computation rules for a 2-input AND gate with output Z and inputs A and B. Z_{obj} is the objective for Z, and A_{obj} and B_{obj} are the objectives computed for the gate inputs. Basically, the output objective propagates to a gate input with x value. For an inverting gate, the value for any input objective would have been the complement of the output objective.

Table 4: Objective computation for an AND gate

Z	Z_{obj}	A	B	A_{obj}	B_{obj}	
0,1	–	–	–	$\varnothing$	$\varnothing$	Z binary
x	1	x	x	1	$\varnothing$	1-objective
		1	x	$\varnothing$	1	
	0	x	x	$\varnothing$	0	0-objective
		x	1	0	$\varnothing$	
x	$\varnothing$	x	x	$\varnothing$	$\varnothing$	no objectives

When two or more gate inputs have value x, a choice exists in selecting the input to continue backtracing from. This choice can be made based on heuristic *controllability measures*[1]. Controllability of a line l for value v measures the relative difficulty of setting $l=v$. A value c is said to be the *controlling value* of a gate, if setting a gate input to c determines the gate output value independent of the other input values. AND and NAND gates have $c=0$, and OR and NOR gates have $c=1$. When the objective value for the gate input is the controlling value c of the gate, one should select the input which is the easiest to set to c, because setting one input is sufficient to set the output. However, when the gate input objective is $\bar{c}$, one should select the most difficult input to set to $\bar{c}$, because eventually all inputs must be set to $\bar{c}$, and it could be wasteful to work on the easier inputs while the most difficult one may be impossible to set. Controllability measures are computed as a preprocessing step in one pass through the model. These measures can be easily embedded in the backward network during the mapping process to determine the order in which the gate inputs with x value are assigned an objective during backtracing. For example, Table 4 gives priority to A over B for 1-objectives, and to B over A for 0-objectives.

For a signal S that fans out to multiple destinations, an objective may propagate back from either one of the gates fed by S. To model this, we treat the source line (the *fanout stem*) and the lines carrying its value to their destinations (the *fanout branches*) as separate entities. Figure 7a) illustrates a stem S with two fanout branches, $B1$ and $B2$, and Figure 7b) shows the circuit used to compute the objective for S from the objectives of $B1$ and $B2$. For a line X, its objective flag is denoted by X_F and its objective value by X_V. At any time, at most one branch may have an objective, which should propagate on the stem. Hence the two flags of the branches are ORed, and the value corresponding to the active flag is selected.

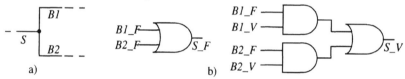

a)   b)

Figure 7. Computing a stem objective

2.6. The PI Decision Block

Figure 8 shows the block diagram of the PI Decision Block. Every heavy line represents a group of n signals, where n is the number of PIs. The PI objectives arrive from the backward network as two separate groups, *Obj_flags* and *Obj_values*. The Control block, whose activity is initiated by *Reset/Start*, receives *PO_is_1* and *PO_is_x* from the forward network, and asserts *Done* on completion; at that time *Justified* shows whether the desired value has been obtained or has been proven unjustifiable. *PI_values*, produced by the PI Value Control block, is the 3-valued representation of the generated vector which feeds the forward network. The Control

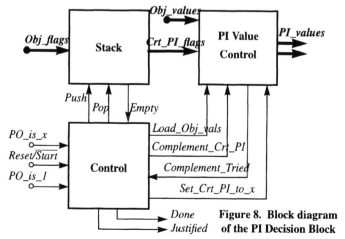

Figure 8. Block diagram of the PI Decision Block

block uses the Stack to implement the decision process. Initially, the *PI_values* are all *x*. The backward network generates one objective as long as *PO_is_x* is asserted, and then *Obj_flags* is vector of 0's with only one bit set. The work of the Control block follows the flowchart given in Figure 4. *PO_is_x* informs the Control block that one PI objective has been found, and *Obj_flags* is pushed on the Stack (*Push*). The top of the Stack is always available as *Crt_PI_flags*, which is used to select the PI whose value is to be changed. The value of the selected PI is taken from *Obj_values* under the control of *Load_Obj_vals*. If no objectives are found (*PO_is_x=0*) and the PO has value 1, the Control block asserts both *Result* and *Justified*. Otherwise, the binary PO value is 0; if the Stack is not *Empty*, the value of the current PI is either complemented (*Complement_Crt_PI*), or, if *Complement_Tried* is asserted, it is set to *x* (*Set_Crt_PI_to_x*) and the Stack is popped (*Pop*). When the Stack is *Empty*, the Control block asserts *Done* and sets *Justified* to 0.

3. Implementation

3.1. Size

Our approach is hardware-intensive. Just the forward network is almost two times larger than the original circuit, because, as shown in Table 2, every original gate is mapped into two gates (except inverters, whose implementation is logic-free). In the backward network, for every stem with k fanout branches we need a circuit with $k+2$ gates (shown in Figure 7b for $k=2$). For every original gate with p inputs, we need a $2p$-gate circuit to implement Table 4. For a circuit with n PIs, the Stack is implemented by a n-word RAM with n bits/word, and PI Value Control has 3 n-bit registers (two for the PI values and one for the *Complement_tried* flags - see Figure 8). Table 5 gives the size of the SAT circuits for the combinational benchmark circuits ISCAS-85[5]. For these examples, the SAT circuit is between 6.9 and 12 times larger than the original circuit. Since the words stored in Stack have only one bit set, the memory requirements for Stack can be significantly reduced by storing only the index of this bit in the word rather than the whole word; the trade-off is the encoding and decoding of the data written into and read from memory.

Table 5: Model sizes

	Original model			SAT model	
Circuit	Gates	PIs	POs	Gates	Increase
c432	160	36	7	2035	12
c499	202	41	32	2368	11
c880	383	60	26	3929	10
c1355	546	41	32	4768	8
c1908	880	33	25	6358	7
c2670	1193	233	140	13108	10.9
c3540	1669	50	22	11591	6.9
c5315	2307	178	123	19361	8.3
c6288	2416	32	32	18160	7.5
c7552	3512	207	108	26908	7.6

Since the SAT circuit is implemented on reconfigurable logic, its size is not important, unless it does not fit in the emulator. The largest capacity available today for emulation is about 6 million gates[15].

3.2. Speed

Figure 9 outlines a timing diagram of the processing done in one clock cycle. After the PI assignment (1), the new PI value propagates in the forward network (2). The signals *PO_is_1* and *PO_is_x* become valid at the end of the interval (2). As a result of values changing in the forward network, new objectives may be generated and old objectives may be cancelled in the backward network; the propagation of objectives occurs during interval (3). The last interval (4) represents the activity in the PI Decision Block - the decision logic and stack manipulation - which prepares the next PI assignment (1).

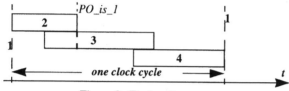

Figure 9. Timing diagram

Let us analyze some of the corresponding work done in a conventional software implementation between two consecutive PI assignments. The tasks in (2) require forward simulation (including scheduling and retrieving events, determining the gates reached by events, and 3-valued gate evaluations). The tasks in (4) are similar, except the propagation follows the fanin instead of fanout, and objectives are computed instead of values. The average length of the backtraced path increases with the size of the circuit. Gate evaluations (both forward and backward) typically involve a loop in which the inputs of a gate are in turn retrieved and analyzed. *The massive parallelism inherent in the SAT circuit allows all this work to be accomplished in only one clock cycle, independent of the size of the circuit.* The clock period should be greater than the sum of the largest path delays in the forward and the backward network.

At the time of this writing, our experimental work is still in progress. We generate and simulate (using a 1MHz emulation clock cycle) SAT circuits for equivalence checking between two identical benchmark circuits. In general, this generates the worst-case behavior of the algorithm, which must exhaust the search space to determine that there is no solution. The problem is that fair comparisons with a software SAT algorithm handling the same circuit are difficult. Our estimate is that our SAT method implemented on reconfigurable hardware will be about 2 to 3 orders of magnitude faster than the software SAT.

4. Conclusions and Future Work

In this paper we have introduced a new approach to SAT, in which a *SAT(C)* circuit is automatically designed to justify the PO value of a given combinational circuit *C*. The architecture of the SAT circuit implements a branch-and-bound justification algorithm, and one of its major innovations is the separate backward network used to propagate objectives. Compared with prior work[12][17] whose efficiency is limited by their fixed-order exploration of the search space, in our approach PIs may be assigned in any order. Moreover, both the ordering of the PIs and their values are dynamically determined based on values currently assigned.

Our approach relies on fine-grain massive parallelism to such an extent that all the operations performed between successive decision steps in the algorithm are completed in a single clock cycle. Although the SAT circuit is about ten times larger than the original circuit, this is not a significant problem with a "virtual hardware" implementation using reconfigurable hardware.

Currently, the backtracing is not guided by any cost function, and this usually results in additional search. Other planned algorithmic improvements are concurrent propagation of objectives on several paths and concurrent assignment of several PIs.

References

[1] M. Abramovici, M. A. Breuer and A. D. Friedman, *Digital Systems Testing and Testable Design*, IEEE Press, 1994

[2] M. Abramovici and P. Menon, "Fault Simulation on Reconfigurable Hardware," *Proc. IEEE Symp. on Field-Programmable Custom Computing Machines*, April, 1997

[3] J. Babb, M. Frank, and A. Agarwal, "Solving graph problems with dynamic computation structures," in *SPIE Photonics East: Reconfigurable Technology for Rapid Product Development & Computing*, pp. 225-236, November, 1996

[4] R. Brayton, G. Hachtel, C. McMullen, and A. Sangiovanni-Vincentelli, *Logic Minimization Algorithms for VLSI Synthesis*, Kluwer Academic Publishers, 1984

[5] F. Brglez and H. Fujiwara, "Neutral netlist of ten combinational benchmark circuits and a target translator in FORTRAN," *Proc. IEEE Intn'l. Symp. on Circuits and Systems*, June 1985.

[6] S. T. Chakradhar, V. D. Agrawal, and M. L. Bushnell, "Neural Net and Boolean Satisfiability Models of Logic Circuits," *IEEE Design & Test of Computers*, vol. 7, no. 3, pp. 54-57, October,1990

[7] S. T. Chakradhar and V. D. Agrawal, "A Novel VLSI Solution to a Difficult Graph Problem," *Proc. Intn.l. Symp. on VLSI Design*, pp. 124-129, January,1990

[8] S. A. Cook, "The Complexity of Theorem-Proving Procedures," *Proc. 3rd Annual ACM Symp. on Theory of Computation*, pp. 151-158, 1971

[9] S. Devadas, K. Keutzer, S. Malik, and A. Wang, "Certified Timing Verification and the Transition Delay of a Logic Circuit," *Proc. Design Automation Conf.*, pp. 549-555, June, 1992

[10] H. Fujiwara and T. Shimono, "On the Acceleration of Test Generation Algorithms," *IEEE Trans. on Computers*, vol. C-32, no 12, pp. 1137-1144, December, 1983.

[11] P. Goel, "An Implicit Enumeration Algorithm to Generate Tests for Combinational Logic Circuits," *IEEE Trans. on Computers*, Vol. C-30, No. 3, pp. 215-222, March, 1981.

[12] F. Hirose, K. Takayama, and N. Kawato, "A Method to Generate Tests for Combinational Logic Circuits Using an Ultrahigh-speed Logic Simulator," *Proc. Intn'l. Test Conf.*, pp 102-107, Oct. 1988

[13] T. Larrabee, "Test Pattern Generation Using Boolean Satisfiability," *IEEE Trans. on CAD*, Vol. 11, No. 1, pp. 4-15, January, 1992

[14] P. C. McGeer *et al.*, "Timing Analysis and Delay-Fault Test Generation Using Path Recursive Functions," *Proc. Intn'l. Symp. on CAD*, pp. 180-183, November 1991

[15] *RPM Emulation System Data Sheet*, Quickturn Systems Inc., 1991

[16] P. R. Stephan, R. K. Brayton, and A. Sangiovanni-Vincentelli, "Combinational Test Generation Using Satisfiability," *IEEE Trans. on CAD*, vol. 15, no. 9, pp. 1167-1176, Sept. 1996.

[17] T. Suyama, M. Yokoo, and H. Sawada, "Solving Satisfiability Problems on FPGAs," *Proc. Intn'l. Workshop on Field-Programmable Logic and Applications*, 1996

[18] K. Takayama, F. Hirose, and N. Kawato, "A Test Generation System Using a Logic Simulation Engine," *Fujitsu Sci. Tech. J.*, vol. 27, no 3, pp. 285-289, Sept. 1991

[19] R.G. Wood and R.A. Rutenbar, "FPGA Routing and Routability Estimation Via Boolean Satisfiability", *Proc. Intn'l. Symp. on FPGAs*, February 1997

[20] Y. Wu and S. Adham, "BIST Fault Diagnosis in Scan-Based VLSI Environments," *Proc. Intn'l. Test Conf.*, pp. 48-57, October 1996

Auto-configurable Array for GCD Computation

Tudor Jebelean

RISC-Linz, A-4040 Linz, Austria
Jebelean@RISC.Uni-Linz.ac.at

Abstract. A novel one-directional pass-through array for the computation of integer greatest common divisor is designed and implemented on Atmel FPGA. The design is based on the plus-minus GCD algorithm and works in LSB pipelined manner. In contrast with previous designs, the length of the new array is independent of the length of the operands: arbitrary long integers can be processed in multiple passes. The array is auto-configurable: at each step, one new cell is configured according to the input from the previous computation. Preliminary experiments show that for 100 bits a speed-up of 4 over software can be obtained using one 6010 Atmel chip.

Introduction

An emerging trend in scientific computing (e.g. symbolic computation) is the need for increased precision arithmetic – either very long or arbitrary long floats or rationals. Cryptography and data-compression applications also need fast long integer arithmetic.

Reconfigurable computing offers a valuable alternative to software solutions, promising a spectacular increase in speed. Typically, the length of the operands in such computations vary from one application to another, or even during the same computation. Thus, a fixed-size hardware coprocessor will fail to accommodate the operands at a certain moment. In the case of systolic arrays (which is the mostly used design technique for such arithmetic circuits) various methods have been proposed for "packing" several processors in one [2], such that the size of the accommodated operands increases correspondingly. However, these methods lead to a significant increase of the area consumption and also need full reprogramming of the FPGA chip.

We present a method which avoids these inconveniences. Namely, we design a *pass-through* array for the computation of the integer greatest common divisor (GCD). The two operands are processed serially in LSB manner, and after one pass through the array either the GCD is obtained, or the operands are reduced to shorter ones and the process is reiterated. The size of the array is completely independent of the lengths of the input, but, of course, a longer array will result in increased speed.

The array is *auto-configurable*, that is, the functions of the cells are selected upon the values of the first digits which pass through. This is a technique which goes beyond usual *configurable* computing (e. g. configure before computation) and even beyond *re-configurable* (e. g. configure part of the chip during computation): at each step (clock cycle) a new cell is configured according to the result of the previous computation. Current FPGA designs do not allow a direct approach to this computing style, because the configuration cannot be changed using the signals produced inside the chip. (By interaction with the host this is possible to some extent in Xilinx XC6200 series.) We use Atmel FPGA and simulate auto-configurability by multiplexors. Preliminary experiments show that 100-bit operands can be processed in one pass by a 6010 Atmel chip, with a speed-up of 4 over the software solution.

GCD Computation

The computation of the greatest common divisor is the most complicated operation over integers and also the most time consuming in arbitrary precision arithmetic. VLSI implementations are based on the *PlusMinus* algorithm of [1] and on its improvements [4]. We adapt here the later, which consists of two phases:

1. **Preprocessing** :
   - the least-significant common null bits are discarded from both operands A and B;
   - after that, if the LSB a_0 of A is null, then the operands are interchanged (after this, a_0 always becomes 1);
2. **Reduction** is applied iteratively as follows, until B becomes null (we denote by a_0, a_1, b_0, b_1 the LSBs of the operands):
   - If $b_0 = 0$ then B is shifted rightward one position ($B \leftarrow B/2$).
   - if $b_0 = 1$ then $(A, B) \leftarrow (B, (A \pm B)/4)$, where $\pm$ is *plus* if $a_1 \neq b_1$ and *minus* otherwise.

Correctness and termination of this algorithm are shown in [4].

The Array

The overall structure of the GCD circuit is shown in Fig. 1.

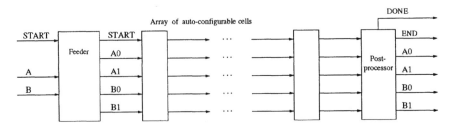

Fig. 1. The overall structure of the GCD circuit.

The *auto-configurability* of the array is realized by the configuring units PCF and ICF shown in Fig. 2. PCF (pre-configure) is used when the data which is used for configuration arrives *one clock before* the data which is used for computation (ENABLE signals the later event). ICF (instant configure) is used when the data which is used for configuration arrives simultaneously with the data which is used for computation (ENABLE has to become 1 one clock later).

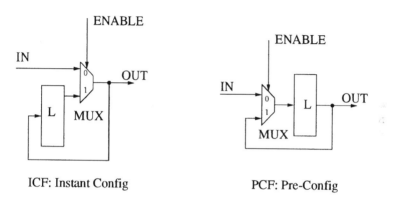

ICF: Instant Config   PCF: Pre-Config

Fig. 2. Configuration units.

The **preprocessing** phase of the GCD algorithm is performed by the *feeder* presented in Fig. 3. This unit receives A, B serially, together with the signal START which remains high for the duration of the data transfer. The OR gate detects the first non-zero pair of bits, its output is AND-ed with START in order to trigger the next computation step, which is instant-configured by a_0.

The feeder is followed by a buffer of latches which allows the simultaneous use of the a_0, b_0 and of a_1, b_1 in the subsequent computation.

The **reduction** phase of the algorithm is implemented as an array of identical configurable cells as in Fig. 4. The signal representing b_0 arrives on the line B1 one clock before START becomes high, and it pre-configures the cell to perform either *shift* or *plus-minus*. The shift is performed by delaying A and START through latches.

If the cell is configured for *plus-minus*, then further instant-configuration is done by the XOR of a_1, b_1, which arrive simultaneously with a_0, b_0 when START becomes high. This configuration signal commands the arithmetic unit (plus or minus, with carry/borrow feedback incorporated). The result of the arithmetic unit is then supplied as B, while A and START are delayed through two latches – this results in shifting B by two bits.

In order to avoid a high delay through lines B0, B1, one inserts a buffer (as in Fig. 3) after a constant number of cells.

We do not detail here the termination detection and the sign detection, which are solved by a simple *postprocessing* circuit, as well as I/O buffering between the host and the coprocessor and between several chips in a multi-chip design, which is easy to perform due to the unidirectional pipelined style of communication.

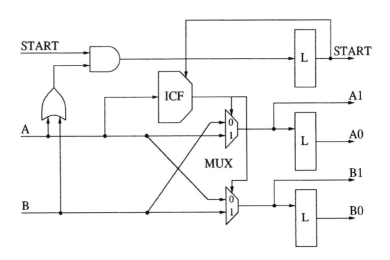

Fig. 3. Feeder and subsequent buffer.

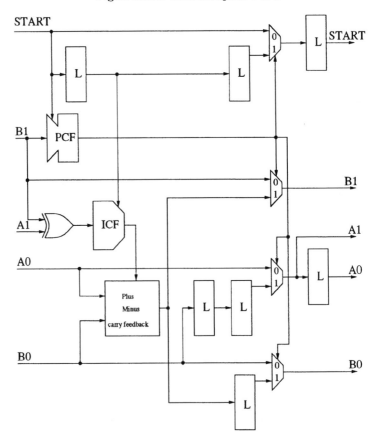

Fig. 4. Configurable cell.

Experiments and Conclusions

The array is being implemented using Atmel 6000 FPGA series. We present here a preliminary assessment of the performance.

Each cell uses 22 Atmel logic components, and is simple enough to be manually placed and routed. Due to local communication of the systolic design, several cells are then placed and routed by simple tiling. At 50% utilization this allows the placement of 150 cells on a 6010 chip, including the preprocessing, postprocessing and I/O buffers. This is enough for the processing in one pass of 100-bit operands. (By software experiments we detected that n-bit operands need on average .75 shift steps and .75 plus-minus steps, which gives an average of $1.5n$ cells and $3n$ clock cycles.) The placed circuit runs at 5 MHz (200 ns clock delay), hence 300 cycles will take 60 μs. (The relatively low speed is also due to the technology-generation involved, on FPGAs of newer technology the performance will increase.)

For comparison, GCD computation in a fast C-implementation (see e.g. [3]) on a DEC workstation 5000/200 (about same generation technology as the Atmel FPGA) takes on average 235 μs, hence the speed-up is 4. Note that by implementing a longer array on a multi-chip module the speed-up will increase linearly for longer operands, because the quadratic time-complexity of the sequential GCD algorithm – in contrast with the linear time-complexity of the systolic device.

References

1. R. P. Brent and H. T. Kung. A systolic algorithm for integer GCD computation. In K. Hwang, editor, *Procs. of the 7th Symp. on Computer Arithmetic*, pages 118–125. IEEE Computer Society, June 1985.
2. P. Dewilde and E. Deprettere. Architectural synthesis of large, nearly regular algorithms. *Ann. Telecom.*, 46(1-2):49 – 59, 1991.
3. T. Jebelean. Comparing several GCD algorithms. In E. Swartzlander, M. J. Irwin, and G. Jullien, editors, *ARITH-11: IEEE Symposium on Computer Arithmetic*, pages 180–185, Windsor, Canada, June 1993.
4. T. Jebelean. Design of a systolic coprocessor for rational addition. In P. Capello, C. Mongenet, G-R. Perrin, P. Quinton, and Y. Robert, editors, *ASAP '95, Strasbourg, France, July 1995*, pages 282–289. IEEE Computer Society Press, 1995.

Structural versus Algorithmic Approaches for Efficient Adders on Xilinx 5200 FPGA

B.Laurent G.Bosco G.Saucier

Institut National Polytechnique de Grenoble/CSI,
46, Avenue Felix Viallet 38031 Grenoble cedex, FRANCE

Abstract. In this paper, classical adder architectures applied to Xilinx XC5200 FPGA are compared. To inherit advantages of both structural and algorithmic approaches, a hybrid solution is proposed such that the optimal trade-off between architectures and technology is reached. The resulting scheme yields optimized performance after the use of Xilinx place and route tools.

1 Introduction

Recently the importance of macroblocks in the automatic synthesis environment has been clearly stated. Macroblocks are called through instantiation or inference in HDL (VHDL or Verilog) synthesis tool. Therefore both I.C. vendors and C.A.D. tools providers offer a large variety of macroblocks. Parameterized macrogenerators commonly synthesize on the fly efficient macroblocks for any bit width and any target technology. The set of macros covered are at least the so called LPM set or library of parameterized modules. We focus in this paper on the performance criterion which aims at minimizing the critical path given by Xilinx place and route tools. Two approaches may be considered to create a library block. It can be handcrafted close to the target technology and we call this a structural approach. It may also be generated automatically using a synthesis algorithm. It is shown in this paper that trade-offs between these two approaches will lead to the best result.

1.1 Technology Independent Algorithm Based Approaches

The addition is the most frequent arithmetic operation, involved not only in simple addition but also in more complex operations like multiplication and division. The key problem of designing adders is to accelerate carry propagation. A straightforward implementation is obtained by abutment of small two bit adders. The carry ripples between each cells and the critical path is structurally quite long. On the other hand, the most commonly scheme for accelarating carry computation is to avoid waiting for the correct carry arrived by calculating all incoming carries in parallel. Many solutions have been proposed for this carry look-ahead addition: Sklansky [1] obtained optimal depth but an increasing fan-out. Brent and Kung Method [2] limits the fan-out but doubles the depth compared to Sklansky's adder. Kogge and Stone [3] reached both optimal

depth and limited fan-out but with an increased complexity. The hybrid computation algorithm invented by Han and Carlson [4] combines Brent and Kung's
algorithm with Kogge and Stone's one. Reasearch has shown that an adder is a
combination of smaller serial and parallel adders. For instance, the carry select
adder computes carry propagation for small consecutive bit length in series in a
first step, and then combines previous schemes in parallel to perform the final
addition.

1.2 Structural Approaches

The Xilinx serie 5200 FPGA proposes technological advantages that may surpass
the benefits of the previous efficient algorithms. Two major devices are used to
perform logic equations. On the one hand, $FMAP$ modules are 4-input Lookup
Tables (LUT) that perform any operations of less than or equal to 4 variables.
On the other hand CY_MUX modules behave like a carry multiplexer [5] (Fig.1).
The time necessary to perform these operations, noted t_{FMAP} and t_{CY_MUX},
are quite different. In a first approximation, $t_{FMAP} = 8.t_{CY\_MUX}$. It is obvious
that the CY_MUX modules provides an opportunity to add digits in a very short
time.

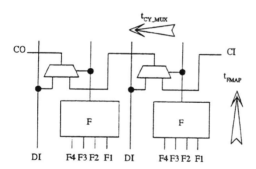

Fig. 1. Xilinx 5200 CLB logic

The paper is organized as follows: in section 2, equations of two operand
adders are presented. Then, structural and technology independent approaches
are derived in section 3. A hybdrid addition scheme, taking into account both
technology and advanced algorithm, is then proposed in section 4 to increase
performance. Finally, experimental results are presented in section 5.

2 Equations of Two-Operand Adders

The problem of performing fast addition and substraction lies in the acceleration
of carry propagation. The outgoing carry can be expressed in several ways. Brent

and Kung [2] introduced in 1982 arithmetic operator Δ, intended to model its generation and propagation according to the input values. The problem statement, presented in [6] and [7], is recalled is section 2.1.

2.1 Carry Generation and Propagation

Let x and y be n-bit integers. Their bits are numbered from 0 to $n-1$. At the bit level, stage i can generate a carry-out regardless of incoming carry; i.e. $g_i = x_i.y_i = 1$ or propagate the incoming carry; i.e. $p_i = x_i \oplus y_i = 1$. At bit-slice level, the generation of a carry-out between bit i and bit j $(i < j)$ noted G_j^i means that a carry is generated inside the slice (defined by bits i and j) and then propagated up to bit j and the propagation of a carry-out between bit i and bit j noted P_j^i means that a carry is propagated from bit i up to bit j. Recursive equations are reduced to compute the generation term G_j^i and propagation term P_j^i for a bit-slice:

- $G_j^i = G_j^k + G_{k-1}^i.P_j^k,\ 0 \leq i \leq k \leq j \leq n-1$
- $P_j^i = P_j^k.P_{k-1}^i,\ 0 \leq i \leq k \leq j \leq n-1$

Let us note: $PG_j^i = (P_j^i, G_j^i)$. Operator Δ is defined as follows: $PG_j^k \Delta PG_{k-1}^i = (P_j^k P_{k-1}^i, G_j^k + P_j^k G_{k-1}^i)$. It has four inputs and two outputs. It is idempotent, associative but not commutative, and is mapped by a virtual Δcell.

The outputs of the adder are then obtained as follows:

- $Out(0) = P_0^0$
- $Out(i) = P_i^i \oplus G_{i-1}^0, 1 \leq i \leq n-1$

2.2 Slices

Let us consider a bit-slice between bit j and bit i $(i < j)$. Assume carry propagation is performed from i to j.

Definition 1 Δslice. The following slice is called a Δslice and noted $\Delta[j, i]$:

$$\Delta[j, i] = PG_i^i, PG_{i+1}^i, PG_{i+2}^i, ..., PG_j^i$$

A bit-slice is an addition performed on a range of bit (i, j) for which all the terms PG_k^i are computed, but does not imply a specific addition scheme. The goal of the final addition is to compute $\Delta[n-1, 0]$. To perform this, internal Δslices, if any, must be also added to form bigger Δslices.

Definition 2 Δtree. A Δ tree is a tree whose nodes and leaves are Δ slices. A node $\Delta[j, i]$ has as subtrees the Δslices $\Delta[k, i]$, $\Delta[l, k+1]$, .., $\Delta[j, r+1]$ with $i \leq k \leq l \leq ... \leq r \leq j$, iff slice $\Delta[j, i]$ is computed using the concatenation of slices $\Delta[k, i]$, $\Delta[l, k+1]$, down to $\Delta[j, r+1]$.

In other words, a *Delta*slice (adder) must be formed by using consecutive smaller *Delta*slices. Otherwise, some PG_k^i may not be computed. Classical adders have a Δtree composed of the single slice $\Delta[n-1,0]$. Figure 2 presents the final $\Delta[15,0]$ obtained by adding $\Delta[15,12]$, $\Delta[11,8]$, $\Delta[7,4]$ and $\Delta[3,0]$, performed in parallel. The addition is a combinaison of two levels of four-bit adders, so-called Δtree.

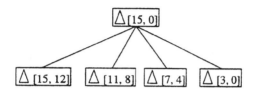

Fig. 2. Δtree of a 16 bit addition using two levels of Δslices

Theorem 3 Depth of a Δtree. *Assume all the Δslices have the same length Δlength. The Final slice $\Delta[n-1,0]$ may be constructed in $\lceil Log_{\Delta length}(n) \rceil$ levels of Δslices.*

To design efficient adders, a good tradeoff must be find between the depth of the Δtree and the Δlength of each Δslice.

3 Direct Implementation

3.1 Serial Addition

The first straightforward implementation of a n bit adder is constructed by abutment of 2 input adders. The terms P_k^0 are useless so that the only operation to be performed is: $G_k^0 = G_k^k + G_{k-1}^0 . P_k^k$, $0 \le k \le n-1$. In fact, $G_k^k = x_k . y_k = x_k \oplus y_k . x_k = \bar{P}_k^k . x_k$. As a consequence, $G_k^0 = \bar{P}_k^k . x_k + P_k^k . G_{k-1}^0$ may be implemented by a 2:1 multiplexer. The design is provided in Fig. 3

The Δtree associated with this adder is the single slice $\Delta[n-1,0]$ implemented in series. The depth of the Δtree is minimal but the Δlength maximal. The delay of such an adder is the sum of the input and output $FMAP$ as well as the carry chain of the n CY_MUX; hence the delay is:

$$t_{serial} = 2t_{FMAP} + n.t_{CY\_MUX} \tag{1}$$

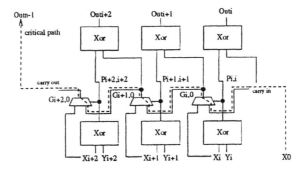

Fig. 3. serial adder using CY_MUX modules

3.2 Parallel Addition

The second implementation consists in translating a logarithmic tree of Δcells (see section 2.1) into $FMAP$ modules. The Sklanski adder has the fastest structural approach with a delay of $\lceil Log_2(n) \rceil$ Δcells. A recursive power-of-two division allows the construction of parallel adders. Figure 4 provides a 16 bit adder using Δcells (squares); the lines represent terms PG_j^is.

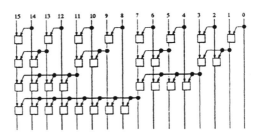

Fig. 4. 16 bit Sklanski adder using Δcells

The critical path of such an architecture goes through $\lceil Log_2(n) \rceil$ $FMAP$s without taking into account input and output Xor cells. Those modules are unsaturated since the equations of P_i^j and G_j^i have two and three inputs respectively. The adder construction is then followed by a selective collapse aiming at reinjecting small sub-functions in order to form four input sub-functions. This allows the reduction of the depth of the adder of one or two $FMAP$. Finally, duplication may be required to reduce the increasing fanout.

The Δtree is a balanced tree of two bit adders. The depth is logarithmic ($\lceil Log_2(n) \rceil$) and the Δlength minimal (2) (Fig.5).

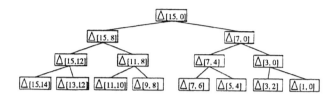

Fig. 5. 16 bit Sklanski Δtree

The delay of such an adder is roughly given by the depth of $FMAP$. In a first approximation:

$$t_{parallel} \leq (\lceil Log_2(n) \rceil + 2).t_{FMAP} \tag{2}$$

3.3 Comparison

The comparison between the two last schemes is given after place and route in section 5. However, it is obvious to see that the serial scheme delay is linear with respect to t_{CY_MUX} and the parallel scheme delay is logarithmic with respect to t_{FMAP}. Even if $t_{FMAP} >> t_{CY\_MUX}$, the parallel option is more efficient for big addition whereas the serial (or technological) option is better for smaller operands. The idea is then to combine both solutions to inherit the advantages of both of them.

4 Hybrid Solution

4.1 Principle

Serial and parallel additions may be combined by creating a Δtree using serial Δslices. As a consequence, the depth of the Δtree is logarithmic and internal additions are performed using fast CY_MUX modules. As an example, a 9 bit addition may be performed with 3-bit serial adders in Fig.6.

The critical path goes through two 3 bit serial adders and three $FMAPs$ modules instead of one 9 bit serial adder and two $FMAPs$ modules. Let us note that this solution is not efficient for small operand sizes.

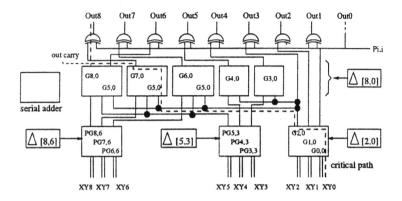

Fig. 6. performing a 9 bit addition with 3 bit serial adders

4.2 Equations

Both terms P_j^i and G_j^i have to be computed with CY_MUX modules. The equations of the first level slices are:

$$- P_j^i = P_j^j.P_{j-1}^i = P_j^j.P_{j-1}^i + \bar{P}_j^j.0$$
$$- G_j^i = P_j^j.x_j + P_j^j.G_{j-1}^i$$

and the other slices:

$$- P_j^i = P_j^k.P_{k-1}^i = P_j^k.P_{k-1}^i + \bar{P}_j^k.0$$
$$- G_j^i = G_j^k + P_j^k.G_{k-1}^i = \bar{P}_j^k.G_j^i + P_j^k.G_{k-1}^i \text{ since } P_j^k.G_j^k = 0.$$

As a consequence, all the addition equations may be expressed using a 2:1 multiplexer.

4.3 Optimal Slice Length

The delay of such an adder is estimated as follows: the critical path goes through as many Δslices as there are levels in the corresponding Δtree ($\lceil Log_{\Delta length}(n) \rceil$). The Δslices have Δlength CY_MUX modules in serie except the last which may be unsaturated, so a delay of $\Delta length.t_{CY\_MUX} + t_{FMAP}$ for the first $\lceil Log_{\Delta length}(n) \rceil - 1$ slices. Each of them but the last adds $\Delta length^{\lceil Log_{\Delta length}(n) \rceil - 1}$ inputs. As a consequence, the last Δslice has $\lceil \frac{n}{\Delta length^{\lceil Log_{\Delta length}(n) \rceil - 1}} \rceil$ inputs to add. Hence the delay can be expressed as follows:

$$t_{hybrid} = t_{FMAP} + (\lceil Log_{\Delta length}(n) \rceil - 1)(\Delta length.t_{CY\_MUX} + t_{FMAP}) + t_{last} \quad (3)$$

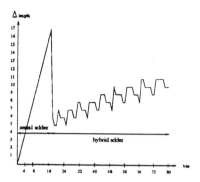

Fig. 7. optimal Δlength for classical adder size

$$t_{last} = \lceil \frac{n}{\Delta length^{\lceil Log_{\Delta length}(n)\rceil - 1}} \rceil . t_{CY\_MUX} + t_{FMAP} \qquad (4)$$

The optimal $\Delta length$ which minimizes this delay, for $\Delta length \in [2, n]$, has to be determined empirically. In figure Fig.7, the optimal $\Delta length$ is given up to 80 bit addition. Typical values of t_{FMAP} and t_{CY_MUX} are derived from previous experiments in order to take into account interconnect delays, which may amount to 2-3 times the logic block delay in such a FPGA. When $\Delta length$ equals to n (n $\leq$ 17), the hybrid solution is the serial scheme. Only for $n > 17$, the logarithmic depth is better.

5 Experimental Results

The three solutions have been implemented on our Xilinx 5200 synthesis tool. All the results are obtained after place and route. The delay of the I/O buffers are included in the figures so that these delays may be substracted to get the adder critical path in the rest of the design. The part type used for experiment is $5215PG299$ with speed grade 3.

The serial and hybrid solutions have the shortest execution time. The parallel algorithm is longer since a Boolean post-processing step (reinjection and fanout reduction) is performed. As an example, a 48 bit adder has an execution time of 1.43s, 4.88s for the serial and hybrid algorithm respectively and 64.9s for the parallel one on a Sparc20 workstation. Equations (1)-(4), derived from the adder delays, are obtained with *KaleidaGraph* tool. They are given below; The hybrid solution is about 20% better than the parallel one for 80 word length (see the coefficient of the logarithmic component of the experimental delays). Let us notice that the parallel architecture turns out to be impracticable for higher bits because of execution time and memory requirements although the serial and hybrid solutions are feasible.

$$- \; t_{serial} = 25.6 + 2.1bit$$
$$- \; t_{parallel} = \text{-}12.6 + 65.9.log_{10}(bit)$$
$$- \; t_{hybrid} = \text{-}0.9 + 53.7log_{10}(bit)$$

The serial adder is linear but the parallel and hybrid adder are logarithmic. For the bit length tested, only two levels of slices are used; more may be required for bigger operands. For significant sizes, the hybrid solution is more than two times as fast as the serial one (ratio $\frac{1}{2.25}$). The parallel solution is better than the serial only after 32 bits.

The area of the most efficient solution (hybrid) is bigger than the parallel scheme. This is because the CY_MUX output must be used only by the next CY_MUX of the carry chain. As a consequence, if the output is needed for other CY_MUX modules, duplication is required. The area estimation is the number of occupied CLBs given by the place and route report file. However, a lot of CLBs are not saturated in parallel and hybrid schemes so that actual adder area, the one which could be obtained using $RLOC$ assignments, is less than the given figures. Up to three out of four logic modules ($FMAP, CY\_MUX$) may be free. They are then two and three times bigger than the serial architecture respectively (Fig.8).

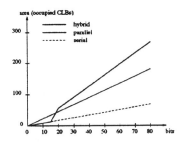

Fig. 8. adder area

6 Conclusion

The main purpose of this paper was to apply classical addition algorithm on a specific FPGA target, the Xilinx XC5200. To take advantage not only on the logarithmic delay of the parallel algorithms but also on the fast serial carry multiplexers (CY_MUX modules), a hybrid solution is proposed. Performance is increased with respect to classical approaches.

References

[1] J.Sklansky, Conditional-Sum Addition Logic, IRE Transactions on Electronic Computers, Vol. EC-9, No.2, pp. 226-231, June 1960.

[2] R.T.Brent and H.Kung, A Regular Layout for Parallel Adders, IEEE Transactions on Computers, Vol. C-31, No.3, pp. 260-264, March 1982.

[3] P.M.Kogge and H.S.Stone, A Parallel Algorithm for the Efficient Solution of a General Class of Recurrence Equations, IEEE Transactions on Computers, Vol. C-22, No.8, pp. 786-792, August 1973.

[4] T.Han and D.A.Carlson, fast Area-Efficient VLSI Adders, 8th Symposium on Computer Arithmetic, pp. 49-56, May 1987.

[5] The Programmable Logic Data Book, Xilinx 1996.

[6] A. Guyot, M. Belrhiti, G. Bosco, Adders synthesis, IFIP Workshop on logic and architecture synthesis, pp. 107-117, December 1994.

[7] A. Bosco, M. Belrhiti, B. Laurent, Adder Solution Space Exploration for Speed/Area Trade-offs, IFIP Workshop on logic and architecture synthesis, December 1996.

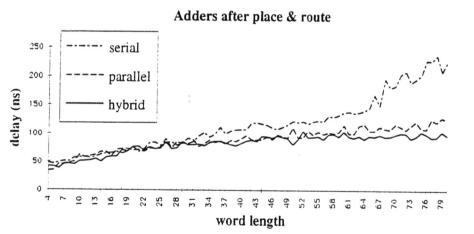

FPGA Implementation of Real-Time Digital Controllers Using On-Line Arithmetic

Arnaud Tisserand[1], Martin Dimmler[2]

[1] Laboratoire de l'Informatique du Parallélisme. Ecole Normale Supérieure de Lyon
46 allée d'Italie, 69364 LYON Cedex 07 France
Email Arnaud.Tisserand@lip.ens-lyon.fr
[2] Institut d'Automatique. Ecole Polytechnique Fédérale de Lausanne
CH-1015 Lausanne, Switzerland.
Email dimmler@ia.epfl.ch

Abstract. The implementation of real-time digital controllers for low-power compact high performance mechanisms is a challenging task in the design of mechatronic micro-systems such as robots, fine positioning devices and intelligent drives. While ASIC and DSP solutions lead to very expensive design processes and high-power consumption realizations, respectively, FPGA implementations using parallel arithmetic are limited by size constraints. Serial arithmetic with most significant digit first, called *on-line arithmetic*, offers the necessary size reduction and acceleration of speed. In this approach traditional functions, such as analog to digital conversions and control data calculations can be combined. FPGA implementation of digital control algorithms using on-line arithmetic leads to efficient real-time realizations with small size, high speed and low power consumption.

1 Introduction

The growing capacity of programmable logic devices offers new possibilities for compact high precision mechanisms. Their controllers are mostly based on a very restricted instruction set processor. In this context FPGAs offer a cheap and compact (no external memory) alternative over a costly and time-consuming development of an *Application Specific Integrated Circuit* (ASIC) or a *digital signal processor* (DSP) environment. Especially for aerospace applications the reduced number of individual components plays a key role because of the very strict specifications. However, the trend to smaller dimensions, improved dynamic behavior and lower power consumption leads to future controller requirements which are difficult to fulfill with ordinary parallel arithmetic implementations on FPGAs.

The problems arising with the adaptation of the classical scheme to the needs of micro-system control are very apparent for *multiple inputs multiple outputs* (MIMO) systems (see 3). They require a large number of calculations of type *Matrix × Vector*. A further acceleration of the calculation speed becomes necessary, but is only possible either by using the strong parallelism or by the acceleration of single arithmetic units. However, these measures lead to increase space and speed requirements, respectively, and are therefore inconvenient for systems with limited size and low power consumption (see Fig. 1a,b).

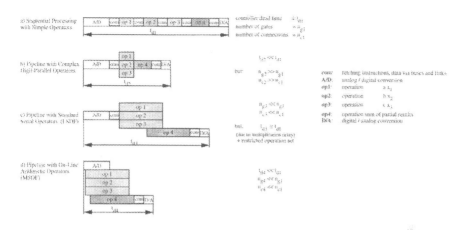

Fig. 1. Timing and size aspects of the computation $ax_1 + bx_2 + cx_3$

In most mechatronic systems the controller is a stand-alone unit which repeats cyclically some operations with very few user interactions. Therefore, a *pipelined* architecture of different operators could compute the appropriate control signals. Pipeline in this context should mean that the whole control algorithm is realized in hardware as a complex operator, e.g. on a FPGA. Digit-serial processing schemes [1] offer hereby an interesting alternative to digit-parallel computation. In spite of slower single operators, their overlapped operations keep the overall calculation time low (see Fig. 1c,d) and their digit-wise processing offers the necessary reduction of the size of the operators and a low connection number. Data transmission in a digit-serial fashion may hereby be done either with *least significant digit first* (LSDF) or with *most significant digit first* (MSDF). Algorithms for LSDF transmission seem to be more natural (standard "paper and pencil" methods for addition and multiplication), but they have some major disadvantages:

- A/D conversion and calculation cannot be overlapped (different digit order).
- Multiplications cause additional delay (result has twice as much digits than operands).
- Not all arithmetic operations are available in LSDF, e.g. division (see [2]).

In order to overcome the above mentioned problems while still keeping the advantageous features of digit-serial computations, like low gate number and interconnection number, we propose the use of a processing scheme in the MSDF direction, the so called on-line arithmetic (see 2). It allows to overlap A/D conversion and computation (see Fig. 1d). This leads to a reduced processing time and therefore to a lower necessary clock frequency than for the classical schemes. Because of the lower gate number, the consequence is also a much lower power consumption of the controller. The above mentioned additional delay for multiplications disappears and algorithms for all arithmetic operations (including division) are available (see for example [3, 4]). Changing the calculation direction is possible thanks to a *redundant number system* (see 2.1).

The mathematical aspects of on-line algorithms have been studied by several authors (see e.g. [3, 4]) during the last twenty years. However, they were rarely implemented and some important extension to their mathematical description have to be made before they can be used in control systems. In the next sections we will give a short introduction to the main ideas of on-line algorithms and digital control algorithms followed by a detailed description of the necessary modification of the original operators for implementations on FPGA. This offers the opportunity of building an on-line operator library for the modular development of micro-system controllers.

2 Introduction to On-Line Arithmetic

In this section we will present some of the basic concepts behind on-line arithmetic. For more details the reader is referred to the literature (e.g. [3, 4]). In on-line arithmetic the operands, as well as the results, flow through arithmetic units in a digit-serial fashion starting with the *most significant digit first* (MSDF). Important characteristics of on-line operators are their *delay* and *period*. As illustrated in figure 2, the delay is the number δ such that i digits of the result can be deduced from $i + \delta$ digits of the operands. Usually, the on-line delay is a small integer (e.g. 1 to 4). The clock *period* τ is the time needed by the signal to cross through the longest path of the circuit.

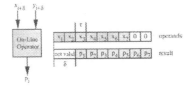

Fig. 2. Delay and Clock Period of On-Line Operations

Using a redundant number system (see 2.1) additions can be performed without any carry propagation. Therefore, computations of additions and multiplications can be performed serially with most significant digits first. This special digit-serial arithmetic has been introduced by Ercegovac and Trivedi in 1977 [5] and was further studied by several authors (see e.g. [6, 4]). Nowadays, there are on-line algorithms for all common arithmetic operations available, in the fixed-point representation as well as in the floating-point representation.

2.1 Redundant Number Systems

In a usual number system, a positive fractional number $A \in \mathbb{R}^+$ is written using a radix r $(r > 0)$ as $\sum_{k=1}^{\infty} a_k r^{-k}$, where for all k, $a_k \in \mathcal{D} = \{0, 1, \ldots, r-1\}$. $\mathcal{D}$ is called the digit set. In [7], a number system is proposed in which radix r numbers are represented with a signed digit set $\mathcal{D}_r = \{-a, -a+1, \ldots, a-1, a\}$ where $a \leq r - 1$. If the number of elements in $\mathcal{D}_r$ is larger than r then some numbers have several representations: the system is redundant. The sign assignment is done on the digit level. Thus, negative numbers are treated in a similar way.

It is dealt here with radix 2. On-line arithmetic in radix 2 has been studied in [3, 4]. The digit set is $\{-1, 0, 1\}$, and we use a bit level representation of the digits, called *borrow-save*, defined as follows: the i-th digit a_i of a number a is represented by two bits, a_i^+ and a_i^-, such that $a_i = a_i^+ - a_i^-$. Thus, digit 1 is represented by $(1, 0)$, digit -1 is represented by $(0, 1)$, while digit 0 has two possible representations, namely $(0, 0)$ and $(1, 1)$. For the conversion between standard and redundant radix 2 numbers see 2.4.

2.2 Example of an On-Line Operation: Addition

It is shown in [4] that the digits s_i of s ($s = a + b$) can be obtained either with the parallel carry free architecture presented in figure 3(left) or with the on-line operator of figure 3(right). The comparison with respect to operator size clearly shows the advantage of the on-line scheme, already for $n = 6$.

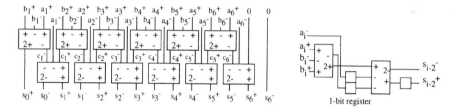

Fig. 3. Borrow-Save parallel and on-line adders (ranks are indicated by indexes)

The main building blocks for both algorithms are PPM cells (*plus plus minus*, see [4]), which reduce 3 bits, x_i, y_i and z_i of same rank to 2 bits, t_{i-1} and u_i, one of the same rank and one of previous rank, so that $x_i + y_i - z_i = 2t_{i-1} - u_i$. Subtractions are realized by exchanging positive (a^+) and negative bits (a^-) of the subtracted operand. See [4, 3] for the multiplication, division and square-root operators.

2.3 Speed and Size of Redundant Arithmetic

Table 1 shows the time and area complexity of the main arithmetic operators using parallel, LSDF and on-line arithmetic(the time complexity of the LSDF and on-line arithmetic is obviously $O(n)$).

2.4 Conversions

The conversion from a standard radix 2 number $s = \sum_{i=1}^{n} s_i 2^{-i}$ to a borrow-save number $b = \sum_{i=1}^{n} b_i 2^{-i}$ is obvious ($b_i^+ - b_i^- = s_i$ with $b_i^+ = s_i$ and $b_i^- = 0$ for example). In the case of a 2's complement number the most significant digit has a negative weight. The conversion from a borrow-save number to standard radix 2 number (or in 2's complement notation) needs the time of a usual LSDF addition (with carry propagation) $time(a = a^+ - a^-) = O(\log_2 n)$ or two D/A converters $(voltage(a) = voltage(a^+) - voltage(a^-))$, where $a^+ = \sum_{i=1}^{n} a_i^+ 2^{-i}$ and $a^- = \sum_{i=1}^{n} a_i^- 2^{-i}$.

	parallel		LSDF	on-line
	time	area	area	area
$\pm$	$O(1)$	$O(n)$	$O(1)$	$O(1)$
$\times$	$O(\log_2 n)$	$O(n^2)$	$O(n)$	$O(n)$
$\div$	$O(n)$	$O(n^2)$	impossible	$O(n)$
$\sqrt{}$	$O(n)$	$O(n^2)$	impossible	$O(n)$
$AX + B$	$O(\log_2 n)$	$O(n^2)$	$O(n)$	$O(n)$

Table 1. Main Arithmetic Operators Time-Area Complexity

3 Digital Control Algorithms

In this section we will give a basic idea of digital control algorithms and how they are used to improve the dynamic behavior of micro-systems. A dynamical system is characterized by its input-output behavior. Here we will treat linear systems, i.e. systems where amplification and superposition principle hold. For micro-systems linear behavior can often be already obtained by design (small dimensions, precise manufacturing methods). However, even the best model is only a simplified description of the real system. Due to model mismatch as well as unmodelled perturbations the model and the real system, driven by the same input signals would differ in the output values. An open-loop strategy (no measurement feedback) would therefore lead to low precision. Also, unstable plants cannot be controlled by an open-loop strategy. In digital control systems these problems are avoided by taking measurements of output and state values (intermediate system values like currents and speeds) of the physical system and subsequently calculating the appropriate input values in order to follow a given trajectory. The analog measurements are digitalized by A/D converters and D/A converters are used to convert the numerical results back to analog input values. The corresponding control scheme is shown in figure 3.

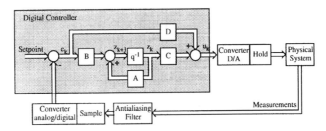

Fig. 4. Digital Control System

Usually, algorithms in the state-space form are applied for linear tracking problems: $z_{k+1} = Az_k + Be_k$ and $u_k = Cz_k + De_k$, where A, B, C are constant matrices, z_k the controller states, u_k the control variables and e_k the errors between setpoints and measurements. In the case where e_k and u_k are scalars the system is called a *single input single output* (SISO) system, otherwise a *multiple inputs multiple outputs* (MIMO) system. The closed-loop system dynamics are determined by the choice of the constant matrices of the previous equations. The only mathematical operations are multiplications and additions. Calculation complexity is dependent on controller order and the input/output dimensions.

In this paper we will use the classical *proportional integral differential* (PID) control algorithm as a simple example for implementation. It is the most frequently used algorithm in todays industrial applications. The state-space representation of this controller is given as:

$$z_{k+1} = Az_k + Be_k \tag{1}$$

$$u_k = de_k + z_{1,k}$$

where $\quad z_{1,k} = \begin{pmatrix} 1 & 0 \end{pmatrix} z_k \quad$ and $\quad A = \begin{pmatrix} a_{11} & 1 \\ a_{21} & 0 \end{pmatrix} \quad B = \begin{pmatrix} b_1 \\ b_2 \end{pmatrix} \tag{2}$

where e_k is the error between set point and measurement, z_k the controller states, u_k the controller output and a_{11}, a_{21}, b_1, b_2, d constants.

4 Implementation of On-Line Arithmetic for Control Algorithms

Since the upcoming of programmable logic, such as *field programmable gate arrays* (FPGA), efficient implementations of algorithms in hardware can be made with software-like design principles (for examples see [8]). In order to simplify the controller design for the end-user modular operator libraries become necessary. These are obtained by addition of some important extensions to the existing mathematical on-line algorithms. In this section we will describe the necessary concepts of initialization, synchronization and normalization and give some implementation guidelines for their realization in programmable logic circuits.

4.1 Initialization

The cyclic repetition of the control algorithm in combination with the serial processing scheme demands an initialization of each operator before every new computation. This partial initialization can be done either by a global or distributed control scheme (see figure 5).

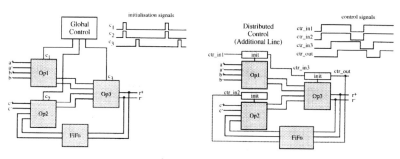

Fig. 5. Global and Distributed Control of Initialization

In the former solution a central control unit supervises the whole digit flow in the control algorithm and sends initialization signals to the operators if necessary. It is mainly based on counters and comparators which detect the important events. It is rather costly in logic and design effort, because it has to be especially adapted to the current operator architecture. On the contrary, a distributed control scheme realizes the initialization by an additional line synchronized to the numbers. This control signal has a value of 1 if there are valuable digits and 0 otherwise (see figure 5). An initialization signal is sent to the operator if the control signals at in- and output are 0 at the same time. This concept was chosen for our implementation because it improves the modularity of the design. Both concepts use at least one clock cycle for the reset operation.

4.2 Intermediate Zeros

In both initialization schemes the digits of the result must have left the operator entirely before initialization. Otherwise, the last δ (on-line delay) digits would be lost. Some number of intermediate zeros between the operands, which is at least equal to the biggest on-line delay of a single operator in the pipeline (δ_{max}) plus one for the initialization, is therefore necessary (see figure 5 right side). The additional digits increase the amount of needed clock cycles for the cyclic computation. However, due to the sampler, the A/D converter already requires a certain period of time for stabilization ($t_{sampler}$) before the converted number can be serially sent to the controller. This time is often in the same order as $(\delta_{max} + 1) \times clock\, period$ and has to be considered in form of intermediate zeros anyway.

In order to maintain the intermediate zeros even after several operations in the pipeline, multiplexers are needed which avoid the interference of subsequent numbers via the static logic of the operator. This output switching can be realized with the aid of the additional control line which was added for initialization of operators.

4.3 Synchronization in Loops

Algebraic loops, like for the controller states z in (1), require some logic for synchronization of forward and backward flow. Here we will present two solutions corresponding to the concepts in figure 5. In the case of distributed control, appropriate shift registers are included in the backward branch which take into account the operator delays as well as the number of digits and intermediate zeros. This leads to area saving but specific designs. An exchange of the A/D converter can lead to a necessary re-adaptation of register sizes (if $t_{sampler} > \delta_{max} \times clockperiod$). On the contrary, in the global control scheme shift register size are independent of intermediate zeros. Synchronization is obtained by deactivation of the clock in the backward branch until the new operand arrives at the forward operator (Op3 in figure 5). This results in very general solutions which can be used with different converters. However, the drawback of the global solution is the high number of necessary gates and buffers for the generation of "wait" cycles. Usually, the smaller distributed control scheme is preferred because the specification of the A/D converters is known in advance or register sizes can be adapted by reprogramming of the logic circuit (on FPGAs).

4.4 Normalization in Loops

A special property of redundant arithmetic has already been implicitly mentioned in the addition example of 2.2. There the sum was represented by $n + 1$ valuable digits whereas the operands only by n. This is not only due to the possible overflow (which can be avoided by appropriate scaling) but also due to several possible representations for fractional numbers (e.g. $1.0(-1) = 0.11$). In order to avoid a continuously growing number of digits after additions, especially in algebraic loops (infinite number of additions), an on-fly conversion to a limited representation is necessary. This could be done by the complete on-fly conversion algorithm presented by Ercegovac and Lang [9] which converts the redundant numbers into conventional binary representations. However, this would cause an additional delay of n clock cycles. We propose an on-fly conversion algorithm without delay which generates a redundant fractional number with zero unit part (i.e. $0.s_1 s_2 s_3 \ldots$). With an appropriate scaling ($|s| < 1$) only two types of results need conversion:

- $1.0 \ldots 0(-1)s_{i+1} \ldots s_n \rightarrow 0.1 \ldots 1 s_{i+1} \ldots s_n$
- $(-1).0 \ldots 01 s_{i+1} \ldots s_n \rightarrow 0.(-1) \ldots (-1)s_{i+1} \ldots s_n$

In order to obtain a zero unit digit the 1 (-1) has to be propagated to the right until the first (-1) (1) appears, i.e. all digits are changed to 1 (-1) including the final (-1) (1). In case of overflow the closest possible value appears on the output ($0.1 \ldots 1$ or $0.(-1) \ldots (-1)$, respectively). The corresponding scheme is shown in figure 6. It needs the space of 10 Actel 2 cells which is not important if only used 2 or 3 times in a design (operator combinations reduce occurrence, see 4.5).

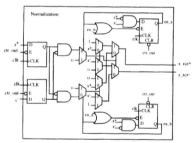

Fig. 6. On-Fly Conversion Operator

4.5 Combinations of Different Operators

For computations of type $Matrix \times Vector$, on-line arithmetic offers an efficient simplification. The static logic part of all multiplications can be executed in parallel without any on-line delay and the final addition can be performed by a simplified final adder with a much smaller delay than a binary tree of adders. The necessary simplification algorithm is given in [4]. This simplification keeps the overall delay of the control algorithm and the number of individual operators very low and reduces therefore the controller dead time and the number of necessary on-fly conversion units. Additionally, scaling is limited to a few intermediate results.

The resulting operator structure for our example is shown in figure 7. After simplification only 3 complex operators are left.

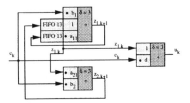

Fig. 7. Resulting Arithmetic Structure for the Controller Example, the Coefficients Correspond to the Matrices in (2)

Under consideration of the above mentioned measures, a library of basic operators can easily be constructed and subsequently used to construct controller algorithms in a common way: either by graphical modules or by formal description in a high level language (e.g. VHDL). A formal description is mostly preferred because of its better possibilities for validation and implementation on different hardware platforms. As shown in figure 8 each arithmetic operator is thereby composed by 4 main building blocks: initialization, mathematical algorithm, normalization and output switch.

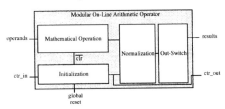

Fig. 8. Modular On-Line Arithmetic Operator

Once the basic arithmetic modules are available implementing a controller using on-line arithmetic is very similar to writing a control program for a microcontroller or DSP. The only difference is due to the compiler: from high level language into specific gate array code (stored in gate array structure) instead of machine code (stored in memory).

5 Experimental Results

The resulting operational scheme for the above mentioned PID example was shown in figure 7 (without explicit indication of normalization, initialization and output switch). It includes 2 inner loops with appropriate registers and 3 *Matrix* × *Vector* operators with simplified final adders. The overall delay of the whole algorithm is $\delta_{alg} = 3$, i.e. 3 clock cycles after the first digit left the A/D converter the result begins to appear on the exit of the operator. The largest on-line delay of a simple operator in the pipeline is $\delta = 3$. This leads to 4 intermediate zeros to permit initialization in the above described manner.

The scheme was implemented with a resolution of 12 bits in an Actel 1240 device, which is a low cost FPGA in anti-fuse technology (for specification details see [10]). The circuit can be driven with a maximal clock frequency of about 20 MHz. An implementation in parallel arithmetic with the same resolution would have already the size of 4 FPGAs of same type for this simple algorithm.

6 Conclusion

In this paper we have shown that a serial processing scheme based on *on-line arithmetic* is particularly well suited for the implementation of real-time control on FPGAs. It was pointed out that the small operator size allows an implementation in programmable logic while still providing high precision and speed as well as a low power consumption. An introduction to on-line arithmetic operators as well as digital control algorithms was given and guidelines for their control-specific implementation were presented. The shown concepts allow a creation of a modular on-line library which takes away the complexity of controller design from the end user and which can be used for fast prototyping of digital controller on FPGAs.

Acknowledgment

The authors would like to thank the CSEM, Neuchatel, Switzerland, for the financial support , Dr. Moerschell and Dr. Holmberg for many fruitful discussions.

References

1. R. Hartley and P. Corbett, "Digit-serial processing techniques", *IEEE Transactions on Circuits and Systems*, vol. 37, no. 6, pp. 707–719, June 1990.
2. M.D. Ercegovac and T. Lang, "Module to perform multiplication, division and square root in systolic arrays for matrix computations", *Journal of Parallel and Distributed Computing*, vol. 11, pp. 212–221, 1991.
3. M.D. Ercegovac, "On-line arithmetic: an overview.", in *SPIE, Real Time Signal Processing VII*, SPIE, Ed., 1984, pp. 86–93.
4. J.C. Bajard, J. Duprat, S. Kla, and J.M. Muller, "Some operators for on-line radix-2 computations", *Journal of Parallel and Distributed Computing*, vol. 22, pp. 336–345, 1994.
5. M.D. Ercegovac and K.S. Trivedi, "On-line algorithms for division and multiplication", *IEEE Trans. Comp.*, vol. C-26, no. 7, pp. 681–687, 1977.
6. M.J. Irwin and R.M. Owens, "On-line algorithms for the design of pipeline architecture", in *4th Symposium on Computer Architecture*. 1979, IEEE Computer Society Press.
7. A. Avizienis, "Signed-digit number representations for fast parallel arithmetic", *IRE Transactions on Electronic Computers*, vol. 10, pp. 389–400, 1961, Reprinted in E.E. Swartzlander, Computer Arithmetic, Vol. 2, IEEE Computer Society Press Tutorial, 1990.
8. Doug Conner, "Reconfigurable logic: Hardware speed with software flexibility", *EDN Europe*, pp. 15–23, July 1996.
9. M.D. Ercegovac and T. Lang, "On-the-fly conversion of redundant into conventional representations", *IEEE Transactions on Computers*, vol. C-36, no. 17, pp. 895–897, July 1987.
10. Actel, *ACT Family FPGA Databook*, Actel Corporation, 1996.

A Prototyping Environment for Fuzzy Controllers

Thomas Hollstein and Andreas Kirschbaum and Manfred Glesner

Darmstadt University of Technology,
Institute of Microelectronic Systems,
Karlstr. 15, D-64283 Darmstadt, Germany,
{thomas|andreask|glesner}@mes.th-darmstadt.de

Abstract. In this paper we present a complete environment for hardware prototyping of generic fuzzy controller architectures. The system specification is based on a generic VHDL module library which is handled by an integrated source- and documentation administration tool. The targeted hardware emulator has been enhanced with off-the-shelf memory to overcome the inherent memory limitations of the FPGA-based platform. The implementation of a Midpoint-of-Area-Defuzzification unit (MOA) demonstrates the feasibility of this approach.

1   Introduction

There exist different approaches for the realization of fuzzy controllers depending on the computational power required. Pure software solutions are suitable for application areas with relaxed timing constraints such as consumer electronics (e.g. washing-machines, vacuum cleaners) or low-dynamic mechanical processes (e.g. oil-hydraulic). Existing implementations range from standard microcontrollers to controllers with fuzzy-specific instruction sets. In contrast, applications with high system dynamics (e.g. in the automotive or avionic area) require low latency controllers which can be realized in hardware only. Comparing different implementation techniques with respect to design costs, programmable digital solutions turn out to be most suitable. Analog hardware realizations as well as dedicated fuzzy-datapaths provide high operational speed but suffer from their non-programmability. This results in high development costs compared to a versatile architecture which can be reused for different applications. Our proposed approach combines the programming flexibility with the advantages of specialized datapath architectures.

Providing a generic fuzzy controller toolkit, the architecture of the controller can be customized by the designer depending on the applications' constraints. Initially the parallelism of execution units (e.g. number of rule evaluators, defuzzification units etc.) as well as the representation of the fuzzy membership functions (degree of overlap, format of storage) have to be determined. Estimates of chip area and performance data for the generated fuzzy controller allow the designer to target a specific area in the possible design space. Application specific programming of membership functions and fuzzy premises assure the controllers flexibility.

In order to provide validation of the system functionality in an early design step rapid-prototyping has to be incorporated into the design flow. This allows the fuzzy controller to be tested in a hardware-in-the-loop simulation in its real environment.

The remainder of the paper is organized as follows: The concept of the generic architecture of the fuzzy controller is presented in Sect. 2. Design specification, hardware configuration and simulation is covered in Sect. 3. In Sect. 4 the rapid-prototyping design flow and our memory enhanced prototyping target hardware is discussed. Section 5 presents a MOA-Defuzzification unit as an application example. Section 6 concludes the paper.

2 Generic Fuzzy Controller

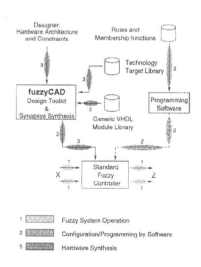

Fig. 1. Design and Configuration Overview

Several approaches for hard-wired or coprocessor realizations of fuzzy systems have been published in the recent years [Hun95] [SU95]. The concept of our fuzzy controller design kit **fuzzyCAD** is different, since it permits the user to configure a controller according to his specific requirements. Additionally a certain degree of flexibility is maintained, because fuzzy premises and membership functions of linguistic variables have not to be fixed beforehand. Furthermore the design kit described in this section is based on a generic approach, which allows the designer to target a desired realization in the space of possible solutions (performance vs. area). The functionality of the controller is related to a classical fuzzy system, introduced by Mamdani [MA75]. In contrast to our initial full hardwired prototyping approach described in [HHKG94], rules can be reprogrammed and added during the use of the hardware. This is essential, since finally an ASIC realization is targeted, which has to be flexible concerning subsequent modifications of fuzzy premises. Figure 1 depicts the design and programming structure.

An overview on the basic architecture of the controller is presented in Fig. 2. The fuzzification unit maps the values of n_{inp} crisp input variables on membership degrees of associated fuzzy sets. It is assumed, that for any crisp input value at maximum n_{ov} membership functions return a non-zero membership degree. Therefore membership functions are stored in n_{ov} RAM benches in order to increase the parallelism of processing. The parameter n_{ov} can be fixed by the designer as well as n_{inp} and the bitwidth of input variables. Membership functions are parameterized by bitwidth and resolution. The rule evaluation unit consists of n_{pe} processing elements for parallel calculation of the results of fuzzy premises. Three types of processing elements can be implemented with respect to different rule complexities (trivial, and/or bracket hierarchies). The results are collected in local output RAMs before transferring them to the defuzzification unit.

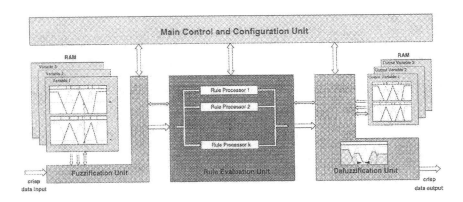

Fig. 2. Structure of the Fuzzy Controller

Concerning computational effort, the inference and defuzzification unit is the bottleneck of the system. A defuzzification unit may be structured with respect to sequential processing of multiple output variables (Fig. 2). The technique for the storage of membership function is identical to the method used in the fuzzification unit. For the defuzzification itself we have developed a hardware related efficient method, called Midpoint of Area (MOA) . The center of the defuzzification area is computed by simultaneous partial integration, starting at both boundaries of the total area (Fig. 3). The objective during integration is to keep the size of the two growing partial areas almost identical. The crisp defuzzification value is simply obtained by the meeting point of the partial integrations.

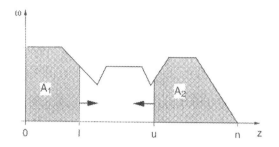

Fig. 3. MOA Defuzzification Method

For a more detailed discussion of the architecture and a comparison of defuzzification strategies, we refer to [HHG96]. The operation of the datapath as well as the configuration and download of membership function shapes and fuzzy rules is coordinated by a central main control unit.

3 Design Specification and Environment

The fuzzy controller has been specified as a hierarchical generic VHDL module library. In order to maintain numerous VHDL source files, a HTML-based design management system has been set up, providing transparent hierarchical handling of program sources and related documentation. VHDL configuration files can be parsed automatically resulting in a HTML page which allows convenient access to instantiated sources files. Simulation stimuli can be generated using
a configuration editor for membership functions and input waveforms. The editor for membership functions is depicted in Fig. 4. Generic parameters are also set be the design configuration software according to designer definitions.

4 Hardware Prototyping

Design validation is a key issue in nowadays complex system design. There exist different validation techniques at different steps of the design process ranging from formal verification, simulation and emulation up to simple intuition [Lee96]. Due to complexity reasons formal verification is only feasible for selected system components and therefore validation techniques relying on simulation or emulation turn out to play a major role.

There are two main reasons for emulating designs instead of using the more flexible simulation technique: speed and availability of simulation models. Whenever a huge amount of test vectors is necessary for validating the design simulation speed becomes a bottleneck. Emulation can speed up validation typically

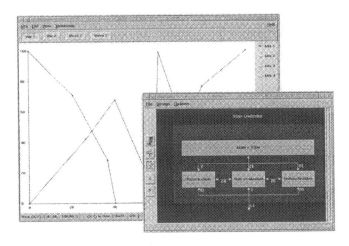

Fig. 4. Editor for Membership Functions and Configuration Tool

with a factor of 10^6 compared to simulation. Usually the system under design has to be tested in its real process environment for which not necessarily simulation models exist. In case of a fuzzy controller, which is part of a closed control loop, it is very likely that there are no simulation models available for e.g. mechanical processes which are to be controlled. This implies the necessity of a hardware-in-the-loop simulation of the control loop including an emulated version of the fuzzy controller.

Nevertheless migrating designs for emulation often introduce changes in the design itself as well as in its environment. As emulation speed is typically a factor of ten slower than the final system it might be necessary to slow down the environment for correct interfacing with the emulated design. Also parts of the system such as memories or macrocells have to be replaced for emulation purpose because there do not exist equivalent modules in the prototyping hardware. Instead, either off-the-shelf components such as RAM devices are integrated in the prototyping hardware or the lacking parts are replaced using exclusively emulator resources e.g. re-synthesis of multiplier-macrocells. In general, for a better acceptance of this prototyping methodology design changes should be guided by CAD tools to minimize the necessary user interference.

For prototyping of the fuzzy controller a commercial FPGA-based emulator system (*Quickturn Logic Animator*) is used. The system can emulate designs with up to 50 K gates with an average frequency of 8 to 16 MHz [Qui94]. All nodes of the emulated design, i.e. interface pins as well as internal signals and registers are accessible through 448 bidirectional probes. A CAD-environment allows for almost automatic configuration of the emulator. A key problem which is still to be solved by the designer is memory implementation. In a straight forward realization memory blocks can be implemented by allocating hardware resources on the FPGAs itself. Doing this, the look-up tables in the configurable logic blocks of the XILINX FPGAs are interpreted as 16x1 RAM blocks instead of being used as function generators. Nevertheless this approach realizes memory at the expense of logic emulation capacity, which reduces the maximum allowable gate size of the design to be emulated enormously. Taking into account all resources of the emulator a maximum of 55 KByte of memory may be emulated; i.e. allocation of 27.5 KByte of memory will halve the total logic capacity of the emulator. Therefore memory intensive designs can only be realized with external RAM components. For these applications a dedicated Extension Interface Module (EIM) has been developed as memory enhancement card for the emulator (Fig. 5).

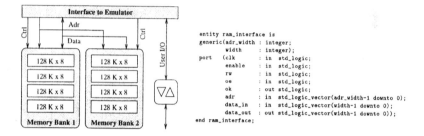

Fig. 5. Memory Card and Unified Interface.

The memory card provides two 32-bit wide memory banks with 128K words each [Dar96]. Depending on the application the memory banks can either be equipped with fast SRAM devices (15 ns) or FLASH-ROMs (55 ns) or a combination of both. For the fuzzy controller the membership functions as well as the fuzzy rules can be kept in the permanent FLASH memory so that they do not have to be reprogrammed when the system is powered up. In order to provide a smooth integration of the different memory implementations (e.g. FPGA look-up tables, SRAM and FLASH devices on the EIM, RAM macrocells) in the overall design flow a unified memory interface has been developed (Fig. 5). The fuzzy controller communicates through this interface with the memory devices which internal behavior is hidden by dedicated VHDL interface routines. This allows for easy retargeting of the controller not only for emulation but also for later standard cell implementations where RAM macrocells are used.

According to Fig. 6 the hardware prototype of the fuzzy controller is generated as follows: The controller is completely specified using the WWW-based **fuzzyCAD** design environment (see Sect. 3). Especially for the purpose of emulation different RAM implementations (internal look-up tables, external off-the-shelf components) and their appropriate interfaces can be chosen within **fuzzyCAD**. The resulting RTL-level VHDL code of the fuzzy controller is fully synthesizable. A standard RTL-synthesis tool (*Synopsys Design Compiler*) is applied to get the gate-level netlist of the design in the targeted technology (stan-

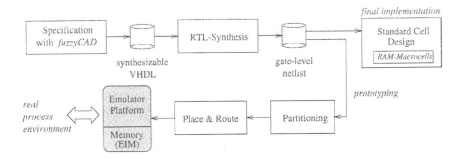

Fig. 6. Rapid-Prototyping Design flow.

dard cells or FPGAs for prototyping). After partitioning the design to multiple FPGAs and placement and routing of each FPGA, the generated bitstreams can be downloaded immediately to the emulator platform. Memory devices located on the memory extension card are automatically integrated. After a few hours (varying with the chosen generic parameters of the design) the prototype of the fuzzy controller will be ready for testing in its real process environment.

5 Application Example

In order to demonstrate the feasibility of the described prototyping approach a MOA-defuzzification unit has been prototyped. The generic VHDL model of this unit is parameterizable with respect to the number of defuzzification variables, bitwidths of membership functions and weights, and the degree of overlap of membership functions. The latter determines the number of independent memory banks for the defuzzification unit. Membership functions $\mu_n(x)$ which do not overlap can be placed in one memory, if a function number is added in order to correlate it later with the appropriate rule weight α. Instead of storing each functional value the membership functions are defined in a vector format to save extra memory.

The results α_n of the evaluation of the rule premises as well as the membership functions itself are stored in separate memory banks M_{α_n} resp. $M_{\mu_n(x)}$ for each group of non-overlapping membership functions (Fig. 7). The following calculation module executes inference and defuzzification and determines the crisp output value for each variable according to the algorithm described in Sect. 2. The different memory blocks are accessed only via the unified interface and therefore the design is independent of the concrete memory implementation.

The defuzzification unit discussed here can be characterized as follows: one output variable, overlap=2, bitwidth=12 for membership functions and rule weights, resolution=2400 points per membership function. Two different prototypes have been realized. One implements memories with the FPGA look-up

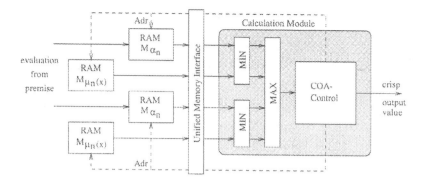

Fig. 7. MOA-Defuzzification Unit for one variable with overlap=2.

tables and the other uses RAM and FLASH devices located on the External Interface Module. Both prototypes can be clocked with approx. 5.5 MHz which results in 5 ms latency for inference and defuzzification independent of the number of rules. A speed up of 5-10 can be expected for the final standard cell implementation. The design comprises 4819 gates without the memory elements.

6 Conclusions

In this paper we have presented a prototyping approach for fuzzy controllers. The developed **fuzzyCad** environment covers all design phases from specification up to automatic prototype generation. The generated fuzzy controller combines programming flexibility with high-speed operation of a specialized datapath architecture. The generic VHDL description of the controller allows for application driven configuration of main parts of the generated architecture, e.g. number of rule evaluation / defuzzification units, bitwidth and resolution of membership functions etc. A standard logic emulator has been enhanced with a memory extension card to overcome the inherent memory limitations of the FPGA-based platform and make it usable for prototyping fuzzy controllers. The feasibility of our approach has been demonstrated by presenting the prototype generation of a MOA-Defuzzification unit.

Currently we assemble the whole fuzzy controller consisting of fuzzification, rule evaluation and defuzzification unit and use it for a fuzzy controlled driver assistance for trucks. In the future we plan to extend the emulator with analog interfaces in order to simplify the integration of the platform in a hardware-in-the-loop simulation. Additionally the structural VHDL description of the fuzzy controller will be replaced with a behavioral description in order to allow for high-level optimization (scheduling) of the timing properties of the resulting architecture.

References

[Dar96] Darmstadt University of Technology: Memory Extension Interface Module (M-EIM) User's Guide. V 1.0 (1996)

[HHG96] Hollstein, T. and Halgamuge, S. and Glesner, M.: Computer Aided Design of Fuzzy Systems based on generic VHDL Specifications. IEEE Transactions on Fuzzy Systems Vol.4 No.4 (1996) 403–417

[HHKG94] Halgamuge, S. K. and Hollstein, T. and Kirschbaum, A. and Glesner, M.: Automatic Generation of Application Specific Fuzzy Controllers for Rapid Proto-typing. IEEE International Conference on Fuzzy Systems, Orlando (1994)

[Hun95] Hung, D. L.: Dedicated Digital Fuzzy Hardware. IEEE MICRO Vol.15 No.4 (1995)

[L+93] Lipsett, R. et al.: VHDL: Hardware Description and Design. Kluwer Academic Publishers, Boston (1993)

[Lee96] Lee, A.: Comparing Models of Computation. Embedded Tutorial given at International Conference on Computer Aided Design (1996)

[MA75] Mamdani, E. H. and Assilian, S.: An experiment in linguistic synthesis with a fuzzy logic controller. IJMMS 7 (1975)

[Qui94] Quickturn Design Systems: Quickturn Logic Animator Design Compilation User's Guide. V 1.0 (1994)

[SU95] Surmann, H and Ungering, A. P.: Fuzzy Rule-Based Systems on General-Purpose Processors. IEEE MICRO Vol.15 No.4 (1995)

A Reconfigurable Sensor-Data Processing System for Personal Robots

Kazumasa Nukata, Yuichiro Shibata, Hideharu Amano, Yuichiro Anzai

Department of Computer Science, Keio University
3-14-1, Hiyoshi Yokohama 223 Japan
{nukata,shibata,hunga,anzai}@aa.cs.keio.ac.jp

Abstract. A reconfigurable sensor-data processing system for personal robots is proposed. This system consists of two reconfigurable units including Altera's CPLDs which are directly connected with sensors, and is implemented on the interface board of a personal robot prototype *ASPIRE-II*. By treating digital raw data directly, flexible and precise processing can be achieved. In this system, the correlation function of data from ultrasonic sensor is computed about 20 times faster than that with the RISC processor in *ASPIRE-II*.

1 Introduction

In order to realize a computing system which can handle not only electronic information treated by conventional computer science but also physical information, personal robots have been researched. Personal robots, like personal computers, are owned by individuals and can naturally interact with humans[2][3].

A sophisticated and high speed sensor system including ultrasonic sensors, heat-source sensors and sound-source sensors is required for such personal robots to achieve real time interactions with humans. However, traditional sensor systems for autonomous robot systems are not suitable for personal robots, since they use analog processing such as filtering in the first stage, and convert the processed information into digital data. This approach accompanies the loss of information by the filtering or the threshold function of the analog circuits. The accuracy and performance deeply depend on the analog circuit, and the strict calibration is required for each analog circuit.

In order to cope with this problem, a novel sensor system called *Chaser-II* is proposed and implemented in a personal robot prototype *ASPIRE-II*. In this system, since raw data from sensors is processed in digital units directly, flexible and precise processing can be achieved. For high speed digital processing, a powerful RISC processor is provided in *ASPIRE-II*.

However, even with a powerful RISC processor or DSP, the processing power is often insufficient for real time processing of sensor data such as pattern-matching, filtering and computation of the correlation function. Although enough performance can be obtained by dedicated ASIC chips, such a hardware is expensive and the flexibility is limited. In this paper, a reconfigurable sensor-data processing system for such personal robots is proposed. This system consists of two reconfigurable units including Altera's CPLDs which are directly connected with sensors, and is implemented on an interface board in the personal robot prototype *ASPIRE-II*. In this system, the correlation function of data from ultrasonic sensor is computed about 20 times faster than that with the RISC processor in *ASPIRE-II*.

In Section 2 of this paper we introduce the prototype of personal robot *ASPIRE-II* and its sensor system *Chaser-II*. In Section 3 we propose the reconfigurable sensor-data processing system. The computation system for the correlation function of data from the ultrasonic sensor is designed and evaluated on the reconfigurable system in Section 4.

2 *ASPIRE-II* and *Chaser-II*

2.1 *ASPIRE-II*

ASPIRE-II is a personal robot prototype based on the experience of the first version called *ASPIRE-I(Asynchronous, Parallel, Interrupt-based and REsposive architecture)*. As shown in Figure 1, it is implemented as a heterogeneous parallel computer, which is connected by the VME bus and has distributed shared memory. Each module has its own processing unit to process information and to decide its behavior for itself.

Each module consists of a mother board which provides the CPUs and a functional board for motor control, sensor, and image processing. Two types of mother board: the SPARClite board and the µSPARC board are provided. The high performance µSPARC board is attached to the board which requires high computational power such as image processing, while the SPARClite mother board which provides quick interrupt function using register windows is used for other purpose.

Figure 2 shows the diagram of the SPARClite board. In order to allow quick inter-module communication, communication memory and dedicated interrupt lines are provided in each module. Since the

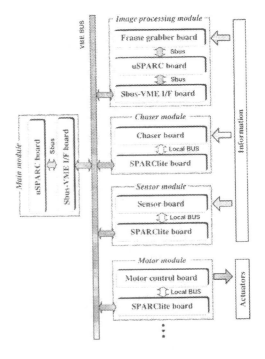

Fig. 1. Diagram of *ASPIRE-II*

VME bus is used as an inter-module connection on *ASPIRE-II*, communication memory is implemented as distributed shared memory. That is, each module shares a single address space and can access the memory on other modules through the bus. The seven-level interrupt in the VME bus can be fully utilized by the software for generating inter-module interrupt signals.

In the current implementation, the following modules are operational:

- Main module: μ-SPARC, Ethernet, SCSI-2, Keyboard, Mouse, SIO, PIO, and SBus interface;
- Motor module: SPARClite, DC motor and SIO;
- Sensor module (Chaser module): SPARClite, Ultrasonic sensors, Sound source direction sensors, Heat source sensors, and SIO;
- General purpose I/O module: SPARClite, Local bus, and SIO.

The general appearance of the personal robot *ASPIRE-II* is shown in Figure 3. It has two main wheels on its right and left side and has two auxiliary wheels at the front and back.

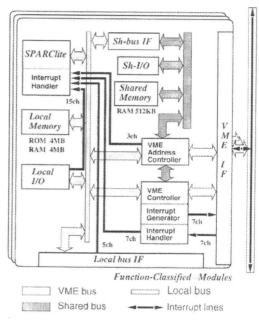

Fig. 2. Diagram of SPARClite Board

2.2 Chaser-II

Chaser-II is a practical human detection sensor system with fully digital processing in real-time. Analog processing is excluded as much as possible, and the raw data is utilized. The following advantages can be obtained with this approach:

- Calibration of each sensor can be achieved by the software.
- A threshold can be changed dynamically by software according to a situation.
- Additional information can be extracted from a sensor.
- All information from a sensor can be utilized because there is no extra operation such as filtering.

Chaser-II is implemented on Sensor module (Chaser module) of *ASPIRE-II*. As shown in Figure 4, two kind of A/D converters: a fast and slow one are connected directly with the ultrasonic sensor, sound source direction sensor and heat

Fig. 3. Personal Robot *ASPIRE-II*

source sensor. The digitized sensor data is directly transferred to the connected SPARClite mother board through the local bus, and processed in the mother board.

3 Reconfigurable sensor-data processing system

3.1 Overview of the reconfigurable system

Compared with other autonomous robot systems, *ASPIRE-II* provides a powerful computation ability. However, even with a powerful μ-SPARC or SPARClite processor in the mother board, the computating power is sometimes insufficient for real time sensor-data processing including a pattern-matching, filtering and computation of the correlation function. Although the performance of a dedicated ASIC chip is enough high, such a hardware is expensive and less flexible. To address this problem, we propose a reconfigurable sensor-data processing system for personal robots. Reconfigurable systems have been researched and developed by making the best use of recent advanced technologies on programmable devices[4]. In this approach, since the system can map each target algorithm to a real hardware directly, a powerful performance of dedicated processors can be achieved as well as a high degree of flexibility of general purpose processors. Although several systems for sensor-data processing have been

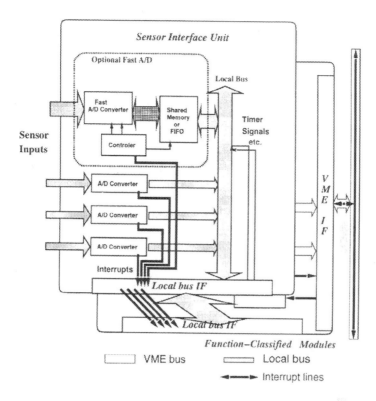

Fig. 4. Interface board for Sensor(Chaser) unit

implemented[5][6][7], almost no system is installed and evaluated in an actual robot environment.

The key to the design of reconfigurable systems is a selection of a programmable device. For our purpose, the function of processing system must be changed depending on the request from the main module. Thus, the circuit on the reconfigurable system should be quickly configured on demand. On the other hand, since the clock frequency of the system is hard to be changed, the reconfigurable unit should work at a constant speed even after the configuration is replaced. To cope with these requirements, we selected Altera's CPLD as a device of the reconfigurable system. Although the maximum number of gates is not large, the operational speed of the CPLD is relatively higher than that of FPGAs, and is not very sensitive to the change of circuits.

3.2 Reconfigurable Units

As shown in Figure 5, this system is implemented on the sensor interface board and connected with the mother board through the local bus. Two Reconfigurable

Units (RUs) are connected with two A/D converters (AD7892: 500kSPS, 12bits data) each of which receives analog data from the ultrasonic and heat source sensor respectively. The sound source sensor can be also used by changing jumper pins. The RUs are also connected with each other by the direct links and the local bus.

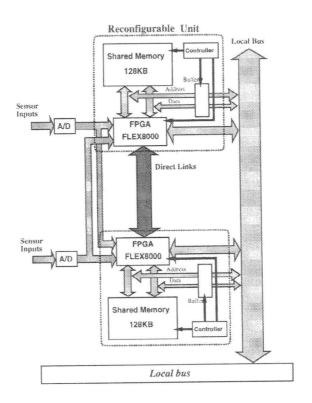

Fig. 5. Diagram of the reconfigurable system

An RU consists of the CPLD (Altera EPF81500A-2) corresponding to 16,000 gates, 128Kbyte SRAM shared memory, and a small controller. The shared memory can be accessed both from the CPLD and the CPU in the mother board, and used for storing configuration data and/or processing data. The arbitration of the shared memory, configuration of the CPLD, and data transfer from/to the mother board are managed by a small controller implemented in Lattice's isp2032. When the robot is starting up, the configuration data and initial execution data are downloaded from the CPU in the mother board through the local bus. After the configuration, the hardware system configured on the CPLD processes the data from the A/D converters. Execution results are directly transferred to the mother board through the local bus, or stored temporarily in the shared memory when required.

Since the size of shared memory is enough for storing multiple configuration data, the controller can select the one of the configurations and replace the function of the CPLD on demand. Although two RUs can be used independently for processing data from each A/D converter, they can also constitute a single large circuit when the size of required hardware exceeds the capacity of a single CPLD. In this case, two controllers are synchronized each other, and start the configuration at the same time. During the execution, direct links between two RUs are used for the data communication.

4 Evaluation

4.1 Correlation function calculator

We designed a circuit for computing the correlation function and implemented on the reconfigurable system. The correlation function is used to measure the similarity of two signal waves. By computing the correlation function of data from ultrasonic sensor and reference data, the distance from target object can be measured more precisely than the method based on a simple threshold function by analog circuits[1]. Computation results also can be used to conjecture the shape of the target object.

The correlation function of two signal waves $x(t)$ and $y(t)$ can be computed with the following equation:

$$R_{xy}(\tau) = \lim_{T \to \infty} \frac{1}{T} \int_0^T x(t + \tau)y(t)dt. \tag{1}$$

In this case, the sensor data is expressed discrete quantity, and the correlation function $R_{xy}(i)$ between the sample data from sensor $x(i); i = 0, 1, 2, \cdots, N$ and the reference data $y(j)$; $j = 1, 2, \cdots M$ ($M \leq N$) is computed. Since the real time computation between the sample data which comes every sample clock and fixed reference data is required, the following equation is used:

$$R_{xy}(i) = \frac{1}{M} \sum_{j=1}^{M} x(i + j) \cdot y(j) : i = 0, 1, \cdots, N - M. \tag{2}$$

Figure 6 shows the structure of correlation function calculator implemented on the reconfigurable system. It consists of two shift-registers with feedback lines (**SR1, SR2**), a multiplier (**Mul**), and an adder (**Add**). The length of the shift register is corresponding to the length of reference data, and set to be 32 in this design. The width of both the sampling data and the reference data is set to be 8 bits, while results are 16 bits.

Figure 7 shows the operation of the calculator (Figure 7(a)). The reference data $(r0 \ldots r31)$ is stored in **SR1** in advance. The sample data $(s0)$ from the A/D converter is inserted to **Mul** (Figure 7(a)), and multiplied results are transferred

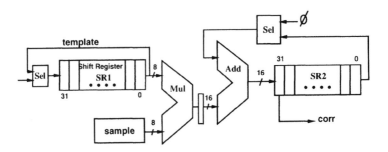

Fig. 6. Reconfigurable correlation function calculator

into **SR2** (Figure 7(b)). After every reference data is multiplied by $s0(t=32)$, the first result is output(Figure 7(c)). The first result is overwritten by the data from the adder $(r0s0+r1s1)$ in the next clock (t=33, Figure 7(d)), and the next sample data $s1$ is inserted. The same operation from Figure 7(a) to Figure 7(d) is iterated, and the result is calculated in every 32 clocks. Figure 7(e) shows the contents of **SR2** after 1024 clocks.

4.2 Evaluation

The above design is described in Altera's hardware description language AHDL. The modification of the size of data, and the number of reference data are quite easy. The synthesis and generation of the configuration data take only 10 minutes on a common personal computer (Pentium 133 MHz and 32MByte main memory).

The operational speed and the number of gates are shown in Table 1. Since the results can be obtained in every 32 clocks, it can be used with the A/D converter whose sample rate is 1.1MHz. This speed is sufficient for the current fast A/D converter used in *ASPIRE-II* whose maximum clock rate is 500kHz.

Table 1. Efficiency of designed circuit

Clock Cycle	28.5 [nsec]
Maximum Operating Frequency	35.08[MHz]
Maximum Sample Rate	1.10[MHz]
Computation time of 1024 samples	936.0[μsec]
Maximum Sample Rate by software	55.2[kHz]
Computation time of 1024 samples by software	18561.0[μsec]
Used LEs	975
Corresponding number of Gates	12,000

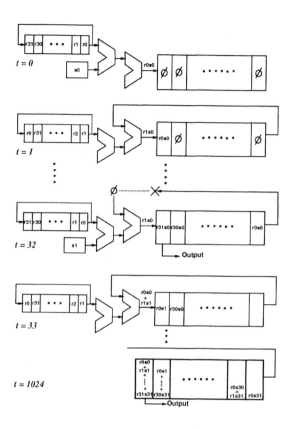

Fig. 7. Operations of the calculator

Table 1 also shows the operational speed when the function of 1024 sampling data is computed with the high speed RISC processor on the mother board of *Aspire-II* (μ-SPARC with 50MHz clock). The program is written in C language, and compiled with options -O2 -funroll-loops -mv8. As shown in this table, the maximum sampling rate is only 55.2kHz in this case, and far less than the sampling rate of the A/D converter. The same number of sampling data is computed with the reconfigurable system with 1056 clocks, and thus, 963.0μsec. It shows that the reconfigurable system works almost 20 times faster than that of the high speed RISC module.

The amount of hardware of the CPLD is evaluated with the number of the required LEs(Logic Elements). As shown in Table 1, this circuit requires 975 LEs (12,000 gates), and so about 75% of total LEs in EPF81500A are used. When the data size or the number of reference data is extended, the circuit is divided into two blocks: a block with **SR1** and **Add**, and another one with **Mul** and **SR2**. Since only 16bits signal lines are required between two blocks, they can be implemented on two RUs of the reconfigurable system.

5 Conclusion

In this paper, a reconfigurable sensor-data processing system for personal robots is proposed. Presently, the system based on Altera's CPLDs is operational on the sensor module in the personal robot prototype *ASPIRE-II*. As an example of processing, the correlation function calculator is implemented on this system, and approximately 20 times operational speed is achieved over the software approach on a RISC processor. Further detailed evaluation with a lot of other processing applications is the future work.

Acknowledgement

The authors would like to thank Dr. Nobuyuki Ymasaki at Electrotechnical Laboratory for his helpful suggestion. We owe this much to his previous work.

References

1. N. Yamasaki, Y. Anzai: *"Active Interface* for Human Robot Interaction", Proc. of IEEE International Conference on Robotics and Automation, Vol. 3, pp.3103–3109, 1995.
2. N. Yamasaki, Y. Anzai: "Applying Personal Robots and *Active Interface* to Video Conference System", Proc. of International Conference on Human-Computer Interaction, Vol. 2, pp.243–248, 1995.
3. D. Stewart, D. Schmitz, P. Khosla, "Implementing Real-Time Robotic System Using CHIMERA II", Proc. of the IEEE International Conference on Robotics and Automation, pp. 598-603, 1990.
4. S. Guccione, M. Gonzalez: "Classification and Performance of Reconfigurable Architectures", Proc. of the International Workshop on Field-Programmable Logic and Applications, LNCS 975, pp.439–448, 1995.
5. D. Woodward, I. Page, D. Levy, R. Harley: "A Programmable I/O System for Real-Time AC Drive Control Applications", Proc. of the International Workshop on Field-Programmable Logic and Applications, LNCS 975, 1995.
6. M. Shand: "Flexible Image Acquisition using Reconfigurable Hardware ", Proc. of the IEEE Symposium on FPGAs for Custom Computing Machines, pp.125–134, 1995.
7. J. Villasenor, B. Schoner, K. Chia, C. Zapata: "Configurable Computing Solutions for Automatic Target Recognition", Proc. of the IEEE Symposium on FPGAs for Custom Computing Machines, pp.70–79, 1996.

Author Index

A

Abramovici, M. 448
Acock, S.J.B. 255
Aggoun, A. 354
Aguirre, M.A. 1
Almeida, C.B. 274
Amano, H. 491
Amos, D. 324
Anzai, Y. 491
Athanas, P. 101
Athikulwongse, K. 374

B

Becker, J. 294
Betz, V. 213
Bhatia, D. 141
Boemo, E. 69
Bosco, G. 462
Bramer, B. 354
Brebner, G. 173
Burge, P. 121
Burns, J. 428

C

Carrabina, J. 392
Chaudhuri, A.S. 344
Chauham, D. 354
Cheung, P.Y.K. . 91, 111, 344
Chichkov, A.V. 274
Coghill, G.G. 265
Corke, P. 400
Crome, C. 324
Cuadrado, F. 392

D

Dagless, E. 245
Dales, M. 428

D

Dandalis, A. 314
Dick, H. 41
Diessel, O. 131
Dimmler, M. 472
Dimond, K.R. 255
Do, T. 51
Duncan, A. 382
Dunn, P. 400

E

ElGindy, H. 131
Elst, G. 235
Emmert, J.M. 141

F

Fattouche, M. 364
Faura, J. 1
Feske, K. 235
Fukami, K. 11

G

Gibb, S.G. 364
Glesner, M. 482
Graumann, P.J.W. 364
Greenfield, D. 324
Guccione, S.A. 61, 284
Guo, S.R. 111

H

Hadžić, I. 438
Hartenstein, R.W. .. 294, 304
Hayashi, K. 11
Herz, M. 294
Hollstein, T. 482
Hutchings, B.L. 193

I

Ibrahim, M.K. 354
Ichimori, T. 11
Insenser, J.M. 1
Inuani, M.K. 223

J

Jebelean, T. 457
Jong, C.C. 410

K

Kahne, B. 101
Katayama, M. 11
Kean, T. 382
Kirschbaum, A. 482
Koegst, M. 235
Kostarnov, I. 79
Kress, R. 304
Kropp, H. 51

L

Lam, Y.Y.H. 410
Laurent, B. 462
Lechner, E. 284
Leeser, M. 21
Leonard, J. 151
Lin, X. 245
Lisa, F. 392
López-Buedo, S. 69
Lu, A. 245
Luk, W. 91, 111, 344
Lysaght, P. 31, 41, 183

M

Mackinlay, P.I. 91
Mangione-Smith, W.H. .. 151
Mathews, T. 364
Maunder, R.B. 265
McGregor, G. 31, 41
Meleis, W.M. 21
Miyazaki, T. 11
Moreno, J.M. 1
Morley, S. 79
Mulka, S. 235
Murooka, T. 11

N

Nageldinger, U. 294, 304
Ng, L.S. 410
Nisbet, S. 61
Nukata, K. 491

O

Osmany, J. 79

P

Page, I. 418
Patterson, J. 428
Payne, R. 161
Pirsch, P. 51
Prasanna, V.K. 314

R

Rexachs, D. 392
Rising, B. 121
Robinson, D. 41
Rose, J. 213

S

Saab, D. 448
Salcic, Z.A. 265
Sandiford, R. 91
Saucier, G. 462
Saul, J. 223
Schwiegershausen, M. 51
Servít, M.Z. 203
Shand, M. 333
Shawe-Taylor, J. 121
Shibata, Y. 491
Shirazi, N. 111
Singh, S. 428
Smith, J.M. 438
Solomon, C. 79

T

Takahara, A. 11
Teerapanyawatt, S. 374
Tisserand, A. 472
Tsutsui, A. 11
Turner, L.E. 364

V

Vai, M.M. 21
van Daalen, M. 121
van Duong, P. 1

W

Won, M.S. 324

Y

Yi, K. 203

Z

Zavracky, P. 21

Springer
and the
environment

At Springer we firmly believe that an international science publisher has a special obligation to the environment, and our corporate policies consistently reflect this conviction.

We also expect our business partners – paper mills, printers, packaging manufacturers, etc. – to commit themselves to using materials and production processes that do not harm the environment. The paper in this book is made from low- or no-chlorine pulp and is acid free, in conformance with international standards for paper permanency.

 Springer

Lecture Notes in Computer Science

For information about Vols. 1–1229

please contact your bookseller or Springer-Verlag

Vol. 1230: J. Duncan, G. Gindi (Eds.), Information Processing in Medical Imaging. Proceedings, 1997. XVI, 557 pages. 1997.

Vol. 1231: M. Bertran, T. Rus (Eds.), Transformation-Based Reactive Systems Development. Proceedings, 1997. XI, 431 pages. 1997.

Vol. 1232: H. Comon (Ed.), Rewriting Techniques and Applications. Proceedings, 1997. XI, 339 pages. 1997.

Vol. 1233: W. Fumy (Ed.), Advances in Cryptology — EUROCRYPT '97. Proceedings, 1997. XI, 509 pages. 1997.

Vol 1234: S. Adian, A. Nerode (Eds.), Logical Foundations of Computer Science. Proceedings, 1997. IX, 431 pages. 1997.

Vol. 1235: R. Conradi (Ed.), Software Configuration Management. Proceedings, 1997. VIII, 234 pages. 1997.

Vol. 1236: E. Maier, M. Mast, S. LuperFoy (Eds.), Dialogue Processing in Spoken Language Systems. Proceedings, 1996. VIII, 220 pages. 1997. (Subseries LNAI).

Vol. 1238: A. Mullery, M. Besson, M. Campolargo, R. Gobbi, R. Reed (Eds.), Intelligence in Services and Networks: Technology for Cooperative Competition. Proceedings, 1997. XII, 480 pages. 1997.

Vol. 1239: D. Sehr, U. Banerjee, D. Gelernter, A. Nicolau, D. Padua (Eds.), Languages and Compilers for Parallel Computing. Proceedings, 1996. XIII, 612 pages. 1997.

Vol. 1240: J. Mira, R. Moreno-Díaz, J. Cabestany (Eds.), Biological and Artificial Computation: From Neuroscience to Technology. Proceedings, 1997. XXI, 1401 pages. 1997.

Vol. 1241: M. Akşit, S. Matsuoka (Eds.), ECOOP'97 – Object-Oriented Programming. Proceedings, 1997. XI, 531 pages. 1997.

Vol. 1242: S. Fdida, M. Morganti (Eds.), Multimedia Applications, Services and Techniques – ECMAST '97. Proceedings, 1997. XIV, 772 pages. 1997.

Vol. 1243: A. Mazurkiewicz, J. Winkowski (Eds.), CONCUR'97: Concurrency Theory. Proceedings, 1997. VIII, 421 pages. 1997.

Vol. 1244: D. M. Gabbay, R. Kruse, A. Nonnengart, H.J. Ohlbach (Eds.), Qualitative and Quantitative Practical Reasoning. Proceedings, 1997. X, 621 pages. 1997. (Subseries LNAI).

Vol. 1245: M. Calzarossa, R. Marie, B. Plateau, G. Rubino (Eds.), Computer Performance Evaluation. Proceedings, 1997. VIII, 231 pages. 1997.

Vol. 1246: S. Tucker Taft, R. A. Duff (Eds.), Ada 95 Reference Manual. XXII, 526 pages. 1997.

Vol. 1247: J. Barnes (Ed.), Ada 95 Rationale. XVI, 458 pages. 1997.

Vol. 1248: P. Azéma, G. Balbo (Eds.), Application and Theory of Petri Nets 1997. Proceedings, 1997. VIII, 467 pages. 1997.

Vol. 1249: W. McCune (Ed.), Automated Deduction – CADE-14. Proceedings, 1997. XIV, 462 pages. 1997. (Subseries LNAI).

Vol. 1250: A. Olivé, J.A. Pastor (Eds.), Advanced Information Systems Engineering. Proceedings, 1997. XI, 451 pages. 1997.

Vol. 1251: K. Hardy, J. Briggs (Eds.), Reliable Software Technologies – Ada-Europe '97. Proceedings, 1997. VIII, 293 pages. 1997.

Vol. 1252: B. ter Haar Romeny, L. Florack, J. Koenderink, M. Viergever (Eds.), Scale-Space Theory in Computer Vision. Proceedings, 1997. IX, 365 pages. 1997.

Vol. 1253: G. Bilardi, A. Ferreira, R. Lüling, J. Rolim (Eds.), Solving Irregularly Structured Problems in Parallel. Proceedings, 1997. X, 287 pages. 1997.

Vol. 1254: O. Grumberg (Ed.), Computer Aided Verification. Proceedings, 1997. XI, 486 pages. 1997.

Vol. 1255: T. Mora, H. Mattson (Eds.), Applied Algebra, Algebraic Algorithms and Error-Correcting Codes. Proceedings, 1997. X, 353 pages. 1997.

Vol. 1256: P. Degano, R. Gorrieri, A. Marchetti-Spaccamela (Eds.), Automata, Languages and Programming. Proceedings, 1997. XVI, 862 pages. 1997.

Vol. 1258: D. van Dalen, M. Bezem (Eds.), Computer Science Logic. Proceedings, 1996. VIII, 473 pages. 1997.

Vol. 1259: T. Higuchi, M. Iwata, W. Liu (Eds.), Evolvable Systems: From Biology to Hardware. Proceedings, 1996. XI, 484 pages. 1997.

Vol. 1260: D. Raymond, D. Wood, S. Yu (Eds.), Automata Implementation. Proceedings, 1996. VIII, 189 pages. 1997.

Vol. 1261: J. Mycielski, G. Rozenberg, A. Salomaa (Eds.), Structures in Logic and Computer Science. X, 371 pages. 1997.

Vol. 1262: M. Scholl, A. Voisard (Eds.), Advances in Spatial Databases. Proceedings, 1997. XI, 379 pages. 1997.

Vol. 1263: J. Komorowski, J. Zytkow (Eds.), Principles of Data Mining and Knowledge Discovery. Proceedings, 1997. IX, 397 pages. 1997. (Subseries LNAI).

Vol. 1264: A. Apostolico, J. Hein (Eds.), Combinatorial Pattern Matching. Proceedings, 1997. VIII, 277 pages. 1997.

Vol. 1265: J. Dix, U. Furbach, A. Nerode (Eds.), Logic Programming and Nonmonotonic Reasoning. Proceedings, 1997. X, 453 pages. 1997. (Subseries LNAI).

Vol. 1266: D.B. Leake, E. Plaza (Eds.), Case-Based Reasoning Research and Development. Proceedings, 1997. XIII, 648 pages. 1997 (Subseries LNAI).

Vol. 1267: E. Biham (Ed.), Fast Software Encryption. Proceedings, 1997. VIII, 289 pages. 1997.

Vol. 1268: W. Kluge (Ed.), Implementation of Functional Languages. Proceedings, 1996. XI, 284 pages. 1997.

Vol. 1269: J. Rolim (Ed.), Randomization and Approximation Techniques in Computer Science. Proceedings, 1997. VIII, 227 pages. 1997.

Vol. 1270: V. Varadharajan, J. Pieprzyk, Y. Mu (Eds.), Information Security and Privacy. Proceedings, 1997. XI, 337 pages. 1997.

Vol. 1271: C. Small, P. Douglas, R. Johnson, P. King, N. Martin (Eds.), Advances in Databases. Proceedings, 1997. XI, 233 pages. 1997.

Vol. 1272: F. Dehne, A. Rau-Chaplin, J.-R. Sack, R. Tamassia (Eds.), Algorithms and Data Structures. Proceedings, 1997. X, 476 pages. 1997.

Vol. 1273: P. Antsaklis, W. Kohn, A. Nerode, S. Sastry (Eds.), Hybrid Systems IV. X, 405 pages. 1997.

Vol. 1274: T. Masuda, Y. Masunaga, M. Tsukamoto (Eds.), Worldwide Computing and Its Applications. Proceedings, 1997. XVI, 443 pages. 1997.

Vol. 1275: E.L. Gunter, A. Felty (Eds.), Theorem Proving in Higher Order Logics. Proceedings, 1997. VIII, 339 pages. 1997.

Vol. 1276: T. Jiang, D.T. Lee (Eds.), Computing and Combinatorics. Proceedings, 1997. XI, 522 pages. 1997.

Vol. 1277: V. Malyshkin (Ed.), Parallel Computing Technologies. Proceedings, 1997. XII, 455 pages. 1997.

Vol. 1278: R. Hofestädt, T. Lengauer, M. Löffler, D. Schomburg (Eds.), Bioinformatics. Proceedings, 1996. XI, 222 pages. 1997.

Vol. 1279: B. S. Chlebus, L. Czaja (Eds.), Fundamentals of Computation Theory. Proceedings, 1997. XI, 475 pages. 1997.

Vol. 1280: X. Liu, P. Cohen, M. Berthold (Eds.), Advances in Intelligent Data Analysis. Proceedings, 1997. XII, 621 pages. 1997.

Vol. 1281: M. Abadi, T. Ito (Eds.), Theoretical Aspects of Computer Software. Proceedings, 1997. XI, 639 pages. 1997.

Vol. 1282: D. Garlan, D. Le Métayer (Eds.), Coordination Languages and Models. Proceedings, 1997. X, 435 pages. 1997.

Vol. 1283: M. Müller-Olm, Modular Compiler Verification. XV, 250 pages. 1997.

Vol. 1284: R. Burkard, G. Woeginger (Eds.), Algorithms — ESA '97. Proceedings, 1997. XI, 515 pages. 1997.

Vol. 1285: X. Jao, J.-H. Kim, T. Furuhashi (Eds.), Simulated Evolution and Learning. Proceedings, 1996. VIII, 231 pages. 1997. (Subseries LNAI).

Vol. 1286: C. Zhang, D. Lukose (Eds.), Multi-Agent Systems. Proceedings, 1996. VII, 195 pages. 1997. (Subseries LNAI).

Vol. 1287: T. Kropf (Ed.), Formal Hardware Verification. XII, 367 pages. 1997.

Vol. 1288: M. Schneider, Spatial Data Types for Database Systems. XIII, 275 pages. 1997.

Vol. 1289: G. Gottlob, A. Leitsch, D. Mundici (Eds.), Computational Logic and Proof Theory. Proceedings, 1997. VIII, 348 pages. 1997.

Vol. 1290: E. Moggi, G. Rosolini (Eds.), Category Theory and Computer Science. Proceedings, 1997. VII, 313 pages. 1997.

Vol. 1292: H. Glaser, P. Hartel, H. Kuchen (Eds.), Programming Languages: Implementations, Logigs, and Programs. Proceedings, 1997. XI, 425 pages. 1997.

Vol. 1294: B.S. Kaliski Jr. (Ed.), Advances in Cryptology — CRYPTO '97. Proceedings, 1997. XII, 539 pages. 1997.

Vol. 1295: I. Prívara, P. Ružička (Eds.), Mathematical Foundations of Computer Science 1997. Proceedings, 1997. X, 519 pages. 1997.

Vol. 1296: G. Sommer, K. Daniilidis, J. Pauli (Eds.), Computer Analysis of Images and Patterns. Proceedings, 1997. XIII, 737 pages. 1997.

Vol. 1297: N. Lavrač, S. Džeroski (Eds.), Inductive Logic Programming. Proceedings, 1997. VIII, 309 pages. 1997. (Subseries LNAI).

Vol. 1298: M. Hanus, J. Heering, K. Meinke (Eds.), Algebraic and Logic Programming. Proceedings, 1997. X, 286 pages. 1997.

Vol. 1299: M.T. Pazienza (Ed.), Information Extraction. Proceedings, 1997. IX, 213 pages. 1997. (Subseries LNAI).

Vol. 1300: C. Lengauer, M. Griebl, S. Gorlatch (Eds.), Euro-Par'97 Parallel Processing. Proceedings, 1997. XXX, 1379 pages. 1997.

Vol. 1302: P. Van Hentenryck (Ed.), Static Analysis. Proceedings, 1997. X, 413 pages. 1997.

Vol. 1303: G. Brewka, C. Habel, B. Nebel (Eds.), KI-97: Advances in Artificial Intelligence. Proceedings, 1997. XI, 413 pages. 1997. (Subseries LNAI).

Vol. 1304: W. Luk, P.Y.K. Cheung, M. Glesner (Eds.), Field-Programmable Logic and Applications. Proceedings, 1997. XI, 503 pages. 1997.

Vol. 1305: D. Corne, J.L. Shapiro (Eds.), Evolutionary Computing. Proceedings, 1997. X, 313 pages. 1997.

Vol. 1308: A. Hameurlain, A M. Tjoa (Eds.), Database and Expert Systems Applications. Proceedings, 1997. XVII, 688 pages. 1997.

Vol. 1310: A. Del Bimbo (Ed.), Image Analysis and Processing. Proceedings, 1997. Volume I. XXI, 722 pages. 1997.

Vol. 1311: A. Del Bimbo (Ed.), Image Analysis and Processing. Proceedings, 1997. Volume II. XXII, 794 pages. 1997.

Vol. 1312: A. Geppert, M. Berndtsson (Eds.), Rules in Database Systems. Proceedings, 1997. VII, 213 pages. 1997.

Vol. 1314: S. Muggleton (Ed.), Inductive Logic Programming. Proceedings, 1996. VIII, 397 pages. 1997. (Subseries LNAI).

Vol. 1315: G. Sommer, J.J. Koenderink (Eds.), Algebraic Frames for the Perception-Action Cycle. Proceedings, 1997. VIII, 395 pages. 1997.